Coronaviruses

Molecular and Cellular Biology

Edited by Volker Thiel

Caister Academic Press

Caister Academic Press
Norfolk, UK

www.caister.com

British Library Cataloguing-in-Publication Data
A catalogue record for this book is available from the British Library

ISBN: 978-1-904455-16-5

Printed and bound in Great Britain

Contents

Contributors

Ralph S. Baric
Departments of Epidemiology and Microbiology & Immunology
University of North Carolina
Chapel Hill, NC
USA

rockx@email.unc.edu

Cornelia C. Bergmann
Cleveland Clinic
Lerner Research Institute
Cleveland, OH
USA

bergmac@ccf.org

Paul Britton
Division of Microbiology
Institute for Animal Health
Compton Laboratory
Compton
Newbury
Berkshire
UK

Paul.britton@bbsrc.ac.uk

Dave Cavanagh
Institute for Animal Health
Compton Laboratory
Compton
Newbury
Berkshire
UK

dave.cavanagh@bbsrc.ac.uk

Luis Enjuanes
National Center of Biotechnology (CNB)
CSIC
Campus de la Universidad Autonoma
Cantoblanco
Madrid
Spain

L.Enjuanes@cnb.uam.es

Raoul J. de Groot
Institute of Virology
Department of Infectious Diseases and Immunology
Faculty of Veterinary Medicine
Utrecht University
Utrecht
The Netherlands

r.groot@vet.uu.nl

Bert Jan Haijema
Institute of Virology
Department of Infectious Diseases and Immunology
Faculty of Veterinary Medicine
Utrecht University
Utrecht
The Netherlands

Lia van der Hoek
Laboratory of Experimental Virology
Department of Medical Microbiology
Center for Infection and Immunity
Amsterdam (CINIMA)
Academic Medical Center
Amsterdam
The Netherlands

c.m.vanderhoek@amc.uva.nl

Kathryn V. Holmes
Dept. of Microbiology
University of Colorado Health Sciences
Center
Aurora, CO

kathryn.holmes@uchsc.edu

Els Keyaerts
Laboratory of Clinical and Epidemiological
Virology
Department of Microbiology and
Immunology
Rega Institute for Medical Research
University of Leuven
Leuven
Belgium

Shinji Makino
Department of Microbiology and
Immunology
The University of Texas Medical Branch
Galveston, TX
USA

shmakino@utmb.edu

Paul S. Masters
David Axelrod Institute
Wadsworth Center
New York State Department of Health
Albany, NY
USA

masters@wadsworth.org

Jose Luis Moreno Carboneros
National Center of Biotechnology (CNB)
CSIC
Campus de la Universidad Autonoma
Cantoblanco
Madrid
Spain

jlmoreno@cnb.uam.es

Krishna Narayanan
Department of Microbiology and
Immunology
University of Texas Medical Branch
Galveston, TX

krnaraya@utmb.edu

John M. Nicholls
The University of Hong Kong
Department of Pathology
Queen Mary Hospital
Hong Kong

nicholls@pathology.hku.hk

Joseph S.M. Peiris
The University of Hong Kong
Department of Microbiology
Queen Mary Hospital
Hong Kong

Leo Lit Man Poon
Department of Microbiology
The University of Hong Kong
Queen Mary Hospital
Pokfulam
Hong Kong

llmpoon@hkucc.hku.hk

Krzysztof Pyrc
Laboratory of Experimental Virology
Department of Medical Microbiology
Center for Infection and Immunity
Amsterdam (CINIMA)
Academic Medical Center
Amsterdam
The Netherlands

k.a.pyrc@gmail.com

Barry Rockx
Post-Doctoral Research Associate
Department of Epidemiology
University of North Carolina
Chapel Hill, NC
USA

rockx@email.unc.edu

Peter J.M. Rottier
Institute of Virology
Department of Infectious Diseases and Immunology
Faculty of Veterinary Medicine
Utrecht University
Utrecht
The Netherlands

p.rottier@vet.uu.nl

Eric J. Snijder
Department of Medical Microbiology
Leiden University Medical Center
The Netherlands

e.j.snijder@lumc.nl

Isabel Sola
National Center of Biotechnology (CNB)
CSIC
Campus de la Universidad Autonoma
Cantoblanco
Madrid
Spain

isola@cnb.uam.es

Martin Spiegel
Institut für Virologie
Georg-August-Universität
Bereich Humanmedizin
Göttingen
Germany

mspiege@gwdg.de

Stephen A Stohlman
Staff
Cleveland Clinic
Lerner Research Institute
Cleveland, OH
USA

Stohlms2@ccf.org

Volker Thiel
Kantonal Hospital St.Gallen
Research Department
St.Gallen
Switzerland
Volker.Thiel@kssg.ch

Marc Van Ranst
Laboratory of Clinical and Epidemiological Virology
Department of Microbiology and Immunology
Rega Institute for Medical Research
University of Leuven
Leuven
Belgium

marc.vanranst@uz.kuleuven.ac.be

Leen Vijgen
Laboratory of Clinical and Epidemiological Virology
Department of Microbiology and Immunology
Rega Institute for Medical Research
University of Leuven
Leuven
Belgium

Friedemann Weber
Department of Virology
Institute for Medical Microbiology and Hygiene
Freiburg
Germany

friedemann.weber@uniklinik-freiburg.de

David E. Wentworth
Department of Biomedical Sciences
School of Public Health
The University at Albany
and
Wadsworth Center
NYSDOH
Albany, NY
USA

dwentwor@wadsworth.org

John Ziebuhr
School of Biomedical Sciences
Queen's University Belfast
Centre for Cancer Research and Cell Biology
Belfast
United Kingdom

j.ziebuhr@qub.ac.uk

Sonia Zuñiga Lucas
National Center of Biotechnology (CNB)
CSIC
Campus de la Universidad Autonoma
Cantoblanco
Madrid
Spain

szuniga@cnb.uam.es

Preface

Coronaviruses are enveloped, positive-stranded RNA viruses that can infect a variety of vertebrates and are mainly associated with respiratory and enteric diseases. They have long been known as important pathogens of lifestock and companion animals where they lead to high economical losses. In humans, however, coronavirus infections manifest usually as mild respiratory tract disease (common cold) and only occasionally more severe symptoms, particularly in elderly or immune-compromised individuals, are observed. This picture of coronavirus infections has been dramatically changed in 2003, as it became clear that the newly recognized severe acute respiratory syndrome is caused by a formerly unknown coronavirus, the SARS-Coronavirus. This event not only highlighted the potential of coronaviruses to seriously impact on human health, but also gave a strong impetus on coronavirus research.

Four years after the SARS epidemic this book provides a comprehensive and contemporary view of coronavirus molecular biology and diseases and review literally every aspect of cutting edge coronavirus research. The first part of this book describes the coronavirus replication cycle, from cell entry to exit, on the molecular level and provides new insights into the molecular evolution of group 2 coronaviruses. Topics on molecular biology include coronavirus binding and entry, replicase gene function, *cis*-acting RNA elements, coronavirus discontinuous transcription, reverse genetics, genome packaging and molecular evolution. The second part focuses on coronavirus diseases and immune responses which are reviewed for long known prototype viruses of all three coronavirus groups, namely feline coronavirus, mouse hepatitis virus, and avian infectious bronchitis virus. Reviews on SARS-Coronavirus include virus pathogenesis, antiviral cytokine responses and vaccine development. Finally, "new" coronaviruses, such as the newly recognized bat coronaviruses and human coronavirus NL63, and the current state of antiviral coronavirus research, are presented.

I'm glad that so many renowned coronavirus experts made their time available and contributed with excellent articles to this volume. I would like to thank all the authors and the staff of Caister Academic Press for their help in the preparation of this book.

Volker Thiel
June 2007

Part I

Coronavirus Binding and Entry

1

David E. Wentworth and Kathryn V. Holmes

Abstract

Coronaviruses are a diverse family of viruses that bind to host cells primarily through interactions between viral spike glycoproteins and specific host cell surface glycoproteins. Some coronaviruses also bind to sialic acids on glycoproteins and glycolipids via their spike and/or hemagglutinin esterase glycoproteins. The interactions between coronaviruses and host cell receptors are critical determinants of species-specificity, tissue tropism, and virulence. This chapter summarizes how coronaviruses bind to host cells and how the viral envelope fuses with host cell membranes to initiate infection. It will focus on the viral spike and hemagglutinin esterase glycoproteins and their interactions with known host cell receptor proteins, sialyloligosaccharides, and lectins. The functions of the other coronavirus structural proteins, non-structural and accessory proteins, as well as the fascinating replication strategy of coronaviruses will be discussed in later chapters.

Introduction

Coronaviruses are enveloped viruses studded with the characteristic large, petal-shaped spikes forming the "corona" seen in negatively stained preparations, which gave the family *Coronaviridae* their name. Three major groups of coronaviruses were initially distinguished based on serologic cross-reactivity, and more recently by genome phylogeny (Gorbalenya *et al.*, 2006). Each group contains viruses that infect a wide variety of hosts. Coronaviruses in groups 1 and 2 infect many mammalian species ranging from bats to humans, and group 3 coronaviruses have, so far, only been found in avian species. The discovery of a distinct group 2 coronavirus (SARS-CoV) as the cause of the 2002–2003 outbreak of Severe Acute Respiratory Syndrome led to subdivision of group 2 coronaviruses into two subgroups 2a and 2b (Gorbalenya *et al.*, 2004; Gorbalenya *et al.*, 2006).

The coronavirus virion (Figure 1.1A) is the apparatus designed to deliver the large positive sense RNA genome (27.3–31.3 kb), and its associated nucleocapsid (N) protein, into host cells. Like most eukaryotic mRNAs, the coronavirus RNA genome has a 5′ cap and a 3′ poly-A tail. Furthermore, the positive sense genomic RNA is infectious when transfected into host cells. Coronavirus virions (Figure 1.1A), contain 4 or 5 structural proteins: S, the spike glycoprotein that binds to receptors and mediates membrane fusion; M, the membrane glycoprotein that spans the membrane three times and forms an internal core; N, the nucleocapsid phosphoprotein that wraps the viral RNA genome in a long,

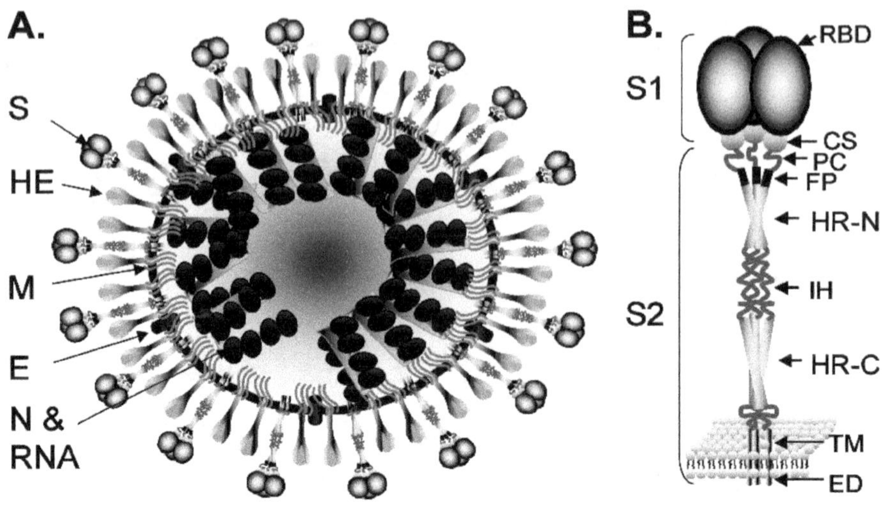

Figure 1.1 Illustrations depicting common structural components of a coronavirus virion and the functional domains of spike glycoproteins. A, Diagram of the coronavirus virion depicting the spike (S), hemagglutinin esterase (HE), membrane (M), envelope (E) and the RNA (ribbon) bound by nucleoprotein (N, oval). B, Diagram of a coronavirus S trimer depicting the S1 and S2 domains. The S1 forms the globular head at the amino terminus that contains the receptor binding domain (RBD). The S2 stalk region which mediates fusion of the viral envelope with host cell membranes. A potential protease cleavage site (CS) separates the S1 and S2, and this is followed by the pre-coil (PC) region, and a putative fusion peptide (FP) is shown in black. The amino-heptad repeat (HR-N) and carboxy-heptad repeat (HR-C) that collapse to form a six helix bundle critical for fusion are separated by an inter-helical (IH) domain. The hydrophobic transmembrane (TM) region is shown going through a lipid bilayer followed by a short endodomain (ED) at the carboxy-terminus.

flexible, helical nucleocapsid; E, the envelope protein, a minor virion component that has ion channel activity important for virus entry and assembly; and HE, the hemagglutinin-esterase glycoprotein found only in group 2a coronaviruses, that binds to sialoglycoconjugates on the host cell membrane and facilitates entry.

Structures and functions of coronavirus surface glycoproteins

Spike (S) glycoprotein

Infection of host cells is initiated by virus binding to receptors on the cell surface and subsequent fusion of the lipid bilayer of the viral envelope with the host cell plasma membrane (e.g. mouse hepatitis virus [MHV]-A59) or endosomal membrane (e.g. human coronavirus [HCoV]-229E). The large, petal shaped spikes on the coronavirus envelope are composed of trimers of the viral S glycoprotein monomers that project ~20 nanometers from the envelope (Figure 1.1A) (Masters, 2006; de Groot, 2006). S is the viral attachment and fusion glycoprotein common to all coronaviruses, and, in different coronaviruses the S ranges from approximately 1100 to 1500 amino acids in length. An amino terminal signal peptide directs the nascent S polypeptide to the endoplasmic reticulum (ER), where the signal peptide is cleaved and a 128–160 kDa monomeric precursor is synthesized (Masters, 2006). The S monomers are extensively glycosylated during maturation, resulting in sizes ranging from ~150–200 kDa. The S monomers form trimers in the ER, and the S trimers are incorporated into budding virions in the ER–Golgi intermediate compartment.

Variations in the properties of S proteins of different coronaviruses modulate virus entry and influence tissue tropism, persistence, virulence and host range. These properties include the specific host receptor(s) bound, ability to bind sialoglycoconjugates, hypervariable regions in the S1 protein, and susceptibility of the S protein to cleavage by proteases, and/or to acid-induced conformational changes.

The S protein has a heavily glycosylated ectodomain that consists of a large amino-terminal head region and a long stalk region (Figure 1.1B). The S proteins of some coronavirus strains are cleaved by host proteases to yield two non-covalently associated subunits, designated S1 and S2 (Figure 1.1B). As a result all coronavirus spikes are said to have S1 and S2 domains. Coronavirus S proteins are variants of class 1 viral fusion glycoproteins, which mediate membrane fusion by orthomyxoviruses, retroviruses, paramyxoviruses, and filoviruses. The structural elements of class 1 fusion proteins of these virus groups are shown in Figure 1.2A (Bosch *et al.*, 2003). These class 1 fusion proteins are all found as trimers on the envelopes of their respective virions, and they usually contain two major domains, e.g. S1 and S2, which are generally responsible for receptor interactions and membrane fusion, respectively. In general, a protease cleavage site separates receptor binding and membrane fusion domains of class 1 viral fusion proteins, and cleavage by furin, trypsin, cathepsins, or other proteases may be required for efficient virus entry. Class 1 viral fusion proteins are anchored in the viral envelope by a transmembrane domain and a short cysteine-rich, acylated endodomain/cytoplasmic tail.

Although the coronavirus S glycoproteins have all the elements of other class 1 fusion proteins, they also have some unique features including long pre-coil and interhelical domains. The coronavirus S1 domain forms a globular head and its major role is to bind host cell receptor glycoproteins and, in some cases, to sialoglycoconjugates. S2, the stalk region of the spike, is responsible for fusion of the viral envelope with host cell membranes. Some coronavirus S proteins (e.g. MHV-A59) also induce fusion of infected cells with uninfected cells resulting in impressive syncytia formation (Frana *et al.*, 1985; Sturman *et al.*, 1985). The critical roles that the S1 and S2 domains, the protease cleavage motif and the endodomain play in the complex process of coronavirus entry are described below.

S1 domain

The S1 domain is the major antigenic determinant of coronaviruses that elicits neutralizing antibodies. Some anti-S1 antibodies neutralize infection by inhibiting binding of S1 to host receptors, and these are important in protection from infection *in vivo* (Wege and Dorries, 1984; Cavanagh *et al.*, 1988; Corapi *et al.*, 1992; Enjuanes *et al.*, 1992; Vautherot *et al.*, 1992; Kubo *et al.*, 1993; Godet *et al.*, 1994; Sestak *et al.*, 1996). Many coronaviruses have evolved variations in S1 to escape neutralization by host immune responses (Cavanagh *et al.*, 2005; Rekik and Dea, 1994; Wang *et al.*, 1994; Simkins *et al.*, 1992). S1 domains of some coronaviruses contain hypervariable regions that contain neutralization

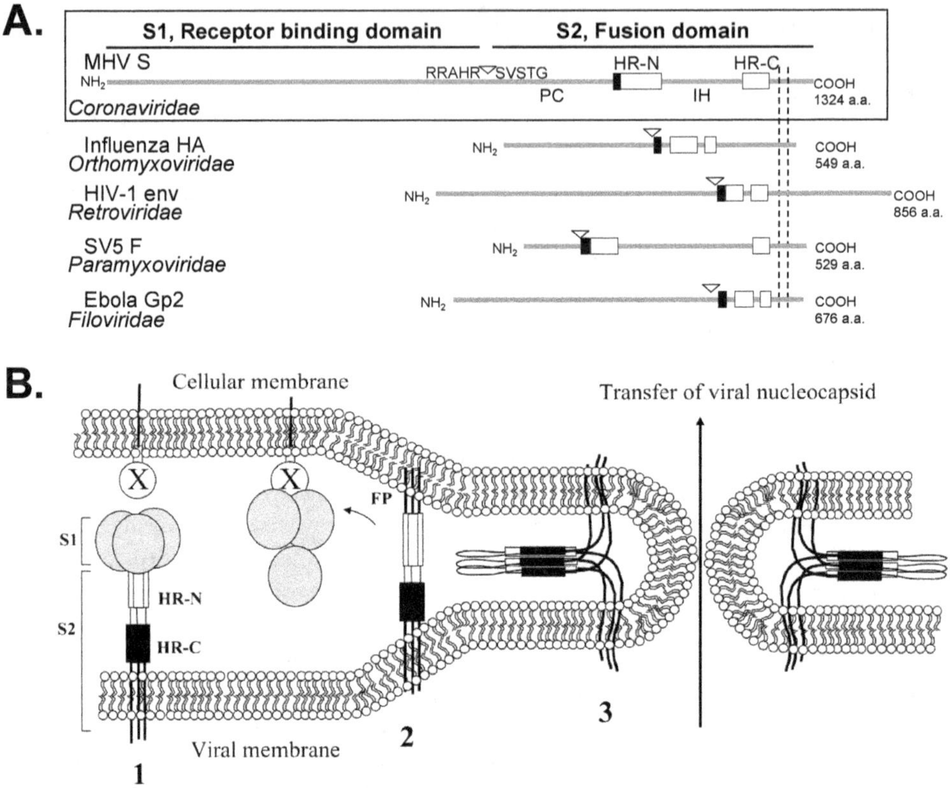

Figure 1.2 Schematic illustrations of class 1 viral fusion proteins and the conformational changes in the coronavirus spike that occur during entry. A, Representations (approximately to scale) of the viral fusion proteins of five different virus families (modified from Bosch *et al.*, 2003 with permission). These include, MHV-A59 S (Coronaviridae), influenza virus HA (Orthomyxoviridae), HIV-1 gp160 (Retroviridae), SV5 F (Paramyxoviridae), and Ebola Gp2 (Filoviridae). Cleavage sites are indicated by triangles; the solid bars represent the (putative) fusion peptides, the vertically hatched bars represent the HR-N (HR1) domains, and the horizontally hatched bars represent the HR-C (HR2) domains. Transmembrane domains are indicated by the vertical dashed lines. The MHV S also shows the S1 and S2 domains, the amino acids that juxtapose the cleavage site, a pre-coil (PC) region that lies between the cleavage site and fusion peptide and the inter-helical (IH) region that lies between the HR-N and HR-C. For each polypeptide, the total length is given at the right. B, "schematic representation of the three different states of the coronavirus S fusion protein during viral entry" (Obtained from Tripet *et al.*, 2004 with permission). "state 1 (native), the S protein is denoted S1 and S2 for its N- and C-terminal domains; state 2 (intermediate state), the N-terminal (S1) domain is dissociated (shedding) to expose the fusion peptide (FP) region, and state 3 (collapsed 6-helix bundle or fusion active state), the collapsed S2 domain draws the viral and cellular membranes together causing fusion and release of the viral nucleocapsid into the host cell. HR-N and HR-C denote coiled-coils at the N terminus and C terminus of the S2 domain. x denotes the host cell surface receptor."

epitopes (Cavanagh *et al.*, 1988; Gallagher *et al.*, 1990; La Monica N. *et al.*, 1991). For example, MHV strains differ markedly in the length of S1 due to a hypervariable region (La Monica N. *et al.*, 1991; Gallagher *et al.*, 1990). Longer S1 proteins contain additional

B cell epitopes recognized by neutralizing monoclonal antibodies (Gallagher *et al.*, 1990). In addition, virus variants with shorter S1 proteins that lack a T cell epitope found in the hypervariable domain are selected *in vivo* (Pewe *et al.*, 1997). Mutations in S1 can also alter interactions of a coronavirus with its receptor, dramatically influencing the species specificity of virus infection (Li *et al.*, 2005c; Baric *et al.*, 1999; Schickli *et al.*, 1997; Thackray and Holmes, 2004).

Specific glycoprotein receptors required for entry of different coronaviruses have been identified. These receptors include: mouse CEACAM1a for MHV strains (Dveksler *et al.*, 1991); aminopeptidase N (APN) orthologs for group 1 coronaviruses HCoV-229E (Yeager *et al.*, 1992), porcine transmissible gastroenteritis virus (TGEV) (Delmas *et al.*, 1992), porcine respiratory coronavirus (PRCoV) (Delmas *et al.*, 1993), and feline coronavirus (FCoV) (Tresnan *et al.*, 1996); and human angiotensin-converting enzyme 2 (ACE2) for SARS-CoV (Li *et al.*, 2003) in group 2b and HCoV-NL63 (Hofmann *et al.*, 2005) in group 1. The specific interactions of these viral S glycoproteins with their host receptors will be discussed later in the chapter. In this section, we will compare the putative receptor-binding domains (RBD) of the S1 proteins of these coronaviruses.

The RBDs have been mapped to the S1 domain using truncated, chimeric, and mutant S proteins, pseudotyped retroviruses, neutralizing antibodies, and reverse genetics. Interestingly, the RBDs of different coronaviruses are not in homologous regions of the S1. In the group 1 coronaviruses, the minimal RBD of HCoV-229E S1 was shown to be amino acids 407–547 (Bonavia *et al.*, 2003; Breslin *et al.*, 2003). However, retrovirus pseudotypes containing chimeras between S proteins of HCoV-NL63 and HCoV-229E required additional amino acids near the amino terminus of HCoV-229E S1 to bind to human APN (Hofmann *et al.*, 2006). The minimal RBD of TGEV S1 was localized to amino acids 579–655 (Godet *et al.*, 1994) a region that is highly conserved in S1 proteins of PRCoV, FCoV, and canine coronavirus (CCoV) which also use APN receptors. The putative RBD of HCoV-NL63 which uses a different receptor (human ACE2), was found between amino acids 232–684 using truncated S1 proteins (Hofmann *et al.*, 2006). In addition, chimeras between S glycoproteins of HCoV-NL63 and HCoV-229E expressed as IgG heavy chain fusion proteins suggest that amino acids 510–549 of HCoV-NL63 S1 are important for its interaction with human ACE2 or for the proper folding and conformation of the fusion protein (Hofmann *et al.*, 2006). Intriguingly, although the amino acid sequences of S1 proteins of HCoV-NL63 and HCoV-229E are conserved they use different receptor proteins.

The RBD of MHV strains (Group 2a) that use mouse CEACAM1a as a receptor is at the amino-terminus of the S glycoprotein (amino acids 1–330) (Kubo *et al.*, 1994; Suzuki and Taguchi, 1996; Taguchi, 1995). Reverse genetics by targeted RNA recombination in MHV-A59 demonstrated that amino acids S33, L79, T82, Y162 and K183 within this amino terminal RBD are important for interaction with CEACAM1a (Thackray *et al.*, 2005). In contrast, the RBD of SARS-CoV (group 2b) that binds to human ACE2 is near the carboxy-terminus of S1 (amino acids 318–510) (Babcock *et al.*, 2004; Wong *et al.*, 2004; Xiao *et al.*, 2003) and residues N479 and T487 of S make critical contacts with human ACE2 (Li *et al.*, 2005a). The SARS-CoV RBD has little amino acid sequence similarity with other coronaviruses, including HCoV-NL63 which also uses human ACE2 as its receptor.

It should be noted that most of the studies that define various RBDs are subject to experimental limitations, and that regions outside of the minimal RBD are very likely to be important for efficient entry. For example, truncation mutagenesis, fusion with other proteins, and the use of chimeras between different viral spikes may result in improper protein folding and incorrect conformation of the putative RBD. Furthermore, although these S1 reagents can be used to identify a region that binds strongly to its receptor, other regions of the S that enhance binding strength or specificity may be missed in these experiments. The apparent plasticity of the S1 proteins and their RBDs that have allowed coronaviruses to adapt to different receptor proteins or receptor orthologs in different species may continue to allow these viruses to emerge in new host species.

Protease cleavage and membrane fusion activity of the S glycoprotein

In the pre-fusion state, class 1 fusion proteins on most enveloped viruses have a single protease cleavage site exposed. Cleavage at this site by proteases (e.g. furin) in the infected cell, or by trypsin or other serine proteases in the extracellular milieu, induces a minor conformational change in the trimer. In general, a hydrophobic fusion peptide immediately follows the cleavage site of class 1 fusion proteins (Figure 1.2A). The fusion peptide inserts into host cell membranes after conformational changes which occur as a result of engagement of the host receptor, a change in the pH, or both. Only coronaviruses in groups 2a and 3 have readily cleavable S glycoproteins. These viruses have a basic amino acid motif in the S that precedes the cleavage site. For example, the S1/S2 cleavage site of the group 2a coronavirus MHV-A59 is preceded by an RRAHR motif (Luytjes *et al.*, 1987), and in IBV strains Beaudette and M41 (group 3), the cleavage site is preceded by an RRFRR motif (Cavanagh *et al.*, 1986). S proteins with mutations that interfere with protease cleavage at this site are less cytotoxic and have less membrane fusion activity. Viruses that express such mutant or variant S proteins enter cells inefficiently, form smaller plaques than wild type viruses and are attenuated *in vivo* (Phillips *et al.*, 2002; de Groot *et al.*, 2001). S proteins of some strains of MHV, such as the OBLV variant derived from MHV-JHM, require not only the murine CEACAM1a receptor, but also acid pH to trigger fusion of the viral envelope with host cell membranes to initiate infection (Nash and Buchmeier, 1997). Entry of MHV-OBLV therefore occurs in acidified endosomes instead of at the plasma membrane where MHV strains with readily cleavable S proteins enter the cell (Nash and Buchmeier, 1997). In coronaviruses of groups 1 and 2b there are no obvious trypsin cleavage sites between the S1 and S2 domains, but these viruses efficiently infect susceptible cells. HCoV-229E (Blau and Holmes, 2001) and TGEV (Hansen *et al.*, 1998) in group 1 and SARS-CoV (Qinfen *et al.*, 2004; Yang *et al.*, 2004b) in group 2b also enter cells via receptor-mediated endocytosis. The SARS-CoV S protein is cleaved in the endosomes by acid-activated cathepsin proteases (Huang *et al.*, 2005; Simmons *et al.*, 2005). Although IBV in group 3 has a readily cleavable S, virus entry appears to occur via receptor-mediated endocytosis rather than at the plasma membrane (Chu *et al.*, 2006). IBV entry is inhibited by inhibitors of acidification of endosomes, and fusion of virions with host cells occurs in an acid pH-dependent and protease-independent manner (Chu *et al.*, 2006).

Some coronavirus infections lead to syncytia formation *in vivo*. *In vitro* expression of recombinant coronavirus S proteins that are readily cleaved by furin or trypsin in cells that express the appropriate virus receptor causes syncytia formation and cell death (Frana *et al.*, 1985; Sturman *et al.*, 1985). The S protein of FCoV in group 1, which lacks a cleavage motif between S1 and S2, can also cause cell–cell fusion at neutral pH (Vennema *et al.*, 1990) by an unknown mechanism.

Protease cleavage, cell–cell fusion and a phenomenon known as "receptor-independent spread" of coronavirus infection are interrelated (Gallagher *et al.*, 1992). The S glycoprotein of MHV-JHM (also known as MHV-4) plays a primary role in the tissue tropism and neurovirulence of this strain (Phillips *et al.*, 1999; Phillips *et al.*, 2002). MHV-JHM has more basic amino acids (RRARR) at its S1/S2 cleavage site than MHV-A59, resulting in more efficient cleavage of the S glycoprotein. Binding of MHV-JHM S to its mCEACAM1a receptor is required to initiate infection, however this strain can spread from an infected cell to cells that lack the receptor or to receptor-bearing cells treated with an antireceptor monoclonal antibody (Mab CC1) (Gallagher *et al.*, 1992; Nash and Buchmeier, 1996). Cell-to-cell fusion is probably very important *in vivo* because it allows virus infection to spread efficiently to naïve cells that express specific receptors, generating syncytia, or transmits viral infection to cells that lack a specific receptor, and/or may permit spread of infection from cell to cell in the presence of anti-S neutralizing antibody (Fiscus and Teramoto, 1987; Nash and Buchmeier, 1996). Although cleavage of S by host proteases is critical for infection and pathogenesis of specific coronavirus strains in group 2a, protease cleavage is apparently not essential for entry of all coronaviruses.

Pre-coil domain and fusion peptide of coronavirus S2 proteins

In most class 1 viral fusion proteins, a hydrophobic fusion peptide (FP) immediately follows the protease cleavage site. However, in coronavirus S2 and filovirus Gp2 glycoproteins the hydrophobic fusion peptide is not adjacent to the cleavage site, but lies within the S2 or fusion domain. A large pre-coil domain (PC) unique to coronaviruses lies between the cleavage site and the fusion peptide (Figure 1.2A). The role of the pre-coil domain in membrane fusion and virus entry is unknown. Within the fusion domain are two heptad repeat regions, here called HR-N and HR-C, the membrane distal and proximal HR regions, respectively (Figure 1.2A). Several lines of evidence suggest that the fusion peptide of coronavirus S proteins immediately precedes the HR-N. Peptide scanning of SARS-CoV spike protein identified a hydrophobic peptide at this site that causes leakage of liposomes (Sainz, Jr. *et al.*, 2005; Guillen *et al.*, 2005). In addition, mutational analysis implicated this region as a coronavirus fusion peptide (Luo and Weiss, 1998; Petit *et al.*, 2005; Bosch *et al.*, 2003). How this internal fusion peptide is exposed to host cell membranes during virus entry is not completely understood.

S2 heptad repeat regions, inter-helical domains and membrane fusion

In coronavirus S2 proteins, a large (~140 amino acid), unique inter-helical domain (IH) lies between the two heptad repeats (Figure 1.2A). The class 1 viral fusion proteins of orthomyxovirus HA, retrovirus env and filovirus gp2 are much shorter than this large coronavirus IH domain (Figure 1.2A). Only paramyxovirus F proteins have a larger IH domain than coronavirus S protein (Figure 1.2A). The function of the coronavirus IH domain is not understood. However, based on the important role of the tiny IH domain of influenza HA glycoprotein in conformational changes leading to membrane fusion, it is possible that the IH domain of coronaviruses also plays a role in the receptor- and/or low pH-induced conformational changes in the S glycoprotein. The presence of the hydrophobic, membrane active peptide in the IH domain of coronaviruses (Sainz, Jr. *et al.*, 2005) suggests that the IH may undergo complex conformational changes to facilitate movement of HR-N and HR-C to form the 6 helix bundle found in the post-fusion state of the S glycoprotein.

The HR-N and HR-C heptad repeat domains in the trimeric coronavirus S2 protein form coiled-coil structures. They are largely responsible for the conformational changes in the S2 which take place during conversion from the pre-fusion to post-fusion conformations of the S protein (Figure 1.2B) (Tripet *et al.*, 2004; Petit *et al.*, 2005; Bosch *et al.*, 2003). For some coronaviruses (e.g. MHV-A59), binding to its soluble receptor glycoprotein at 37°C triggers conformational changes both in S1 and in S2 that makes the entire virion hydrophobic (Lewicki and Gallagher, 2002; Zelus *et al.*, 2003). For other coronaviruses low pH may be required for membrane fusion activity of S, since these viruses enter via acidified endosomes and do not cause syncytia formation at neutral pH (Blau and Holmes, 2001; Hansen *et al.*, 1998; Qinfen *et al.*, 2004; Yang *et al.*, 2004b; Chu *et al.*, 2006). Tripet *et al.*, 2004, proposed a model of membrane fusion by coronavirus S after receptor binding alone includes conformational changes in S, the S1 protein changes orientation or disassociates from S2, exposing the fusion peptide (FP) in S2 which then associates with the lipids in the host cell membrane (Figure 1.2B). The membranes of the host cell and the viral envelope are drawn together by a subsequent massive conformational change in S2 proteins of a number of spikes acting in concert. A fusion pore is formed, allowing the viral nucleocapsid to enter the host cell (Figure 1.2B). The HR-N and HR-C regions are critical for the conversion of S from its native, pre-fusion conformation on the virion to the post-fusion conformation that is implicated in membrane fusion activity (Figure 1.2B) (Tripet *et al.*, 2004). The solution structure of the pre-fusion state of the SARS-CoV HR-C region (amino acids 1141–1193) was determined by NMR (Hakansson-McReynolds *et al.*, 2006). In its pre-fusion conformation, SARS-CoV HR-C forms a trimeric coiled-coil structure, consisting of a bundle of alpha helices that create a left-handed superhelix (Hakansson-McReynolds *et al.*, 2006). In the post-fusion conformation, coronavirus HR-N and HR-C peptides form a six helix bundle similar to those of other class 1 viral fusion proteins (Xu *et al.*, 2004a; Xu *et al.*, 2004b). The six helix bundle of SARS-CoV and MHV-A59 were elucidated by biophysical studies and X-ray crystallography of complexes formed between the purified HR-N and HR-C peptides (Tripet *et al.*, 2004; Xu *et al.*, 2004a; Xu *et al.*, 2004b). The MHV-A59 and SARS-CoV six helix bundle structures are similar to the fusion core of influenza A HA, and HIV gp41, which suggest that the fusion process mediated by S2 of coronaviruses utilizes mechanisms common to all three virus families (Xu *et al.*, 2004a; Xu *et al.*, 2004b). Peptides that correspond with heptad repeat domains of HIV-1 env glycoprotein prevent membrane fusion and inhibit virus entry (Kilby *et al.*, 1998; Bianchi *et al.*, 2005). An HR-C peptide called T20 is used in concert with other antiretroviral drugs in highly active retrovirus therapy (HAART) to treat HIV/AIDS patients. Similarly, HR peptides of coronavirus S proteins inhibit membrane fusion and virus infectivity (Bosch *et al.*, 2003; Yan *et al.*, 2006; Sainz, Jr. *et al.*, 2006; Liu *et al.*, 2004). In addition, antibodies that specifically bind to the pre-fusion coiled coil conformation of HR-C inhibit the infectivity of SARS-CoV virions (Tripet *et al.*, 2006).

Aromatic-, transmembrane- and endo-domains of S proteins

A short region rich in aromatic amino acids (KWPW[Y/W]VWL) preceding the hydrophobic transmembrane domain is conserved in all coronavirus S glycoproteins. A peptide containing this aromatic-rich domain, like the fusion peptide and a peptide in the IH domain, can cause liposome leakage (Guillen *et al.*, 2005). Aromatic-rich regions are found proximal to the transmembrane domains of class 1 viral fusion proteins of other viruses, and they may also play a role in pore formation and membrane fusion.

The transmembrane domain is followed by a short ~26 amino acid endo-domain that lies within the virion or in the cytoplasm of infected cells. The transmembrane and endo-domain are important for the incorporation of S trimers into budding coronavirus virions (Masters, 2006). This region of the trimeric spike protein interacts with the viral M protein during budding, and contains important motifs that direct intracellular transport of the S glycoprotein (Masters, 2006). For incorporation into coronavirus virions, chimeric spike glycoproteins can contain ecto-domains of any other coronavirus, but they must contain the endo-domains from the same coronavirus as the M protein to form infectious virions (Bosch *et al.*, 2005). A cysteine-rich domain in the endodomain proximal to the membrane is essential for virus maturation and infectivity, and contains the site of acylation of the S protein (Masters, 2006; Ye *et al.*, 2004; Thorp *et al.*, 2006). A novel di-lysine ER-retrieval motif at the C terminus of IBV S protein retains the newly formed protein in the ER to facilitate incorporation of S proteins into budding virions (Youn *et al.*, 2005). Adjacent to this motif in the endodomain of S is a tyrosine-based endocytosis motif that promotes recycling of the S protein from the plasma membrane (Youn *et al.*, 2005).

Hemagglutinin-esterase (HE) glycoprotein

Binding of viral envelope glycoproteins to carbohydrate moieties on cell membranes plays a significant role in infection by many viruses. Some coronaviruses in group 2a express a second surface glycoprotein, HE, that has hemagglutination and acetyl-esterase activities. Electron microscopy shows that coronaviruses which express HE glycoproteins are morphologically distinct from those that lack HE. The HE appears as small peplomers (5–7 nanometers long) that do not extend as far from the virion envelope as the S glycoprotein trimers (Figure 1.1A) (de Groot, 2006; Masters, 2006). HE monomers produced in infected cells form disulfide-bonded homo-dimers and are anchored in the viral envelope by a carboxy-terminal hydrophobic domain (de Groot, 2006). Coronaviruses that encode an HE gene include most MHV strains, rat-CoV, bovine-CoV (BCoV), HCoV-OC43, equine-CoV, porcine hemagglutinating encephalomyelitis virus (PHEV), and HKU1-CoV (de Groot, 2006). Although HE genes are not found in the genomes of coronaviruses in groups 1, 2b, or 3, orthologs of coronavirus group 2a HE genes are found in genomes of toroviruses and influenza C virus (de Groot, 2006). Thus, the HE gene was probably acquired by an ancestral group 2a coronavirus from a viral or cellular gene, by RNA recombination (de Groot, 2006). HE binds to sialic acids causing hemagglutination, and its acetyl esterase activity cleaves off specific O-acetyl groups from these sialic acids on glycoproteins or glycolipids. Thus, HE binds to specific sialoglycoconjugates, and also specifically cleaves these sialoglycoconjugates, acting as a "receptor-destroying enzyme" similar to the neuraminidases of orthomyxoviruses and paramyxoviruses. Coronaviruses have evolved specific interactions with glycoconjugates that can be very selective and affect viral tropism. Like the influenza C HE, the HE of BCoV and HCoV-OC43 show a binding preference for 5-N-acetyl-9-O-acetylneuraminic acid (de Groot, 2006; Schwegmann-Wessels and Herrler, 2006; Vlasak *et al.*, 1988). Variation in the acetyl esterase activity of different coronavirus HEs shows that their substrate specificity is strain dependent. The HE glycoproteins of BCoV, HCoV-OC43, PHEV, and MHV-DVIM are sialate-9-O-acetylesterases, whereas HE proteins of MHV-S, MHV-JHM, and rat-CoV are sialate-4-O-acetylesterases (de Groot, 2006). Additionally, MHV-S and JHM appear to specifically bind 5-N-acetyl-4-O-acetylneuraminic acid in solid phase assays (de Groot, 2006).

The HE is largely regarded as an accessory protein and it appears to be a selective disadvantage *in vitro*, yet the gene is maintained by multiple strains of coronavirus in the natural setting and is a pathogenic determinant *in vivo* (de Groot, 2006; Masters, 2006). There may be many roles for HE in coronavirus binding and replication. In the respiratory tract, HE, like influenza HA and NA, may allow reversible binding of virions to sialoglycoconjugates in the mucus, and its esterase activity may allow the virus to penetrate through the mucus layer to reach the epithelial cell surface. HE binding to 5-N-acetyl-O-acetyl-neuraminic acids may serve the same function as binding of herpes virus gB glycoprotein to heparan sulfate residues on cell membranes that concentrates virions on the cell surface where other viral glycoproteins come into close proximity to more specific receptors that are essential for virus entry (Spear, 2004). Another possible function of HE would be to facilitate release of virions from infected cells, as the neuraminidase NA glycoprotein does for influenza A virus. As coronavirus HE is synthesized and transported through the ER and Golgi, its acetylesterase activity cleaves O-acetyl groups from cellular membrane glycoconjugates that are also in transit through these organelles. Virions that are released from the infected cell would then be less likely to bind to previously infected cells than to neighboring uninfected cells. For the group 2a coronaviruses, this acetyl esterase activity of HE, like the neuraminidase activity of influenza virus, is likely to aid in spread of virus infection *in vivo*.

Coronavirus receptor glycoproteins and their interactions with viral S proteins

Receptor glycoproteins for some coronaviruses have been identified, and the specificity of spike-receptor interactions is a major determinant of viral host range, tissue tropism, and pathogenesis. Related coronaviruses generally utilize the same glycoprotein or orthologous glycoproteins as receptors (Figure 1.3). However this is not always the case. Some closely related coronaviruses use unrelated receptors, and conversely some divergent coronaviruses use the same receptor (Figure 1.3). This section of the chapter focuses on the interaction of different coronavirus spikes, particularly the S1 domain, with CEACAM1a, APN, or ACE2 and illustrate how different receptor isoforms, post-translational modifications, and amino acid variations between species influence coronavirus receptor activity.

Carcinoembryonic antigen cell adhesion molecule 1 (CEACAM1 or CD66a)

The receptors for MHV strains are mouse glycoproteins in the carcinoembryonic antigen family of glycoproteins in the immunoglobulin (Ig) superfamily. The carcinoembryonic antigen cell adhesion molecule 1 (CEACAM1) gene encodes the principal receptor glycoproteins for MHV (Dveksler *et al.*, 1991; Williams *et al.*, 1991; Dveksler *et al.*, 1993b).

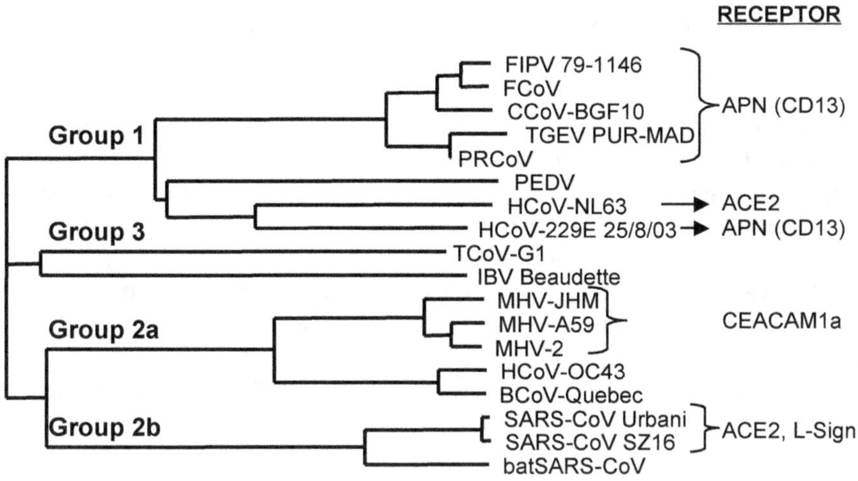

Figure 1.3 Amino acid relationship between the S proteins of coronaviruses representing groups 1–3. Known protein receptors for specific viruses are listed on the right. The tree was generated with Vector NTI (Invitrogen Inc.) using a PAM 250 matrix, Kimura's correction and ignoring the positions with gaps. The viruses used for this alignment include; FIPV 79–1146 (AY994055), FCoV (X80799), CCoV-BGF10 (AY342160), TGEV-PUR46-MAD (NC 002306), PEDV1 (NC003436), PRCoV (Z24675), HCoV-NL63 (AY567487), HCoV-229E 25/8/03 (DQ243986), TCoV-G1 (AY342357), IBV-Beaudette (P11223), MHV-JHM (NC 006852), MHV-A59 (AY700211), MHV-2-Pen97 (AF208066), HCoV-OC43 (NC 005147), BCoV-Quebec (AF220295), SARS-CoV-Urbani (AY278741), SARS-CoV-SZ16 (AY304488), batSARS-CoV (DQ022305).

CEACAM1 (CD66a) glycoproteins also play important roles in embryology, immunology, cancer, and angiogenesis. CEACAM1 glycoproteins are expressed on apical membranes of epithelial cells in the gastrointestinal and respiratory tracts, by vascular endothelial cells, macrophages, B cells, neutrophils, activated T cells, colonic M cells and on dendritic cells (Godfraind *et al.*, 1995; Obrink, 1991; McCuaig *et al.*, 1992; Zhou *et al.*, 2001; Baranov and Hammarstrom, 2004; Turner *et al.*, 2004). In inbred mouse strains, two alleles of CEACAM1 determine susceptibility or resistance to MHV infection *in vitro* and *in vivo*. Highly susceptible mice, such as BALB/c and C57BL/6, express the CEACAM1a allele, whereas a SJL/J mice, which are highly resistant to MHV, are homozygous for the CEACAM1b allele (Knobler *et al.*, 1981; Knobler *et al.*, 1984; Dveksler *et al.*, 1993a). Homozygous CEACAM1a knockout mice are completely resistant to infection with MHV-A59 by both intranasal and intracerebral routes of inoculation (Hemmila *et al.*, 2004).

Alternative splicing of the mouse CEACAM1a RNA transcript yields four isoforms that share the amino-terminal domain (D1) and all have MHV receptor activity (Dveksler *et al.*, 1993b). The amino acid sequences of CEACAM1a and CEACAM1b primarily differ in the amino-terminal domain (D1), which is important for binding MHV S glycoprotein. The two domain (D1, D4) CEACAM1a isoform is commonly expressed by cell lines used to propagate MHV and it is an efficient receptor *in vitro*. A recombinant soluble form of CEACAM1a (D1, D4) binds to MHV-A59 S and has strong MHV neutralizing activity because this interaction leads to conformational changes in S (Zelus *et al.*, 2003). The crystal structure of this soluble form of CEACAM1a (D1, D4) was solved at 3.32 Å (Figure 1.4) (Tan *et al.*, 2002). This structure reveals unique regions in the CC″ loop in D1 (Figure 1.4, light gray) that were shown by mutational analysis to be critical for MHV receptor activity (Rao *et al.*, 1997; Wessner *et al.*, 1998). The amino acid sequence of the CC″ loop differs significantly between CEACAM1a and 1b. A projecting hydrophobic residue, isoleucine 41, in murine CEACAM1a (Figure 1.4) is a "hot spot" in this loop that may provide the major binding energy for interaction with the S of MHV-A59 (Tan *et al.*, 2002). Isoleucine 41 of CEACAM1a may interact with Y162 of MHV S, which is part of a putative hydrophobic domain in the RBD (Thackray *et al.*, 2005). Nearby on the virus-binding face of CEACAM1a are charged residues; R96, E93 and R47 that may contribute to the specificity of the interaction between the spike and receptor (Figure 1.4). In CEACAM1b, I-41 is replaced by T, and R47 is replaced by H. Substitution of these residues in murine CEACAM1a markedly reduces its MHV receptor activity (Turner and Holmes, unpublished data).

The roles that domains 2, 3, and 4 of CEACAM1 play in coronavirus entry are poorly understood. Yet, studies indicate that these other domains are also important in the entry process. For example, *in vitro* studies show that soluble CEACAM1a containing

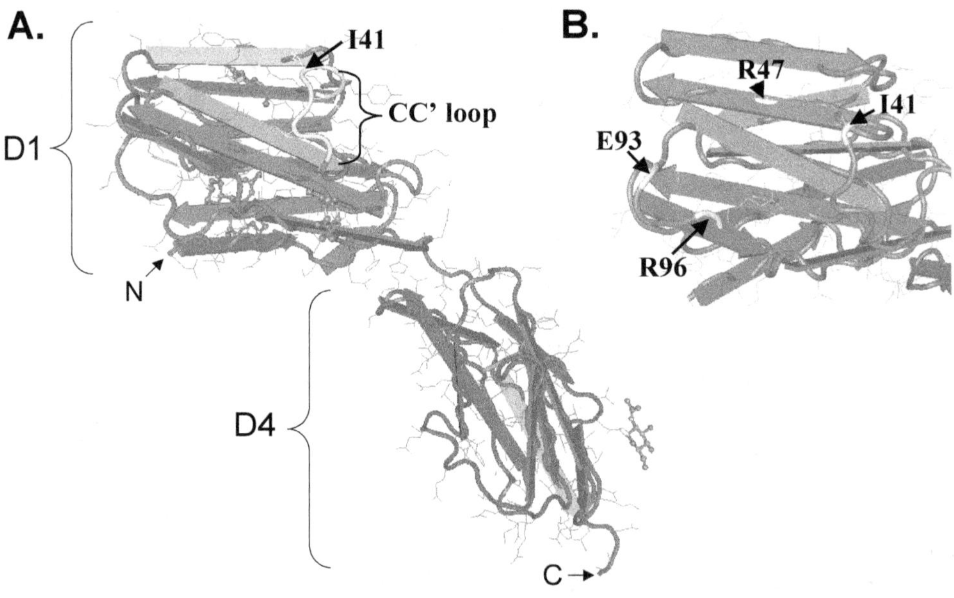

Figure 1.4 Crystal structure of a two domain isoform of the MHV receptor CEACAM1a. A, Drawing of a monomer of mouse CEACAM1a [1,4], illustrating the two Ig-like domains, D1 and D4 oriented such that the cell membrane is on the bottom of the figure. The amino (N) and carboxy (C) termini are indicated and amino acids 36–43 that make up the CC" loop in D1 are highlighted in light gray. The protein backbone is shown as tubes, the amino acid side chains as thin wire and the strands are indicated with flat arrows. B, A top view of the virus binding face on D1. Arrows indicate isoleucine 41, arginine 47, glutamic acid 93, and arginine 96 which are also highlighted in light gray. This figure was generated using Cn3d 4.1 software from the National Center for Biotechnology Information.

domains 1 and 2 has little MHV neutralizing activity, whereas soluble CEACAM1a containing domains 1 and 4 has strong MHV neutralizing activity (Zelus *et al.*, 1998). The CEACAM1a isoforms expressed *in vivo* are also important determinants of MHV pathogenesis. Transgenic mice with a targeted disruption in the CEACAM1a allele express low levels of the CEACAM1a four domain (D1-D4) isoform, and unusually high levels of the naturally occurring two domain isoform (D1,D4). Compared to wild type BALB/C mice, the transgenic animals that expressed altered isoform ratios of CEACAM1a, had fewer and much smaller lesions in their livers, and much higher LD_{50} titers after inoculation with MHV-A59 (Blau *et al.*, 2001).

Aminopeptidase N (APN or CD13)

The serogroup 1 coronaviruses HCoV-229E, FCoV, and TGEV all use APN as receptors (Delmas *et al.*, 1992; Yeager *et al.*, 1992; Tresnan *et al.*, 1996). APN is a ~150kDa type II glycoprotein, composed of 967 amino acids, and is a zinc-containing amino-peptidase also known as CD13 (Look *et al.*, 1989; Noren *et al.*, 1997). APN is expressed as a heavily glycosylated dimer on the surface of epithelial cells of the kidney, intestine and respiratory tract (Look *et al.*, 1989; Noren *et al.*, 1997) APN is also expressed by granulocytes, fibroblasts, endothelial cells, cerebral pericytes at blood-brain barrier, synaptic junctions, macrophages, and dendritic cells (Riemann *et al.*, 1999; Dong *et al.*, 2000). APN aids in removal of amino acids from peptides in the lumen of the small intestine, cleaves peptides bound to MHC class II molecules on antigen presenting cells, and degrades neuropeptides in synaptic junctions (Riemann *et al.*, 1999; Dong *et al.*, 2000). The high levels of APN surface expression on respiratory and intestinal epithelium are probably important in coronavirus infection, tissue tropism, and transmission.

Human, feline and porcine APN orthologs show strong amino acid conservation and have ~83% similarity and ~78% identity. Yet, species-specific amino acid and/or post-translational differences in APN glycoproteins influence receptor activity for HCoV-229E, FCoV, and TGEV. For example, porcine APN is a receptor for TGEV, but not for FCoV or HCoV-229E (Delmas *et al.*, 1994; Benbacer *et al.*, 1997). Similarly, human APN is a receptor for HCoV-229E but not TGEV, CCoV or FCoV. In contrast, feline APN functions as a receptor for FCoV, TGEV, HCoV-229E and CCoV *in vitro* (Tresnan *et al.*, 1996). Although canine APN has not been examined directly, studies show that both feline APN (Tresnan *et al.*, 1996) and chimeric human/canine APN proteins have CCoV receptor activity (Benbacer *et al.*, 1997), suggesting that canine APN serves as a CCoV receptor in canine cells. Chimeras of human, porcine, feline, and canine APN proteins were used to identify regions of APN critical for receptor activity for these group 1 coronaviruses (Benbacer *et al.*, 1997; Kolb *et al.*, 1997). Amino acids 260–353 of human APN are required for HCoV-229E entry, and within this region, amino acids 288–295 strongly influence coronavirus receptor activities of porcine/human chimeras (Kolb *et al.*, 1996; Kolb *et al.*, 1997). Insertion of a single asparagine linked glycosylation sequon from porcine APN at amino acids 291–293 of human APN blocked HCoV-229E infection (Wentworth and Holmes, 2001). In contrast to HCoV-229E, amino acids 670–840 of APN are important for the entry of TGEV, FCoV, and CCoV (Hegyi and Kolb, 1998). An antifeline APN monoclonal antibody (RG4) blocks feline APN-mediated entry of all four of these coronaviruses, suggesting that the discontinuous regions of APN required for entry of the human verses animal group 1 coronaviruses are close together on the surface of APN (Hohdatsu *et al.*, 1998; Tusell *et al.*, 2006). The observation that feline APN functions as a common receptor for these four coronaviruses suggests that there may be a conserved spike-binding domain in APN glycoproteins, and that minor species-specific differences in amino acids or post-translational modifications in or near this domain may inhibit interactions of APN with spikes of coronaviruses that infect other species. Although mouse APN has no coronavirus receptor activity, amino acid substitutions in mouse APN that removed the asparagine-glycosylation sequon at amino acids 288–290, that blocks HCoV-229E receptor activity in porcine APN (Wentworth and Holmes, 2001), caused a gain of receptor activity for HCoV-229E, but not for FCoV, CCoV or TGEV (Tusell *et al.*, 2006). In feline APN, the hydroxyl group of T742 is required for entry of FCoV, CCoV and TGEV, since T742V has no receptor activity while T742S has receptor activity for these viruses (Tusell *et al.*, 2006). Substitution of T742R in feline APN, which is found at the homologous site (741) in human APN, also destroyed receptor activity for FCoV, TGEV, and CCoV. Conversely, substitution of R741T in human APN caused a gain of receptor activity for TGEV, but not FCoV or CCoV (Tusell *et al.*, 2006). In mouse APN, substitution of two short discontinuous amino acid stretches with orthologous regions from feline APN, amino acids 732–746 and 754–788, conferred receptor activity for FCoV, and CCoV (Tusell *et al.*, 2006). Whereas, substitution of amino acids 732–746 from feline APN into mouse APN conferred TGEV receptor activity (Tusell *et al.*, 2006). Thus in the broad interface between the 200 kDa spike protein and its 150 kDa APN receptor, single amino acids or small groups of discontinuous amino acids determine the species specificity of coronavirus infection (Wentworth and Holmes, 2001; Tusell *et al.*, 2006; Benbacer *et al.*, 1997). Structural information for APN, group 1 coronavirus spikes, and the S-APN complex will undoubtedly provide more insight into the molecular determinants of group 1 coronavirus species specificity.

Angiotensin converting enzyme 2 (ACE2)

In 2002–2003, an outbreak of SARS spread from Southern China to 28 other countries and infected more than 8000 people, resulting in 774 deaths (WHO, 2004). SARS-CoV was identified as the causative agent of the outbreak (Drosten *et al.*, 2003; Ksiazek *et al.*, 2003; Peiris *et al.*, 2003). The identification of RNA from SARS-CoV-like viruses and anti-SARS-CoV antibodies in several species in live animal markets, suggests that the SARS outbreak resulted from zoonotic transmission(s) (Guan *et al.*, 2003; Tu *et al.*, 2004; Li *et al.*, 2006). Specifically, viruses genetically related to SARS-CoV and antibodies to SARS-CoV were detected in Himalayan palm civets (*Paguma larvata*) and raccoon dogs (*Nyctereutes procyonoides*) (Guan *et al.*, 2003; Tu *et al.*, 2004). Rapid evolution of SARS-CoV in palm civets and humans (Song *et al.*, 2005) suggests that the virus was adapting to its new hosts from SARS-CoV-like viruses enzootic in another species, and that palm civets were an intermediate or amplifying host, that lead to zoonosis (Li *et al.*, 2006). The reservoir of SARS-CoV-like coronaviruses appears to be several species of horseshoe bats (*Rhinolophus sinicus, R. pearsoni, R. pussilus,* and *R. macrotis*) (Lau *et al.*, 2005; Li *et al.*, 2005b). Unlike many other CoVs, SARS-CoV isolates from humans have a broad host range and infect cells from a wide variety of species including human, African green monkey, Rhesus macaque, Cynomolgus macaque, mink, and pig, *in vitro* (Gillim-Ross *et al.*,

2004; Mossel *et al.*, 2005). SARS-CoV also experimentally infects diverse animal species including non-human primates, ferrets, cats, mice, and hamsters which can be used to model aspects of SARS-CoV infection (Fouchier *et al.*, 2003; Martina *et al.*, 2003; Czub *et al.*, 2005; Roberts *et al.*, 2005; Subbarao *et al.*, 2004; Wentworth *et al.*, 2004).

Although SARS-CoV was only recently discovered, there has been tremendous progress toward the understanding of important spike-receptor interactions. Two receptors have been identified and the crystal structures of the most efficient receptor and fragments of the SARS-CoV spike have been solved. Human ACE2 is the principal receptor for SARS-CoV in group 2b (Li *et al.*, 2003) and it is also a receptor for HCoV-NL63 (Hofmann *et al.*, 2005; Smith *et al.*, 2006) in group 1 (Figure 1.3). ACE2 is an important regulator of the renin–angiotensin system that maintains blood pressure and is critical to cardiovascular and renal function (Donoghue *et al.*, 2000; Danilczyk *et al.*, 2003; Oudit *et al.*, 2003; Warner *et al.*, 2004). ACE2 is a zinc-containing carboxypeptidase that cleaves angiotensin 1 into angiotensin (1–9) and angiotensin II, a vasoconstrictor, into angiotensin (1–7) (Donoghue *et al.*, 2000; Warner *et al.*, 2004). The ACE2 gene was previously defined as a quantitative trait locus on the X chromosome and its function is important in diseases such as hypertension, cardiovascular disease, and diabetes (Danilczyk *et al.*, 2003; Danilczyk *et al.*, 2004).

The predicted open reading frame for human ACE2 encodes an 805 amino acid glycoprotein. Compared to human ACE2 the predicted ACE2 glycoproteins from other mammals have strong amino acid conservation, which ranges from 95% identity for African green monkey to 82% identity for mouse (Figure 1.5) (Heller *et al.*, 2006; Olivieri *et al.*, 2006). Analysis of ACE2 orthologs revealed some specific amino acids that may be important for receptor activity, and mapping the differences on the crystal structure of ACE2 (Towler *et al.*, 2004) was used as a guide for an informative ACE2 mutagenesis study (Li *et al.*, 2005c). Mouse and rat ACE2 are inefficient receptors for epidemic strains

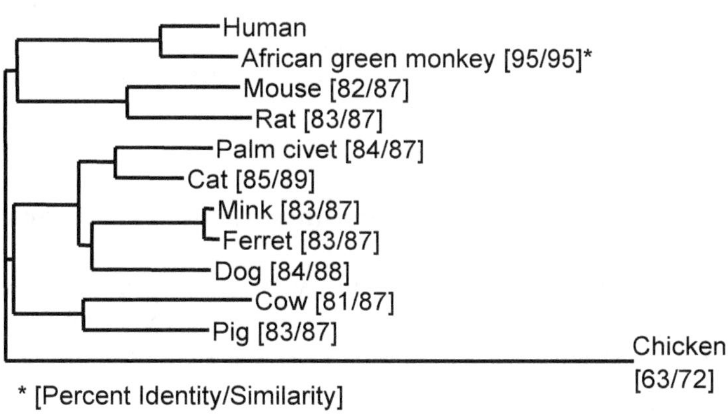

Figure 1.5 Amino acid comparison of angiotensin converting enzyme 2 orthologs. The deduced amino acid sequence of ACE2 from human, African green monkey, mouse, rat, Himalayan palm civet, domestic cat, mink, ferret, dog, cow, pig and chicken were compared by multiple sequence alignment and the percent amino acid identity or similarity with human ACE2 is shown in brackets. The tree depicting their amino acid relationship was generated by Vector NTI (Invitrogen, Inc.) using a blosum62mt2 matrix and Kimura's correction.

of SARS-CoV (e.g. Urbani, Tor2), whereas human and palm civet ACE2 have very strong receptor activity (Li *et al.*, 2004; Li *et al.*, 2005c). Chimeras of human and rat ACE2 were used to identify residues critical for binding to a SARS-CoV S1 domain fused to the Fc domain of human IgG1 (S1-Ig) and for infection by an HIV-1 based vector pseudotyped with SARS-CoV spikes (Li *et al.*, 2005c). Substitution of amino acid residues N82M, F83Y, and S84P, that encode a glycosylation site in rat ACE2, for those of human ACE2, as well as substitution of H353K dramatically improved SARS-CoV S1-Ig binding to this mutant rat ACE2 (Li *et al.*, 2005c). The putative RBD fragment of SARS-CoV S1 (amino acids 318–510) (Babcock *et al.*, 2004) was also fused to the Fc domain of human IgG1 (RBD-Ig) and used to analyze ACE2 variants and S mutations that were identified in SARS-CoV-like viruses found in animals (Li *et al.*, 2005c). Comparison of SARS-CoV strains that transmitted well from human-to-human (e.g. Urbani or TOR2) with SARS-CoV-like strains found in animals (e.g. SZ3 and SZ16) identified amino acids within the RBD of S1 that are important in binding to human, palm civet and rat ACE2 (Li *et al.*, 2005c; Li *et al.*, 2006). The RBD-Ig variants examined, demonstrated that amino acids 479 and 487 of SARS-CoV S play essential roles in binding and species-specificity (Li *et al.*, 2005c). Co-crystallization of baculovirus-expressed peptidase domain (amino acids 19–615) of ACE2 with an RBD fragment (amino acids 306–527) of SARS-CoV S yielded structure at 2.9 Å resolution (Li *et al.*, 2005a) (Figure 1.6). This structure confirms the importance of K479N and S487T substitutions in the spike protein that differ between SARS-CoV-like animal viruses (e.g. SZ3 and SZ16) and epidemic strains of SARS-CoV (e.g. Urbani), respectively (Li *et al.*, 2005a; Li *et al.*, 2005c). The co-crystal structure (Li *et al.*, 2005a) shows that N479 in the SARS-CoV S contacts H34 of human ACE2, and T487 of S interacts with both Y41 and K353 of human ACE2 (Figure 1.6). Other specific amino acids within S and outside of RBD may also be important in binding/entry into human cells. Alignment of the predicted amino acids of bat-SARS-CoV S with SARS-CoV/Urbani illustrates that many of the amino acids critical for interaction of Urbani-RBD with human ACE2 have substitutions in the bat SARS-CoV S (Figure 1.6). Additionally, there are amino acid deletions in this region of the bat-SARS-CoV S that are likely to alter the structure of this region of the S1 (Figure 1.6). This observation suggests that amino acids outside of the "RBD" may also be critical for binding to ACE2 or that bat SARS-CoV uses a receptor other than ACE2.

Accessory factors or alternative receptors in coronavirus entry

Interactions of coronavirus S glycoproteins with sialoglycoconjugates

The S glycoproteins of some coronaviruses in group 1 (TGEV), group 2a (BCoV and HCOV-OC43) and group 3 (IBV) also bind specifically to sialoglycoconjugates on cell membranes and can cause hemagglutination (Winter *et al.*, 2006; Schwegmann-Wessels and Herrler, 2006). The interaction of these coronaviruses with sialic acids on glycoproteins and glycolipids is important in virus entry and tissue tropism (Winter *et al.*, 2006; Schwegmann-Wessels and Herrler, 2006).

Among the group 1 CoVs, both enteropathogenic TGEV and the attenuated variant PRCoV use APN as a receptor *in vitro* (Delmas *et al.*, 1992; Delmas *et al.*, 1993). TGEV infects both respiratory and enteric epithelial cells which causes diarrhea and severe disease

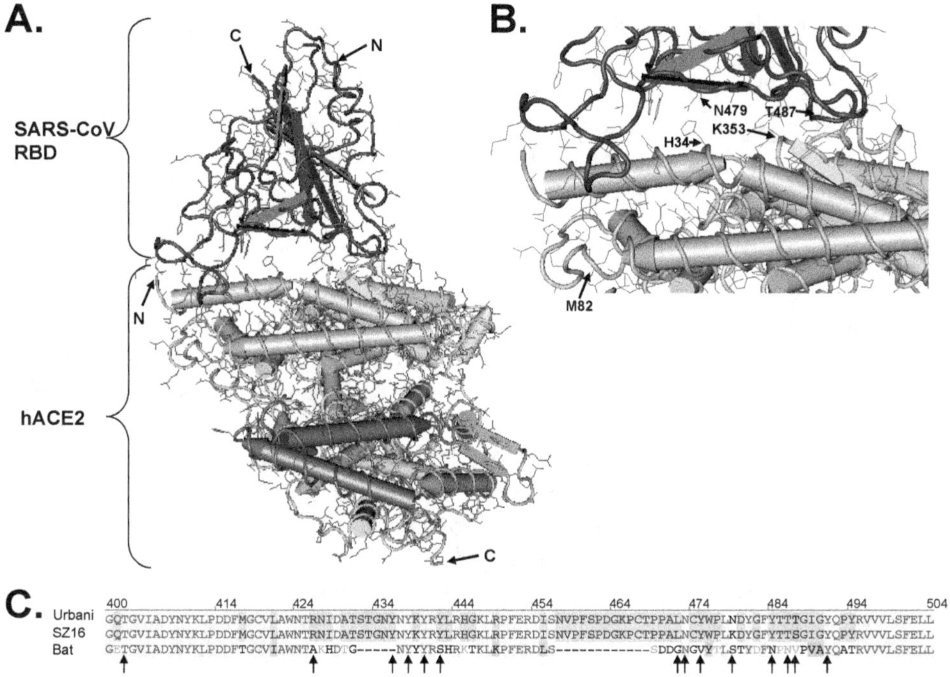

Figure 1.6 Representation of the SARS-CoV RBD co-crystallized with human angiotensin converting enzyme 2. A, The RBD is shown in dark gray binding to the top of ACE2 (light gray). The amino (N) and carboxy (C) termini of both structures are indicated. B, A magnified region of the binding face indicating ACE2 amino acids M82, H34, and K353, and SARS-CoV S amino acids N479 and T487. The protein backbone is shown as tubes, the amino acid side chains as thin wire, the strands are indicated with flat arrows and helix are shown as cylinder arrows. C, Alignment of the deduced amino acids of SARS-CoV-Urbani (human) spike, from residue 400–504, with the deduced sequence of spikes from SARS-CoV-SZ16 (raccoon dog) and batSARS-CoV (Chinese horseshoe bat). Dashes indicate predicted gaps and the arrows at the bottom indicate residues involved in binding of Urbani S with human ACE2. A, and B, were generated using Cn3d 4.1 software from the National Center for Biotechnology Information, and C, was generated using Vector NTI (Invitrogen Inc.).

in suckling pigs (Sanchez *et al.*, 1992; Zhang *et al.*, 2006). Whereas, PRCoV tropism is limited primarily to the respiratory tract and these viruses generally causes mild disease (Sanchez *et al.*, 1992; Zhang *et al.*, 2006). PRCoV strains lack the hemagglutinating activity of TGEV strains, and have a 224 or 227 amino acid deletion near the amino terminus of the spike glycoprotein that is not required for binding to APN (Sanchez *et al.*, 1992; Jackwood *et al.*, 1993). The 224/227 amino acid domain near the N-terminus the TGEV spike glycoprotein is required for binding preferentially to sialoglycoconjugates containing N-glycolyl neuraminic acid (Schultze *et al.*, 1996; Ballesteros *et al.*, 1997). Residues 145, 147, 155, 195, 196, 208 and 209 in this domain of the TGEV S glycoprotein are critical for binding to N-glycolyl neuraminic acid and substitutions at these residues also reduce enteropathogenicity (Bernard and Laude, 1995; Krempl *et al.*, 1997; Krempl *et al.*, 2000). Taken together the data show that naturally occurring deletions or point mutations in

TGEV S can result in loss of sialic acid binding activity and produce respiratory tropic virus variants.

In group 2a, the S glycoproteins of BCoV and HCoV-OC43, like their HE glycoproteins, bind to N-acetyl-9-O-acetylated neuraminic acid moieties on host cells (Kunkel and Herrler, 1993). The S proteins of BCoV and HCoV-OC43 are considered the major sialic acid binding proteins on these viruses (Schwegmann-Wessels and Herrler, 2006) and the receptor destroying acetyl esterase activity of their HE proteins may be critical for dissemination of virus from infected cells. Binding of BCoV/HCoV-OC43 S to N-acetyl-9-O-acetylated neuraminic acid moieties may lead to subsequent interaction with an unknown protein receptor and this initiates fusion of the viral envelope with the cell membrane (Schultze and Herrler, 1994). Thus, in contrast to TGEV, sialic acid plays an essential role in the entry of BCoV and related viruses (Schwegmann-Wessels and Herrler, 2006).

The group 3 coronaviruses are very diverse and infect a variety of avian species, causing respiratory, reproductive and renal disease in poultry (Jia *et al.*, 1995; Jonassen *et al.*, 2005; Cavanagh *et al.*, 2002; Cavanagh *et al.*, 2005; Gough *et al.*, 2006). The most widely studied group 3 virus is IBV-Beaudette, which infects diverse cells (e.g. chicken embryonic kidney and African Green Monkey kidney [Vero]) *in vitro*. S protein of IBV binds preferentially to sialic acid moieties with an alpha 2,3 linkage (Winter *et al.*, 2006). Treatment of these cells with neuraminidase dramatically reduces their susceptibility to infection by the Beaudette and M41 strains of IBV indicating that alpha 2,3 sialic acid may be the primary binding determinant for these strains of IBV (Winter *et al.*, 2006). Like BCoV and HCoV-OC43 a more specific host receptor required for IBV entry remains to be identified.

L-SIGN (CD-209L), DC-SIGN (CD209) and L-SECtin in coronavirus infection

Related C-type lectin family members known as specific ICAM-3 grabbing non-integrins expressed by dendritic cells, DC-SIGN (CD-209), or by sinusoidal endothelial cells in liver and lymph nodes, L-SIGN (CD209L or DC-SIGNR), bind to many bacterial and viral pathogens. These two lectins bind high-mannose oligosaccharides and play important roles in pathogen capture and antigen presentation, which are key to immune recognition by the host (Koppel *et al.*, 2005). However, some pathogens such as HIV-1, hepatitis C virus, and Ebola virus have exploited these molecules to facilitate infection (Simmons *et al.*, 2003; Takada *et al.*, 2004; Liu *et al.*, 2005; Cocquerel *et al.*, 2006; Lai *et al.*, 2006; Barreiro and Quintana-Murci, 2006). Both DC-SIGN and L-SIGN enhance SARS-CoV S pseudotyped retrovirus infection of cells that also express human ACE2 (Gramberg *et al.*, 2005; Marzi *et al.*, 2004; Yang *et al.*, 2004a). L-SECtin, another C-type lectin that is co-expressed with L-SIGN on liver and lymph node sinusoidal endothelial cells, can also promote entry of retrovirus pseudotypes expressing SARS-CoV spikes into cells that also express human ACE2 (Gramberg *et al.*, 2005). However, SARS-CoV S pseudotyped retrovirus infection of cells co-expressing human ACE2 shows more pronounced enhancement by DC-SIGN and L-SIGN than by L-SECtin (Gramberg *et al.*, 2005).

L-SIGN can also function as an alternative receptor for wild type SARS-CoV (Jeffers *et al.*, 2004) or HCoV-229E (Jeffers *et al.*, 2006), although it has much less receptor activity for these viruses than human ACE2 or human APN, respectively. L-SIGN was identified as a functional receptor for SARS-CoV by screening a human lung cDNA library for virus infection (Jeffers *et al.*, 2004). Expression of L-SIGN by normally non-permissive cells that lack human ACE2 confers susceptibility to SARS-CoV (Jeffers *et al.*, 2004). L-SIGN is a type II protein containing an amino-terminal cytoplasmic domain, a neck region that extends from the cell membrane which is followed by a carboxy-terminal carbohydrate recognition domain which binds high-mannose oligosaccharides. Different isoforms of L-SIGN vary because of polymorphisms in tandem repeats of a conserved 23 amino acid sequence in the extracellular neck region of the protein. Repeats of 5 and 7 are commonly found in humans. The 7 repeat isoform predominates in humans (Bashirova *et al.*, 2001; Liu *et al.*, 2006) and it was the isoform identified as a functional SARS-CoV receptor (Jeffers *et al.*, 2004). The majority of the polymorphic isoforms form both homo- or hetero-tetramers and many people are heterozygous and express L-SIGN 5/7 (Guo *et al.*, 2006). Variation in the L-SIGN isoforms expressed by individuals with different genetic backgrounds may also be a factor in their susceptibility to HIV-1, Hepatitis C and SARS-CoV (Bashirova *et al.*, 2001; Chan *et al.*, 2006; Feinberg *et al.*, 2005; Nattermann *et al.*, 2006). The definitive roles played by DC-SIGN, L-SIGN, and L-SECtin in coronavirus pathogenesis remain elusive. However, taken together the data indicate that these three lectins enhance infection mediated by SARS-CoV S *in vitro* (Jeffers *et al.*, 2006; Gramberg *et al.*, 2005; Marzi *et al.*, 2004; Yang *et al.*, 2004a), and examination of the lung tissue from people who died from SARS demonstrated that ACE2 and L-SIGN were co-expressed by the majority of cells that were infected by SARS-CoV *in vivo* (Chan *et al.*, 2006).

Concluding remarks

As coronaviruses evolved and spread to infect many different host species, causing a variety of diseases, they developed several ways to recognize and infect specific cell types in their hosts. Coronavirus binding and entry are important determinants of viral host range and tissue tropism. Binding of the viruses to host cells depends on the viral spike glycoprotein and, for group 2a coronaviruses, on a viral hemagglutinin-esterase glycoprotein. HE and some S glycoproteins bind to specific sialoglyco-conjugates on the host cell surface, concentrating virions on the host cell membrane. However binding of the virus to carbohydrate moieties alone apparently cannot mediate fusion of the viral envelope with host cell membranes. To initiate infection, the large coronavirus S proteins on virions bind to specific receptor glycoproteins on their host cells, which triggers conformational changes in S that result in membrane fusion and virus infection. Lectins on host cell membranes that recognize glycans on viral glycoproteins may also bind virions to cell membranes, where fusion can be mediated by interactions of S with its cognate receptor. Thus, coronavirus entry is a multistep process that requires specific recognition of cellular moieties by the virus. Mutations that alter the receptor-binding domain of S or species-specific differences in the spike-binding domain of the receptor can alter viral host range and tissue tropism as well as pathogenesis. Mutations in S that alter receptor recognition and/or species-specificity may lead to the emergence of novel coronaviruses with epidemic or epizootic potential. Selection of S mutants that recognize receptors on a new host species can cause epidemics of emerging coronavirus diseases in the new human or animal host. Understanding of coronavirus evolution depends upon further investigation of the S and HE glycoproteins of different human and animal coronaviruses, identification of their receptors, and knowledge of their mechanisms of binding and membrane fusion. Virus binding and conformational changes in S that lead to membrane fusion are good targets for development of antiviral drugs and vaccines to prevent coronavirus infection.

Acknowledgments

Work in the authors' laboratories is supported by the National Institutes of Health, NIAID grants P01-AI-0595760, R21-AI-05794102, and R01-AI-25231, and by a Public Health Preparedness and Response to Bioterrorism cooperative agreement between the Department of Health and Human Services, and the Centers for Disease Control and Prevention.

References

WHO (2004). Summary of probable SARS cases with onset of illness from 1 November 2002 to 31 July 2003. World Health Organization http://www.who.int/csr/sars/country/table2004_04_21/en/index.html.

Babcock, G.J., Esshaki, D.J., Thomas, W.D., Jr., and Ambrosino, D.M. (2004). Amino acids 270 to 510 of the severe acute respiratory syndrome coronavirus spike protein are required for interaction with receptor. J. Virol. *78*, 4552–4560.

Ballesteros, M.L., Sanchez, C.M., and Enjuanes, L. (1997). Two amino acid changes at the N-terminus of transmissible gastroenteritis coronavirus spike protein result in the loss of enteric tropism. Virology *227*, 378–388.

Baranov, V., and Hammarstrom, S. (2004). Carcinoembryonic antigen (CEA) and CEA-related cell adhesion molecule 1 (CEACAM1), apically expressed on human colonic M cells, are potential receptors for microbial adhesion. Histochem. Cell Biol. *121*, 83–89.

Baric, R.S., Sullivan, E., Hensley, L., Yount, B., and Chen, W. (1999). Persistent infection promotes cross-species transmissibility of mouse hepatitis virus. J. Virol. *73*, 638–649.

Barreiro, L.B., and Quintana-Murci, L. (2006). DC-SIGNR Neck-Region Polymorphisms and HIV-1 Susceptibility: From Population Stratification to a Possible Advantage of the 7/5 Heterozygous Genotype. J. Infect. Dis. *194*, 1184–1185.

Bashirova, A.A., Geijtenbeek, T.B., van Duijnhoven, G.C., Van Vliet, S.J., Eilering, J.B., Martin, M.P., Wu, L., Martin, T.D., Viebig, N., Knolle, P.A., KewalRamani, V.N., van Kooyk, Y., and Carrington, M. (2001). A dendritic cell-specific intercellular adhesion molecule 3-grabbing nonintegrin (DC-SIGN)-related protein is highly expressed on human liver sinusoidal endothelial cells and promotes HIV-1 infection. J. Exp. Med. *193*, 671–678.

Benbacer, L., Kut, E., Besnardeau, L., Laude, H., and Delmas, B. (1997). Interspecies aminopeptidase-N chimeras reveal species-specific receptor recognition by canine coronavirus, feline infectious peritonitis virus, and transmissible gastroenteritis virus. J. Virol. *71*, 734–737.

Bernard, S., and Laude, H. (1995). Site-specific alteration of transmissible gastroenteritis virus spike protein results in markedly reduced pathogenicity. J. Gen. Virol. *76*, 2235–2241.

Bianchi, E., Finotto, M., Ingallinella, P., Hrin, R., Carella, A.V., Hou, X.S., Schleif, W.A., Miller, M.D., Geleziunas, R., and Pessi, A. (2005). Covalent stabilization of coiled coils of the HIV gp41 N region yields extremely potent and broad inhibitors of viral infection. Proc. Natl. Acad. Sci. USA *102*, 12903–12908.

Blau, D.M., and Holmes, K.V. (2001). Human coronavirus HCoV-229E enters susceptible cells via the endocytic pathway. Adv. Exp. Med. Biol. *494*, 193–198.

Blau, D.M., Turbide, C., Tremblay, M., Olson, M., Letourneau, S., Michaliszyn, E., Jothy, S., Holmes, K.V., and Beauchemin, N. (2001). Targeted disruption of the Ceacam1 (MHVR) gene leads to reduced susceptibility of mice to mouse hepatitis virus infection. J. Virol. *75*, 8173–8186.

Bonavia, A., Zelus, B.D., Wentworth, D.E., Talbot, P.J., and Holmes, K.V. (2003). Identification of a receptor-binding domain of the spike glycoprotein of human coronavirus HCoV-229E. J. Virol. *77*, 2530–2538.

Bosch, B.J., de Haan, C.A., Smits, S.L., and Rottier, P.J. (2005). Spike protein assembly into the coronavirion: exploring the limits of its sequence requirements. Virology *334*, 306–318.

Bosch, B.J., van der Zee, R., de Haan, C.A., and Rottier, P.J. (2003). The coronavirus spike protein is a class I virus fusion protein: structural and functional characterization of the fusion core complex. J. Virol. *77*, 8801–8811.

Breslin, J.J., Mork, I., Smith, M.K., Vogel, L.K., Hemmila, E.M., Bonavia, A., Talbot, P.J., Sjostrom, H., Noren, O., and Holmes, K.V. (2003). Human coronavirus 229E: receptor binding domain and neutralization by soluble receptor at 37 degrees C. J. Virol. *77*, 4435–4438.

Cavanagh, D., Davis, P.J., and Mockett, A.P. (1988). Amino acids within hypervariable region 1 of avian coronavirus IBV (Massachusetts serotype) spike glycoprotein are associated with neutralization epitopes. Virus Res. *11*, 141–150.

Cavanagh, D., Davis, P.J., Pappin, D.J., Binns, M.M., Boursnell, M.E., and Brown, T.D. (1986). Coronavirus IBV: partial amino terminal sequencing of spike polypeptide S2 identifies the sequence Arg-Arg-Phe-Arg-Arg at the cleavage site of the spike precursor propolypeptide of IBV strains Beaudette and M41. Virus Res. *4*, 133–143.

Cavanagh, D., Mawditt, K., Welchman, D.B., Britton, P., and Gough, R.E. (2002). Coronaviruses from pheasants (Phasianus colchicus) are genetically closely related to coronaviruses of domestic fowl (infectious bronchitis virus) and turkeys. Avian Pathol. *31*, 81–93.

Cavanagh, D., Picault, J.P., Gough, R., Hess, M., Mawditt, K., and Britton, P. (2005). Variation in the spike protein of the 793/B type of infectious bronchitis virus, in the field and during alternate passage in chickens and embryonated eggs. Avian Pathol. *34*, 20–25.

Chan, V.S., Chan, K.Y., Chen, Y., Poon, L.L., Cheung, A.N., Zheng, B., Chan, K.H., Mak, W., Ngan, H.Y., Xu, X., Screaton, G., Tam, P.K., Austyn, J.M., Chan, L.C., Yip, S.P., Peiris, M., Khoo, U.S., and Lin, C.L. (2006). Homozygous L-SIGN (CLEC4M) plays a protective role in SARS coronavirus infection. Nat. Genet. *38*, 38–46.

Chu, V.C., McElroy, L.J., Chu, V., Bauman, B.E., and Whittaker, G.R. (2006). The Avian Coronavirus Infectious Bronchitis Virus Undergoes Direct Low-pH-Dependent Fusion Activation during Entry into Host Cells. J. Virol. *80*, 3180–3188.

Cocquerel, L., Voisset, C., and Dubuisson, J. (2006). Hepatitis C virus entry: potential receptors and their biological functions. J. Gen. Virol. *87*, 1075–1084.

Corapi, W.V., Olsen, C.W., and Scott, F.W. (1992). Monoclonal antibody analysis of neutralization and antibody- dependent enhancement of feline infectious peritonitis virus. J. Virol. *66*, 6695–6705.

Czub, M., Weingartl, H., Czub, S., He, R., and Cao, J. (2005). Evaluation of modified vaccinia virus Ankara based recombinant SARS vaccine in ferrets. Vaccine *23*, 2273–2279.

Danilczyk, U., Eriksson, U., Crackower, M.A., and Penninger, J.M. (2003). A story of two ACEs. J. Mol. Med. *81*, 227–234.

Danilczyk, U., Eriksson, U., Oudit, G.Y., and Penninger, J.M. (2004). Physiological roles of angiotensin-converting enzyme 2. Cell Mol. Life Sci. *61*, 2714–2719.

de Groot, L., Pinon, J.D., Phillips, J., Lavi, E., and Weiss, S.R. (2001). Pathogenesis of fusion deficient recombinant mouse hepatitis viruses. Adv. Exp. Med. Biol. *494*, 159–162.

de Groot, R.J. (2006). Structure, function and evolution of the hemagglutinin-esterase proteins of corona- and toroviruses. Glycoconj. J. *23*, 59–72.

Delmas, B., Gelfi, J., Kut, E., Sjostrom, H., Noren, O., and Laude, H. (1994). Determinants essential for the transmissible gastroenteritis virus-receptor interaction reside within a domain of aminopeptidase-N that is distinct from the enzymatic site. J. Virol. *68*, 5216–5224.

Delmas, B., Gelfi, J., L'Haridon, R., Vogel, L.K., Sjostrom, H., Noren, O., and Laude, H. (1992). Aminopeptidase N is a major receptor for the entero-pathogenic coronavirus TGEV. Nature *357*, 417–420.

Delmas, B., Gelfi, J., Sjostrom, H., Noren, O., and Laude, H. (1993). Further characterization of aminopeptidase-N as a receptor for coronaviruses. Adv. Exp. Med. Biol. *342*, 293–298.

Dong, X., An, B., Salvucci, K.L., Storkus, W.J., Amoscato, A.A., and Salter, R.D. (2000). Modification of the amino terminus of a class II epitope confers resistance to degradation by CD13 on dendritic cells and enhances presentation to T cells. J. Immunol. *164*, 129–135.

Donoghue, M., Hsieh, F., Baronas, E., Godbout, K., Gosselin, M., Stagliano, N., Donovan, M., Woolf, B., Robison, K., Jeyaseelan, R., Breitbart, R.E., and Acton, S. (2000). A novel angiotensin-converting enzyme-related carboxypeptidase (ACE2) converts angiotensin I to angiotensin 1–9. Circ. Res. *87*, E1–E9.

Drosten, C., Gunther, S., Preiser, W., van der Werf, S., Brodt, H.R., Becker, S., Rabenau, H., Panning, M., Kolesnikova, L., Fouchier, R.A., Berger, A., Burguiere, A.M., Cinatl, J., Eickmann, M., Escriou,

N., Grywna, K., Kramme, S., Manuguerra, J.C., Muller, S., Rickerts, V., Sturmer, M., Vieth, S., Klenk, H.D., Osterhaus, A.D., Schmitz, H., and Doerr, H.W. (2003). Identification of a novel coronavirus in patients with severe acute respiratory syndrome. N. Engl. J. Med. *348*, 1967–1976.

Dveksler, G.S., Basile, A.A., Cardellichio, C.B., Beauchemin, N., Dieffenbach, C.W., and Holmes, K.V. (1993a). Expression of MHV-A59 receptor glycoproteins in susceptible and resistant strains of mice. Adv. Exp. Med. Biol. *342*, 267–272.

Dveksler, G.S., Dieffenbach, C.W., Cardellichio, C.B., McCuaig, K., Pensiero, M.N., Jiang, G.S., Beauchemin, N., and Holmes, K.V. (1993b). Several members of the mouse carcinoembryonic antigen-related glycoprotein family are functional receptors for the coronavirus mouse hepatitis virus-A59. J. Virol. *67*, 1–8.

Dveksler, G.S., Pensiero, M.N., Cardellichio, C.B., Williams, R.K., Jiang, G.S., Holmes, K.V., and Dieffenbach, C.W. (1991). Cloning of the mouse hepatitis virus (MHV) receptor: expression in human and hamster cell lines confers susceptibility to MHV. J. Virol. *65*, 6881–6891.

Enjuanes, L., Sune, C., Gebauer, F., Smerdou, C., Camacho, A., Anton, I.M., Gonzalez, S., Talamillo, A., Mendez, A., Ballesteros, M.L., and *et al.* (1992). Antigen selection and presentation to protect against transmissible gastroenteritis coronavirus. Vet. Microbiol. *33*, 249–262.

Feinberg, H., Guo, Y., Mitchell, D.A., Drickamer, K., and Weis, W.I. (2005). Extended neck regions stabilize tetramers of the receptors DC-SIGN and DC-SIGNR. J. Biol. Chem. *280*, 1327–1335.

Fiscus, S.A., and Teramoto, Y.A. (1987). Functional differences in the peplomer glycoproteins of feline coronavirus isolates. J. Virol. *61*, 2655–2657.

Fouchier, R.A., Kuiken, T., Schutten, M., Van Amerongen, G., Van Doornum, G.J., Van Den Hoogen, B.G., Peiris, M., Lim, W., Stohr, K., and Osterhaus, A.D. (2003). Aetiology: Koch's postulates fulfilled for SARS virus. Nature *423*, 240.

Frana, M.F., Behnke, J.N., Sturman, L.S., and Holmes, K.V. (1985). Proteolytic cleavage of the E2 glycoprotein of murine coronavirus: host-dependent differences in proteolytic cleavage and cell fusion. J. Virol. *56*, 912–920.

Gallagher, T.M., Buchmeier, M.J., and Perlman, S. (1992). Cell receptor-independent infection by a neurotropic murine coronavirus. Virology *191*, 517–522.

Gallagher, T.M., Parker, S.E., and Buchmeier, M.J. (1990). Neutralization-resistant variants of a neurotropic coronavirus are generated by deletions within the amino-terminal half of the spike glycoprotein. J. Virol. *64*, 731–741.

Gillim-Ross, L., Taylor, J., Scholl, D.R., Ridenour, J., Masters, P.S., and Wentworth, D.E. (2004). Discovery of novel human and animal cells infected by the severe acute respiratory syndrome coronavirus by replication-specific multiplex reverse transcription-PCR. J. Clin. Microbiol. *42*, 3196–3206.

Godet, M., Grosclaude, J., Delmas, B., and Laude, H. (1994). Major receptor-binding and neutralization determinants are located within the same domain of the transmissible gastroenteritis virus (coronavirus) spike protein. J. Virol. *68*, 8008–8016.

Godfraind, C., Langreth, S.G., Cardellichio, C.B., Knobler, R., Coutelier, J.P., Dubois-Dalcq, M., and Holmes, K.V. (1995). Tissue and cellular distribution of an adhesion molecule in the carcinoembryonic antigen family that serves as a receptor for mouse hepatitis virus. Lab. Invest. *73*, 615–627.

Gorbalenya, A.E., Enjuanes, L., Ziebuhr, J., and Snijder, E.J. (2006). Nidovirales: evolving the largest RNA virus genome. Virus Res. *117*, 17–37.

Gorbalenya, A.E., Snijder, E.J., and Spaan, W.J. (2004). Severe acute respiratory syndrome coronavirus phylogeny: toward consensus. J. Virol. *78*, 7863–7866.

Gough, R.E., Drury, S.E., Culver, F., Britton, P., and Cavanagh, D. (2006). Isolation of a coronavirus from a green-cheeked Amazon parrot (Amazon viridigenalis Cassin). Avian Pathol. *35*, 122–126.

Gramberg, T., Hofmann, H., Moller, P., Lalor, P.F., Marzi, A., Geier, M., Krumbiegel, M., Winkler, T., Kirchhoff, F., Adams, D.H., Becker, S., Munch, J., and Pohlmann, S. (2005). LSECtin interacts with filovirus glycoproteins and the spike protein of SARS coronavirus. Virology *340*, 224–236.

Guan, Y., Zheng, B.J., He, Y.Q., Liu, X.L., Zhuang, Z.X., Cheung, C.L., Luo, S.W., Li, P.H., Zhang, L.J., Guan, Y.J., Butt, K.M., Wong, K.L., Chan, K.W., Lim, W., Shortridge, K.F., Yuen, K.Y., Peiris, J.S., and Poon, L.L. (2003). Isolation and characterization of viruses related to the SARS coronavirus from animals in southern China. Science *302*, 276–278.

Guillen, J., Perez-Berna, A.J., Moreno, M.R., and Villalain, J. (2005). Identification of the membrane-active regions of the severe acute respiratory syndrome coronavirus spike membrane glycoprotein using a 16/18-mer peptide scan: implications for the viral fusion mechanism. J. Virol. *79*, 1743–1752.

Guo, Y., Atkinson, C.E., Taylor, M.E., and Drickamer, K. (2006). All but the shortest polymorphic forms of the viral receptor DC-SIGNR assemble into stable homo- and heterotetramers. J. Biol. Chem. *281*, 16794–16798.

Hakansson-McReynolds, S., Jiang, S., Rong, L., and Caffrey, M. (2006). Solution structure of the severe acute respiratory syndrome-coronavirus heptad repeat 2 domain in the prefusion state. J. Biol. Chem. *281*, 11965–11971.

Hansen, G.H., Delmas, B., Besnardeau, L., Vogel, L.K., Laude, H., Sjostrom, H., and Noren, O. (1998). The coronavirus transmissible gastroenteritis virus causes infection after receptor-mediated endocytosis and acid-dependent fusion with an intracellular compartment. J. Virol. *72*, 527–534.

Hegyi, A., and Kolb, A.F. (1998). Characterization of determinants involved in the feline infectious peritonitis virus receptor function of feline aminopeptidase N. J. Gen. Virol. *79*, 1387–1391.

Heller, L.K., Gillim-Ross, L., Olivieri, E.R., and Wentworth, D.E. (2006). *Mustela vison* ACE2 functions as a receptor for SARS-coronavirus. Adv. Exp. Med. Biol. *581*, 507–510.

Hemmila, E., Turbide, C., Olson, M., Jothy, S., Holmes, K.V., and Beauchemin,N. (2004). Ceacam1a–/– mice are completely resistant to infection by murine coronavirus mouse hepatitis virus A59. J. Virol. *78*, 10156–10165.

Hofmann, H., Pyrc, K., van der Hoek, L., Geier, M., Berkhout, B., and Pohlmann, S. (2005). Human coronavirus NL63 employs the severe acute respiratory syndrome coronavirus receptor for cellular entry. Proc. Natl. Acad. Sci. USA *102*, 7988–7993.

Hofmann, H., Simmons, G., Rennekamp, A.J., Chaipan, C., Gramberg, T., Heck, E., Geier, M., Wegele, A., Marzi, A., Bates, P., and Pohlmann, S. (2006). Highly conserved regions within the spike proteins of human coronaviruses 229E and NL63 determine recognition of their respective cellular receptors. J. Virol. *80*, 8639–8652.

Hohdatsu, T., Izumiya, Y., Yokoyama, Y., Kida, K., and Koyama, H. (1998). Differences in virus receptor for type I and type II feline infectious peritonitis virus. Arch. Virol. *143*, 839–850.

Huang, I.C., Bosch, B.J., Li, F., Li, W., Lee, K.H., Ghiran, S., Vasilieva, N., Dermody, T.S., Harrison, S.C., Dormitzer, P.R., Farzan, M., Rottier, P.J., and Choe, H. (2005). SARS coronavirus, but not human coronavirus NL63, utilizes cathepsin L to infect ACE2-expressing cells. J. Biol. Chem. *281*, 3198–3203.

Jackwood, D.J., Bae, I., Jackwood, R.J., and Saif, L.J. (1993). Transmissible gastroenteritis virus and porcine respiratory coronavirus: molecular characterization of the S gene using cDNA probes and nucleotide sequence analysis. Adv. Exp. Med. Biol. *342*, 43–48.

Jeffers, S.A., Hemmila, E.M., and Holmes, K.V. (2006). Human coronavirus 229E can use CD209L (L-SIGN) to enter cells. Adv. Exp. Med. Biol. *581*, 265–269.

Jeffers, S.A., Tusell, S.M., Gillim-Ross, L., Hemmila, E.M., Achenbach, J.E., Babcock, G.J., Thomas, W.D., Jr., Thackray, L.B., Young, M.D., Mason, R.J., Ambrosino, D.M., Wentworth, D.E., Demartini, J.C., and Holmes, K.V. (2004). CD209L (L-SIGN) is a receptor for severe acute respiratory syndrome coronavirus. Proc. Natl. Acad. Sci. USA *101*, 15748–15753.

Jia, W., Karaca, K., Parrish, C.R., and Naqi, S.A. (1995). A novel variant of avian infectious bronchitis virus resulting from recombination among three different strains. Arch. Virol. *140*, 259–271.

Jonassen, C.M., Kofstad, T., Larsen, I.L., Lovland, A., Handeland, K., Follestad, A., and Lillehaug, A. (2005). Molecular identification and characterization of novel coronaviruses infecting graylag geese (Anser anser), feral pigeons (Columbia livia) and mallards (Anas platyrhynchos). J. Gen. Virol. *86*, 1597–1607.

Kilby, J.M., Hopkins, S., Venetta, T.M., DiMassimo, B., Cloud, G.A., Lee, J.Y., Alldredge, L., Hunter, E., Lambert, D., Bolognesi, D., Matthews, T., Johnson, M.R., Nowak, M.A., Shaw, G.M., and Saag, M.S. (1998). Potent suppression of HIV-1 replication in humans by T-20, a peptide inhibitor of gp41-mediated virus entry. Nat. Med. *4*, 1302–1307.

Knobler, R.L., Haspel, M.V., and Oldstone, M.B. (1981). Mouse hepatitis virus type 4 (JHM strains). induced fatal central nervous system disease. I. genetic control and murine neuron as the susceptible site of disease. J. Exp. Med. *153*, 832–843.

Knobler, R.L., Tunison, L.A., and Oldstone, M.B. (1984). Host genetic control of mouse hepatitis virus type 4 (JHM strain) replication. I. Restriction of virus amplification and spread in macrophages from resistant mice. J. Gen. Virol. *65*, 1543–1548.

Kolb, A.F., Hegyi, A., and Siddell, S.G. (1997). Identification of residues critical for the human coronavirus 229E receptor function of human aminopeptidase N. J. Gen. Virol. *78*, 2795–2802.

Kolb, A.F., Maile, J., Heister, A., and Siddell, S.G. (1996). Characterization of functional domains in the human coronavirus HCV 229E receptor. J. Gen. Virol. *77*, 2515–2521.

Koppel, E.A., van Gisbergen, K.P., Geijtenbeek, T.B., and van Kooyk, Y. (2005). Distinct functions of DC-SIGN and its homologues L-SIGN (DC-SIGNR) and mSIGNR1 in pathogen recognition and immune regulation. Cell Microbiol. *7*, 157–165.

Krempl, C., Ballesteros, M.L., Zimmer, G., Enjuanes, L., Klenk, H.D., and Herrler, G. (2000). Characterization of the sialic acid binding activity of transmissible gastroenteritis coronavirus by analysis of haemagglutination-deficient mutants. J. Gen. Virol. *81*, 489–496.

Krempl, C., Schultze, B., Laude, H., and Herrler, G. (1997). Point mutations in the S protein connect the sialic acid binding activity with the enteropathogenicity of transmissible gastroenteritis coronavirus. J. Virol. *71*, 3285–3287.

Ksiazek, T.G., Erdman, D., Goldsmith, C.S., Zaki, S.R., Peret, T., Emery, S., Tong, S., Urbani, C., Comer, J.A., Lim, W., Rollin, P.E., Dowell, S.F., Ling, A.E., Humphrey, C.D., Shieh, W.J., Guarner, J., Paddock, C.D., Rota, P., Fields, B., DeRisi, J., Yang, J.Y., Cox, N., Hughes, J.M., LeDuc, J.W., Bellini, W.J., and Anderson, L.J. (2003). A novel coronavirus associated with severe acute respiratory syndrome. N. Engl. J. Med. *348*, 1953–1966.

Kubo, H., Takase-Yoden, S., and Taguchi, F. (1993). Neutralization and fusion inhibition activities of monoclonal antibodies specific for the S1 subunit of the spike protein of neurovirulent murine coronavirus JHMV cl-2 variant. J. Gen. Virol. *74*, 1421–1425.

Kubo, H., Yamada, Y.K., and Taguchi, F. (1994). Localization of neutralizing epitopes and the receptor-binding site within the amino-terminal 330 amino acids of the murine coronavirus spike protein. J. Virol. *68*, 5403–5410.

Kunkel, F., and Herrler, G. (1993). Structural and functional analysis of the surface protein of human coronavirus OC43. Virology *195*, 195–202.

La Monica, N., Banner, L.R., Morris, V.L., and Lai, M.M. (1991). Localization of extensive deletions in the structural genes of two neurotropic variants of murine coronavirus JHM. Virology *182*, 883–888.

Lai, W.K., Sun, P.J., Zhang, J., Jennings, A., Lalor, P.F., Hubscher, S., McKeating, J.A., and Adams, D.H. (2006). Expression of DC-SIGN and DC-SIGNR on human sinusoidal endothelium: a role for capturing hepatitis C virus particles. Am. J. Pathol. *169*, 200–208.

Lau, S.K.P., Woo, P.C.Y., Li, K.S.M., Huang, Y., Tsoi, H.W., Wong, B.H.L., Wong, S.S.Y., Leung, S.Y., Chan, K.H., and Yuen, K.Y. (2005). Severe acute respiratory syndrome coronavirus-like virus in Chinese horseshoe bats. Proc. Natl. Acad. Sci. USA *102*, 14040–14045.

Lewicki, D.N., and Gallagher, T.M. (2002). Quaternary structure of coronavirus spikes in complex with carcinoembryonic antigen-related cell adhesion molecule cellular receptors. J. Biol. Chem. *277*, 19727–19734.

Li, F., Li, W., Farzan, M., and Harrison, S.C. (2005a). Structure of SARS coronavirus spike receptor-binding domain complexed with receptor. Science *309*, 1864–1868.

Li, W., Greenough, T.C., Moore, M.J., Vasilieva, N., Somasundaran, M., Sullivan, J.L., Farzan, M., and Choe, H. (2004). Efficient replication of severe acute respiratory syndrome coronavirus in mouse cells is limited by murine angiotensin-converting enzyme 2. J. Virol. *78*, 11429–11433.

Li, W., Moore, M.J., Vasilieva, N., Sui, J., Wong, S.K., Berne, M.A., Somasundaran, M., Sullivan, J.L., Luzuriaga, K., Greenough, T.C., Choe, H., and Farzan, M. (2003). Angiotensin-converting enzyme 2 is a functional receptor for the SARS coronavirus. Nature *426*, 450–454.

Li, W., Shi, Z., Yu, M., Ren, W., Smith, C., Epstein, J.H., Wang, H., Crameri, G., Hu, Z., Zhang, H., Zhang, J., McEachern, J., Field, H., Daszak, P., Eaton, B.T., Zhang, S., and Wang, L.F. (2005b). Bats Are Natural Reservoirs of SARS-Like Coronaviruses. Science *310*, 676–679.

Li, W., Zhang, C., Sui, J., Kuhn, J.H., Moore, M.J., Luo, S., Wong, S.K., Huang, I.C., Xu, K., Vasilieva, N., Murakami, A., He, Y., Marasco, W.A., Guan, Y., Choe, H., and Farzan, M. (2005c). Receptor and viral determinants of SARS-coronavirus adaptation to human ACE2. EMBO J. *24*, 1634–1643.

Li, W., Wong, S.K., Li, F., Kuhn, J.H., Huang, I.C., Choe, H., and Farzan, M. (2006). Animal Origins of the Severe Acute Respiratory Syndrome Coronavirus: Insight from ACE2-S-Protein Interactions. J. Virol. *80*, 4211–4219.

Liu, H., Carrington, M., Wang, C., Holte, S., Lee, J., Greene, B., Hladik, F., Koelle, D.M., Wald, A., Kurosawa, K., Rinaldo, C.R., Celum, C., Detels, R., Corey, L., McElrath, M.J., and Zhu, T. (2006). Repeat-region polymorphisms in the gene for the dendritic cell-specific intercellular adhesion molecule-3-grabbing nonintegrin-related molecule: effects on HIV-1 susceptibility. J. Infect. Dis. *193*, 698–702.

Liu, H., Hladik, F., Andrus, T., Sakchalathorn, P., Lentz, G.M., Fialkow, M.F., Corey, L., McElrath, M.J., and Zhu, T. (2005). Most DC-SIGNR transcripts at mucosal HIV transmission sites are alternatively spliced isoforms. Eur. J. Hum. Genet. *13*, 707–715.

Liu, S., Xiao, G., Chen, Y., He, Y., Niu, J., Escalante, C.R., Xiong, H., Farmar, J., Debnath, A.K., Tien, P., and Jiang, S. (2004). Interaction between heptad repeat 1 and 2 regions in spike protein of SARS-associated coronavirus: implications for virus fusogenic mechanism and identification of fusion inhibitors. Lancet *363*, 938–947.

Look, A.T., Ashmun, R.A., Shapiro, L.H., and Peiper, S.C. (1989). Human myeloid plasma membrane glycoprotein CD13 (gp150) is identical to aminopeptidase N. J. Clin. Invest. *83*, 1299–1307.

Luo, Z., and Weiss, S.R. (1998). Roles in cell-to-cell fusion of two conserved hydrophobic regions in the murine coronavirus spike protein. Virology *244*, 483–494.

Luytjes, W., Sturman, L.S., Bredenbeek, P.J., Charite, J., van der Zeijst, B.A., Horzinek, M.C., and Spaan, W.J. (1987). Primary structure of the glycoprotein E2 of coronavirus MHV-A59 and identification of the trypsin cleavage site. Virology *161*, 479–487.

Martina, B.E.E., Haagmans, B.L., Kuiken, T., Fouchier, R.A.M., Rimmelzwaan, G.F., Van Amerongen, G., Peiris, J.S.M., Lim, W., and Osterhaus, A.D. (2003). Virology: SARS virus infection of cats and ferrets. Nature *425*, 915.

Marzi, A., Gramberg, T., Simmons, G., Moller, P., Rennekamp, A.J., Krumbiegel, M., Geier, M., Eisemann, J., Turza, N., Saunier, B., Steinkasserer, A., Becker, S., Bates, P., Hofmann, H., and Pohlmann, S. (2004). DC-SIGN and DC-SIGNR interact with the glycoprotein of Marburg virus and the S protein of severe acute respiratory syndrome coronavirus. J. Virol. *78*, 12090–12095.

Masters, P.S. (2006). The molecular biology of coronaviruses. Adv. Virus Res. *66*, 193–292.

McCuaig, K., Turbide, C., and Beauchemin, N. (1992). mmCGM1a: a mouse carcinoembryonic antigen gene family member, generated by alternative splicing, functions as an adhesion molecule. Cell Growth Differ. *3*, 165–174.

Mossel, E.C., Huang, C., Narayanan, K., Makino, S., Tesh, R.B., and Peters, C.J. (2005). Exogenous ACE2 expression allows refractory cell lines to support severe acute respiratory syndrome coronavirus replication. J. Virol. *79*, 3846–3850.

Nash, T.C., and Buchmeier, M.J. (1996). Spike glycoprotein-mediated fusion in biliary glycoprotein-independent cell-associated spread of mouse hepatitis virus infection. Virology *223*, 68–78.

Nash, T.C., and Buchmeier, M.J. (1997). Entry of mouse hepatitis virus into cells by endosomal and nonendosomal pathways. Virology *233*, 1–8.

Nattermann, J., Ahlenstiel, G., Berg, T., Feldmann, G., Nischalke, H.D., Muller, T., Rockstroh, J., Woitas, R., Sauerbruch, T., and Spengler, U. (2006). The tandem-repeat polymorphism of the DC-SIGNR gene in HCV infection. J. Viral Hepat. *13*, 42–46.

Noren, O., Sjostrom, H., and Olsen, J. (1997). Aminopeptidase N. In Cell-Surface Peptidases in Health and Disease, A.J. Kenney and C.M. Boustead, eds. (Oxford: BIOS Scientific Publishers), pp. 175–191.

Obrink, B. (1991). C-CAM (cell-CAM 105)--a member of the growing immunoglobulin superfamily of cell adhesion proteins. Bioassays *13*, 227–234.

Olivieri, E.R., Heller, L.K., Gillim-Ross, L., and Wentworth, D.E. (2006). Analysis of SARS-CoV receptor activity of ACE2 orthologs. Adv. Exp. Med. Biol. *581*, 277–280.

Oudit, G.Y., Crackower, M.A., Backx, P.H., and Penninger, J.M. (2003). The role of ACE2 in cardiovascular physiology. Trends Cardiovasc. Med. *13*, 93–101.

Peiris, J.S., Lai, S.T., Poon, L.L., Guan, Y., Yam, L.Y., Lim, W., Nicholls, J., Yee, W.K., Yan, W.W., Cheung, M.T., Cheng, V.C., Chan, K.H., Tsang, D.N., Yung, R.W., Ng, T.K., and Yuen, K.Y. (2003). Coronavirus as a possible cause of severe acute respiratory syndrome. Lancet *361*, 1319–1325.

Petit, C.M., Melancon, J.M., Chouljenko, V.N., Colgrove, R., Farzan, M., Knipe, D.M., and Kousoulas, K.G. (2005). Genetic analysis of the SARS-coronavirus spike glycoprotein functional domains involved in cell-surface expression and cell-to-cell fusion. Virology *341*, 215–230.

Pewe, L., Xue, S., and Perlman, S. (1997). Cytotoxic T-cell-resistant variants arise at early times after infection in C57BL/6 but not in SCID mice infected with a neurotropic coronavirus. J. Virol. *71*, 7640–7647.

Phillips, J.J., Chua, M.M., Lavi, E., and Weiss, S.R. (1999). Pathogenesis of chimeric MHV4/MHV-A59 recombinant viruses: the murine coronavirus spike protein is a major determinant of neurovirulence. J. Virol. *73*, 7752–7760.

Phillips, J.J., Chua, M.M., Rall, G.F., and Weiss, S.R. (2002). Murine coronavirus spike glycoprotein mediates degree of viral spread, inflammation, and virus-induced immunopathology in the central nervous system. Virology *301*, 109–120.

Qinfen, Z., Jinming, C., Xiaojun, H., Huanying, Z., Jicheng, H., Ling, F., Kunpeng, L., and Jingqiang, Z. (2004). The life cycle of SARS coronavirus in Vero E6 cells. J. Med. Virol. *73*, 332–337.

Rao, P.V., Kumari, S., and Gallagher, T.M. (1997). Identification of a contiguous 6-residue determinant in the MHV receptor that controls the level of virion binding to cells. Virology *229*, 336–348.

Rekik, M.R., and Dea, S. (1994). Comparative sequence analysis of a polymorphic region of the spike glycoprotein S1 subunit of enteric bovine coronavirus isolates. Arch. Virol. *135*, 319–331.

Riemann, D., Kehlen, A., and Langner, J. (1999). CD13—not just a marker in leukemia typing. Immunology Today *20*, 83–88.

Roberts, A., Vogel, L., Guarner, J., Hayes, N., Murphy, B., Zaki, S., and Subbarao, K. (2005). Severe acute respiratory syndrome coronavirus infection of golden Syrian hamsters. J Virol. *79*, 503–511.

Sainz, B., Jr., Mossel, E.C., Gallaher, W.R., Wimley, W.C., Peters, C.J., Wilson, R.B., and Garry, R.F. (2006). Inhibition of severe acute respiratory syndrome-associated coronavirus (SARS-CoV) infectivity by peptides analogous to the viral spike protein. Virus Res. *120*, 146–155.

Sainz, B., Jr., Rausch, J.M., Gallaher, W.R., Garry, R.F., and Wimley, W.C. (2005). Identification and characterization of the putative fusion peptide of the severe acute respiratory syndrome-associated coronavirus spike protein. J. Virol. *79*, 7195–7206.

Sanchez, C.M., Gebauer, F., Sune, C., Mendez, A., Dopazo, J., and Enjuanes, L. (1992). Genetic evolution and tropism of transmissible gastroenteritis coronaviruses. Virology *190*, 92–105.

Schickli, J.H., Zelus, B.D., Wentworth, D.E., Sawicki, S.G., and Holmes, K.V. (1997). The murine coronavirus mouse hepatitis virus strain A59 from persistently infected murine cells exhibits an extended host range. J. Virol. *71*, 9499–9507.

Schultze, B., and Herrler, G. (1994). Recognition of cellular receptors by bovine coronavirus. Arch. Virol. Suppl. *9*, 451–459.

Schultze, B., Krempl, C., Ballesteros, M.L., Shaw, L., Schauer, R., Enjuanes, L., and Herrler, G. (1996). Transmissible gastroenteritis coronavirus, but not the related porcine respiratory coronavirus, has a sialic acid (N-glycolylneuraminic acid) binding activity. J. Virol. *70*, 5634–5637.

Schwegmann-Wessels, C., and Herrler, G. (2006). Sialic acids as receptor determinants for coronaviruses. Glycoconj. J. *23*, 51–58.

Sestak, K., Lanza, I., Park, S.K., Weilnau, P.A., and Saif, L.J. (1996). Contribution of passive immunity to porcine respiratory coronavirus to protection against transmissible gastroenteritis virus challenge exposure in suckling pigs. Am. J. Vet. Res. *57*, 664–671.

Simkins, R.A., Weilnau, P.A., Bias, J., and Saif, L.J. (1992). Antigenic variation among transmissible gastroenteritis virus (TGEV) and porcine respiratory coronavirus strains detected with monoclonal antibodies to the S protein of TGEV. Am. J. Vet. Res. *53*, 1253–1258.

Simmons, G., Gosalia, D.N., Rennekamp, A.J., Reeves, J.D., Diamond, S.L., and Bates, P. (2005). Inhibitors of cathepsin L prevent severe acute respiratory syndrome coronavirus entry. Proc. Natl. Acad. Sci. USA *102*, 11876–11881.

Simmons, G., Reeves, J.D., Grogan, C.C., Vandenberghe, L.H., Baribaud, F., Whitbeck, J.C., Burke, E., Buchmeier, M.J., Soilleux, E.J., Riley, J.L., Doms, R.W., Bates, P., and Pohlmann, S. (2003). DC-SIGN and DC-SIGNR bind ebola glycoproteins and enhance infection of macrophages and endothelial cells. Virology *305*, 115–123.

Smith, M.K., Tusell, S., Travanty, E.A., Berkhout, B., van der Hoek, L., and Holmes, K.V. (2006). Human angiotensin-converting enzyme 2 (ACE2) is a receptor for human respiratory coronavirus NL63. Adv. Exp. Med. Biol. *581*, 285–288.

Song, H.D., Tu, C.C., Zhang, G.W., Wang, S.Y., Zheng, K., Lei, L.C., Chen, Q.X., Gao, Y.W., Zhou, H.Q., Xiang, H., Zheng, H.J., Chern, S.W., Cheng, F., Pan, C.M., Xuan, H., Chen, S.J., Luo, H.M., Zhou, D.H., Liu, Y.F., He, J.F., Qin, P.Z., Li, L.H., Ren, Y.Q., Liang, W.J., Yu, Y.D., Anderson, L., Wang, M., Xu, R.H., Wu, X.W., Zheng, H.Y., Chen, J.D., Liang, G., Gao, Y., Liao, M., Fang, L., Jiang, L.Y., Li, H., Chen, F., Di, B., He, L.J., Lin, J.Y., Tong, S., Kong, X., Du, L., Hao, P., Tang, H., Bernini, A., Yu, X.J., Spiga, O., Guo, Z.M., Pan, H.Y., He, W.Z., Manuguerra, J.C., Fontanet, A., Danchin, A., Niccolai, N., Li, Y.X., Wu, C.I., and Zhao, G.P. (2005). Cross-host evolution of severe acute respiratory syndrome coronavirus in palm civet and human. Proc. Natl. Acad. Sci. USA *102*, 2430–2435.

Spear, P.G. (2004). Herpes simplex virus: receptors and ligands for cell entry. Cell Microbiol. *6*, 401–410.

Sturman, L.S., Ricard, C.S., and Holmes, K.V. (1985). Proteolytic cleavage of the E2 glycoprotein of murine coronavirus: activation of cell-fusing activity of virions by trypsin and separation of two different 90K cleavage fragments. J. Virol. *56*, 904–911.

Subbarao, K., McAuliffe, J., Vogel, L., Fahle, G., Fischer, S., Tatti, K., Packard, M., Shieh, W.J., Zaki, S., and Murphy, B. (2004). Prior Infection and Passive Transfer of Neutralizing Antibody Prevent Replication of Severe Acute Respiratory Syndrome Coronavirus in the Respiratory Tract of Mice. J. Virol. *78*, 3572–3577.

Suzuki, H., and Taguchi, F. (1996). Analysis of the receptor-binding site of murine coronavirus spike protein. J. Virol. *70*, 2632–2636.

Taguchi, F. (1995). The S2 subunit of the murine coronavirus spike protein is not involved in receptor binding. J. Virol. *69*, 7260–7263.

Takada, A., Fujioka, K., Tsuiji, M., Morikawa, A., Higashi, N., Ebihara, H., Kobasa, D., Feldmann, H., Irimura, T., and Kawaoka, Y. (2004). Human macrophage C-type lectin specific for galactose and N-acetylgalactosamine promotes filovirus entry. J. Virol. *78*, 2943–2947.

Tan, K., Zelus, B.D., Meijers, R., Liu, J.H., Bergelson, J.M., Duke, N., Zhang, R., Joachimiak, A., Holmes, K.V., and Wang, J.H. (2002). Crystal structure of murine sCEACAM1a[1,4]: a coronavirus receptor in the CEA family. EMBO J. *21*, 2076–2086.

Thackray, L.B., and Holmes, K.V. (2004). Amino acid substitutions and an insertion in the spike glycoprotein extend the host range of the murine coronavirus MHV-A59. Virology *324*, 510–524.

Thackray, L.B., Turner, B.C., and Holmes, K.V. (2005). Substitutions of conserved amino acids in the receptor-binding domain of the spike glycoprotein affect utilization of murine CEACAM1a by the murine coronavirus MHV-A59. Virology *334*, 98–110.

Thorp, E.B., Boscarino, J.A., Logan, H.L., Goletz, J.T., and Gallagher, T.M. (2006). Palmitoylations on murine coronavirus spike proteins are essential for virion assembly and infectivity. J. Virol. *80*, 1280–1289.

Towler, P., Staker, B., Prasad, S.G., Menon, S., Tang, J., Parsons, T., Ryan, D., Fisher, M., Williams, D., Dales, N.A., Patane, M.A., and Pantoliano, M.W. (2004). ACE2 structures reveal a large hinge-bending motion important for inhibitor binding and catalysis. J. Biol. Chem. *279*, 17996–18007.

Tresnan, D.B., Levis, R., and Holmes, K.V. (1996). Feline aminopeptidase N serves as a receptor for feline, canine, porcine, and human coronaviruses in serogroup I. J. Virol. *70*, 8669–8674.

Tripet, B., Howard, M.W., Jobling, M., Holmes, R.K., Holmes, K.V., and Hodges, R.S. (2004). Structural characterization of the SARS-coronavirus spike S fusion protein core. J. Biol. Chem. *279*, 20836–20849.

Tripet, B., Kao, D.J., Jeffers, S.A., Holmes, K.V., and Hodges, R.S. (2006). Template-based coiled-coil antigens elicit neutralizing antibodies to the SARS-coronavirus. J. Struct. Biol. *155*, 176–194.

Tu, C., Crameri, G., Kong, X., Chen, J., Sun, Y., Yu, M., Xiang, H., Xia, X., Liu, S., Ren, T., Yu, Y., Eaton, B.T., Xuan, H., and Wang, L.F. (2004). Antibodies to SARS coronavirus in civets. Emerg. Infect. Dis. *10*, 2244–2248.

Turner, B.C., Hemmila, E.M., Beauchemin, N., and Holmes, K.V. (2004). Receptor-dependent coronavirus infection of dendritic cells. J. Virol. *78*, 5486–5490.

Tusell, S.M., Schittone, S, and Holmes, K.V.(2006) Mutational analysis of aminopeptidase N, a receptor for several group 1 coronaviruses, identifies key determinants of viral host range. J. Virol. doi:10.1128/JVI.01510-06.

Vautherot, J.F., Madelaine, M.F., Boireau, P., and Laporte, J. (1992). Bovine coronavirus peplomer glycoproteins: detailed antigenic analyses of S1, S2 and HE. J. Gen. Virol. *73*, 1725–1737.

Vennema, H., Heijnen, L., Zijderveld, A., Horzinek, M.C., and Spaan, W.J. (1990). Intracellular transport of recombinant coronavirus spike proteins: implications for virus assembly. J. Virol. *64*, 339–346.

Vlasak, R., Luytjes, W., Spaan, W., and Palese, P. (1988). Human and bovine coronaviruses recognize sialic acid-containing receptors similar to those of influenza C viruses. Proc. Natl. Acad. Sci. USA *85*, 4526–4529.

Wang, L., Junker, D., Hock, L., Ebiary, E., and Collisson, E.W. (1994). Evolutionary implications of genetic variations in the S1 gene of infectious bronchitis virus. Virus Res. *34*, 327–338.

Warner, F.J., Smith, A.I., Hooper, N.M., and Turner, A.J. (2004). Angiotensin-converting enzyme-2: a molecular and cellular perspective. Cell Mol. Life Sci. *61*, 2704–2713.

Wege, H., and Dorries, R. (1984). Hybridoma antibodies to the murine coronavirus JHM: characterization of epitopes on the peplomer protein (E2). J. Gen. Virol. *65*, 1931–1942.

Wentworth, D.E., Gillim-Ross, L., Espina, N., and Bernard, K.A. (2004). Mice susceptible to SARS coronavirus. Emerg. Infect. Dis. *10*, 1293–1296.

Wentworth, D.E., and Holmes, K.V. (2001). Molecular determinants of species specificity in the coronavirus receptor aminopeptidase N (CD13): influence of N-linked glycosylation. J. Virol. *75*, 9741–9752.

Wessner, D.R., Shick, P.C., Lu, J.H., Cardellichio, C.B., Gagneten, S.E., Beauchemin, N., Holmes, K.V., and Dveksler, G.S. (1998). Mutational analysis of the virus and monoclonal antibody binding sites in MHVR, the cellular receptor of the murine coronavirus mouse hepatitis virus strain A59. J. Virol. *72*, 1941–1948.

Williams, R.K., Jiang, G.S., and Holmes, K.V. (1991). Receptor for mouse hepatitis virus is a member of the carcinoembryonic antigen family of glycoproteins. Proc. Natl. Acad. Sci. USA *88*, 5533–5536.

Winter, C., Schwegmann-Wessels, C., Cavanagh, D., Neumann, U., and Herrler, G. (2006). Sialic acid is a receptor determinant for infection of cells by avian Infectious bronchitis virus. J. Gen. Virol. *87*, 1209–1216.

Wong, S.K., Li, W., Moore, M.J., Choe, H., and Farzan, M. (2004). A 193-amino acid fragment of the SARS coronavirus S protein efficiently binds angiotensin-converting enzyme 2. J. Biol. Chem. *279*, 3197–3201.

Xiao, X., Chakraborti, S., Dimitrov, A.S., Gramatikoff, K., and Dimitrov, D.S. (2003). The SARS-CoV S glycoprotein: expression and functional characterization. Biochem. Biophys. Res. Commun. *312*, 1159–1164.

Xu, Y., Liu, Y., Lou, Z., Qin, L., Li, X., Bai, Z., Pang, H., Tien, P., Gao, G.F., and Rao, Z. (2004a). Structural basis for coronavirus-mediated membrane fusion. Crystal structure of mouse hepatitis virus spike protein fusion core. J. Biol. Chem. *279*, 30514–30522.

Xu, Y., Lou, Z., Liu, Y., Pang, H., Tien, P., Gao, G.F., and Rao, Z. (2004b). Crystal structure of severe acute respiratory syndrome coronavirus spike protein fusion core. J. Biol. Chem. *279*, 49414–49419.

Yan, Z., Tripet, B., and Hodges, R.S. (2006). Biophysical characterization of HRC peptide analogs interaction with heptad repeat regions of the SARS-coronavirus Spike fusion protein core. J. Struct. Biol *155*, 162–175.

Yang, Z.Y., Huang, Y., Ganesh, L., Leung, K., Kong, W.P., Schwartz, O., Subbarao, K., and Nabel, G.J. (2004b). pH-dependent entry of severe acute respiratory syndrome coronavirus is mediated by the spike glycoprotein and enhanced by dendritic cell transfer through DC-SIGN. J. Virol. *78*, 5642–5650.

Yang, Z.Y., Huang, Y., Ganesh, L., Leung, K., Kong, W.P., Schwartz, O., Subbarao, K., and Nabel, G.J. (2004a). pH-Dependent Entry of Severe Acute Respiratory Syndrome Coronavirus Is Mediated by the Spike Glycoprotein and Enhanced by Dendritic Cell Transfer through DC-SIGN. J. Virol. *78*, 5642–5650.

Ye, R., Montalto-Morrison, C., and Masters, P.S. (2004). Genetic analysis of determinants for spike glycoprotein assembly into murine coronavirus virions: distinct roles for charge-rich and cysteine-rich regions of the endodomain. J. Virol. *78*, 9904–9917.

Yeager, C.L., Ashmun, R.A., Williams, R.K., Cardellichio, C.B., Shapiro, L.H., Look, A.T., and Holmes,K.V. (1992). Human aminopeptidase N is a receptor for human coronavirus 229E. Nature *357*, 420–422.

Youn, S., Collisson, E.W., and Machamer, C.E. (2005). Contribution of trafficking signals in the cytoplasmic tail of the infectious bronchitis virus spike protein to virus infection. J. Virol. *79*, 13209–13217.

Zelus, B.D., Schickli, J.H., Blau, D.M., Weiss, S.R., and Holmes, K.V. (2003). Conformational changes in the spike glycoprotein of murine coronavirus are induced at 37 degrees C either by soluble murine CEACAM1 receptors or by pH 8. J. Virol. *77*, 830–840.

Zelus, B.D., Wessner, D.R., Williams, R.K., Pensiero, M.N., Phibbs, F.T., deSouza, M., Dveksler, G.S., and Holmes, K.V. (1998). Purified, soluble recombinant mouse hepatitis virus receptor, Bgp1(b), and Bgp2 murine coronavirus receptors differ in mouse hepatitis virus binding and neutralizing activities. J. Virol. *72*, 7237–7244.

Zhang, X., Hasoksuz, M., Spiro, D., Halpin, R., Wang, S., Stollar, S., Janies, D., Hadya, N., Tang, Y., Ghedin, E., and Saif, L. (2006). Complete genomic sequences, a key residue in the spike protein and deletions in nonstructural protein 3b of US strains of the virulent and attenuated coronaviruses, transmissible gastroenteritis virus and porcine respiratory coronavirus. Virology doi:10.1016/j.virol.2006.08.051.

Zhou, G.Q., Zhang, Y., and Hammarstrom, S. (2001). The carcinoembryonic antigen (CEA) gene family in non-human primates. Gene *264*, 105–112.

The Coronavirus Replicase Gene: Special Enzymes for Special Viruses

2

John Ziebuhr and Eric J. Snijder

Abstract

Coronaviruses have single-stranded, positive-sense RNA genomes of about 30 kilobases, by far the largest non-segmented RNA virus genomes currently known. The key functions required for coronavirus RNA synthesis are encoded by the viral replicase gene. The gene comprises more than 20000 nucleotides and encodes two replicase polyproteins, pp1a and pp1ab, that are proteolytically processed by viral proteases. Over the past years, it has become clear that the unique size of the coronavirus genome and the special mechanism that coronaviruses (and several other nidoviruses) have evolved to produce an extensive set of subgenome-length RNAs is linked to the production of a number of non-structural proteins (nsps) that is unprecedented among RNA viruses. Many of these replicase cleavage products in fact are multidomain proteins themselves, thus further increasing the complexity of protein functions and interactions. Structural studies suggest that several nsps, following their release from larger precursor molecules, form dimers or even multimers. The various pp1a/pp1ab precursors and processing products are thought to assemble into large, membrane-associated complexes that, in a temporally coordinated manner, catalyze the reactions involved in RNA replication and transcription and, most probably, serve yet other functions in the viral life cycle. This article reviews the expression, maturation and the functional and structural properties and peculiarities of coronavirus replicase gene-encoded enzymic and other functions, including protease, polymerase, helicase, ADP-ribose 1″-phosphatase and RNase activities.

An outline of positive-strand RNA virus replication

The nucleic acid contained in a positive-strand RNA (+RNA) virus particle fulfils an unusual double role in the viral life cycle: it is not only the template for genome replication, but also acts as the first viral mRNA. Upon its release from the incoming +RNA virion, the successful translation of the genome RNA is a critical first step in infection. It produces the viral enzymes (non-structural proteins or "*replicase*," always including an RNA-dependent RNA polymerase [RdRp]) that will amplify the genomic mRNA and, thus, boost their own synthesis. As a result, +RNA virus replication can be an explosive process, which can usurp a major part of the host cell's infrastructure and resources within a matter of hours, with obvious adverse consequences for the host itself.

Positive-strand RNA viruses that infect eukaryotes express and replicate their genome entirely in the cytoplasm of the host cell. The lack of a nuclear stage implies that the virus needs to synthesize and/or recruit all factors required to synthesize mRNA and ensure its recognition by the cellular translation apparatus. The key proteins directing RNA synthesis are encoded by the viral replicase gene itself. They coordinate the formation of a complex for viral RNA synthesis ("*replication complex*") that is commonly associated with the cytoplasmic face of a host organelle or with specific, virus-induced membrane structures (see below). Mammalian +RNA viruses commonly rely on the "polyprotein strategy:" their genome translation product is a multidomain precursor that is proteolytically cleaved. When the genome contains just a single open reading frame (as in the case of e.g. picornaviruses and flaviviruses), polyprotein processing yields both non-structural and structural proteins, in equal amounts. Viruses having a genome with more than one cistron (e.g. togaviruses, coronaviruses and arteriviruses) are able to regulate the ratio of non-structural to structural protein synthesis. In such viruses, genome translation again yields the viral replicase. The structural proteins, encoded in downstream reading frame(s) that are not accessible to ribosomes engaged in genome translation, are commonly produced via synthesis of one or more subgenomic mRNAs.

Coronaviruses are special

The key stages of the early phase in the replication cycle of coronaviruses, and all other mammalian +RNA viruses for that matter, can be summarized as follows.

1 translation of the incoming genome,
2 maturation of the resulting viral enzymes,
3 formation of a membrane-bound complex for viral RNA synthesis, and
4 initiation of viral RNA synthesis.

In this chapter, these events will be summarized for coronaviruses. Against the background of the (necessarily incomplete) summary of +RNA virus genome expression strategies in the first paragraph, coronaviruses (and other members of the order *Nidovirales*) stand out for several reasons (Lai and Holmes, 2001; Siddell *et al.*, 2005; Snijder *et al.*, 2005; Ziebuhr, 2005; Gorbalenya *et al.*, 2006).

- First, at 27–31 kilobases, the polycistronic coronavirus genome RNA (Figure 2.1) is unusually large and by far the largest among known +RNA viruses, which commonly have a genome size of between 5 and 15 kilobases (Gorbalenya *et al.*, 2006).
- Second, the two reading frames that make up the replicase gene (ORF1a and ORF1b) occupy about two-thirds of the genome and specify two precursor polyproteins (pp1a and pp1ab) of unprecedented size and complexity (more than 4000 and 7000 residues, respectively). Pp1ab is the C-terminally extended version of pp1a and its synthesis is regulated by a highly conserved ribosomal frameshifting mechanism (see below), yielding a fixed ratio between ORF1a- and ORF1b-encoded subunits. Also, the total number of 15–16 cleavage products (or nsps) generated from the coronavirus pp1a/pp1ab precursors through the activity of 2 or 3 internal proteinases is without precedent in +RNA virology.
- Third, compared to other groups of +RNA viruses, the RNA synthesis directed by the coronavirus replicase is more complex. Not just because of the sheer size of the genome, and the issue of the high mutation frequency that is commonly associated with

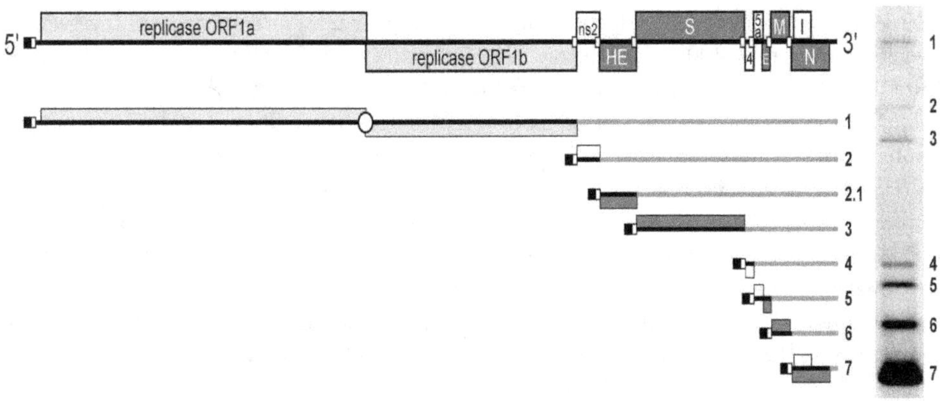

Figure 2.1 Coronavirus genomic structure and genome expression strategy, using *Mouse hepatitis virus* as example. The open reading frames in the viral genomes are indicated at the top of the scheme. The names of the different genes are indicated. Below the genome organization, the structural relationship of the genome and sg mRNAs is depicted. The leader sequence and TRSs found at the 5′ end of all mRNAs are indicated as black and white boxes, respectively. The ribosomal frameshifting element (RFS) found in the genome-length mRNA1 (circle) and the translated region of each mRNA are indicated. The right-hand panels show a typical pattern of MHV mRNAs isolated from infected mouse L cells, visualized by hybridization to a probe complementary to the 3′ end of the genome and therefore recognizing all viral mRNA species and reflecting their relative abundance. The HE gene of the MHV laboratory strain used here (MHV-A59) is no longer functional and mRNA2.1 is not produced.

RNA virus RNA polymerases, but also because of another hallmark of coronavirus biology: the production of an extensive nested set of 5 to 10 subgenomic messenger RNAs (Figure 2.1). These transcripts, which are discussed in more detail below, serve to express the structural and accessory protein genes of coronaviruses, which are located in the 3′-proximal third of the genome. Subgenomic RNA synthesis depends on a discontinuous step, which presumably operates during minus strand synthesis and is required to provide all subgenomic mRNAs with a common 5′, leader sequence (for reviews, see also Sawicki and Sawicki, 1995; Sawicki and Sawicki, 2005; Siddell *et al.*, 2005; Snijder *et al.*, 2005; Pasternak *et al.*, 2006).

The complexity of coronavirus molecular biology

Since most +RNA viruses can survive by relying on non-structural proteins of "only" 2000–4000 residues, the size of the coronavirus genome and replicase has often been interpreted as an omen of the complexity of coronavirus molecular biology that remained to be unraveled. Indeed, the amplification of such a large RNA genome raises questions about the existence of proofreading or repair mechanism that might compensate for the intrinsic high error frequency of the viral RdRp enzyme (Holmes and Rambaut, 2004; Moya *et al.*, 2004). In this context, also high-frequency (similarity-assisted) RNA recombination (Nagy and Simon, 1997), which was regularly described in connection with coronavirus replication (Makino *et al.*, 1986; Lai, 1996; Brian and Spaan, 1997), should be mentioned as a possible factor in the replication (and evolution) of this virus group. RNA recombination, in turn, shows some clear mechanistic similarities with the current model for subgenomic RNA production, in which a template switch is guided by a crucial basepairing interaction between nascent strand and "acceptor template" (for recent reviews, see Sawicki and Sawicki, 2005; Siddell *et al.*, 2005; Snijder *et al.*, 2005; Pasternak *et al.*, 2006). Quite likely, as illustrated by studies on fellow nidoviruses (van Dinten *et al.*, 1997; Tijms *et al.*, 2001), the replicase gene encodes protein functions with a specific role in the process of subgenomic RNA synthesis. Finally, the (re)analysis of the coronavirus replicase that was triggered by the 2003 SARS-Coronavirus outbreak has identified many novel enzyme functions (Snijder *et al.*, 2003; Ivanov *et al.*, 2004a; Putics *et al.*, 2005; Minskaia *et al.*, 2006), several of them unique to coronaviruses (and their closest relatives). Although their functional characterization is still in full progress, it has become clear that at least some of these enzymatic activities may be dispensable for replication in cell culture, suggesting that they may exert their function at the level of the interface between virus and host organism (Schwarz *et al.*, 1990; Putics *et al.*, 2005).

Over the past decade, the molecular virological and biochemical characterization of the coronavirus replicase has been profoundly influenced by in-depth bioinformatics studies (Gorbalenya *et al.*, 1989a; Gorbalenya *et al.*, 1991; Gorbalenya, 2001; Ziebuhr *et al.*, 2001; Gonzalez *et al.*, 2003; Snijder *et al.*, 2003; Gorbalenya *et al.*, 2004), the advent of systems for coronavirus reverse genetics (Almazán *et al.*, 2000; Yount *et al.*, 2000; Casais *et al.*, 2001; Thiel *et al.*, 2001; Yount *et al.*, 2003; Coley *et al.*, 2005), and the increasing availability of structural information concerning individual replicase domains (Anand *et al.*, 2002; Anand *et al.*, 2003; Yang *et al.*, 2003; Egloff *et al.*, 2004; Sutton *et al.*, 2004; Bartlam *et al.*, 2005; Peti *et al.*, 2005; Saikatendu *et al.*, 2005; Yang *et al.*, 2005; Zhai *et al.*, 2005). Still, a major challenge for the future remains: the integration of functional and structural information into a coherent picture of the overall "replication complex," and the way in which this macromolecular assembly operates in and interacts with the coronavirus-infected host cell. Clearly, this step will also be crucial for the rational design of antiviral strategies aimed at combating coronavirus infections. In this chapter, we will summarize our current understanding of the structure, expression and function of the proteins encoded by the coronavirus replicase gene. Also, we will summarize what is known about the "natural working environment" of these enzymes, the replication complex inside the coronavirus-infected cell.

The conserved core of the nidovirus replicase gene

The order *Nidovirales* (Cavanagh, 1997; Spaan *et al.*, 2005b) unites three virus families: *Arteriviridae, Roniviridae,* and *Coronaviridae* (consisting of the genera *Torovirus* and *Coronavirus,* the latter again being subdivided into three groups). The currently known representatives of these families mostly infect mammals, but can also have avian (coronaviruses) or invertebrate (roniviruses) hosts, and they can cause a wide range of diseases. Also several molecular biological differences seem to argue against the grouping of these virus families into one taxon: genome sizes range from 12–16 kb for arteriviruses to 26–31 kb for coronaviruses and roniviruses, nucleocapsid structures are very different, and the same is true for the overall virion morphology.

Despite these major differences, the largest gene in the nidovirus genome—the replicase gene—provided convincing grounds for their taxonomic unification. In phylogenetic analyses of key replicase domains, including the RdRp and helicase, the replicase genes of the different nidovirus subgroups were found to cluster, suggesting that they share a common ancestor (Gorbalenya *et al.*, 1989a; Snijder *et al.*, 1990; den Boon *et al.*, 1991; Cavanagh, 1997; de Vries *et al.*, 1997; Cowley *et al.*, 2000; Gonzalez *et al.*, 2003; Snijder *et al.*, 2005; Gorbalenya *et al.*, 2006). Furthermore, the common genetic plan and common genome expression strategy supported the notion that nidoviruses, or—more precisely—nidovirus replicase genes, form a separate lineage in +RNA virus evolution. Important molecular biological characteristics shared by all nidoviruses (Figure 2.1 and 2.2) are

1 the polycistronic nature of the genome,
2 the presence of two large replicase open reading frames (1a and 1b) connected by a ribosomal frameshift site,
3 the conserved array of key domains within the replicase gene (see below), and
4 the synthesis of a nested set of subgenomic mRNAs to mediate structural protein gene expression.

Nevertheless, a detailed comparison between nidovirus replicase genes also reveals major differences, in particular when arteriviruses are compared with nidoviruses that have much bigger genomes and replicase genes (Figure 2.2). This aspect was recently reviewed in detail by Alexander Gorbalenya and colleagues (Gorbalenya *et al.*, 2006), who concluded that in fact only two "true genetic markers" (i.e. domains shared by all nidoviruses and not found in other RNA virus lineages) can be identified in nidovirus replicases:

1 the (putative) multinuclear zinc-binding domain (ZBD) (Gorbalenya *et al.*, 1989a; Snijder *et al.*, 1990; Cowley *et al.*, 2000; van Dinten *et al.*, 2000; Seybert *et al.*, 2005) that forms the N-terminal part of the nidovirus helicase protein and,
2 the uridylate-specific endoribonuclease (NendoU) domain (Snijder *et al.*, 1990; Snijder *et al.*, 2003; Bhardwaj *et al.*, 2004; Ivanov *et al.*, 2004a; Posthuma *et al.*, 2006), encoded in the 3′-proximal part of ORF1b.

ZBD and NendoU are part of a more extensive set of conserved domains, which do have counterparts in other virus groups, although the conserved array of domains as it occurs in nidoviruses is unique for this lineage. These additional domains include three ORF1a-encoded (putative) transmembrane domains (see below), the chymotrypsin-like main proteinase and the ORF1b-encoded RdRp and superfamily 1 helicase (HEL1) domains. Together with ZBD and NendoU, they form the conserved backbone of the nidovirus replicase gene, roughly occupying the 3′ half of ORF1a and ORF1b. The viral main proteinase is responsible for all but one of the proteolytic cleavages required to release the individual nsps from the conserved replicase backbone (see below).

Major variations between nidovirus replicase genes

By and large, three areas of variation inside nidovirus replicase genes can be discriminated. The first concerns two apparent insertions in the generally conserved ORF1b-encoded part of the replicase, which are found in all nidoviruses with a large genome (corona-, toro-, and roniviruses). These are a 3′→5′ exonuclease (ExoN), located between the HEL1 and

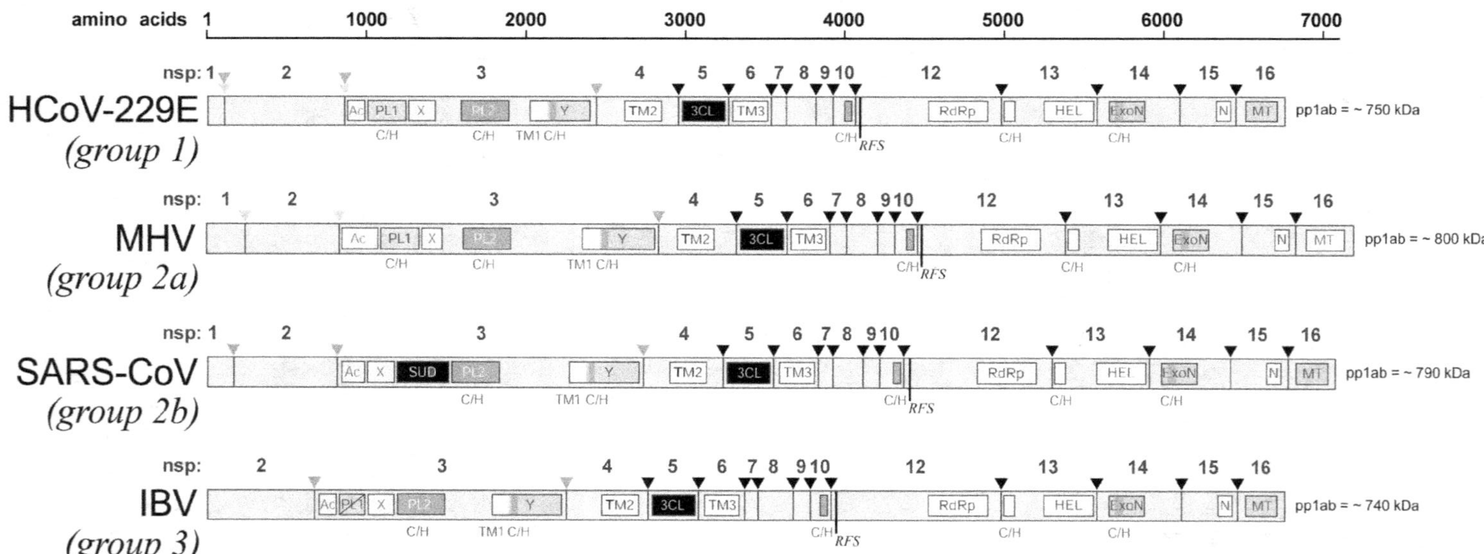

Figure 2.2 The replicase gene organization of representatives of coronavirus groups 1, 2a, 2b and 3 (Spaan *et al.*, 2005a). The replicase gene organization is depicted in the form of the polyprotein pp1ab. The delineation of amino acids encoded in ORF1a and ORF1b is indicated as RFS (ribosomal frameshift) and arrows represent sites in pp1ab that are cleaved by papain-like proteinases (light and dark grey) or the main (3CL) proteinase (black). The proteolytic cleavage products are numbered (Snijder *et al.*, 2003) and within the cleavage products, the location of domains that have been identified as structurally or functionally related are highlighted. These include domains with conserved Cys and His residues (C/H), putative transmembrane domains (TM), domains with conserved features (Ac, X and Y), and domains that have been associated with proteolysis (PL1, PL2 and 3CL), RNA-dependent RNA synthesis (RdRp), helicase (HEL), exoribonuclease (ExoN), uridylate-specific endoribonuclease (N), and methyltransferase (MT) activities. The figure was adapted from Siddell *et al.* (2005).

NendoU domains, and a putative ribose-2′-O methyl transferase (MT), residing in the most C-terminal replicase subunit (Feder *et al.*, 2003; Snijder *et al.*, 2003; von Grotthuss *et al.*, 2003; Minskaia *et al.*, 2006) (Figure 2.2 and see below). It has been postulated by Alexander Gorbalenya and colleagues that in particular the acquisition of the ExoN domain may have been a key event in the expansion of the nidovirus genome to the size range of the present-day corona-, toro-, and roniviruses (Gorbalenya *et al.*, 2006).

Secondly, the 5′ half of ORF1a is a region of major variation, both between and within nidovirus subgroups. Proteolytic processing of this region, the N-proximal domain of both replicase polyproteins, is performed by so-called "accessory proteinases," which belong to the superfamily of papain-like cysteine proteinases (PL^{pro}s; reviewed in Gorbalenya *et al.*, 1991; Ziebuhr *et al.*, 2000; Ziebuhr, 2005). Depending on the nidovirus, there can be one to four of such PL^{pro} domains, but coronaviruses typically contain one or two accessory proteinases, which reside in the large nsp3 protein and are linked to zinc-binding domains (Herold *et al.*, 1999) (see below). In addition to various conserved domains that appear to be specific for nidovirus groups or subgroups (and sometimes even single species), the 5′ half of ORF1a encodes one domain that is conserved in all coronaviruses and toroviruses, but is lacking in arteriviruses and roniviruses: the ADP-ribose 1″-phosphatase (ADRP) or X domain (Gorbalenya *et al.*, 1991; Snijder *et al.*, 2003; Putics *et al.*, 2005; Saikatendu *et al.*, 2005; Draker *et al.*, 2006; Putics *et al.*, 2006). Substitution of putative ADRP active-site residues did not significantly affect virus replication in cell culture, suggesting that the enzyme plays a role at e.g. the level of virus-host interactions (Putics *et al.*, 2005).

The third area/type of variation, at least when the three nidovirus families are compared as a whole, concerns the (sometimes large) regions separating the domains that form the conserved replicase backbone. Examples are the proteins encoded between the last putative transmembrane domain and ribosomal frameshift site, and the N-terminal two-thirds of the large RdRp-containing cleavage product. Clearly those domains, and several others, can still be considered "white spots" on the nidovirus replicase map, although recent studies have yielded a very interesting perspective for the set of small coronavirus proteins, known as nsp7 to nsp10, encoded in the 3′ end of ORF1a. Genetic analysis implicated nsp10 in RNA synthesis (Sawicki *et al.*, 2005) and recent structural studies showed that nsp9 (Egloff *et al.*, 2004; Sutton *et al.*, 2004) and a hexadecameric complex formed by nsp7 and nsp8 (Peti *et al.*, 2005; Zhai *et al.*, 2005) have RNA-binding properties and may be accessory factors for the RdRp complex (see below).

Expression of the coronavirus replicase gene

The initiation of nidovirus RNA synthesis depends on the expression of the replicase gene, which is regulated at both the co-translational (ribosomal frameshifting) and post-translational (proteolytic processing) levels. Genome translation depends on the common 5′-cap-dependent mechanism of ribosome binding, which, in principle, allows the subsequent translation of the most 5′-located reading frame of an mRNA, in this case replicase ORF1a. Subgenomic mRNAs are produced to ensure the expression of all downstream genes, with one exception: replicase ORF1b. To express this part of the genome, which encodes key enzymatic functions to initiate and sustain viral RNA synthesis (Figure 2.2), nidoviruses rely on a ribosomal frameshift mechanism.

Ribosomal frameshifting

Several other RNA virus groups employ ribosomal frameshifting (for a recent review, see Brierley and Dos Ramos, 2005, and references therein). It is not uncommon that this strategy is used for the relative downregulation of the expression of the viral RNA polymerase compared to replicase subunits encoded upstream of the frameshift site. Due to a combination of primary and higher order structural features of the short genome region where nidovirus ORF1a and ORF1b overlap, a fraction of the ribosomes translating ORF1a from the genome RNA shift into the −1 reading frame. Instead of terminating at the ORF1a stop codon (which would yield pp1a), they continue translation in ORF1b, thus producing the full-length pp1ab replicase polyprotein. Nidovirus frameshift signals, which were first characterized for the coronavirus IBV by Brierley *et al.* (1989), consist of two RNA elements:

1 the so-called "slippery" sequence, i.e. the actual site where the ribosome changes frame, and
2 a higher-order RNA structure that is positioned just downstream and presumably promotes frameshifting by stalling the ribosome when translating the slippery sequence (Jacks *et al.*, 1988; Brierley *et al.*, 1989; ten Dam *et al.*, 1990).

In nidoviruses, the downstream RNA structure is an RNA pseudoknot, a stable RNA hairpin of which the loop base-pairs with a downstream sequence element, thus expanding the physical obstacle for the elongating ribosome. The slippery sequence is a heptanucleotide stretch that usually has the sequence XXXYYYZ (X can be any nucleotide, Y is A/U, Z is not G). Eukaryotic frameshifting presumably occurs by "simultaneous slippage" of the tRNAs bound to the XXY and YYZ codons (ORF1a) into the −1 frame (XXX YYY; ORF1b) (Jacks *et al.*, 1988; Brierley *et al.*, 1989; Baranov *et al.*, 2005). The specific codon usage is thought to allow two out of three anticodon nucleotides of these tRNAs to stay base-paired with the template during the shift. For a recent review on the mechanistic details of ribosomal frameshifting and the RNA structures regulating the process, the reader is referred to Brierley and Dos Ramos (2005) and references therein.

Although direct data from nidovirus-infected cell systems are not available, studies with reporter plasmids encoding nidovirus ribosomal frameshift signals suggested a relatively high frameshift efficiency of 15–40% (Brierley and Dos Ramos, 2005). This level determines the stoichiometry of the replicase subunits encoded in ORF1a and ORF1b, which has apparently been conserved in nidovirus evolution. Among the ORF1a-encoded proteins/domains that are "relatively overexpressed" in this manner are the viral proteinases and the (putative) *trans*-membrane domains that are thought to anchor the viral replication complex to (modified) intracellular membranes (see below).

Processing of the N-terminal replicase region by accessory proteinases

Like many other +RNA viruses (Kräusslich and Wimmer, 1988; Dougherty and Semler, 1993), coronaviruses employ proteolytic processing of their replicase polyproteins as a crucial regulatory mechanism at the post-translational level. Following (and possibly also during) their translation, pp1a and pp1ab are processed by two or three autoproteinases encoded by ORF1a. The (predicted) number of mature replicase products (non-structural proteins or nsps) is 15 or 16 (see Ziebuhr *et al.*, 2000; Snijder *et al.*, 2003; Ziebuhr, 2005, and references therein), but it should be noted that—as in other systems—processing intermediates may be functional subunits themselves, whose activity could in fact be modulated or terminated by additional proteolytic events.

The proteinases integrated into the coronavirus replicase were among the first enzymes to be studied experimentally and characterized by using bioinformatics methods. Upon translation of the MHV genome in rabbit reticulocyte lysates, a 28-kDa N-terminal cleavage product (nsp1) was identified (Denison and Perlman, 1986). It was shown that the proteinase responsible for this cleavage resides in a domain of (what is now referred to as) nsp3, a domain that was characterized as a distant relative of the cellular cysteine proteinase papain (Baker *et al.*, 1989; Gorbalenya *et al.*, 1991; Baker *et al.*, 1993). A variety of studies on coronavirus replicase proteolysis (for reviews, see Ziebuhr *et al.*, 2000; Ziebuhr, 2005, and references therein) have revealed additional PL^{pro} cleavages in the nsp1–4 region and have started to define factors influencing PL^{pro} activity, including the interplay between $PL1^{pro}$ and $PL2^{pro}$ in viruses containing two active PL^{pro} enzymes (Ziebuhr *et al.*, 2001; Ziebuhr, 2005).

Processing of the conserved replicase backbone by the main proteinase

The proteolysis of the conserved core of the coronavirus replicase is controlled by the proteinase domain that has become known as "3C-like cysteine proteinase" ($3CL^{pro}$, a name based on a relationship with the picornavirus 3C proteinase) or "main proteinase" (in view of its central role in the expression of the major replicative proteins) (Ziebuhr *et al.*, 2000). This activity resides in nsp5 and was initially predicted by Alexander Gorbalenya and coworkers on the basis of comparative sequence analysis using the first coronavirus replicase sequence (IBV; Boursnell *et al.*, 1987) that became available (Gorbalenya *et al.*, 1989a). In the decade following this groundbreaking analysis, comparisons with additional coronavirus replicase sequences and experimental studies have resulted in the current scheme of $3CL^{pro}$-mediated processing, which includes 11 proteolytic cleavages in the nsp4-nsp16 region of all coronaviruses and was reviewed by Ziebuhr *et al.* (Ziebuhr *et al.*, 2000; Ziebuhr, 2005) (Figure 2.2).

For several coronavirus $3CL^{pro}$s, peptide cleavage data revealed differential processing kinetics for specific sites. The order of cleavages was found to be conserved among coronaviruses and appears to depend on the accessibility and structural context of these sites in the context of the polyprotein (Piñon *et al.*, 1999; Hegyi and Ziebuhr, 2002; Fan *et al.*, 2004). It is assumed that coronavirus polyprotein processing occurs in a temporally coordinated manner, but information on the kinetics of the process in infected cells is rather limited. Proteinase inhibitors can block coronavirus replication, illustrating the essential role of pp1a/pp1ab processing for viral RNA synthesis (Yang *et al.*, 2005). Still, the importance of individual cleavages has not been addressed in great detail. Reverse genetics studies demonstrated that specific cleavages (e.g. of the nsp1|nsp2 site) and even specific domains in the N-terminal part of the replicase are dispensable for MHV replication (Denison *et al.*, 2004; Brockway and Denison, 2005; Graham *et al.*, 2005). By analogy with arteriviruses (van Dinten *et al.*, 1999; van Aken, Zevenhoven, Gorbalenya, and Snijder, unpublished) it is assumed that probably all $3CL^{pro}$ cleavages in the conserved backbone of the replicase will prove to be important, if not essential for (efficient) virus replication.

Enzymic activities encoded by the coronavirus replicase gene

Papain-like proteinases

As summarized above, the N-proximal regions of the coronavirus replicase polyproteins are processed by PL^{pro}s that reside in nsp3 (Ziebuhr *et al.*, 2000). Recent evidence suggests that at least some of these proteases might have additional functions by acting as deubiquitinating enzymes and thus be involved in pathways outside the replicase polyprotein processing (Lindner *et al.*, 2005; Sulea *et al.*, 2005), for example, in counteracting the host cell's antiviral response. Except for IBV and SARS-CoV, which only possess one (active) PL^{pro} (Lim *et al.*, 2000; Ziebuhr *et al.*, 2000; Thiel *et al.*, 2003; Harcourt *et al.*, 2004), coronaviruses employ two PL^{pro} domains called $PL1^{pro}$ and $PL2^{pro}$ that cleave at 2 or 3 sites. Based on their position in nsp3 (downstream of the ADRP) the single PL^{pro} domains of IBV and SARS-CoV are referred to as orthologs of the $PL2^{pro}$s of other coronaviruses (Snijder *et al.*, 2003; Thiel *et al.*, 2003).

Coronavirus PL^{pro}s employ a catalytic Cys-His-Asp triad (Barretto *et al.*, 2005; Sulea *et al.*, 2005) and are predicted to adopt a papain-like $\alpha+\beta$ structure in which the α- and β-domains are connected by a zinc ribbon structure (Herold *et al.*, 1999). SARS-CoV PL^{pro} structural informatics data suggest that this intervening domain has a circularly permutated (rather than classical) zinc ribbon structure (Sulea *et al.*, 2006). Most coronavirus PL^{pro}s require small residues, Gly or Ala, at both the P1 and P2 positions of the cleavage sites (Ziebuhr *et al.*, 2000). Exceptions are the nsp1|nsp2 and nsp2|nsp3 sites cleaved by the $PL1^{pro}$ of MHV and other group II coronaviruses, which have a small residue (Gly or Ala) at the P1 position, but significantly larger residues (Arg, Cys) at the P2 position. Furthermore, these sites have basic residues at their P4 and P2′ positions. Recently, a refined structure model was reported for the SARS-CoV $PL2^{pro}$ that, together with previous PL^{pro} cleavage site mutagenesis data (Dong and Baker, 1994; Bonilla *et al.*, 1997), was used to reconcile the available information on coronavirus PL^{pro} substrate specificities (Sulea *et al.*, 2006). The structure model is essentially based on (i) sequence comparisons with other papain-like proteases and (ii) structure information derived from two template structures, the herpesvirus-associated ubiquitin-specific protease (HAUSP) bound to ubiquitin aldehyde (Hu *et al.*, 2002) and the foot-and-mouth disease virus (FMDV) leader protease (L^{pro}) (Guarné *et al.*, 1998; Guarné *et al.*, 2000). The study revealed possible determinants and structural constraints of the coronavirus PL^{pro} substrate specificities, including key specificity determinants, which then formed the basis to propose a new (binding site-based) classification of coronavirus PL^{pro}s into two major groups, called R-group and O-group (Sulea *et al.*, 2006). The R-group encompasses all coronavirus $PL2^{pro}$s (including those of SARS-CoV and IBV) and the $PL1^{pro}$s of group I coronaviruses (HCoV-229E, PEDV, TGEV and HCoV-NL63). Enzymes that belong to this group are suggested to have a catalytic core domain resembling that of HAUSP and other ubiquitin-specific proteases (USP) as aligned in the MEROPS database (Rawlings *et al.*, 2004). Their significantly restricted (R), USP-like (HAUSP-like) binding site is characterized by a conserved signature of key substrate-binding residues, with an Asn being absolutely conserved in the S1 binding site and two Tyr residues (one of them being conservatively substituted by Phe in a few cases) in the S2 subsite. Like in USPs, the restricted binding mode of coronavirus R-group PL^{pro}s is suggested to result from occluded S1 and S2 subsites that are suitable to accommodate a P1-P2 di-Gly but no side chains larger than that of Ala. By contrast, coronaviral O-group PL^{pro}s share a S2 subsite that is distinct from the R-group proteases but similar to papain in that the two S2 subsite Tyr residues mentioned above are substituted by smaller residues. This, together with other subsitutions, opens the S2 site and allows for binding of larger P2 side chains. The distinct features of the R- versus O-group S1 and S2 subsites appear to be associated with further differences in other binding sites and even extend to the presumed oxyanion hole (Sulea *et al.*, 2006).

The classification of group II coronaviral PL^{pro}s into two distinct groups receives further support by previous MHV PL^{pro} inhibition data, which had shown that E-64d inhibits PL1^{pro} but not PL2^{pro} (Kanjanahaluethai and Baker, 2000). It is also consistent with the (partially) overlapping substrate specificities reported for the PL^{pro}s of HCoV-229E (but not for those of MHV) (Ziebuhr *et al.*, 2001; Kanjanahaluethai *et al.*, 2003). The fact that (most) group II coronaviruses appear to encode two types of PL^{pro}s, whereas, for example, group I coronaviruses encode only one type suggests differences in the origin and/or evolution of PL^{pro} activities between these groups of coronaviruses.

3C-like (main) proteinase

The coronavirus "main proteinase," 3CL^{pro} (nsp5), is a protein of about 300 residues. It releases itself from the flanking hydrophobic domains (nsp4 and nsp6) by autoproteolytic cleavage and is responsible for cleaving as many as 11 sites in pp1a/pp1ab (Figure 2.2). Using selective inhibitors, the proteolytic activity of 3CL^{pro} was demonstrated to be essential for coronavirus replication (Yang *et al.*, 2005). Furthermore, genetic studies using MHV *ts* mutants implicate the enzyme in a specific step in viral RNA synthesis, namely the switch from negative-strand to (efficient) plus-strand RNA synthesis (Sawicki *et al.*, 2005).

Essentially, coronavirus main proteinases comprise a (chymotrypsin-like) two-β-barrel core domain and an additional, C-terminal domain called "domain III," which is formed by five α-helices that are arranged in a unique fold (Anand *et al.*, 2002; Anand *et al.*, 2003; Yang *et al.*, 2003) (Figure 2.3A). Coronavirus main proteases form stable dimers (Anand *et al.*, 2002; Chou *et al.*, 2004; Shi *et al.*, 2004; Hsu *et al.*, 2005), with the two protomers being oriented perpendicular to one another (Figure 2.3A). Dimerization mainly occurs through (i) hydrogen-bonding interactions between the protein's N terminus ("N-finger") and residues near the S1 subsite of the other monomer (Anand *et al.*, 2002; Yang *et al.*, 2003) as well as interactions between the carboxyl-terminal domains III. Besides its role in dimerization, the "N-finger" helps to keep the active site of the other monomer in a catalytically competent conformation (Yang *et al.*, 2003; Tan *et al.*, 2005). Also domain III has recently been implicated in the regulation of the catalytic activity of the proteinase (Shi and Song, 2006).

There is strong evidence to suggest that it is the dimeric form of 3CL^{pro} that cleaves most of the downstream sites in pp1a and pp1ab in *trans* (Fan *et al.*, 2004). Furthermore, it has been proposed that 3CL^{pro} domains might dimerize (albeit much less efficiently) when still part of the polyprotein, and a four-step maturation process has been proposed which starts off with two unprocessed polyproteins that interact with each other, proceeds through several processing intermediates to finally yield the fully processed (and dimeric) form of the 3CL^{pro} that is seen in the available crystal structures (Hsu *et al.*, 2005). The model implies that the release of fully activated, dimeric 3CL^{pro} may be a rather slow

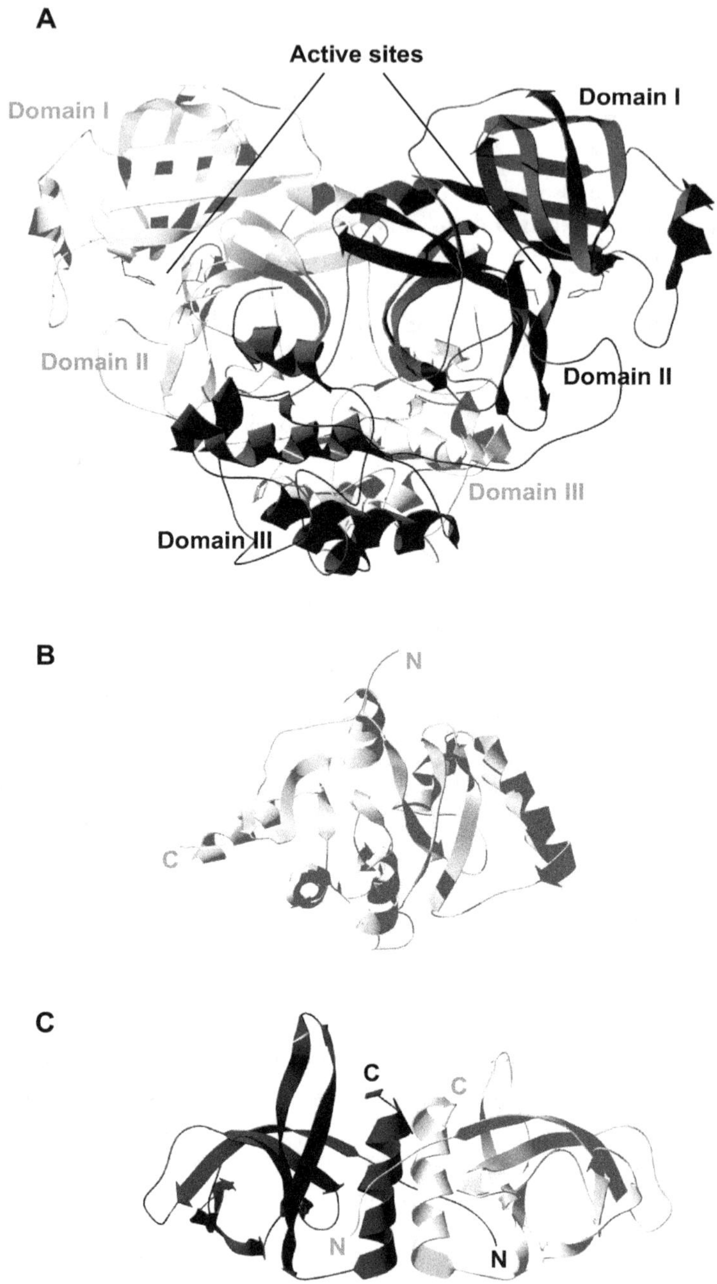

Figure 2.3 Structures of the 3CL^{pro}, ADRP, and nsp9. (A) Crystal structure of the TGEV 3CL^{pro} dimer (PDB code 1LVO). Protomers A and B are shown in ribbon representation and are colored black and grey, respectively. The N-terminal β-barrel domains I and II and the C-terminal α-helical domain III are indicated. Also shown are the catalytic Cys and His residues in ball-and-stick representation. (B) Crystal structure of the SARS-CoV ADRP (2ACF) shown in ribbon representation. (C) Crystal structure of the SARS-CoV nsp9 dimer (PDB code 1QZ8). Protomers A and B are shown in ribbon representation and are colored black and grey, respectively.

process, which indeed is consistent with the observation of quite stable 3CLpro-containing precursors in coronavirus-infected cells (Schiller *et al.*, 1998). The model by Hsu *et al.* (Hsu *et al.*, 2005) is further supported by MHV ORF1a *ts* mutants data demonstrating that the nsp4-nsp10 region of pp1a/pp1ab forms a single large cistron, indicating that, early in infection, this part of the replicase might act as a large (essentially unprocessed) precursor molecule (Sawicki *et al.*, 2005). As soon as sufficient dimeric 3CL^{pro} has been produced (i.e. later in infection), pp1a/pp1ab could be processed extremely effectively (Ziebuhr and Siddell, 1999; Prentice *et al.*, 2004b), thus providing the fully processed nsps that are needed to form the large, multimeric protein assemblies identified in structural studies and believed to be required for maximum plus-strand RNA synthesis (see below).

It should be noted, however, that this model of the release (and gradual activation) of 3CL^{pro} is still based on a quite limited set of data. Furthermore, it is worth noting that coronavirus 3CL^{pro} domains (and their counterparts in other nidoviruses) are flanked in the polyprotein by transmembrane domains that seem to affect the structure and/or function of the proteinase domain (Tibbles *et al.*, 1996; Piñon *et al.*, 1997). Also, in most of the structural and functional studies published to date, either fully processed 3CL^{pro}s or 3CL^{pro}s with artificial N- and/or C-terminal extensions rather than authentic neighboring regions (nsp4 and nsp6) were used (Anand *et al.*, 2002; Anand *et al.*, 2003; Yang *et al.*, 2003; Bartlam *et al.*, 2005; Yang *et al.*, 2005). It remains to be seen as to whether the flanking membrane domains will be able to trigger a (partial) refolding of the protease, thus inducing self-cleavage in *cis*. Early intra- and intermolecular cleavages by 3CL^{pro}-containing precursors might also occur at more distant sites, for example, in the nsp12–16 region. This idea is supported by genetic data showing that the nsps 12 to 16 belong to different complementation groups (Sawicki *et al.*, 2005), suggesting that, quite early in infection, they already act as separate functional subunits.

The distant relationship between coronavirus main proteinases and picornavirus 3C proteinases (3C^{pro}s), which in the first place had allowed the identification of the coronavirus homolog in the avian coronavirus IBV polyprotein, inspired the name "3C-like" for this type of coronavirus proteinases (Gorbalenya *et al.*, 1989a). However, despite their similar substrate specificities (see below) and the conservation of a two-β-barrel fold, the two groups of proteinases differ significantly from each other in many structural details and even their catalytic residues. Thus, for example, whilst picornavirus 3C proteinases employ a Cys-His-Asp(Glu) catalytic triad (Allaire *et al.*, 1994; Matthews *et al.*, 1994; Mosimann *et al.*, 1997; Seipelt *et al.*, 1999; Yin *et al.*, 2005), the coronavirus homologs employ a Cys-His catalytic dyad (Liu and Brown, 1995; Lu *et al.*, 1995; Ziebuhr *et al.*, 1995; Ziebuhr *et al.*, 1997; Anand *et al.*, 2002; Hegyi *et al.*, 2002; Anand *et al.*, 2003; Yang *et al.*, 2003; Tan *et al.*, 2005). A stable water molecule that occupies the position of the usual third member of the triad has been suggested to stabilize the protonated His in the intermediate state during proteolytic cleavage (Anand *et al.*, 2002; Anand *et al.*, 2003), thus possibly compensating for the lack of an acidic catalytic residue in coronavirus main proteases.

Despite only moderate sequence conservation among coronavirus main proteinases, their substrate specificities are extremely well conserved (Hegyi and Ziebuhr, 2002; Thiel *et al.*, 2003). Gln is invariably present at the P1 position, whereas small residues (Ala, Ser, and Gly) are conserved at the P1′ position of most cleavage sites (Ziebuhr *et al.*, 2000). The presence of larger residues, such as Asn and Leu, at the P1′ position has detrimental effects on cleavage efficiency (Hegyi and Ziebuhr, 2002; Fan *et al.*, 2004; Fan *et al.*, 2005). At the P2 position, Leu is typically conserved, but the enzyme's S2 subsite seems to be large enough to accept, at least in some cases, bulkier hydrophobic residues, such as Phe (Yang *et al.*, 2003). Furthermore, the S4 subsite contributes to the enzyme's specificity by accommodating Val, Thr, Ala or Ser residues (Ziebuhr *et al.*, 2000; Thiel *et al.*, 2003; Fan *et al.*, 2005). The well conserved specificity correlates with a highly conservative substrate-recognition pocket as revealed by crystal structure analyses of representative $3CL^{pro}$s from all three genetic groups of coronaviruses (Yang *et al.*, 2005), which makes the enzyme an attractive target for the development of irreversible inhibitors with clinical potential against a broad range of existing and potentially (re-)emerging coronavirus-associated diseases.

RNA polymerase and helicase

The coronavirus RNA-dependent RNA polymerase (RdRp) domain occupies the C-terminal part of nsp12 and has been classified as an outgroup of superfamily 1 RdRps (Koonin, 1991). Its predicted activity has recently been established for SARS-CoV (Cheng *et al.*, 2005) and a structural model of the protein has been reported (Xu *et al.*, 2003). MHV data suggest that nsp12 interacts with ORF1a-encoded proteins, including nsp5 ($3CL^{pro}$), nsp8, and nsp9 (Brockway *et al.*, 2003). When expressed separately as a green fluorescent protein-tagged fusion protein in virus-infected cells, the MHV nsp12 assembled with the viral replicase, suggesting that nsp12 (in contrast to most ORF1a functions) is a "diffusible" component of the replicase. This idea is strongly supported by complementation studies using MHV *ts* mutants (Sawicki *et al.*, 2005).

The coronavirus helicase resides in nsp13 and has been classified as belonging to the superfamily 1 of helicases (Gorbalenya *et al.*, 1989b; Gorbalenya *et al.*, 1989a) (Figure 2.2). At their N-termini, coronavirus helicases and their nidovirus homologs have a complex zinc-binding domain (ZBD) including 12 conserved Cys and His residues that are predicted to form a binuclear cluster (Gorbalenya *et al.*, 1989a; van Dinten *et al.*, 2000; Bernini *et al.*, 2006). The ZBD has been demonstrated to be critically involved in the enzymatic activities of both coronavirus and arterivirus helicases (Seybert *et al.*, 2005), which explains some of the defects in RNA synthesis and virus reproduction reported for arterivirus EAV ZBD mutants. Furthermore, the arterivirus reverse genetics data suggest that a protein segment, called "hinge spacer," that links ZBD to the C-terminal core helicase domain is involved in subgenomic RNA synthesis (van Dinten *et al.*, 1997; van Dinten *et al.*, 2000). Based on the arterivirus data, it seems reasonable to suggest that coronavirus helicases have diverse functions in the viral life cycle. Biochemical studies have indeed demonstrated multiple enzymatic activities for coronavirus helicases, including nucleic acid-stimulated NTPase and dNTPase as well as RNA and DNA duplex-unwinding activities (Seybert *et al.*, 2000a; Seybert *et al.*, 2000b; Tanner *et al.*, 2003; Thiel *et al.*, 2003; Ivanov *et al.*, 2004b; Ivanov and Ziebuhr, 2004). Furthermore, nsp13 was shown to have RNA 5′-triphosphatase activity which might be involved in the 5′-capping of coronavirus RNAs (Ivanov *et al.*, 2004b; Ivanov and Ziebuhr, 2004).

Coronavirus helicases unwind their (partial-) duplex substrates with 5′-to-3′ polarity, that is, they move in a 5′-to-3′ direction along the strand to which they initially bind (Seybert *et al.*, 2000a). The polarity of the unwinding reaction differs from that of the well characterized (superfamily 2) helicases of the *Flaviviridae* (Kadaré and Haenni, 1997), indicating differences between these two groups of enzymes regarding their biological functions in viral replication. Coronavirus helicases act very effectively on both DNA and RNA substrates, a property that is rarely found in helicases (Seybert *et al.*, 2000a; Thiel *et al.*, 2003; Ivanov *et al.*, 2004a; Ivanov and Ziebuhr, 2004). Still, this specific feature could be exploited to devise robust (DNA- rather than RNA-based) activity assays suitable for high-throughput screening of potential inhibitors of these enzymes.

Exo- and endoribonuclease activities

Coronaviruses and other nidoviruses with genome sizes approaching 30 kb (i.e. toroviruses, roniviruses) encode a 3′→5′ exoribonuclease activity (ExoN) that occupies the N-terminal part of the coronavirus nsp14 (Snijder *et al.*, 2003; Minskaia *et al.*, 2006) (Figure 2.2). ExoN is related to the DEDD superfamily of exonucleases (Moser *et al.*, 1997; Zuo and Deutscher, 2001) which also includes enzymes involved in DNA replication-associated proofreading mechanisms. Arteriviruses (featuring much smaller genomes than other nidoviruses) lack an ExoN homolog, indicating that nidovirus exoribonucleases might be involved in pathways related to proofreading and repair which could be essential in the replication of RNA genomes whose sizes exceed a (postulated) critical threshold, which may be the case in the nidovirus genera mentioned above (Gorbalenya *et al.*, 2006). However, a definitive proof for this speculation is missing as is information on the fidelities (error frequencies) of coronavirus and other nidovirus RNA polymerase activities.

A recombinant form of the SARS-CoV ExoN has recently been demonstrated to mediate metal ion-dependent 3′→5′ exoribonuclease (ExoN) activity that acts on both single stranded (ss) and double-stranded (ds) RNA (Minskaia *et al.*, 2006). Although the precise specificity of the enzyme remains to be determined, it seems clear that RNA structure is an important factor in substrate binding and/or activity. The coronavirus ExoN is specific for RNA, as DNA and ribose-2′-O-methylated RNA proved to be resistant to cleavage. Reverse genetics data obtained for HCoV-229E demonstrated that ExoN activity is essential for the production of infectious virus. Substitutions of predicted ExoN active-site residues (e.g. Asp, Glu, His) resulted in severe defects in viral RNA synthesis (Minskaia *et al.*, 2006). Thus, besides a drastically reduced genome replication, distinct defects in subgenome-length RNA synthesis, such as aberrant sizes of specific RNAs and changes in the molar ratios of individual RNAs, were observed. Although the molecular mechanisms underlying the diverse defects in RNA synthesis remain to be investigated, it seems reasonable to suggest that, at some stage of genome-length and/or subgenome-length RNA synthesis, ExoN-mediated 3′→5′ exonuclease ("trimming") activity might be required, for example to remove nucleotide misincorporations during RNA synthesis or nucleotide mismatches at leader-body junctions during minus-strand RNA synthesis.

Coronaviruses encode a second ribonuclease activity that is called NendoU and, in contrast to ExoN, is conserved in all nidoviruses (but no other RNA virus), making it a genetic marker of the order *Nidovirales* (Snijder *et al.*, 2003; Ivanov *et al.*, 2004a). NendoU is related to a family of cellular proteins that is prototyped by the poly(U)-specific endoribonuclease, XendoU, which is involved in the processing of intron-encoded small nucleolar (sno) RNAs in *Xenopus laevis* (Laneve *et al.*, 2003). Reverse genetics data obtained for both corona- and arteriviruses have implicated NendoU in a critical (but currently unknown) function in the life cycle of these viruses (Ivanov *et al.*, 2004a; Posthuma *et al.*, 2006). Whereas substitutions of several conserved Asp codons proved to be lethal when introduced in the arterivirus EAV and coronavirus HCoV-229E genomes (Ivanov *et al.*, 2004a; Posthuma *et al.*, 2006), other substitutions including those of presumed active-site His residues caused only moderate defects in viral RNA synthesis, with subgenomic RNA synthesis being consistently more affected than genome replication (Posthuma *et al.*, 2006). Interestingly, with drops in virus titers of up to 5 logs, production of infectious virus was shown to be affected much more profoundly than viral RNA synthesis (Posthuma *et al.*, 2006), suggesting additional functions (apart from viral RNA synthesis) for NendoU in the viral replication cycle. Thus, for example, NendoU might process specific viral and/or cellular RNAs in order to promote the production of fully infectious viral genomes or regulate (counteract?) specific cellular pathways, thereby supporting the efficient production and release of infectious virus particles. As pointed out earlier (Snijder *et al.*, 2003), the conservation (and co-expression) of endoribonuclease, exoribonuclase and ribose-2′-O methyltransferase in coronaviruses has a striking parallel in the cellular pathway of snoRNA processing and site-specific RNA methylation, which depends on a very similar set of enzymic activities (Kiss, 2001; Filipowicz and Pogacic, 2002). It remains to be seen if the coronavirus (nidovirus) ORF1b-encoded enzymes cooperate in similarly complex metabolic pathways or if they interfere with those of the host cell at some stage of the viral replication cycle.

Uridylate-specific endoribonuclease activities have recently been established and characterized for the NendoU homologs from SARS-CoV, HCoV-229E and IBV (Bhardwaj *et al.*, 2004; Ivanov *et al.*, 2004a). The recombinant enzymes were demonstrated to require Mn^{2+} ions for activity (Bhardwaj *et al.*, 2004; Ivanov *et al.*, 2004a) and retained only little activity in the presence of Mg^{2+} or Ca^{2+} (Ivanov *et al.*, 2004a). NendoU was confirmed to cleave both single-stranded and double-stranded RNA (Bhardwaj *et al.*, 2004; Ivanov *et al.*, 2004a). Using primer extension studies, the enzyme was shown to cleave 3′ to uridylate residues (Ivanov *et al.*, 2004a). Initial studies showed that the cleavage activity is affected by both the local RNA structure (basepaired or non-basepaired uridylate) and the sequence context of the uridylate residue (Ivanov *et al.*, 2004a). Further studies are required to get a more complete picture of the enzyme's substrate specificity and possibly identify the biologically relevant substrates. NendoU cleavages produce molecules with 2′-3′ cyclic phosphate ends (Ivanov *et al.*, 2004a), causing an aberrant migration of NendoU cleavage products in polyacrylamide gels that led to erroneous conclusions on the precise site of cleavage in initial studies (Bhardwaj *et al.*, 2004). The generation of 2′-3′ cyclic phosphates in the cleavage reaction is consistent with the fact that NendoU does not cleave ribose-2′-O methylated RNAs (Ivanov *et al.*, 2004b). Crystal structure information and mutagenesis data suggest that, similar to RNase A, the NendoU catalytic center involves two conserved His and one Lys residue (Ivanov *et al.*, 2004a; Ricagno *et al.*, 2006; Ricagno, Egloff, Ulferts, Coutard, Nurizzo, Campanacci, Cambillau, Ziebuhr, and Canard, unpublished). Both in the crystal and in solution, NendoU forms hexamers (Guarino *et al.*, 2005; Ricagno, Egloff, Ulferts, Coutard, Nurizzo, Campanacci, Cambillau, Ziebuhr, and Canard, unpublished). The functional consequences of hexamerization on the activities of the NendoU domain and, more generally, nsp15 remain to be studied in detail.

ADP-ribose-1″-phosphatase

The ADP-ribose-1″-phosphatase (Appr-1″-pase, ADRP) domain is located upstream of PL2^{pro} in the coronavirus nsp3 and was originally identified as a conserved domain (called "X domain") in coronaviruses and viruses of the alphavirus-like supergroup (Gorbalenya *et al.*, 1991; Koonin *et al.*, 1992). It is also conserved in toroviruses (Draker *et al.*, 2006), but not in arteri- and roniviruses (Snijder *et al.*, 2003). The domain is related to a large family of so-called *macro* domain proteins (with currently more than 400 proteins in the SMART database; Letunic *et al.*, 2004) that are named after the non-histone domain of macroH2A histones. Meanwhile, it is generally accepted that *macro* domains are associated with ADP-ribose binding or with the processing of ADP-ribose derivatives, such as Appr-1″-p (Karras *et al.*, 2005). The enormous diversity of this protein family (which probably is associated with differences in substrate specificities and/or enzymatic activities) suggests that members of the *macro* domain family may be involved in diverse biological processes that are controlled by (or involve) binding and/or processing of ADP-ribose or structurally related molecules (Karras *et al.*, 2005; Kustatscher *et al.*, 2005).

Characterization of the HCoV-229E ADRP revealed that the protein dephosphorylates Appr-1″-p to ADP-ribose in a very specific manner (Putics *et al.*, 2005) as had been shown previously for several cellular homologs (Kumaran *et al.*, 2005; Shull *et al.*, 2005). Similar data were also obtained for the ADRPs from SARS-CoV and TGEV (Putics *et al.*, 2005; Saikatendu *et al.*, 2005; Putics *et al.*, 2006). Recently, the ADRP of SARS-CoV was crystallized and its structure was solved to 1.4 Å resolution (Saikatendu *et al.*, 2005). The protein was shown to have a three-layered $\alpha/\beta/\alpha$ core that is very similar to the fold of cellular *macro* domains (Allen *et al.*, 2003; Kumaran *et al.*, 2005). It however carries an N-terminal extension by two β-strands, one of which forms part of the central β-sheet (Figure 2.4B) (Saikatendu *et al.*, 2005). Mutagenesis and structural data implicate residues from four conserved regions of the ADRP in the Appr-1″-p binding site and several key residues that are probably involved in the binding to adenine, the α and β phosphates and the two ribose moieties could be identified (Saikatendu *et al.*, 2005). By contrast, the precise catalytic mechanism used by the ADRP is less clear and requires further studies to be fully understood.

Site-directed mutagenesis of the HCoV-229E full-length infectious clone revealed that inactivation of the Appr-1″-pase activity does not significantly effect viral replication in cell culture (Putics *et al.*, 2005). Both the accumulation of viral RNA and the growth characteristics of the mutant virus proved to be indistinguishable from that of the wild-type virus. The data suggest that the ADRP may have a critical function *in vivo*, i.e. the infected host.

Other predicted enzyme activities

In several group II coronaviruses (excluding SARS-CoV), a putative cyclic nucleotide phosphodiesterase (CPDase) domain has been identified. The domain is also conserved in bovine and equine toroviruses (Snijder *et al.*, 2003; Draker *et al.*, 2006), albeit at a different position in the genome. The activity of the protein has not been characterized to date. It is known, however, that the CPDase domain is not essential for MHV replication in cell culture (Schwarz *et al.*, 1990). It has been proposed that the (putative) CPDase operates in conjunction with the ADRP in the same metabolic pathway (Snijder *et al.*,

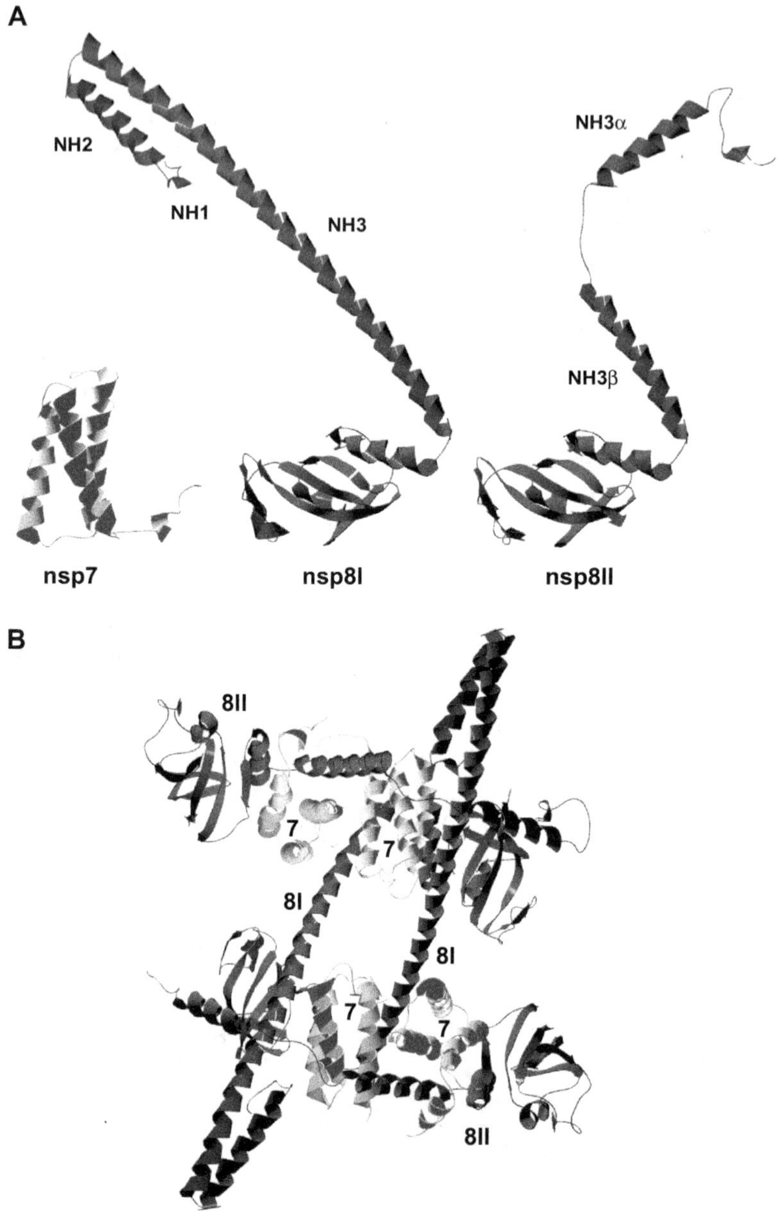

Figure 2.4 Structures of nsp7 and nsp8. (A) Three-dimensional structures of SARS-CoV nsp7 and the two alternative nsp8 structures designated nsp8I and nsp8II (PDB code 2AHM) (Zhai *et al.*, 2005). The structures are shown in ribbon representation and are colored grey (nsp7) and black (nsp8), respectively. The α-helices in the N-terminal shaft domains of nsp8I and nsp8II are indicated according to the nomenclature used by Zhai *et al.* (Zhai *et al.*, 2005). (B) Structure of the octameric complex formed by 4 molecules of nsp7 (grey) and two molecules of each nsp8I and nsp8II (Zhai *et al.*, 2005). In the crystal structure, there is one such octameric complex per asymmetric unit. The cylinder-like hexadecameric supercomplex described in the text comprises two asymmetric units that are related by the crystallographic two-fold *c*-axis (not shown).

2003), where it is predicted to convert ADP-ribose 1″,2″ cyclic phosphate (Appr>p), a side product of tRNA splicing (Culver *et al.*, 1994), to Appr-1″-p, which is subsequently dephosphorylated by the ADRP. Both the conservation pattern and the currently available (but limited) data suggest that the ADRP and CPDase activities provide a selective advantage only in the host (Putics *et al.*, 2005; Sperry *et al.*, 2005). At the moment, it is not clear as to whether Appr>p and Appr-1″-p really represent the relevant substrates of these two activities, and also the biological meaning of this pathway to coronavirus replication has remained obscure.

Another activity that has not yet been characterized biochemically is the predicted ribose-2′-O methyltransferase (MT) residing in nsp16. In contrast to the ADRP, the MT domain proved to be essential for viral RNA synthesis as shown by reverse genetics experiments using HCoV-229E (Hertzig, Schelle and Ziebuhr, unpublished). The MT is probably involved in the maturation of the 5′ cap structure of viral RNAs but may also be involved in other RNA processing pathways, where it might cooperate with NendoU and/or ExoN (see above). Thus, for example, endonucleolytic cleavage of (viral or cellular?) RNA by NendoU could generate small RNAs that then act as cofactors for subsequent, site-specific RNA ribose-2′-O methylation reactions by the MT (similar to the cellular snoRNA processing pathways discussed above) or, vice versa, the MT could affect/control the activity of NendoU by methylating specific RNAs to protect them from NendoU-mediated cleavage (Ivanov *et al.*, 2004a).

Non-structural proteins 7 to 11

The functions of the five proteins, nsp7 to nsp11, encoded downstream of the third transmembrane domain (nsp6) in ORF1a are slowly beginning to emerge and significant progress has recently been made towards the determination of the structures of these proteins (Egloff *et al.*, 2004; Sutton *et al.*, 2004; Peti *et al.*, 2005; Zhai *et al.*, 2005). In a fascinating study, Zihe Rao and colleagues have shown that nsp7 and nsp8 form a hexadecameric supercomplex comprised of eight copies each of nsp7 and nsp8 (Zhai *et al.*, 2005) (Figure 2.4). The complex has a cylinder-like structure with an inner diameter of approximately 30 Å. On the basis of (i) the diameter of the cylinder, (ii) the positively charged inner surface and (iii) biochemical (RNA-binding) data obtained for nsp7 and nsp8, it has been proposed that the supercomplex may encircle double-stranded RNA and thereby act as a processivity factor of the viral polymerase (nsp12). The multimeric structure of the nsp7-nsp8 supercomplex, which is likely to be of functional significance, provides a plausible explanation for the conservation of a ribosomal frameshifting mechanism in coronaviruses and their relatives: the relative overexpression of ORF1a over ORF1b proteins would provide a sufficient supply of the subunits of multimeric structures formed by, e.g. nsp7, 8 and 9, and it seems reasonable to think that these supercomplexes represent cofactors of the (ORF1b-encoded) core replicative enzymes, for example, by acting as their processivity or loading factors, accessory enzymes or essential scaffold proteins.

The structure of nsp7 is comprised of three α-helices that form an N-terminal triple-stranded antiparallel coiled coil ("helical bundle") that is linked to a fourth (C-terminal) α-helix (Peti *et al.*, 2005; Zhai *et al.*, 2005) (Figure 2.4A). Nsp8, by contrast, folds into a "golf club"-like structure with N-terminal "shaft" and C-terminal "head" domains. Interestingly, this "golf club"-like structure adopts two different conformations (featuring either a straight or a bent shaft) that are called nsp8I and nsp8II, respectively (Zhai *et al.*, 2005) (Figure 2.4A). The bent shaft in nsp8II becomes possible because the long α-helix NH3 found in nsp8I bends into two helices, NH3α and NH3β. The two different conformations of nsp8 are essential to form the octameric and, ultimately, hexadecameric supercomplex structures seen in the crystal structure (Zhai *et al.*, 2005) (Figure 2.4B). Whereas nsp8I and nsp8II form the framework of this supercomplex, the eight nsp7 molecules further stabilize it by filling some of the remaining space. This remarkable and truly unique subunit arrangement in the supercomplex has been compared with the roles of bricks (nsp8) and mortar (nsp7) (Zhai *et al.*, 2005).

The nsp7-nsp8 supercomplex is thought to interact with other replicase subunits and one of them has been confirmed to be nsp9, a dimeric, single-stranded RNA-binding protein that specifically interacts with nsp8 (Egloff *et al.*, 2004; Sutton *et al.*, 2004). Crystal structures of nsp9 have been determined to 2.7 and 2.8 Å resolution, respectively, by two laboratories in 2004 (Egloff *et al.*, 2004; Sutton *et al.*, 2004) (Figure 2.3C). The structural studies revealed that nsp9 has an open six-stranded β-barrel fold that is most closely related to the second β-barrel domain of coronavirus 3C-like and the first β-barrel domain of rhinovirus 3C proteases but it can also be superimposed onto the related subdomains of other chymotrypsin-like proteases (Sutton *et al.*, 2004). Interestingly, both laboratories noticed a distant similarity (but no structural homology) of the nsp9 structure with the oligosaccharide/oligonucleotide-binding (OB) fold. The β-barrel core of nsp9 is flanked by an N-terminal extension and a C-terminal helix. Zhai *et al.* (Zhai *et al.*, 2005) suggested that the region formed by the N-terminal 50 residues of nsp8II, which is located at the entrance of the channel formed by the nsp7-nsp8 supercomplex, would be the most likely binding site for nsp9.

Nsp10, which immediately precedes the RdRp (nsp12) in pp1ab and nsp11 in pp1a, is a protein of ~140 residues with multiple conserved Cys residues (Figure 2.2). Based on an MHV *ts* mutant study, Sawicki *et al.* (Sawicki *et al.*, 2005) have suggested that nsp10 has an essential role in minus-strand RNA synthesis but detailed studies into the precise functions of this protein are still missing. Nsp11 is a small peptide of about 20 residues that shares its N-terminal residues (encoded upstream of the ribosomal frameshift site) with nsp12. Thus far, the peptide has not been detected in coronavirus-infected cells and its biological relevance is generally questioned.

The coronavirus replication complex

The association of the viral replication complex with (modified) intracellular membranes is a common feature of positive-strand RNA viruses replicating in eukaryotic cells (see e.g. Mackenzie *et al.*, 1999; Rust *et al.*, 2001; Schwartz *et al.*, 2002; Ahlquist *et al.*, 2003; Salonen *et al.*, 2003, and references therein). It is assumed to be advantageous for creating a suitable (micro)environment for viral RNA synthesis, recruiting membrane-bound host factors and, possibly, for hiding the double-stranded RNA intermediates of replication from host defense mechanisms that can be triggered by double-stranded RNA. +RNA viruses can hijack membranes from a variety of organelles including the endoplasmic reticulum, lysosomes and endosomes, mitochondria, and chloroplasts. Typically, these membranes are modified to produce either specific vesicles or spherules (membrane invaginations).

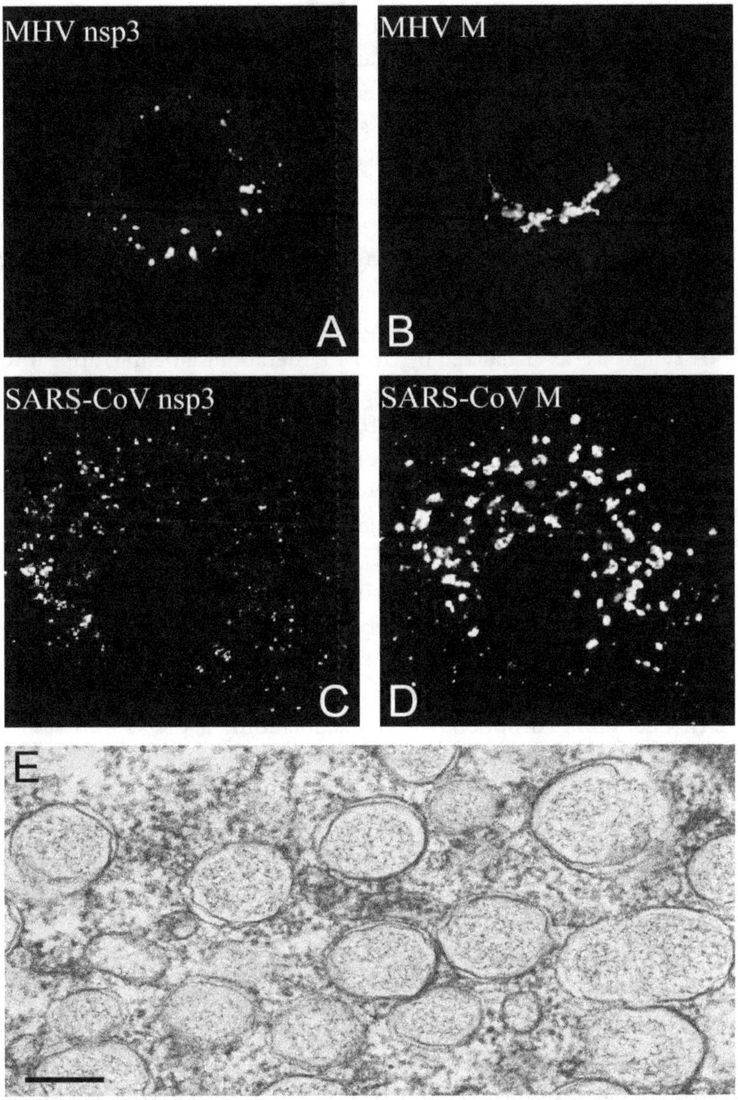

Figure 2.5 Subcellular localization of the coronavirus replication complex. Panels A and B: immunofluorescence double labeling assay of MHV-infected L cells at 6 hours p.i., showing the localization of the MHV replicase, represented by nsp3 (rabbit antiserum kindly provided by Dr. Susan Baker, Loyola University, Chicago, USA), and the structural protein M (mouse monoclonal antibody kindly provided by Dr. Michael Buchmeier, The Scripps Research Institute, La Jolla, USA), which labels membranes presumably involved in virus assembly. The overlay of the two images (not shown) revealed a complete separation of the two labeling patterns. Panels C and D: the same double labeling, with the same result, for SARS-coronavirus in Vero-E6 cells at 6 h.p.i. (Snijder *et al.*, 2006). Bottom panel: electron microscopy of SARS-CoV-infected Vero-E6 cells (9 hours p.i.), cryofixed by high-speed plunge freezing, a step followed by freeze substitution (Snijder *et al.*, 2006). Bar is 100 nm. The image shows the virus-induced double-membrane structures with which the coronavirus replication complex is thought to be associated (for details, see Snijder *et al.*, 2006, and references therein). Images courtesy of Dr. Yvonne van der Meer (panels A-D) and Dr. Mieke Mommaas (panel E), respectively, Leiden University Medical Center, the Netherlands.

For nidoviruses that have been studied in some detail (EAV, MHV, SARS-CoV), immunofluorescence microscopy showed that most replicase subunits (including 3CL^{pro}, RdRp and helicase) localize to (the same) membrane structures, often present in the perinuclear region of the infected cell (Figure 2.5; see Lai and Holmes, 2001; Siddell *et al.*, 2005, and references therein). For both coronaviruses and arteriviruses, ORF1a-encoded (putative) multispanning transmembrane domains are thought to form the membrane-bound scaffold with which the other replicase subunits, lacking such hydrophobic domains, can associate to form the functional complex. Another common feature is the fact that both coronaviruses and arteriviruses have been reported to induce double-membrane vesicles (DMVs; diameter for coronaviruses 200 to 350 nm; Figure 2.5) with which the replicase and viral RNA synthesis (as visualized by BrUTP metabolic labeling) appear to be associated (see, e.g. Pedersen *et al.*, 1999; Bost *et al.*, 2001; Gosert *et al.*, 2002; Ng and Liu, 2002; Harcourt *et al.*, 2004; Prentice *et al.*, 2004a; Prentice *et al.*, 2004b; Snijder *et al.*, 2006, and references therein). However, many aspects of these structures remain to be analyzed, including their ultrastructure, the question of whether these structures are actively engaged in RNA synthesis and—if so—the localization of the RNA-synthesizing complex within or on the structure. For the coronavirus TGEV, genome- and subgenome-length minus strands, the presumed templates for replication and transcription, respectively, were predominantly found in nuclease-resistant membranous complexes (Sethna and Brian, 1997). Also studies with MHV identified membrane fractions containing newly made viral RNA and a specific set of viral proteins (Sims *et al.*, 2000). Several MHV pp1a/pp1ab processing products that contain hydrophobic domains, including nsp3 (Gosert *et al.*, 2002) and nsp4–10 (Schiller *et al.*, 1998), were characterized as integral membrane proteins, again suggesting that they serve as a scaffold for the remaining, "more enzymatic" part of the complex.

The origin of the membranes used to form the DMVs has been the subject of several studies. For the arterivirus EAV, the endoplasmic reticulum (ER) was implicated on the basis of morphological evidence and the use of antibodies recognizing cellular marker proteins (van der Meer *et al.*, 1998; Pedersen *et al.*, 1999; Snijder *et al.*, 2001). Also, the expression of just two ORF1a-derived proteins (nsp2 and nsp3, both containing putative *trans*-membrane domains) could induce DMV formation from ER membranes. However, for the coronaviruses MHV and SARS-CoV, studies on the origin of the membranes that carry the replication complex have remained equivocal thus far. This may be partially due to technical reasons (Snijder *et al.*, 2006), including the use of different cell lines, cellular marker proteins, and antisera. Different intracellular compartments were implicated in different studies. For MHV, one study pointed mainly to membranes of (late) endosomal origin as the site of viral RNA synthesis (van der Meer *et al.*, 1998). However, more recent studies implicated ER-derived DMVs (Gosert *et al.*, 2002), resembling those seen for arteriviruses, or the cellular housekeeping process of authophagy, which produces double membrane-containing intermediates of autophagosome formation (Kirkegaard *et al.*, 2004; Prentice *et al.*, 2004a). The suggestion of involvement of the autophagic pathway in DMV formation was also made for SARS-CoV (Prentice *et al.*, 2004a), but other studies, using a combination of immunofluorescence and electron microscopy, again pointed to the endoplasmic reticulum as the most likely membrane donor (Ivanov *et al.*, 2004b; Snijder *et al.*, 2006).

Clearly, at the moment, this emerging area of nidovirus research is producing more questions than answers, including those related to the topology of the viral *trans*-membrane nsps, their interactions with other replicase subunits (and possibly host proteins), the mechanism of membrane modification, and the ultrastructure of the active viral RNA-synthesizing complex.

Conclusion

Collaborative research efforts in the areas of bioinformatics, structural biology, biochemistry, cell biology and molecular virology will be needed to further dissect and understand the molecular mechanisms that coronaviruses have evolved to synthesize, express, and maintain their unusually large RNA genomes. As outlined in this review, both the genome replication and the RNA strand transfer mechanism used during fusion of sg RNA leader and body segments may require specific protein factors that remain to be studied in detail. Interestingly, the distantly related toroviruses (for which the genome sequence analysis has been completed only recently) (Draker *et al.*, 2006) may represent an intermediate stage in nidovirus/coronavirus evolution and, therefore, their characterization may provide valuable insights into the factors and mechanisms involved in the life cycle of the various nidovirus families and genera. Despite the proposed common ancestry of the core enzymes driving nidovirus RNA synthesis, it is likely that future studies will uncover more surprising differences between individual nidovirus subgroups. Thus, the continued functional dissection of the replicative proteins of different nidoviruses will help to clarify some of the outstanding issues concerning the evolution of nidovirus replication and transcription.

Over the past few years, impressive progress into the functions and structures of coronavirus replicative proteins has been made (Ziebuhr, 2004; Bartlam *et al.*, 2005). The information derived from these studies has been used to develop selective enzyme inhibitors (see, e.g., Ghosh *et al.*, 2005; Kaeppler *et al.*, 2005; Martina *et al.*, 2005; Shie *et al.*, 2005; Yang *et al.*, 2005; Wu *et al.*, 2006). Some of these inhibitors have been shown to block viral replication *in vitro* and, potentially, might be further developed to drugs suitable to treat coronavirus infections.

Acknowledgments

For pleasant contacts/collaborations and helpful discussions through the years, we would like to thank many colleagues from the nidovirus field, including our past and present colleagues in Leiden, Würzburg, and Belfast. In particular, we would like to acknowledge the input and support of Dr. Alexander Gorbalenya, Dr. Stuart Siddell, Dr. Volker Thiel, and Dr. Willy Spaan. Our nidovirus research was supported by grants from the Council for Chemical Sciences of the Netherlands Organization for Scientific Research (to E.J.S.), the Deutsche Forschungsgemeinschaft and the Sino-German Center for Science Promotion (to J.Z.), and the European Commission in the context of the activities of the Euro-Asian SARS-DTV Network (SP22-CT-2004–511064).

References

Ahlquist, P., Noueiry, A.O., Lee, W.M., Kushner, D.B., and Dye, B.T. (2003). Host factors in positive-strand RNA virus genome replication. J. Virol. 77, 8181–8186.

Allaire, M., Chernaia, M.M., Malcolm, B.A., and James, M.N. (1994). Picornaviral 3C cysteine proteinases have a fold similar to chymotrypsin-like serine proteinases. Nature 369, 72–76.

Allen, M.D., Buckle, A.M., Cordell, S.C., Lowe, J., and Bycroft, M. (2003). The crystal structure of AF1521 a protein from Archaeoglobus fulgidus with homology to the non-histone domain of macroH2A. J. Mol. Biol. *330*, 503–511.

Almazán, F., González, J.M., Pénzes, Z., Izeta, A., Calvo, E., Plana-Durán, J., and Enjuanes, L. (2000). Engineering the largest RNA virus genome as an infectious bacterial artificial chromosome. Proc. Natl. Acad. Sci. USA *97*, 5516–5521.

Anand, K., Palm, G.J., Mesters, J.R., Siddell, S.G., Ziebuhr, J., and Hilgenfeld, R. (2002). Structure of coronavirus main proteinase reveals combination of a chymotrypsin fold with an extra alpha-helical domain. EMBO J. *21*, 3213–3224.

Anand, K., Ziebuhr, J., Wadhwani, P., Mesters, J.R., and Hilgenfeld, R. (2003). Coronavirus main proteinase ($3CL^{pro}$) structure: basis for design of anti-SARS drugs. Science *300*, 1763–1767.

Baker, S.C., Shieh, C.K., Soe, L.H., Chang, M.F., Vannier, D.M., and Lai, M.M. (1989). Identification of a domain required for autoproteolytic cleavage of murine coronavirus gene A polyprotein. J. Virol. *63*, 3693–3699.

Baker, S.C., Yokomori, K., Dong, S., Carlisle, R., Gorbalenya, A.E., Koonin, E.V., and Lai, M.M. (1993). Identification of the catalytic sites of a papain-like cysteine proteinase of murine coronavirus. J. Virol. *67*, 6056–6063.

Baranov, P.V., Henderson, C.M., Anderson, C.B., Gesteland, R.F., Atkins, J.F., and Howard, M.T. (2005). Programmed ribosomal frameshifting in decoding the SARS-CoV genome. Virology *332*, 498–510.

Barretto, N., Jukneliene, D., Ratia, K., Chen, Z., Mesecar, A.D., and Baker, S.C. (2005). The papain-like protease of severe acute respiratory syndrome coronavirus has deubiquitinating activity. J. Virol. *79*, 15189–15198.

Bartlam, M., Yang, H., and Rao, Z. (2005). Structural insights into SARS coronavirus proteins. Curr. Opin. Struct. Biol. *15*, 664–672.

Bernini, A., Spiga, O., Venditti, V., Prischi, F., Bracci, L., Huang, J., Tanner, J.A., and Niccolai, N. (2006). Tertiary structure prediction of SARS coronavirus helicase. Biochem. Biophys. Res. Commun. *343*, 1101–1104.

Bhardwaj, K., Guarino, L., and Kao, C.C. (2004). The severe acute respiratory syndrome coronavirus Nsp15 protein is an endoribonuclease that prefers manganese as a cofactor. J. Virol. *78*, 12218–12224.

Bonilla, P.J., Hughes, S.A., and Weiss, S.R. (1997). Characterization of a second cleavage site and demonstration of activity in trans by the papain-like proteinase of the murine coronavirus mouse hepatitis virus strain A59. J. Virol. *71*, 900–909.

Bost, A.G., Prentice, E., and Denison, M.R. (2001). Mouse hepatitis virus replicase protein complexes are translocated to sites of M protein accumulation in the ERGIC at late times of infection. Virology *285*, 21–29.

Boursnell, M.E., Brown, T.D., Foulds, I.J., Green, P.F., Tomley, F.M., and Binns, M.M. (1987). Completion of the sequence of the genome of the coronavirus avian infectious bronchitis virus. J. Gen. Virol. *68*, 57–77.

Brian, D.A., and Spaan, W.J.M. (1997). Recombination and coronavirus defective interfering RNAs. Semin. Virol. *8*, 101–111.

Brierley, I., Digard, P., and Inglis, S.C. (1989). Characterization of an efficient coronavirus ribosomal frameshifting signal: requirement for an RNA pseudoknot. Cell *57*, 537–547.

Brierley, I., and Dos Ramos, F.J. (2005). Programmed ribosomal frameshifting in HIV-1 and the SARS-CoV. Virus Res., in press.

Brockway, S.M., Clay, C.T., Lu, X.T., and Denison, M.R. (2003). Characterization of the expression, intracellular localization, and replication complex association of the putative mouse hepatitis virus RNA-dependent RNA polymerase. J. Virol. *77*, 10515–10527.

Brockway, S.M., and Denison, M.R. (2005). Mutagenesis of the murine hepatitis virus nsp1-coding region identifies residues important for protein processing, viral RNA synthesis, and viral replication. Virology *340*, 209–223.

Casais, R., Thiel, V., Siddell, S.G., Cavanagh, D., and Britton, P. (2001). Reverse genetics system for the avian coronavirus infectious bronchitis virus. J. Virol. *75*, 12359–12369.

Cavanagh, D. (1997). Nidovirales: a new order comprising Coronaviridae and Arteriviridae. Arch. Virol. *142*, 629–633.

Cheng, A., Zhang, W., Xie, Y., Jiang, W., Arnold, E., Sarafianos, S.G., and Ding, J. (2005). Expression, purification, and characterization of SARS coronavirus RNA polymerase. Virology *335*, 165–176.

Chou, C.Y., Chang, H.C., Hsu, W.C., Lin, T.Z., Lin, C.H., and Chang, G.G. (2004). Quaternary structure of the severe acute respiratory syndrome (SARS) coronavirus main protease. Biochemistry *43*, 14958–14970.

Coley, S.E., Lavi, E., Sawicki, S.G., Fu, L., Schelle, B., Karl, N., Siddell, S.G., and Thiel, V. (2005). Recombinant mouse hepatitis virus strain A59 from cloned, full-length cDNA replicates to high titers in vitro and is fully pathogenic in vivo. J. Virol. *79*, 3097–3106.

Cowley, J.A., Dimmock, C.M., Spann, K.M., and Walker, P.J. (2000). Gill-associated virus of Penaeus monodon prawns: an invertebrate virus with ORF1a and ORF1b genes related to arteri- and coronaviruses. J. Gen. Virol. *81*, 1473–1484.

Culver, G.M., Consaul, S.A., Tycowski, K.T., Filipowicz, W., and Phizicky, E.M. (1994). tRNA splicing in yeast and wheat germ. A cyclic phosphodiesterase implicated in the metabolism of ADP-ribose 1,"2′-cyclic phosphate. J. Biol. Chem. *269*, 24928–24934.

de Vries, A.A.F., Horzinek, M.C., Rottier, P.J.M., and de Groot, R.J. (1997). The genome organization of the Nidovirales: similarities and differences between arteri-, toro-, and coronaviruses. Semin. Virol. *8*, 33–47.

den Boon, J.A., Snijder, E.J., Chirnside, E.D., de Vries, A.A., Horzinek, M.C., and Spaan, W.J. (1991). Equine arteritis virus is not a togavirus but belongs to the coronaviruslike superfamily. J. Virol. *65*, 2910–2920.

Denison, M.R., and Perlman, S. (1986). Translation and processing of mouse hepatitis virus virion RNA in a cell-free system. J. Virol. *60*, 12–18.

Denison, M.R., Yount, B., Brockway, S.M., Graham, R.L., Sims, A.C., Lu, X., and Baric, R.S. (2004). Cleavage between replicase proteins p28 and p65 of mouse hepatitis virus is not required for virus replication. J. Virol. *78*, 5957–5965.

Dong, S., and Baker, S.C. (1994). Determinants of the p28 cleavage site recognized by the first papain-like cysteine proteinase of murine coronavirus. Virology *204*, 541–549.

Dougherty, W.G., and Semler, B.L. (1993). Expression of virus-encoded proteinases: functional and structural similarities with cellular enzymes. Microbiol. Rev. *57*, 781–822.

Draker, R., Roper, R.L., Petric, M., and Tellier, R. (2006). The complete sequence of the bovine torovirus genome. Virus. Res. *115*, 56–68.

Egloff, M.P., Ferron, F., Campanacci, V., Longhi, S., Rancurel, C., Dutartre, H., Snijder, E.J., Gorbalenya, A.E., Cambillau, C., and Canard, B. (2004). The severe acute respiratory syndrome-coronavirus replicative protein nsp9 is a single-stranded RNA-binding subunit unique in the RNA virus world. Proc. Natl. Acad. Sci. USA *101*, 3792–3796.

Fan, K., Wei, P., Feng, Q., Chen, S., Huang, C., Ma, L., Lai, B., Pei, J., Liu, Y., Chen, J., and Lai, L. (2004). Biosynthesis, purification, and substrate specificity of severe acute respiratory syndrome coronavirus 3C-like proteinase. J. Biol. Chem. *279*, 1637–1642.

Fan, K., Ma, L., Han, X., Liang, H., Wei, P., Liu, Y., and Lai, L. (2005). The substrate specificity of SARS coronavirus 3C-like proteinase. Biochem. Biophys. Res. Commun. *329*, 934–940.

Feder, M., Pas, J., Wyrwicz, L.S., and Bujnicki, J.M. (2003). Molecular phylogenetics of the RrmJ/fibrillarin superfamily of ribose 2′-O-methyltransferases. Gene *302*, 129–138.

Filipowicz, W., and Pogacic, V. (2002). Biogenesis of small nucleolar ribonucleoproteins. Curr. Opin. Cell Biol. *14*, 319–327.

Ghosh, A.K., Xi, K., Ratia, K., Santarsiero, B.D., Fu, W., Harcourt, B.H., Rota, P.A., Baker, S.C., Johnson, M.E., and Mesecar, A.D. (2005). Design and synthesis of peptidomimetic severe acute respiratory syndrome chymotrypsin-like protease inhibitors. J. Med. Chem. *48*, 6767–6771.

Gonzalez, J.M., Gomez-Puertas, P., Cavanagh, D., Gorbalenya, A.E., and Enjuanes, L. (2003). A comparative sequence analysis to revise the current taxonomy of the family Coronaviridae. Arch. Virol. *148*, 2207–2235.

Gorbalenya, A.E., Koonin, E.V., Donchenko, A.P., and Blinov, V.M. (1989a). Coronavirus genome: prediction of putative functional domains in the non-structural polyprotein by comparative amino acid sequence analysis. Nucleic Acids Res. *17*, 4847–4861.

Gorbalenya, A.E., Koonin, E.V., Donchenko, A.P., and Blinov, V.M. (1989b). Two related superfamilies of putative helicases involved in replication, recombination, repair and expression of DNA and RNA genomes. Nucleic Acids Res. *17*, 4713–4730.

Gorbalenya, A.E., Koonin, E.V., and Lai, M.M. (1991). Putative papain-related thiol proteases of positive-strand RNA viruses. Identification of rubi- and aphthovirus proteases and delineation of a novel conserved domain associated with proteases of rubi-, alpha- and coronaviruses. FEBS Lett. *288*, 201–205.

Gorbalenya, A.E. (2001). Big nidovirus genome. When count and order of domains matter. Adv. Exp. Med. Biol. *494*, 1–17.

Gorbalenya, A.E., Snijder, E.J., and Spaan, W.J. (2004). Severe acute respiratory syndrome coronavirus phylogeny: toward consensus. J. Virol. *78*, 7863–7866.

Gorbalenya, A.E., Enjuanes, L., Ziebuhr, J., and Snijder, E.J. (2006). Nidovirales: evolving the largest RNA virus genome. Virus Res. *117*, 17–37.

Gosert, R., Kanjanahaluethai, A., Egger, D., Bienz, K., and Baker, S.C. (2002). RNA replication of mouse hepatitis virus takes place at double-membrane vesicles. J. Virol. *76*, 3697–3708.

Graham, R.L., Sims, A.C., Brockway, S.M., Baric, R.S., and Denison, M.R. (2005). The nsp2 replicase proteins of murine hepatitis virus and severe acute respiratory syndrome coronavirus are dispensable for viral replication. J. Virol. *79*, 13399–13411.

Guarino, L.A., Bhardwaj, K., Dong, W., Sun, J., Holzenburg, A., and Kao, C. (2005). Mutational analysis of the SARS virus Nsp15 endoribonuclease: identification of residues affecting hexamer formation. J. Mol. Biol. *353*, 1106–1117.

Guarné, A., Tormo, J., Kirchweger, R., Pfistermueller, D., Fita, I., and Skern, T. (1998). Structure of the foot-and-mouth disease virus leader protease: a papain-like fold adapted for self-processing and eIF4G recognition. EMBO J. *17*, 7469–7479.

Guarné, A., Hampoelz, B., Glaser, W., Carpena, X., Tormo, J., Fita, I., and Skern, T. (2000). Structural and biochemical features distinguish the foot-and-mouth disease virus leader proteinase from other papain-like enzymes. J. Mol. Biol. *302*, 1227–1240.

Harcourt, B.H., Jukneliene, D., Kanjanahaluethai, A., Bechill, J., Severson, K.M., Smith, C.M., Rota, P.A., and Baker, S.C. (2004). Identification of severe acute respiratory syndrome coronavirus replicase products and characterization of papain-like protease activity. J. Virol. *78*, 13600–13612.

Hegyi, A., Friebe, A., Gorbalenya, A.E., and Ziebuhr, J. (2002). Mutational analysis of the active centre of coronavirus 3C-like proteases. J. Gen. Virol. *83*, 581–593.

Hegyi, A., and Ziebuhr, J. (2002). Conservation of substrate specificities among coronavirus main proteases. J. Gen. Virol. *83*, 595–599.

Herold, J., Siddell, S.G., and Gorbalenya, A.E. (1999). A human RNA viral cysteine proteinase that depends upon a unique Zn^{2+}-binding finger connecting the two domains of a papain-like fold. J. Biol. Chem. *274*, 14918–14925.

Holmes, E.C., and Rambaut, A. (2004). Viral evolution and the emergence of SARS coronavirus. Philos. Trans. R. Soc. Lond. B. Biol. Sci. *359*, 1059–1065.

Hsu, M.F., Kuo, C.J., Chang, K.T., Chang, H.C., Chou, C.C., Ko, T.P., Shr, H.L., Chang, G.G., Wang, A.H., and Liang, P.H. (2005). Mechanism of the maturation process of SARS-CoV 3CL protease. J. Biol. Chem. *280*, 31257–31266.

Hu, M., Li, P., Li, M., Li, W., Yao, T., Wu, J.W., Gu, W., Cohen, R.E., and Shi, Y. (2002). Crystal structure of a UBP-family deubiquitinating enzyme in isolation and in complex with ubiquitin aldehyde. Cell *111*, 1041–1054.

Ivanov, K.A., Hertzig, T., Rozanov, M., Bayer, S., Thiel, V., Gorbalenya, A.E., and Ziebuhr, J. (2004a). Major genetic marker of nidoviruses encodes a replicative endoribonuclease. Proc. Natl. Acad. Sci. USA *101*, 12694–12699.

Ivanov, K.A., Thiel, V., Dobbe, J.C., van der Meer, Y., Snijder, E.J., and Ziebuhr, J. (2004b). Multiple enzymatic activities associated with severe acute respiratory syndrome coronavirus helicase. J. Virol. *78*, 5619–5632.

Ivanov, K.A., and Ziebuhr, J. (2004). Human coronavirus 229E nonstructural protein 13: characterization of duplex-unwinding, nucleoside triphosphatase, and RNA 5′-triphosphatase activities. J. Virol. *78*, 7833–7838.

Jacks, T., Madhani, H.D., Masiarz, F.R., and Varmus, H.E. (1988). Signals for ribosomal frameshifting in the Rous sarcoma virus gag-pol region. Cell *55*, 447–458.

Kadaré, G., and Haenni, A.L. (1997). Virus-encoded RNA helicases. J. Virol. *71*, 2583–2590.

Kaeppler, U., Stiefl, N., Schiller, M., Vicik, R., Breuning, A., Schmitz, W., Rupprecht, D., Schmuck, C., Baumann, K., Ziebuhr, J., and Schirmeister, T. (2005). A new lead for nonpeptidic active-site-directed inhibitors of the severe acute respiratory syndrome coronavirus main protease discovered by a combination of screening and docking methods. J. Med. Chem. *48*, 6832–6842.

Kanjanahaluethai, A., and Baker, S.C. (2000). Identification of mouse hepatitis virus papain-like proteinase 2 activity. J. Virol. *74*, 7911–7921.

Kanjanahaluethai, A., Jukneliene, D., and Baker, S.C. (2003). Identification of the murine coronavirus MP1 cleavage site recognized by papain-like proteinase 2. J. Virol. *77*, 7376–7382.

Karras, G.I., Kustatscher, G., Buhecha, H.R., Allen, M.D., Pugieux, C., Sait, F., Bycroft, M., and Ladurner, A.G. (2005). The macro domain is an ADP-ribose binding module. EMBO J. *24*, 1911–1920.

Kirkegaard, K., Taylor, M.P., and Jackson, W.T. (2004). Cellular autophagy: surrender, avoidance and subversion by microorganisms. Nat. Rev. Microbiol. *2*, 301–314.

Kiss, T. (2001). Small nucleolar RNA-guided post-transcriptional modification of cellular RNAs. EMBO J. *20*, 3617–3622.

Koonin, E.V. (1991). The phylogeny of RNA-dependent RNA polymerases of positive-strand RNA viruses. J. Gen. Virol. *72*, 2197–2206.

Koonin, E.V., Gorbalenya, A.E., Purdy, M.A., Rozanov, M.N., Reyes, G.R., and Bradley, D.W. (1992). Computer-assisted assignment of functional domains in the nonstructural polyprotein of hepatitis E virus: delineation of an additional group of positive-strand RNA plant and animal viruses. Proc. Natl. Acad. Sci. USA *89*, 8259–8263.

Kräusslich, H.G., and Wimmer, E. (1988). Viral proteinases. Annu. Rev. Biochem. *57*, 701–754.

Kumaran, D., Eswaramoorthy, S., Studier, F.W., and Swaminathan, S. (2005). Structure and mechanism of ADP-ribose-1″-monophosphatase (Appr-1″-pase), a ubiquitous cellular processing enzyme. Protein Sci. *14*, 719–726.

Kustatscher, G., Hothorn, M., Pugieux, C., Scheffzek, K., and Ladurner, A.G. (2005). Splicing regulates NAD metabolite binding to histone macroH2A. Nat. Struct. Mol. Biol. *12*, 624–625.

Lai, M.M.C. (1996). Recombination in large RNA viruses: coronaviruses. Semin. Virol. *7*, 381–388.

Lai, M.M.C., and Holmes, K.V. (2001). *Coronaviridae*: the viruses and their replication, In Fields Virology, D. M. Knipe, and P. M. Howley, eds. (Philadelphia: Lippincott Williams & Wilkins), pp. 1163–1185.

Laneve, P., Altieri, F., Fiori, M.E., Scaloni, A., Bozzoni, I., and Caffarelli, E. (2003). Purification, cloning, and characterization of XendoU, a novel endoribonuclease involved in processing of intron-encoded small nucleolar RNAs in Xenopus laevis. J. Biol. Chem. *278*, 13026–13032.

Letunic, I., Copley, R.R., Schmidt, S., Ciccarelli, F.D., Doerks, T., Schultz, J., Ponting, C.P., and Bork, P. (2004). SMART 4.0: towards genomic data integration. Nucleic Acids Res. *32*, D142–144.

Lim, K.P., Ng, L.F., and Liu, D.X. (2000). Identification of a novel cleavage activity of the first papain-like proteinase domain encoded by open reading frame 1a of the coronavirus avian infectious bronchitis virus and characterization of the cleavage products. J. Virol. *74*, 1674–1685.

Lindner, H.A., Fotouhi-Ardakani, N., Lytvyn, V., Lachance, P., Sulea, T., and Menard, R. (2005). The papain-like protease from the severe acute respiratory syndrome coronavirus is a deubiquitinating enzyme. J. Virol. *79*, 15199–15208.

Liu, D.X., and Brown, T.D. (1995). Characterisation and mutational analysis of an ORF 1a-encoding proteinase domain responsible for proteolytic processing of the infectious bronchitis virus 1a/1b polyprotein. Virology *209*, 420–427.

Lu, Y., Lu, X., and Denison, M.R. (1995). Identification and characterization of a serine-like proteinase of the murine coronavirus MHV-A59. J. Virol. *69*, 3554–3559.

Mackenzie, J.M., Jones, M.K., and Westaway, E.G. (1999). Markers for trans-Golgi membranes and the intermediate compartment localize to induced membranes with distinct replication functions in flavivirus-infected cells. J. Virol. *73*, 9555–9567.

Makino, S., Keck, J.G., Stohlman, S.A., and Lai, M.M. (1986). High-frequency RNA recombination of murine coronaviruses. J. Virol. *57*, 729–737.

Martina, E., Stiefl, N., Degel, B., Schulz, F., Breuning, A., Schiller, M., Vicik, R., Baumann, K., Ziebuhr, J., and Schirmeister, T. (2005). Screening of electrophilic compounds yields an aziridinyl peptide as new active-site directed SARS-CoV main protease inhibitor. Bioorg. Med. Chem. Lett. *15*, 5365–5369.

Matthews, D.A., Smith, W.W., Ferre, R.A., Condon, B., Budahazi, G., Sisson, W., Villafranca, J.E., Janson, C.A., McElroy, H.E., Gribskov, C.L., and *et al.* (1994). Structure of human rhinovirus 3C protease reveals a trypsin-like polypeptide fold, RNA-binding site, and means for cleaving precursor polyprotein. Cell *77*, 761–771.

Minskaia, E., Hertzig, T., Gorbalenya, A.E., Campanacci, V., Cambillau, C., Canard, B., and Ziebuhr, J. (2006). Discovery of an RNA virus 3′→5′ exoribonuclease that is critically involved in coronavirus RNA synthesis. Proc. Natl. Acad. Sci. USA *103*, 5108–5113.

Moser, M.J., Holley, W.R., Chatterjee, A., and Mian, I.S. (1997). The proofreading domain of Escherichia coli DNA polymerase I and other DNA and/or RNA exonuclease domains. Nucleic Acids Res. *25*, 5110–5118.

Mosimann, S.C., Cherney, M.M., Sia, S., Plotch, S., and James, M.N. (1997). Refined X-ray crystallographic structure of the poliovirus 3C gene product. J. Mol. Biol. *273*, 1032–1047.

Moya, A., Holmes, E.C., and Gonzalez-Candelas, F. (2004). The population genetics and evolutionary epidemiology of RNA viruses. Nat. Rev. Microbiol. 2, 279–288.

Nagy, P.D., and Simon, A.E. (1997). New insights into the mechanisms of RNA recombination. Virology *235*, 1–9.

Ng, L.F., and Liu, D.X. (2002). Membrane association and dimerization of a cysteine-rich, 16-kilodalton polypeptide released from the C-terminal region of the coronavirus infectious bronchitis virus 1a polyprotein. J. Virol. *76*, 6257–6267.

Pasternak, A.O., Spaan, W.J.M., and Snijder, E.J. (2006). Nidovirus transcription: How to make sense...? J. Gen. Virol. *87*, 1403–1421.

Pedersen, K.W., van der Meer, Y., Roos, N., and Snijder, E.J. (1999). Open reading frame 1a-encoded subunits of the arterivirus replicase induce endoplasmic reticulum-derived double-membrane vesicles which carry the viral replication complex. J. Virol. *73*, 2016–2026.

Peti, W., Johnson, M.A., Herrmann, T., Neuman, B.W., Buchmeier, M.J., Nelson, M., Joseph, J., Page, R., Stevens, R.C., Kuhn, P., and Wuthrich, K. (2005). Structural genomics of the severe acute respiratory syndrome coronavirus: nuclear magnetic resonance structure of the protein nsP7. J. Virol. *79*, 12905–12913.

Piñon, J.D., Mayreddy, R.R., Turner, J.D., Khan, F.S., Bonilla, P.J., and Weiss, S.R. (1997). Efficient autoproteolytic processing of the MHV-A59 3C-like proteinase from the flanking hydrophobic domains requires membranes. Virology *230*, 309–322.

Piñon, J.D., Teng, H., and Weiss, S.R. (1999). Further requirements for cleavage by the murine coronavirus 3C-like proteinase: identification of a cleavage site within ORF1b. Virology *263*, 471–484.

Posthuma, C.C., Nedialkova, D.D., Zevenhoven-Dobbe, J.C., Blokhuis, J.H., Gorbalenya, A.E., and Snijder, E.J. (2006). Site-directed mutagenesis of the Nidovirus replicative endoribonuclease NendoU exerts pleiotropic effects on the arterivirus life cycle. J. Virol. *80*, 1653–1661.

Prentice, E., Jerome, W.G., Yoshimori, T., Mizushima, N., and Denison, M.R. (2004a). Coronavirus replication complex formation utilizes components of cellular autophagy. J. Biol. Chem. *279*, 10136–10141.

Prentice, E., McAuliffe, J., Lu, X., Subbarao, K., and Denison, M.R. (2004b). Identification and characterization of severe acute respiratory syndrome coronavirus replicase proteins. J. Virol. *78*, 9977–9986.

Putics, A., Filipowicz, W., Hall, J., Gorbalenya, A.E., and Ziebuhr, J. (2005). ADP-ribose-1″-monophosphatase: a conserved coronavirus enzyme that is dispensable for viral replication in tissue culture. J. Virol. *79*, 12721–12731.

Putics, A., Gorbalenya, A.E., and Ziebuhr, J. (2006). Identification of protease and ADP-ribose 1″-monophosphatase activities associated with transmissible gastroenteritis virus non-structural protein 3. J. Gen. Virol. *87*, 651–656.

Rawlings, N.D., Tolle, D.P., and Barrett, A.J. (2004). MEROPS: the peptidase database. Nucleic Acids Res. *32*, D160–164.

Ricagno, S., Coutard, B., Grisel, S., Bremond, N., Dalle, K., Tocque, F., Campanacci, V., Lichiere, J., Lantez, V., Debarnot, C., *et al.* (2006). Crystallization and preliminary X-ray diffraction analysis of Nsp15 from SARS coronavirus. Acta Crystallograph. Sect. F Struct. Biol. Cryst. Commun. *62*, 409–411.

Rust, R.C., Landmann, L., Gosert, R., Tang, B.L., Hong, W., Hauri, H.P., Egger, D., and Bienz, K. (2001). Cellular COPII proteins are involved in production of the vesicles that form the poliovirus replication complex. J. Virol. *75*, 9808–9818.

Saikatendu, K.S., Joseph, J.S., Subramanian, V., Clayton, T., Griffith, M., Moy, K., Velasquez, J., Neuman, B.W., Buchmeier, M.J., Stevens, R.C., and Kuhn, P. (2005). Structural basis of severe acute respiratory syndrome coronavirus ADP-ribose-1″-phosphate dephosphorylation by a conserved domain of nsP3. Structure (Camb) *13*, 1665–1675.

Salonen, A., Vasiljeva, L., Merits, A., Magden, J., Jokitalo, E., and Kaariainen, L. (2003). Properly folded nonstructural polyprotein directs the semliki forest virus replication complex to the endosomal compartment. J. Virol. *77*, 1691–1702.

Sawicki, S.G., and Sawicki, D.L. (1995). Coronaviruses use discontinuous extension for synthesis of subgenome-length negative strands. Adv. Exp. Med. Biol. *380*, 499–506.

Sawicki, S.G., and Sawicki, D.L. (2005). Coronavirus transcription: a perspective. Curr. Top. Microbiol. Immunol. *287*, 31–55.

Sawicki, S.G., Sawicki, D.L., Younker, D., Meyer, Y., Thiel, V., Stokes, H., and Siddell, S.G. (2005). Functional and genetic analysis of coronavirus replicase-transcriptase proteins. PLoS Pathog. *1*, e39.

Schiller, J.J., Kanjanahaluethai, A., and Baker, S.C. (1998). Processing of the coronavirus MHV-JHM polymerase polyprotein: identification of precursors and proteolytic products spanning 400 kilodaltons of ORF1a. Virology *242*, 288–302.

Schwartz, M., Chen, J., Janda, M., Sullivan, M., den Boon, J., and Ahlquist, P. (2002). A positive-strand RNA virus replication complex parallels form and function of retrovirus capsids. Mol. Cell *9*, 505–514.

Schwarz, B., Routledge, E., and Siddell, S.G. (1990). Murine coronavirus nonstructural protein ns2 is not essential for virus replication in transformed cells. J. Virol. *64*, 4784–4791.

Seipelt, J., Guarne, A., Bergmann, E., James, M., Sommergruber, W., Fita, I., and Skern, T. (1999). The structures of picornaviral proteinases. Virus Res. *62*, 159–168.

Sethna, P.B., and Brian, D.A. (1997). Coronavirus genomic and subgenomic minus-strand RNAs copartition in membrane-protected replication complexes. J. Virol. *71*, 7744–7749.

Seybert, A., Hegyi, A., Siddell, S.G., and Ziebuhr, J. (2000a). The human coronavirus 229E superfamily 1 helicase has RNA and DNA duplex-unwinding activities with 5′-to-3′ polarity. RNA *6*, 1056–1068.

Seybert, A., van Dinten, L.C., Snijder, E.J., and Ziebuhr, J. (2000b). Biochemical characterization of the equine arteritis virus helicase suggests a close functional relationship between arterivirus and coronavirus helicases. J. Virol. *74*, 9586–9593.

Seybert, A., Posthuma, C.C., van Dinten, L.C., Snijder, E.J., Gorbalenya, A.E., and Ziebuhr, J. (2005). A complex zinc finger controls the enzymatic activities of nidovirus helicases. J. Virol. *79*, 696–704.

Shi, J., Wei, Z., and Song, J. (2004). Dissection study on the SARS 3C-like protease reveals the critical role of the extra domain in dimerization of the enzyme: Defining the extra domain as a new target for design of highly-specific protease inhibitors. J. Biol. Chem. *279*, 24765–24773.

Shi, J., and Song, J. (2006). The catalysis of the SARS 3C-like protease is under extensive regulation by its extra domain. FEBS J. *273*, 1035–1045.

Shie, J.J., Fang, J.M., Kuo, C.J., Kuo, T.H., Liang, P.H., Huang, H.J., Yang, W.B., Lin, C.H., Chen, J.L., Wu, Y.T., and Wong, C.H. (2005). Discovery of potent anilide inhibitors against the severe acute respiratory syndrome 3CL protease. J. Med. Chem. *48*, 4469–4473.

Shull, N.P., Spinelli, S.L., and Phizicky, E.M. (2005). A highly specific phosphatase that acts on ADP-ribose 1″-phosphate, a metabolite of tRNA splicing in Saccharomyces cerevisiae. Nucleic Acids Res. *33*, 650–660.

Siddell, S.G., Ziebuhr, J., and Snijder, E.J. (2005). Coronaviruses, toroviruses, and arteriviruses, In Topley and Wilson's microbiology and microbial infections, B. W. J. Mahy, and V. ter Meulen, eds. (London: Hodder Arnold), pp. 823–856.

Sims, A.C., Ostermann, J., and Denison, M.R. (2000). Mouse hepatitis virus replicase proteins associate with two distinct populations of intracellular membranes. J. Virol. *74*, 5647–5654.

Snijder, E.J., den Boon, J.A., Bredenbeek, P.J., Horzinek, M.C., Rijnbrand, R., and Spaan, W.J. (1990). The carboxyl-terminal part of the putative Berne virus polymerase is expressed by ribosomal frameshifting and contains sequence motifs which indicate that toro- and coronaviruses are evolutionarily related. Nucleic Acids Res. *18*, 4535–4542.

Snijder, E.J., van Tol, H., Roos, N., and Pedersen, K.W. (2001). Non-structural proteins 2 and 3 interact to modify host cell membranes during the formation of the arterivirus replication complex. J. Gen. Virol. *82*, 985–994.

Snijder, E.J., Bredenbeek, P.J., Dobbe, J.C., Thiel, V., Ziebuhr, J., Poon, L.L., Guan, Y., Rozanov, M., Spaan, W.J., and Gorbalenya, A.E. (2003). Unique and conserved features of genome and proteome of SARS-coronavirus, an early split-off from the coronavirus group 2 lineage. J. Mol. Biol. *331*, 991–1004.

Snijder, E.J., Siddell, S.G., and Gorbalenya, A.E. (2005). The order *Nidovirales*. In Topley & Wilson's Microbiology and Microbial Infections, B. W. J. Mahy, and V. ter Meulen, eds. (London: Hodder Arnold), pp. 390–404.

Snijder, E.J., van der Meer, Y., Zevenhoven-Dobbe, J., Onderwater, J.J.M., van der Meulen, J., Koerten, H.K., and Mommaas, A.M. (2006). Ultrastructure and origin of membrane vesicles associated with the severe acute respiratory syndrome coronavirus replication complex. J. Virol. *80*, in press.

Spaan, W.J.M., Brian, D., Cavanagh, D., de Groot, R.J., Enjuanes, L., Gorbalenya, A.E., Holmes, K.V., Masters, P.S., Rottier, P., Taguchi, F., and Talbot, P. (2005a). Family *Coronaviridae*. In Virus Taxonomy, Eighth Report of the International Committee on Taxonomy of Viruses, C. M. Fauquet, M. A. Mayo, J. Maniloff, U. Desselberger, and L. A. Ball, eds. (Amsterdam: Elsevier, Academic Press), pp. 947–964.

Spaan, W.J.M., Cavanagh, D., de Groot, R.J., Enjuanes, L., Gorbalenya, A.E., Snijder, E.J., and Walker, P.J. (2005b). Order *Nidovirales*. In Virus Taxonomy, Eighth Report of the International Committee on Taxonomy of Viruses, C. M. Fauquet, M. A. Mayo, J. Maniloff, U. Desselberger, and L. A. Ball, eds. (Amsterdam: Elsevier, Academic Press), pp. 937–945.

Sperry, S.M., Kazi, L., Graham, R.L., Baric, R.S., Weiss, S.R., and Denison, M.R. (2005). Single-amino-acid substitutions in open reading frame (ORF) 1b-nsp14 and ORF 2a proteins of the coronavirus mouse hepatitis virus are attenuating in mice. J. Virol. *79*, 3391–3400.

Sulea, T., Lindner, H.A., Purisima, E.O., and Menard, R. (2005). Deubiquitination, a new function of the severe acute respiratory syndrome coronavirus papain-like protease? J. Virol. *79*, 4550–4551.

Sulea, T., Lindner, H.A., Purisima, E.O., and Menard, R. (2006). Binding site-based classification of coronaviral papain-like proteases. Proteins *62*, 760–775.

Sutton, G., Fry, E., Carter, L., Sainsbury, S., Walter, T., Nettleship, J., Berrow, N., Owens, R., Gilbert, R., Davidson, A., *et al.* (2004). The nsp9 replicase protein of SARS-coronavirus, structure and functional insights. Structure (Camb) *12*, 341–353.

Tan, J., Verschueren, K.H., Anand, K., Shen, J., Yang, M., Xu, Y., Rao, Z., Bigalke, J., Heisen, B., Mesters, J.R., *et al.* (2005). pH-dependent conformational flexibility of the SARS-CoV main proteinase (M^{pro}) dimer: molecular dynamics simulations and multiple X-ray structure analyses. J. Mol. Biol. *354*, 25–40.

Tanner, J.A., Watt, R.M., Chai, Y.B., Lu, L.Y., Lin, M.C., Peiris, J.S., Poon, L.L., Kung, H.F., and Huang, J.D. (2003). The severe acute respiratory syndrome (SARS) coronavirus NTPase/helicase belongs to a distinct class of 5′ to 3′ viral helicases. J. Biol. Chem. *278*, 39578–39582.

ten Dam, E.B., Pleij, C.W., and Bosch, L. (1990). RNA pseudoknots: translational frameshifting and readthrough on viral RNAs. Virus Genes *4*, 121–136.

Thiel, V., Herold, J., Schelle, B., and Siddell, S.G. (2001). Infectious RNA transcribed in vitro from a cDNA copy of the human coronavirus genome cloned in vaccinia virus. J. Gen. Virol. *82*, 1273–1281.

Thiel, V., Ivanov, K.A., Putics, A., Hertzig, T., Schelle, B., Bayer, S., Weissbrich, B., Snijder, E.J., Rabenau, H., Doerr, H.W., *et al.* (2003). Mechanisms and enzymes involved in SARS coronavirus genome expression. J. Gen. Virol. *84*, 2305–2315.

Tibbles, K.W., Brierley, I., Cavanagh, D., and Brown, T.D. (1996). Characterization in vitro of an autocatalytic processing activity associated with the predicted 3C-like proteinase domain of the coronavirus avian infectious bronchitis virus. J. Virol. *70*, 1923–1930.

Tijms, M.A., van Dinten, L.C., Gorbalenya, A.E., and Snijder, E.J. (2001). A zinc finger-containing papain-like protease couples subgenomic mRNA synthesis to genome translation in a positive-stranded RNA virus. Proc. Natl. Acad. Sci. USA *98*, 1889–1894.

van der Meer, Y., van Tol, H., Krijnse Locker, J., and Snijder, E.J. (1998). ORF1a-encoded replicase subunits are involved in the membrane association of the arterivirus replication complex. J. Virol. *72*, 6689–6698.

van Dinten, L.C., den Boon, J.A., Wassenaar, A.L., Spaan, W.J., and Snijder, E.J. (1997). An infectious arterivirus cDNA clone: identification of a replicase point mutation that abolishes discontinuous mRNA transcription. Proc. Natl. Acad. Sci. USA *94*, 991–996.

van Dinten, L.C., Rensen, S., Gorbalenya, A.E., and Snijder, E.J. (1999). Proteolytic processing of the open reading frame 1b-encoded part of arterivirus replicase is mediated by nsp4 serine protease and Is essential for virus replication. J. Virol. *73*, 2027–2037.

van Dinten, L.C., van Tol, H., Gorbalenya, A.E., and Snijder, E.J. (2000). The predicted metal-binding region of the arterivirus helicase protein is involved in subgenomic mRNA synthesis, genome replication, and virion biogenesis. J. Virol. *74*, 5213–5223.

von Grotthuss, M., Wyrwicz, L.S., and Rychlewski, L. (2003). mRNA cap-1 methyltransferase in the SARS genome. Cell *113*, 701–702.

Wu, C.Y., King, K.Y., Kuo, C.J., Fang, J.M., Wu, Y.T., Ho, M.Y., Liao, C.L., Shie, J.J., Liang, P.H., and Wong, C.H. (2006). Stable benzotriazole esters as mechanism-based inactivators of the severe acute respiratory syndrome 3CL protease. Chem. Biol. *13*, 261–268.

Xu, X., Liu, Y., Weiss, S., Arnold, E., Sarafianos, S.G., and Ding, J. (2003). Molecular model of SARS coronavirus polymerase: implications for biochemical functions and drug design. Nucleic Acids Res. *31*, 7117–7130.

Yang, H., Yang, M., Ding, Y., Liu, Y., Lou, Z., Zhou, Z., Sun, L., Mo, L., Ye, S., Pang, H., *et al.* (2003). The crystal structures of severe acute respiratory syndrome virus main protease and its complex with an inhibitor. Proc. Natl. Acad. Sci. USA *100*, 13190–13195.

Yang, H., Xie, W., Xue, X., Yang, K., Ma, J., Liang, W., Zhao, Q., Zhou, Z., Pei, D., Ziebuhr, J., *et al.* (2005). Design of wide-spectrum inhibitors targeting coronavirus main proteases. PLoS Biol. *3*, e324.

Yin, J., Bergmann, E.M., Cherney, M.M., Lall, M.S., Jain, R.P., Vederas, J.C., and James, M.N. (2005). Dual modes of modification of hepatitis A virus 3C protease by a serine-derived beta-lactone: selective crystallization and formation of a functional catalytic triad in the active site. J. Mol. Biol. *354*, 854–871.

Yount, B., Curtis, K.M., and Baric, R.S. (2000). Strategy for systematic assembly of large RNA and DNA genomes: transmissible gastroenteritis virus model. J. Virol. *74*, 10600–10611.

Yount, B., Curtis, K.M., Fritz, E.A., Hensley, L.E., Jahrling, P.B., Prentice, E., Denison, M.R., Geisbert, T.W., and Baric, R.S. (2003). Reverse genetics with a full-length infectious cDNA of severe acute respiratory syndrome coronavirus. Proc. Natl. Acad. Sci. USA *100*, 12995–13000.

Zhai, Y., Sun, F., Li, X., Pang, H., Xu, X., Bartlam, M., and Rao, Z. (2005). Insights into SARS-CoV transcription and replication from the structure of the nsp7-nsp8 hexadecamer. Nat. Struct. Mol. Biol. *12*, 980–986.

Ziebuhr, J., Herold, J., and Siddell, S.G. (1995). Characterization of a human coronavirus (strain 229E) 3C-like proteinase activity. J. Virol. *69*, 4331–4338.

Ziebuhr, J., Heusipp, G., and Siddell, S.G. (1997). Biosynthesis, purification, and characterization of the human coronavirus 229E 3C-like proteinase. J. Virol. *71*, 3992–3997.

Ziebuhr, J., and Siddell, S.G. (1999). Processing of the human coronavirus 229E replicase polyproteins by the virus-encoded 3C-like proteinase: identification of proteolytic products and cleavage sites common to pp1a and pp1ab. J. Virol. *73*, 177–185.

Ziebuhr, J., Snijder, E.J., and Gorbalenya, A.E. (2000). Virus-encoded proteinases and proteolytic processing in the Nidovirales. J. Gen. Virol. *81*, 853–879.

Ziebuhr, J., Thiel, V., and Gorbalenya, A.E. (2001). The autocatalytic release of a putative RNA virus transcription factor from its polyprotein precursor involves two paralogous papain-like proteases that cleave the same peptide bond. J. Biol. Chem. *276*, 33220–33232.

Ziebuhr, J. (2004). Molecular biology of severe acute respiratory syndrome coronavirus. Curr. Opin. Microbiol. *7*, 412–419.

Ziebuhr, J. (2005). The coronavirus replicase. Curr. Top. Microbiol. Immunol. *287*, 57–94.

Zuo, Y., and Deutscher, M.P. (2001). Exoribonuclease superfamilies: structural analysis and phylogenetic distribution. Nucleic Acids Res. *29*, 1017–1026.

Genomic *Cis*-acting Elements in Coronavirus RNA Replication

3

Paul S. Masters

Abstract

In common with the genomes of all other RNA viruses, coronavirus genomes contain *cis*-acting RNA elements that ensure the specific replication of viral RNA by a virally encoded RNA-dependent RNA polymerase. The embedded *cis*-acting elements devoted to coronavirus replication constitute a surprisingly small fraction of the total genome, but this is probably a reflection of the fact that coronaviruses have the largest genomes of all RNA viruses. The boundaries of *cis*-acting elements essential to replication are fairly well-defined, and an increasingly well resolved picture of the RNA secondary structures of these regions is emerging. However, we are only in the early stages of understanding how these *cis*-acting structures and sequences interact with the viral replicase and host cell components, and much remains to be done before we understand the precise mechanistic roles of such elements in RNA synthesis.

Introduction: coronavirus genomes

Coronavirus genomes are positive-sense, non-segmented, single-stranded RNA molecules. Not only do they have the same polarity as mRNA, but they also exhibit the hallmarks of the typical eukaryotic mRNA: a 5′ monomethylated guanosine cap (Lai and Stohlman, 1981; van Vliet *et al.*, 2002) and a 3′ polyadenylate tail (Schochetman *et al.*, 1976; Lomniczi, 1977; Lai and Stohlman, 1978; Wege *et al.*, 1978). Beyond these features, however, the resemblance to mRNAs breaks down, specifically, in the amounts and types of information harbored by coronavirus genomes. Most conspicuously, coronavirus genomes, with lengths ranging from 27.3 kb (human coronavirus [HCoV]-229E) to 31.3 kb (mouse hepatitis virus [MHV]), fall at the extreme upper end of the size distribution for mature RNA molecules from any source. This makes them the largest of all RNA virus genomes, irrespective of polarity and including consideration of the aggregate lengths of the genomes of segmented RNA viruses. Also at variance with typical eukaryotic mRNA structure is the fact that coronavirus genomes contain multiple open reading frames (ORFs). The gene for the viral replicase occupies the 5′-proximal two-thirds of the genome, with the genes for the four canonical structural proteins situated downstream, in the invariant order 5′- replicase—spike protein (S)—envelope protein (E)—membrane protein (M)—nucleocapsid protein (N) -3′. Additionally, as few as one or as many as eight accessory genes are found interspersed among the structural protein genes.

Beyond its provision of an expanded coding capacity, viral genomic status imposes a heavier burden of duties than those that are required of a typical eukaryotic mRNA. As is true for almost all other positive-sense RNA viruses, coronavirus genomic RNA (gRNA) initiates a productive infection when transfected into permissive host cells. Thus, all of the functions necessary for viral replication must be programmed into the sequence and structure of the gRNA molecule. Much of this program is executed by the encoded viral proteins, but the genome must additionally contain *cis*-acting elements that specify and organize the events of the viral life cycle, through interactions with viral and host machineries (Brian and Baric, 2005).

In one of the earliest steps of infection, the coronavirus genome acts as the mRNA for synthesis of the huge replicase polyprotein. Initiation of replicase expression appears to take place through the conventional cap-dependent scanning mechanism (Denison and Perlman, 1986), but the process of elongation of the encoded protein ventures into non-standard territory. As discussed in Chapter 2, two replicase polyprotein products, coterminal at their amino ends, are produced through ribosomal frameshifting. This translational deviation is dependent upon two coordinated *cis*-acting elements, a heptanucleotide slippery sequence and a closely spaced pseudoknot, embedded in the replicase ORF. In one of the final steps of infection, the assembling viral structural proteins package the progeny genomes into newly forming virions. As discussed in Chapter 6, this selection of gRNA, from among all available viral and cellular RNAs, is likewise dependent upon a *cis*-acting element embedded in the replicase ORF.

Bracketed between the events of replicase translation and virion assembly is the remarkable process that distinguishes RNA viruses in general, and coronaviruses in particular: RNA-dependent RNA synthesis. For coronaviruses, RNA synthesis entails both gRNA replication and subgenomic RNA (sgRNA) transcription. As discussed in Chapter 4, the latter is governed by transcription-regulating sequences (TRSs) within the gRNA and may be assisted by associated host factors. The current chapter will focus on the *cis*-acting genomic elements that govern gRNA replication.

5′ genomic elements acting in RNA replication

Boundary of the 5′ requirement for replication

Systematic analysis of the elements required for coronavirus genome replication was originally enabled by the identification of defective interfering (DI) RNAs, which are remnants of viral gRNAs containing multiple extensive deletions. DI RNAs are not independently viable, but they can propagate by using the RNA synthetic machinery provided by homologous helper virus. Coronavirus DI RNAs were first discovered in cultures of the JHM strain of MHV that had undergone multiple undiluted serial passages (Makino *et al.*, 1984 and 1985). Subsequently, DI RNAs were found and characterized in other strains and species of coronaviruses (van der Most *et al.*, 1991; Hofmann *et al.*, 1990; Chang *et al.*, 1994; Penzes *et al.*, 1994; Mendez *et al.*, 1996). However, although serial passaging may be an important factor, the conditions that generate DI RNAs of MHV or of other coronaviruses are not well understood (Makino *et al.*, 1984). In general, coronavirus DI RNAs do not appear to arise as readily as has been reported for some other RNA virus families. Individual coronavirus DI RNAs exhibit varying degrees of interference with the replication of their helper viruses (Makino *et al.*, 1988a; Penzes *et al.*, 1994). Moreover, some coronavirus DI RNAs retain the capacity to be packaged efficiently, while others have lost the ability to mimic this genomic function.

Since DI RNAs compete with gRNA for access to the viral replicase, DI RNAs must have evolved so as to retain precisely those *cis*-acting sequence elements that are essential for replication. Such elements have been mapped through the cloning and manipulation of both naturally occurring and artificially constructed DI RNAs, which were then assayed for replicative competence upon transfection into helper virus-infected cells. One of the earliest studies of this type, which dissected a DI RNA of MHV-JHM, found that between 638 and 859 nucleotides (nt) from the 5′ end of the genome were necessary for replication (Lin and Lai, 1993). Analysis of successive deletions of an MHV-JHM DI RNA from the same lineage, however, revealed a minimal required 5′ *cis*-acting signal of only 445 to 474 nt (Kim *et al.*, 1993). A caveat is that in this construct, the 5′-most 463 nt of the DI RNA were derived from MHV-A59. Consistent with this result, a length of 467 nt at the 5′ end of an MHV-A59 DI RNA was found to be the minimum that supported replication (Luytjes *et al.*, 1996). In the very closely related group 2 coronavirus, bovine coronavirus (BCoV), analysis of a naturally occurring DI RNA demonstrated that the 5′-most 498 nt of this virus contain the minimal *cis*-acting element for RNA replication (Chang *et al.*, 1994). Similar-sized stretches of 5′-end genomic sequence were found to be required in group 1 and group 3 coronaviruses. Analysis of transmissible gastroenteritis virus (TGEV) DI RNA constructs initially mapped the 5′ *cis*-acting replication signal for this virus to the first 1348 nt of the genome (Izeta *et al.*, 1999); subsequently, the value was refined down to 649 nt (Escors *et al.*, 2003). For infectious bronchitis virus (IBV), dissection of a DI RNA revealed that a 5′-end segment of 544 nt could support replication, whereas a length of 338 nt was not sufficient (Dalton *et al.*, 2001).

In most cases, therefore, the minimal 5′ *cis*-acting element for genome replication extends into a portion of the replicase ORF. In the group 1 and group 2 coronaviruses, the 5′ terminus of the genome contains an untranslated region (UTR) that ranges in size from 210 nt (MHV, BCoV and HCoV-OC43) to 314 nt (TGEV). Consequently, a portion of the coding sequence for the nsp1 segment of the replicase polyprotein plays a dual role, in that it contains sequences or structures that also participate in RNA synthesis. By contrast, the much larger group 3 coronavirus (IBV) 5′ UTR of 528 nt is likely to contain the entirety of the 5′ *cis*-acting replication element; alternatively, the element overlaps the start of the replicase ORF by no more than 16 nt. In this regard, it is noteworthy that the IBV replicase has no homolog for either the group 1 or group 2 nsp1 proteins.

It is necessary to point out two other aspects of the coronavirus genomic 5′ end. First, in all cases, the minimal 5′ *cis*-acting element for genome replication runs well beyond the 5′ leader RNA that is common to gRNA and all sgRNAs. Consequently, sgRNAs do not contain all of the necessary determinants that would allow them to function as independent replicons, as was proposed at the time when negative-strand sgRNAs were discovered (Sethna *et al.*, 1989). Second, within the non-leader portion of the genomic 5′ UTR, there is invariably a very short ORF similar to the upstream ORFs (uORFs) that have been shown to operate in translational regulatory mechanisms in certain eukaryotic mRNAs (Geballe and Sachs, 2000). The various coronavirus uORFs, which range from three to 11 codons in length, have been tabulated by Brian and Baric (2005). It is at present unclear what role the genomic uORF plays in coronavirus translation or RNA replication.

5′ sequences and structures

Almost all experimental definition of the 5′ *cis*-acting element for coronavirus genome replication has been carried out in studies of BCoV. This work has shown that the 5′ genomic region comprises at least five stem–loop structures, numbered consecutively in the 5′ to 3′ direction; the first four of these have been characterized in considerable detail (Figure 3.1) (Brian and Baric, 2005). Stem–loops I and II contain the entire leader region, and the leader copy of the TRS is situated within loop II. The RNA secondary structures of stem–loops I and II were confirmed by enzymatic probing, and both elements were found to be essential for BCoV DI RNA replication (Chang *et al.*, 1994 and 1996). Many mutations that were engineered in these structures in the DI RNA were found to be rapidly repaired, by replacement with a wild-type copy of the leader RNA provided by the helper virus. This phenomenon, often referred to as "leader switching" (Makino and Lai, 1989) and originally envisioned to occur through a leader-priming mechanism, was

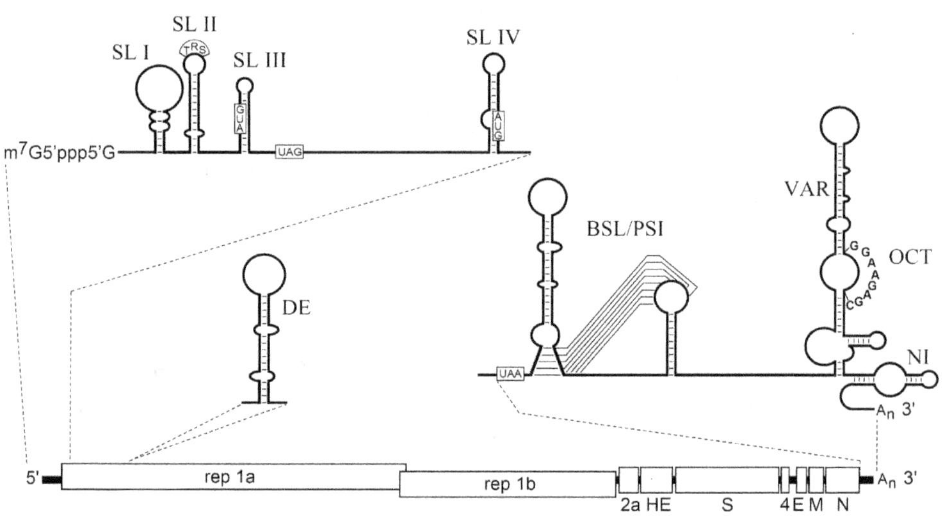

Figure 3.1 *Cis*-acting RNA structures and sequences that affect coronavirus RNA replication, generalized from the group 2 coronaviruses (the best characterized group). The general layout of the genome is shown at the bottom, with the two overlapping ORFs of the replicase (rep 1a and 1b) at the upstream end, and the structural and accessory genes at the downstream end. The schematic of the expanded 5′ terminus summarizes the elements that have been defined in BCoV: stem–loops I through IV (SL I—SL IV). The location of the TRS in stem–loop II is shown. Boxes indicate the start and stop codons of the uORF, which begins in stem–loop III, and the start codon of the replicase ORF, which begins in stem–loop IV. The depiction of the expanded 3′ terminus, which begins at the stop codon of the N gene, summarizes elements that have been defined in MHV and BCoV. These include: the bulged stem–loop and pseudoknot (BSL/PSI), the variable downstream region (VAR), the absolutely conserved octanucleotide motif (OCT), and the minimal element required for negative-strand synthesis initiation (NI). Also shown is the internal discontiguous element (DE) found in JHM DI RNAs (see text for details).

shown to actually be brought about by a high-frequency recombination event that occurs in a 24-nt window downstream of the leader TRS (Chang *et al.*, 1996). Stem–loops I and II are not highly conserved among coronaviruses, even between the closely related MHV and BCoV (Chang *et al.*, 1996). However, structures similar to stem–loop II have been proposed for members of all three groups of coronaviruses, even though the exact location of the TRS varies considerably among them (van den Born *et al.*, 2004). Two alternative 5′ stem–loops have been proposed for the leader-containing region of IBV (Stirrups *et al.*, 2000). Although the first stem–loop is phylogenetically conserved among a number of IBV strains, experimental evidence corroborating the existence of these structures has not yet been shown. In some ways, the coronavirus stem–loop II appears comparable to the well-characterized stem–loop G in the equine arterivirus (EAV) genome, an element which plays essential roles in both replication and transcription (van den Born *et al.*, 2004). However, stem–loop G is the last of seven RNA secondary structures within the much-larger EAV leader. Additionally, the EAV replicase ORF begins within the downstream arm of stem–loop G, in contrast to the placement much further downstream of the replicase ORF in coronavirus genomes. Thus, the functional analogy between these RNA structures in the arterivirus and coronavirus families may break down at the level of specific details regarding their mechanisms of action.

BCoV 5′ stem–loop III is located 13 nt downstream of stem–loop II. Its structure has been confirmed by enzymatic probing and by mutagenesis (Chang *et al.*, 1996; Raman *et al.*, 2003). Mutational analysis has shown that it is the secondary structure of the stem, rather than its primary sequence, that is functionally essential for replication (Raman *et al.*, 2003). A noteworthy feature of stem–loop III is that it contains the above-mentioned uORF of the 5′ UTR, which starts in the upstream arm of the stem and stops 10 nt downstream of the stem. Integrity of the uORF, but not conservation of the uORF peptide sequence, was found to be critical to DI RNA replication. It is not yet clear whether stem–loop III functions as a positive-strand or negative-strand structure, or as both. Stem–loop III is well conserved among the group 2 coronaviruses, and potential homologs have been identified in members of groups 1 and 3. Outside of the group 2 coronaviruses, however, the positioning of the uORF within the structure varies among individual virus species.

BCoV 5′ stem–loop IV follows stem–loop III by some 70 nt and contains the start codon for the replicase ORF in the downstream arm of its stem. Enzymatic probing has been used to confirm that the secondary structure of stem–loop IV conforms to predictions, and the essential nature of this structure has been demonstrated by its deletion in a DI RNA (Raman and Brian, 2005). Mutational analysis showed that the intactness, rather than the primary sequence, of the stem is critical for replication; this analysis also indicated that the stem functions in the positive strand but is not required in the negative strand. Stem–loop IV is well conserved in group 2 coronaviruses that are closely related to BCoV, but SARS-CoV, group 1, and group 3 coronaviruses do not have an obviously homologous structure. Curiously, SARS-CoV and members of the group 1 coronaviruses each have either two or three stem–loops at a similar spacing downstream of stem–loop III. Within each species, all of the candidate stem–loop IV counterparts have hexanucleotide loops with identical, or nearly identical, sequences (Raman and Brian, 2005).

3′ genomic elements acting in RNA replication

Boundary of the 3′ requirement for replication

At the 3′ terminus of the coronavirus genome is a 3′ UTR, which ranges in size from 273 nt (TGEV and feline infectious peritonitis virus [FIPV]) to 505 nt (IBV) and is followed by a polyadenylate tail. Unlike the 5′ UTR, the precise boundary of the 3′ UTR cannot in all cases be unambiguously assigned, because the upstream region is in evolutionary flux. The best example of such a situation is found in the group 3 coronaviruses. Immediately downstream of the N gene, many IBV isolates contain an extremely divergent 200-nt stretch of sequence, which has been designated as a hypervariable region of the 3′ UTR (Williams *et al.*, 1993; Sapats *et al.*, 1996). However, coronavirus sequences closely related to IBV that have been recently characterized in geese and pigeons provide a different perspective on this hypervariable segment (Jonassen *et al.*, 2005). In the goose coronavirus, there is a considerably larger span of sequence (840 nt), containing two additional ORFs, between the N gene and the conserved, distal portion of the 3′ UTR. Similarly, in the pigeon coronavirus sequence, there is one extra ORF relative to the IBV genome. It is thus plausible that the IBV hypervariable region is not properly a part of the 3′ UTR, but, rather, represents a degenerate remnant of one or more ancestral genes.

As with 5′ *cis*-acting elements, the delineation of the replication-essential portions of the 3′ genomic termini of coronaviruses also began with DI RNA studies. Deletion analyses of MHV-JHM DI RNAs found that the minimal length of the 3′ terminus required for replication fell between 378 and 436 nt (Lin and Lai, 1993) or between 416 and 462 nt (Kim *et al.*, 1993). Similarly, the smallest 3′ end sequence on a replication-competent MHV-A59 DI RNA was 461 nt (Luytjes *et al.*, 1996). In all cases it was noted that the required region not only encompassed the entire 301-nt 3′ UTR but also ran into the adjacent N ORF, an arrangement apparently analogous to the overlap between the 5′ replication element and the start of the replicase ORF. By contrast, the 3′ *cis*-acting replication signals that have been mapped for group 1 and group 3 coronavirus DI RNAs do not contain any fragment of the N gene. For TGEV, deletion construction limited the maximal size of the required 3′ element to 492 nt, which includes accessory ORF 7 and the entire 3′ UTR (Izeta *et al.*, 1999). However, this analysis did not rule out the possibility that an even smaller 3′ genomic sequence would suffice for replication. Indeed, for the closely related FIPV, a genomic deletion mutant of ORFs 7a and 7b has been constructed, which would suggest that the group 1 minimal 3′ *cis*-acting replication signal does not exceed 283 nt (Haijema *et al.*, 2004). For IBV, the minimal required element was found to be between 245 and 338 nt in length (Dalton *et al.*, 2001). This essential portion of the IBV 3′ terminus, besides lacking the N gene, also excludes the upstream hypervariable region of the 3′ UTR mentioned above. Mutations engineered in the MHV genome, rather than in DI RNAs, also support the notion that the required 3′ genomic element for viral RNA replication actually lies entirely within the 3′ UTR. Engineered rearrangements of the MHV genome that translocated the N gene to an upstream position had no effect on viral replication (de Haan *et al.*, 2002; Goebel *et al.*, 2004a). Consistent with this observation, the MHV genome has been shown to tolerate multiple point and small-deletion mutations in the downstream portion of the N gene (Kuo and Masters, 2002; Hurst *et al.*,

2005). These findings make it very unlikely that a 3′ *cis*-acting RNA sequence essential for viral genome replication is present in the MHV N gene.

Although the entire 3′ UTR is required for MHV replication, only a fraction of this region appears to be involved in the initiation of negative-strand RNA synthesis. In one study, ribonuclease protection assays were used to examine the production of negative-strand RNA from a series of non-replicating DI RNA mutants that contained successively larger deletions of the upstream portion of the 3′ UTR. This work found that negative-strand RNA synthesis could be supported by as few as 45 to 55 nt at the 3′ end of the genome, in addition to the polyadenylate tail (Lin *et al.*, 1994). It was therefore concluded that the remainder of the 3′ *cis*-acting element must be required for positive-strand, rather than negative-strand, RNA synthesis. In another study, reverse genetic methods were used to construct a non-viable MHV mutant, in which the arms of a putative 4-bp stem in the last 26 nt of the genome were interchanged. Cellular transfection of gRNA of this mutant led to synthesis of negative-strand gRNA, but not negative-strand sgRNA (Johnson *et al.*, 2005). These results reinforce the proposed role of the terminal segment of the 3′ UTR in negative-strand RNA synthesis, but the differential effect observed on gRNA and on sgRNA remains to be explained. Finally, a more detailed examination of the 3′ polyadenylate requirement, using DI RNAs from both MHV and BCoV, found that a minimum of 5 to 10 adenine residues are necessary for replication. Furthermore, DI RNA replication competence was shown to be correlated with the ability to bind poly(A) binding protein (PABP) (Spagnolo and Hogue, 2000).

3′ sequences and structures

A fairly detailed, although not complete, picture of the 3′ *cis*-acting region of the genome has been derived for at least some of the coronaviruses (Figure 3.1). At the upstream end of the 3′ UTR in all group 2 and group 3 coronaviruses, there is a functionally essential RNA bulged stem–loop. This RNA secondary structure was discovered as a result of attempts to replace all or part of the MHV 3′ UTR with corresponding regions of the genome of the most closely related group 2 coronavirus, BCoV (Hsue and Masters, 1997). In these two coronaviruses, the bulged stem–loop comprises the first 68 nt immediately following the N gene stop codon. In IBV and all other group 3 coronaviruses, a similar structure, a nearly perfect 42-nt stem–loop, is found at the upstream end of the conserved region of the 3′ UTR. The IBV stem–loop has been demonstrated to be required for DI RNA replication (Dalton *et al.*, 2001).

Near the upstream end of the 3′ UTR in all group 1 and group 2 coronaviruses, there is a classical hairpin-type RNA pseudoknot encompassing 50 to 54 nt. This structure was discovered in BCoV and was noted to have counterparts in a number of other coronaviruses (Williams *et al.*, 1999). The pseudoknot was demonstrated to be required for BCoV DI RNA replication (Williams *et al.*, 1999), and it was subsequently shown to be essential in the MHV genome (Goebel *et al.*, 2004a). Curiously, only the 3′ UTRs of the group 2 coronaviruses have both the bulged stem–loop and the pseudoknot, with the latter always situated immediately downstream of the former. This suggests that group 1 and group 3 coronaviruses have evolved alternative means to carry out the functions provided by whichever of the two RNA structures is absent in that group. The conservation of both elements in the group 2 coronaviruses has been shown not only for the clusters of viruses that are closely related to MHV and BCoV (Wu *et al.*, 2003), but also for the more phylogenetically distant HCoV-HKU1 (Woo *et al.*, 2005) and SARS-CoV (Goebel *et al.*, 2004a). The structures of the bulged stem–loop and the pseudoknot have been confirmed in MHV and BCoV by chemical and enzymatic probing. Moreover, exhaustive mutational analyses have defined essential and non-essential components of these structures and have established the limits to which essential parts of each element could be destabilized, prior to loss of function (Hsue and Masters, 1997; Williams *et al.*, 1999; Hsue *et al.*, 2000; Goebel *et al.*, 2004a).

A distinctive property of the adjacent RNA elements in group 2 is that the downstream segment of the bulged stem–loop overlaps with the first stem of the pseudoknot. Since the two RNA secondary structures cannot form simultaneously, it has been proposed that they constitute a molecular switch, with one component regulating the folding of the other, possibly so as to govern a transition between different steps of viral replication (Goebel *et al.*, 2004a). Additionally, the two elements are very likely to perform the same role in every group 2 coronavirus, since, in all cases tested, the 3′ UTRs of members of this group have been found to be functionally interchangeable. Initially, it was shown that the 3′ UTR of MHV was able to be entirely replaced by its counterpart from BCoV (Hsue and Masters, 1997). This principle was made more general with the demonstration that any of a number of closely related group 2 helper viruses could support replication of a BCoV DI RNA (Wu *et al.*, 2003). Finally, across a yet greater phylogenetic distance, the SARS-CoV 3′ UTR was found to be able to entirely replace the MHV 3′ UTR, despite there being only 38% identity in primary sequence between these homologous genomic segments of the two viruses (Goebel *et al.*, 2004b). Notably, such reciprocity does not extend across group boundaries: neither the group 1 TGEV 3′ UTR nor the group 3 IBV 3′ UTR was able to functionally replace the 3′ UTR of MHV.

Downstream of the pseudoknot is a tract of roughly 150 nt, which, within the group 1 or group 2 coronaviruses, is much less well conserved than is either the upstream portion of the 3′ UTR or the 3′-terminal extreme of the genome. In MHV, this region has been predicted to fold into a complex RNA secondary structure that is consistent with enzymatic structural probing (Liu *et al.*, 2001). Although the corresponding region of the BCoV 3′ UTR is predicted to form a substantially different structure, there is one stem segment that is homologous with the MHV structure, and mutations made in this stem segment were found to inhibit MHV DI RNA replication (Liu *et al.*, 2001). Remarkably, embedded within this most divergent portion of the 3′ UTR is an invariant octanucleotide motif, 5′-GGAAGAGC-3′. This motif, which is always located 70 to 80 nt from the 3′ end of the genome, was first noted as a commonality in the earliest sequences determined for the otherwise-disparate 3′ UTRs from the three different groups (Boursnell *et al.*, 1985; Lapps *et al.*, 1987; Schreiber *et al.*, 1989); it has since been found to be absolutely conserved in coronaviruses. It is therefore assumed to have a critical role in viral replication, although, as yet, no function has been ascribed to it.

Another motif, a 40- to 45-nt bulged stem–loop, is found in the downstream region of the 3′ UTRs of only the group 3 coronaviruses and, additionally, in that of SARS-CoV. Surprisingly, this element, which is designated s2m, is also found near the 3′ end of the genome in most astroviruses and in one picornavirus (Jonassen *et al.*, 1998). The distribution of this well-conserved RNA sequence suggests that it was independently horizontally acquired by a number of viral families. The structure of the s2m element has recently been solved by X-ray crystallography, which has revealed it to consist of two A-form helices kinked at a 90° angle, in a conformation that is stabilized by a number of tertiary interactions (Robertson *et al.*, 2005). The backbone fold of s2m resembles that of the 530 loop of 16 S ribosomal RNA, which has led to the speculation that s2m could assist viral RNA translation through recruitment of eukaryotic translation initiation factor 1A. Although this is an intriguing possibility, it remains to be demonstrated whether the s2m element plays any role, either essential or modulatory, in those coronavirus genomes in which it appears.

Internal genomic element acting in RNA replication

The earliest deletion analyses of MHV-JHM DI RNAs found that, in addition to sequences derived from the 5′ and 3′ ends of the genome, a short, discontiguous segment of gRNA was required for replication (Lin and Lai, 1993; Kim *et al.*, 1993). This internal genomic element was originally mapped to a region of 135 nt, falling at a distance of between 3.1 and 3.3 kb from the 5′ end of the genome (Figure 3.1). Curiously, the corresponding region of the MHV-A59 genome could not functionally replace its counterpart in MHV-JHM DI RNA, unless the temperature was elevated from 37 to 39.5°C (Kim and Makino, 1995). The internal, discontiguous replication element was shown to function during positive-strand RNA synthesis; its absence had no effect on negative-strand RNA synthesis (Kim and Makino, 1995). In further work, the essential portion of the element was more precisely localized, in MHV-JHM DI RNA constructs, to a 58-nt bulged stem–loop, the structure of which was confirmed by enzymatic probing (Repass and Makino, 1998).

At variance with these findings, a deletion analysis of a naturally occurring MHV-A59 DI RNA yielded many constructs that replicated well without the MHV-A59 counterpart of the internal genomic element (Luytjes *et al.*, 1996). Moreover, a naturally occurring BCoV DI RNA had been isolated that solely comprised terminal genomic segments: the 5′-most 498 nt of the genome fused to the entire N gene and the 3′ UTR (1635 nt) (Chang *et al.*, 1994). A synthetic MHV-A59 DI RNA that was patterned after the BCoV DI RNA was likewise shown to be replication-competent in the absence of any internal discontiguous genomic sequence elements (Masters *et al.*, 1994). In a similar manner, subsequent deletion analyses of TGEV and IBV DI RNAs revealed no requirements for internal discontiguous sequences (Izeta *et al.*, 1999; Dalton *et al.*, 2001).

Consequently, the role of the internal MHV-JHM DI RNA element is unclear, as best illustrated by one paper that paradoxically described the element as "required" and "optional" within a single sentence (Lin *et al.*, 1996). On the whole, the evidence argues that coronaviruses do not require internal *cis*-acting RNA sequences or structures to enable replication of their genomes. Nevertheless, multiple naturally occurring DI RNAs of both MHV-JHM and MHV-A59 have evolved to harbor the same internal element (Makino *et al.*, 1988b; van der Most *et al.*, 1991), suggesting that this element serves a DI RNA-specific function. Interestingly, there has been only one reported direct comparison between the similarly sized DIssE, the smallest MHV-JHM DI RNA that contains the element, and B36, the synthetic MHV-A59 DI RNA that lacks the element (Johnson *et al.*, 2005). In this analysis, DIssE appears to replicate more efficiently than does B36, an advantage perhaps conferred by the presence of the internal element. One possibility is that the internal element provides a barrier that maintains the independent folding of the 5′ and 3′ segments of some DI RNAs. This would prevent the formation of alternative secondary structures within or among regions that are normally separated by great distances in the intact genome. If such is the case, then there may exist numerous other functionally redundant RNA structures in coronavirus genomes, a situation which would explain why not all DI RNAs need to contain the particular element that has been identified in MHV-JHM. We might speculate still further that the internal discontiguous element actually functions as a parasitic RNA that facilitates DI RNA formation in MHV. The 58-nt internal element, which lies near the 5′ end of the nsp3 section of the replicase gene, has no counterpart in the aligned nsp3 sequences of coronavirus species other than MHV (Ziebuhr *et al.*, 2001). It would therefore be very interesting to test, by reverse genetic methods, whether the internal element plays any detectable role in the intact MHV genome.

Host factors implicated in RNA replication

By analogy with a number of other RNA viruses, coronavirus RNA synthesis is presumed to involve both virally encoded activities and host activities that are appropriated to perform virus-specific functions (Shi and Lai, 2005). Currently, all efforts to identify host factors have been carried out with group 2 coronaviruses, either MHV or BCoV. A number of such factors, all of which are members of the heterogeneous nuclear ribonucleoprotein (hnRNP) family, were discovered on the basis of their ability to bind *in vitro* to RNA segments containing either the TRS or other signals of known importance in transcription. This set includes hnRNP A1 (Furuya and Lai, 1993; Zhang and Lai, 1995; Li *et al.*, 1997), pyrimidine tract-binding protein (PTB, or hnRNP I), and synaptotagmin-binding cytoplasmic RNA-interacting protein (SYNCRIP) (Choi *et al.*, 2004). Among these, the most extensively examined has been hnRNP A1, which was initially found among a group of proteins that bound to the negative strand of the MHV TRS.

In addition to its proposed connection with transcription, hnRNP A1, when overexpressed, upregulated the kinetics of MHV RNA synthesis; this activity was taken to indicate that hnRNP A1 also affects genome replication (Shi *et al.*, 2000). Conversely, a dominant-negative effect on MHV replication was obtained through expression of a truncated form of hnRNP A1. The relevance of these observations was placed in doubt by the demonstration that MHV was entirely able to carry out RNA synthesis in a mutant mouse cell line that does not express hnRNP A1 (Shen and Masters, 2001). However, this apparent contradiction was subsequently explained by the finding that other hnRNP A/B family members are functionally redundant with hnRNP A1; moreover, overexpression of hnRNP A/B enhanced MHV RNA synthesis (Shi *et al.*, 2003).

Similarly, although PTB was initially implicated in MHV transcription, owing to its binding near the leader TRS (Li *et al.*, 1999), it was also found to interact with the 3′ *cis*-acting element for replication. Specifically, PTB bound to the negative strand of the region containing the coronavirus-invariant octanucleotide motif, and small deletion mutations in this locus reduced or abolished DI RNA replication (Huang and Lai, 1999). An additional hnRNP A1 binding site was found in the positive-strand RNA complement of the same region (Huang and Lai, 2001).

To date, only one study has investigated potential host factors associated with 5′ *cis*-acting replication structures. In that work, it was found that three separate complexes of six host cell proteins bound to the BCoV 5′ stem–loop IV (Raman and Brian, 2005). These proteins bound specifically to the wild-type stem–loop IV and functional mutants thereof, but did not bind to mutants that were unable to support DI RNA replication. Further work will be needed to determine whether these proteins represent a new set of candidate cellular factors, or whether some of them are identical to factors previously identified with substrate RNAs from other segments of the coronavirus genome.

At the other end of the MHV genome, two related 11-nt sequences in the 3′ *cis*-acting genomic element were shown to be bound by similar complexes of at least four host proteins (Yu and Leibowitz, 1995a and 1995b; Liu *et al.*, 1997). The first motif, at 26 to 36 nt upstream of the polyadenylate tract, maps in the center of the minimal 3′ genomic region required for the initiation of negative-strand RNA synthesis (Lin *et al.*, 1994). The second motif, at 129 to 139 nt from the 3′ end of the genome, falls in the more divergent region of the 3′ UTR, downstream of the pseudoknot. Mutations that abolished protein complex binding to either of the two 11-nt motifs were found to also inhibit DI RNA replication. The largest component of the complex that binds to the further-downstream motif was identified as mitochondrial aconitase (an enzyme which, it must be noted, has a significant extramitochondrial fraction) (Nanda and Leibowitz, 2001). This was the first proposed coronavirus host factor that was not previously known to have RNA-binding activity. Further work found that other protein complex components are the chaperones HSP60, HSP40, and mitochondrial HSP70; these identifications suggest a host factor role in guiding the proper assembly or membrane association of the replicase machinery (Nanda *et al.*, 2004). Finally, downstream of the virus-specific sequence of the 3′ *cis*-acting element, PABP was shown to bind to the polyadenylate tail (Spagnolo and Hogue, 2000). For both MHV and BCoV, the length of oligoadenylate at the 3′ end of a DI RNA was directly correlated with both the extent of PABP binding and the replication competence of the DI RNA.

Conclusion: roles for cis-acting RNA elements

Although the essential nature of *cis*-acting genomic elements in coronavirus RNA replication is well established, the exact roles of these elements are far from clear at this time. One obvious function in which they must participate is the initiation of negative-strand and positive-strand RNA synthesis. However some DI RNA evidence suggests that only relatively small regions of the required elements are directly involved in RNA synthesis initiation. This leaves much mechanistic detail still to be elucidated. Elements within gRNA that do not directly act at the level of initiation may help the virus to establish conditions that need to be in place prior to initiation of either negative-strand or positive-strand RNA synthesis. Thus, these elements could be involved in processes such as regulating the transition from genome translation to initiation of negative-strand RNA synthesis, as well as in switching from negative-strand to positive-strand RNA synthesis. In addition, it is not known how gRNA becomes localized to the membranous compartment in the host cell where coronavirus RNA synthesis occurs (Sethna and Brian, 1997; van der Meer *et al.*, 1999; Bi *et al.*, 1999; Shi *et al.*, 1999). Such targeting may be signaled by an embedded viral RNA element, or, as has been suggested, it could be mediated by the nascent replicase protein while it is being translated from gRNA or DI RNA (Chang and Brian, 1996). There are also ancillary events in RNA synthesis that could be governed by *cis*-acting elements. One of these is the non-templated polyadenylation of positive-strand RNA. Coronavirus negative-strand RNAs have only short (9- to 26-nt) tracts of oligouridylate at their 5′ ends (Hofmann and Brian, 1991). This suggests that the viral replicase adds most of the polyadenylate tail by a "chattering" mechanism similar to that used by some negative-strand RNA viruses. Alternatively, a host cytoplasmic poly(A) polymerase may need to be recruited. Either of these mechanisms could require a 3′ *cis*-acting RNA element. Likewise, nothing is known about the mechanism of 5′ capping of coronavirus gRNA or sgRNA; the potential recruitment of cellular factors for guanylylation or cap methylation could similarly require a 5′ *cis*-acting RNA element.

Some of the events in coronavirus RNA synthesis are proposed to occur through genome circularization. Clearly, translation of the capped and polyadenylated coronavirus gRNA must involve 5′-to-3′ communication via eIF4E, eIF4G, and PABP. In addition, however, coronavirus genome circularization is sometimes invoked, by analogy with mechanisms that have been developed in greater detail for the replication of other positive-strand RNA viruses (Herold and Andino, 2001; Teterina *et al.*, 2001; Barton *et al.*, 2001; Khromykh *et al.*, 2003). Plausible scenarios have been presented for coronavirus genome circularization in transcription and replication (Lin *et al.*, 1996; Huang and Lai, 2001; Sola *et al.*, 2005; Shi and Lai, 2005), but there is, as yet, no evidence that such circularization actually occurs or that it plays a necessary role at any stage of viral RNA synthesis. Unraveling the precise mechanisms of the interplay among coronavirus genomic *cis*-acting elements, viral replicase components, and host cellular factors is likely to remain a challenging area of investigation for some time to come.

References

Barton, D.J., O'Donnell, B.J., and Flanegan, J.B. (2001). 5′ cloverleaf in poliovirus RNA is a *cis*-acting replication element required for negative-strand synthesis. EMBO J. *20*, 1439–1448.

Bi, W., Pinon, J.D., Hughes, S., Bonilla, P.J., Holmes, K.V., Weiss, S.R., and Leibowitz, J.L. (1999). Localization of mouse hepatitis virus open reading frame 1A derived proteins. J. Neurovirol. *4*, 594–605.

Boursnell, M.E.G., Binns, M.M., Foulds, I.J., and Brown, T.D.K. (1985). Sequences of the nucleocapsid genes from two strains of avian infectious bronchitis virus. J. Gen. Virol. *66*, 573–580.

Brian, D.A., and Baric, R.S. (2005). Coronavirus genome structure and replication. Curr. Top. Microbiol. Immunol. *287*, 1–30.

Chang, R.-Y., and Brian, D.A. (1996). *cis* requirement for N-specific protein sequence in bovine coronavirus defective interfering RNA replication. J. Virol. *70*, 2201–2207.

Chang, R.-Y., Hofmann, M.A., Sethna, P.B., and Brian, D.A. (1994). A *cis*-acting function for the coronavirus leader in defective interfering RNA replication. J. Virol. *68*, 8223–8231.

Chang, R.-Y., Krishnan, R., and Brian, D.A. (1996). The UCUAAAC promoter motif is not required for high-frequency leader recombination in bovine coronavirus defective interfering RNA. J. Virol. *70*, 2720–2729.

Choi, K.S., Mizutani, A., and Lai, M.M.C. (2004). SYNCRIP, a member of the heterogeneous nuclear ribonucleoprotein family, is involved in mouse hepatitis virus RNA synthesis. J. Virol. *78*, 13153–13162.

Dalton, K., Casais, R., Shaw, K., Stirrups, K., Evans, S., Britton, P., Brown, T.D.K., and Cavanagh, D. (2001). *cis*-acting sequences required for coronavirus infectious bronchitis virus defective-RNA replication and packaging. J. Virol. *75*, 125–133.

de Haan, C.A.M., Volders, H., Koetzner, C.A., Masters, P.S., and Rottier, P.J.M. (2002). Coronaviruses maintain viability despite dramatic rearrangements of the strictly conserved genome organization. J. Virol. *76*, 12491–12493.

Denison, M.R., and Perlman, S. (1986). Translation and processing of mouse hepatitis virus virion RNA in a cell-free system. J. Virol. *60*, 12–18.

Escors, D., Izeta, A., Capiscol, C., and Enjuanes, L. (2003). Transmissible gastroenteritis coronavirus packaging signal is located at the 5′ end of the virus genome. J. Virol. *77*, 7890–7902.

Furuya, T., and Lai, M.M.C. (1993). Three different cellular proteins bind to complementary sites on the 5′-end-positive and 3′-end-negative strands of mouse hepatitis virus RNA. J. Virol. *67*, 7215–7222.

Geballe, A.P., and Sachs, M.S. (2000). Translational control by upstream open reading frames. In Translational Control of Gene Expression, N. Sononberg, J.W.B. Hershey, and M.B. Mathews, eds. (New York: Cold Spring Harbor Laboratory Press).

Goebel, S.J., Hsue, B., Dombrowski, T.F., and Masters, P.S. (2004a). Characterization of the RNA components of a putative molecular switch in the 3′ untranslated region of the murine coronavirus genome. J. Virol. *78*, 669–682.

Goebel, S.J., Taylor, J., and Masters, P.S. (2004b). The 3′ *cis*-acting genomic replication element of the severe acute respiratory syndrome coronavirus can function in the murine coronavirus genome. J. Virol. *78*, 7846–7851.

Haijema, B.J., Volders, H., and Rottier, P.J.M. (2004). Live, attenuated coronavirus vaccines through the directed deletion of group-specific genes provide protection against feline infectious peritonitis. J. Virol. *78*, 3863–3871.

Herold, J., and Andino, R. (2001). Poliovirus RNA replication requires genome circularization through a protein-protein bridge. Mol. Cell *7*, 581–591.

Hofmann, M.A., and Brian, D.A. (1991). The 5′ end of coronavirus minus-strand RNAs contain a short poly(U) tract. J. Virol. *65*, 6331–6333.

Hofmann, M.A., Sethna, P.B., and Brian, D.A. (1990). Bovine coronavirus mRNA replication continues throughout persistent infection in cell culture. J. Virol. *64*, 4108–4114.

Hsue, B., and Masters, P.S. (1997). A bulged stem–loop structure in the 3′ untranslated region of the genome of the coronavirus mouse hepatitis virus is essential for replication. J. Virol. *71*, 7567–7578.

Hsue, B., Hartshorne, T., and Masters, P.S. (2000). Characterization of an essential RNA secondary structure in the 3′ untranslated region of the murine coronavirus genome. J. Virol. *74*, 6911–6921.

Huang, P., and Lai, M.M.C. (1999). Polypyrimidine tract-binding protein binds to the complementary strand of the mouse hepatitis virus 3′ untranslated region, thereby altering RNA conformation. J. Virol. *73*, 9110–9016.

Huang, P., and Lai, M.M.C. (2001). Heterogeneous nuclear ribonucleoprotein A1 binds to the 3′-untranslated region and mediates potential 5′-3′-end cross talks of mouse hepatitis virus RNA. J. Virol. *75*, 5009–5017.

Hurst, K.R., Kuo, L., Koetzner, C.A., Ye, R., Hsue, B., and Masters, P.S. (2005). A major determinant for membrane protein interaction localizes to the carboxy-terminal domain of the mouse coronavirus nucleocapsid protein. J. Virol. *79*, 13285–13297.

Izeta, A., Smerdou, C., Alonso, S., Penzes, Z., Mendez, A., Plana-Duran, J., and Enjuanes, L. (1999). Replication and packaging of transmissible gastroenteritis coronavirus-derived synthetic minigenomes. J. Virol. *73*, 1535–1545.

Johnson, R.F., Feng, M., Liu, P., Millership, J.J., Yount, B., Baric, R.S., and Leibowitz, J.L. (2005). Effect of mutations in the mouse hepatitis virus 3′(+)42 protein binding element on RNA replication. J. Virol. *79*, 14570–14585.

Jonassen, C.M., Jonassen, T.O., and Grinde, B. (1998). A common RNA motif in the 3′ end of the genomes of astroviruses, avian infectious bronchitis virus and an equine rhinovirus. J. Gen. Virol. *79*, 715–718.

Jonassen, C.M., Kofstad, T., Larsen, I.-L., Lovland, A., Handeland, K., Follestad, A., and Lillehaug, A. (2005). Molecular identification and characterization of novel coronaviruses infecting graylag geese (Anser anser), feral pigeons (Columbia livia) and mallards (Anas platyrhynchos). J. Gen. Virol. *86*, 1597–1607.

Kim, Y.-N., and Makino, S. (1995). Characterization of a murine coronavirus defective interfering RNA internal *cis*-acting replication signal. J. Virol. *69*, 4963–4971.

Kim, Y.-N., Jeong, Y.S., and Makino, S. (1993). Analysis of *cis*-acting sequences essential for coronavirus defective interfering RNA replication. Virology *197*, 53–63.

Khromykh, A.A., Meka, H., Guyatt, K.J., and Westaway, E.G. (2001). Essential role of cyclization sequences in flavivirus RNA replication. J. Virol. *75*, 6719–6728.

Kuo, L., and Masters, P.S. (2002). Genetic evidence for a structural interaction between the carboxy termini of the membrane and nucleocapsid proteins of mouse hepatitis virus. J. Virol. *76*, 4987–4999.

Lai, M.M.C., and Stohlman, S.A. (1978). RNA of mouse hepatitis virus. J. Virol. *26*, 236–242.

Lai, M.M.C., and Stohlman, S.A. (1981). Comparative analysis of RNA genomes of mouse hepatitis viruses. J. Virol. *38*, 661–670.

Lapps, W., Hogue, B.G., and Brian, D.A. (1987). Sequence analysis of the bovine coronavirus nucleocapsid and matrix protein genes. Virology *157*, 47–57.

Li, H.-P., Zhang, X., Duncan, R., Comai, L., and Lai, M.M.C. (1997). Heterogeneous nuclear ribonucleoprotein A1 binds to the transcription-regulatory region of mouse hepatitis virus RNA. Proc. Natl. Acad. Sci. USA *94*, 9544–9549.

Li, H.-P. Huang, P., Park, S., and Lai, M.M.C. (1999). Polypyrimidine tract-binding protein binds to the leader RNA of mouse hepatitis virus and serves as a regulator of viral transcription. J. Virol. *73*, 772–777.

Lin, Y.-J., and Lai, M.M.C. (1993). Deletion mapping of a mouse hepatitis virus defective interfering RNA reveals the requirement of an internal and discontiguous sequence for replication. J. Virol. *67*, 6110–6118.

Lin, Y.-J., Liao, C.-L., and Lai, M.M.C. (1994). Identification of the *cis*-acting signal for minu-strand RNA synthesis of a murine coronavirus: implications for the role of minus-strand RNA in RNA replication and transcription. J. Virol. *68*, 8131–8140.

Lin, Y.-J., Zhang, X., Wu, R.-C., and Lai, M.M.C. (1996). The 3′ untranslated region of coronavirus RNA is required for subgenomic mRNA transcription from a defective interfering RNA. J. Virol. *70*, 7236–7240.

Liu, Q., Yu, W., and Leibowitz, J.L. (1997). A specific host cellular protein binding element near the 3′ end of mouse hepatitis virus genomic RNA. Virology *232*, 74–85.

Liu, Q., Johnson, R.F., and Leibowitz, J.L. (2001). Secondary structural elements within the 3′ untranslated region of mouse hepatitis virus strain JHM genomic RNA. J. Virol. *75*, 12105–12113.

Lomniczi, B. (1977). Biological properties of avian coronavirus RNA. J. Gen. Virol. *36*, 531–533.

Luytjes, W., Gerritsma, H., and Spaan, W.J.M. (1996). Replication of synthetic interfering RNAs derived from coronavirus mouse hepatitis virus-A59. Virology *216*, 174–183.

Makino, S., and Lai, M.M.C. (1989). High-frequency leader sequence switching during coronavirus defective interfering RNA replication. J. Virol. *63*, 5285–5292.

Makino, S., Taguchi, F., and Fujiwara, K. (1984). Defective interfering particles of mouse hepatitis virus. Virology *133*, 9–17.

Makino, S., Fujioka, N., and Fujiwara, K. (1985). Structure of the intracellular defective viral RNAs of defective interfering particles of mouse hepatitis virus. J. Virol. *54*, 329–336.

Makino, S., Shieh, C.-K., Keck, J.G., and Lai, M.M.C. (1988a). Defective interfering particles of murine coronavirus: mechanism of synthesis of defective viral RNAs. Virology *163*, 104–111.

Makino, S., C.-K. Shieh, L.H. Soe, S.C. Baker, and M.M.C. Lai. (1988b). Primary structure and translation of a defective interfering RNA of murine coronavirus. Virology *166*, 550–560.

Masters, P.S., Koetzner, C.A., Kerr, C.A., and Heo, Y. (1994). Optimization of targeted RNA recombination and mapping of a novel nucleocapsid gene mutation in the coronavirus mouse hepatitis virus. J. Virol. *68*, 328–337.

Mendez, A., Smerdou, C., Izeta, A., Gebauer, F., and Enjuanes, L. (1996). Molecular characterization of transmissible gastroenteritis coronavirus defective interfering genomes: packaging and heterogeneity. Virology *217*, 495–507.

Nanda, S.K., and Leibowitz, J.L. (2001). Mitochondrial aconitase binds to the 3′ untranslated region of the mouse hepatitis virus genome. J. Virol. *75*, 3352–3362.

Nanda, S.K., Johnson, R.F., Liu, Q., and Leibowitz, J.L. (2004). Mitochondrial HSP70, HSP40, and HSP60 bind to the 3′ untranslated region of the Murine hepatitis virus genome. Arch. Virol. *149*, 93–111.

Penzes, Z., Tibbles, K., Shaw, K., Britton, P., Brown, T.D.K., and Cavanagh, D. (1994). Characterization of a replicating and packaged defective RNA of avian coronavirus infectious bronchitis virus. Virology *203*, 286–293.

Raman, S., and Brian, D.A. (2005). Stem–loop IV in the 5′ untranslated region is a *cis*-acting element in bovine coronavirus defective interfering RNA replication. J. Virol. *79*, 12434–12446.

Raman, S., Bouma, P., Williams, G.D., and Brian, D.A. (2003). Stem–loop III in the 5′ untranslated region is a *cis*-acting element in bovine coronavirus defective interfering RNA replication. J. Virol. *77*, 6720–6730.

Repass, J.F., and Makino, S. (1998). Importance of the positive-strand RNA secondary structure of a murine coronavirus defective interfering RNA internal replication signal in positive-strand RNA synthesis. J. Virol. *72*, 7926–7933.

Robertson, M.P., Igel, H., Baertsch, R., Haussler, D., Ares, M. Jr., Scott, W.G. (2005). The structure of a rigorously conserved RNA element within the SARS virus genome. PLoS Biol. *3*, e5.

Sapats, S.I., Ashton, F., Wright, P.J., and Ignjatovic, J. (1996). Novel variation in the N protein of avian infectious bronchitis virus. Virology *226*, 412–417.

Schochetman, G., Stevens, R.H., and Simpson, R.W. (1977). Presence of infectious polyadenylated RNA in coronavirus avian bronchitis virus. Virology *77*, 772–782.

Schreiber, S.S., Kamahora, T., and Lai, M.M.C. (1989). Sequence analysis of the nucleocapsid protein gene of human coronavirus 229E. Virology *169*, 142–151.

Sethna, P.B., and Brian, D.A. (1997). Coronavirus genomic and subgenomic minus-strand RNAs copartition in membrane-protected replication complexes. J. Virol. *71*, 7744–7749.

Sethna, P.B., Hung, S.L., and Brian, D.A. (1989). Coronavirus subgenomic minus-strand RNAs and the potential for mRNA replicons. Proc. Natl. Acad. Sci. USA *86*, 5626–5630.

Shen, X., and Masters, P.S. (2001). Evaluation of the role of heterogeneous nuclear ribonucleoprotein A1 as a host factor in murine coronavirus discontinuous transcription and genome replication. Proc. Natl. Acad. Sci. USA *98*, 2717–2722.

Shi, S.T., and Lai, M.M.C. (2005). Viral and cellular proteins involved in coronavirus replication. Curr. Top. Microbiol. Immunol. *287*, 95–131.

Shi, S.T., Schiller, J.J., Kanjanahaluethai, A., Baker, S.C., Oh, J.W., and Lai, M.M.C. (1999). Colocalization and membrane association of murine hepatitis virus gene 1 products and De novo-synthesized viral RNA in infected cells. J. Virol. *73*, 5957–5969.

Shi, S.T., Huang, P., Li, H.-P., and Lai, M.M.C. (2000). Heterogeneous nuclear ribonucleoprotein A1 regulates RNA synthesis of a cytoplasmic virus. EMBO J. *19*, 4701–4711.

Shi, S.T., Yu, G.Y., and Lai, M.M.C. (2003). Multiple type A/B heterogeneous nuclear ribonucleoproteins (hnRNPs) can replace hnRNP A1 in mouse hepatitis virus RNA synthesis. J. Virol. *77*, 10584–10593.

Sola, I., Moreno, J.L., Zuniga, S., Alonso, S., and Enjuanes, L. (2005). Role of nucleotides immediately flanking the transcription-regulating sequence core in coronavirus subgenomic mRNA synthesis. J. Virol. *79*, 2506–2516.

Spagnolo, J.F., and Hogue, B.G. (2000). Host protein interactions with the 3′ end of bovine coronavirus RNA and the requirement of the poly(A) tail for coronavirus defective genome replication. J. Virol. *74*, 5053–5065.

Stirrups, K., Shaw, K., Evans, S., Dalton, K., Cavanagh, D., and Britton, P. (2000). Leader switching occurs during the rescue of defective RNAs by heterologous strains of the coronavirus infectious bronchitis virus. J. Gen. Virol. *81*, 791–801.

Teterina, N.L., Egger, D., Bienz, K., Brown, D.M., Semler, B.L., and Ehrenfeld, E. (2001). Requirements for assembly of poliovirus replication complexes and negative-strand RNA synthesis. J. Virol. *75*, 3841–3850.

van den Born, E., Gultyaev, A.P., and Snijder, E.J. (2004). Secondary structure and function of the 5′-proximal region of the equine arteritis virus RNA genome. RNA *10*, 424–437.

van der Meer, Y., Snijder, E.J., Dobbe, J.C., Schleich, S., Denison, M.R., Spaan, W.J.M., and Locker, J.K. (1999). Localization of mouse hepatitis virus nonstructural proteins and RNA synthesis indicates a role for late endosomes in viral replication. J. Virol. *73*, 7641–7657.

van der Most, R.G., Bredenbeek, P.J., and Spaan, W.J.M. (1991). A domain at the 3′ end of the polymerase gene is essential for encapsidation of coronavirus defective interfering RNAs. J. Virol. *65*, 3219–3226.

van Vliet, A.L., Smits, S.L., Rottier, P.J.M., and de Groot, R.J. (2002). Discontinuous and non-discontinuous subgenomic RNA transcription in a nidovirus. EMBO J. *21*, 6571–6580.

Wege, H., Muller, A., and ter Meulen, V. (1978). Genomic RNA of the murine coronavirus JHM. J. Gen. Virol. *41*, 217–227.

Williams, A.K., Wang, L., Sneed, L.W., and Collisson, E.W. (1993). Analysis of a hypervariable region in the 3′ non-coding end of the infectious bronchitis virus genome. Virus Res. *28*, 19–27.

Williams, G.D., Chang, R.Y., and Brian, D.A. (1999). A phylogenetically conserved hairpin-type 3′ untranslated region pseudoknot functions in coronavirus RNA replication. J. Virol. *73*, 8349–8355.

Woo, P.C.Y., Lau, S.K.P., Chu, C.-M., Chan, K.-H., Tsoi, H.-W., Huang, Y., Wong, B.H.L., Poon, R.W.S., Cai, J.J., Luk, W.-K., Poon, L.L.M., Wong, S.S.Y., Guan, Y., Peiris, J.S.M., and Yuen, K.-Y. (2005). Characterization and complete genome sequence of a novel coronavirus, coronavirus HKU1, from patients with pneumonia. J. Virol. *79*, 884–895.

Wu, H.Y., Guy, J.S., Yoo, D., Vlasak, R., Urbach, E., and Brian, D.A. (2003). Common RNA replication signals exist among group 2 coronaviruses: evidence for in vivo recombination between animal and human coronavirus molecules. Virology *315*, 174–183.

Yu, W., and Leibowitz, J.L. (1995a). Specific binding of host cellular proteins to multiple sites within the 3′ end of mouse hepatitis virus genomic RNA. J. Virol. *69*, 2016–2023.

Yu, W., and Leibowitz, J.L. (1995b). A conserved motif at the 3′ end of mouse hepatitis virus genomic RNA required for host protein binding and viral RNA replication. Virology *214*, 128–38.

Ziebuhr, J., Thiel, V., Gorbalenya, A.E. (2001). The autocatalytic release of a putative RNA virus transcription factor from its polyprotein precursor involves two paralogous papain-like proteases that cleave the same peptide bond. J. Biol. Chem. *276*, 33220–33232.

Zhang, X., and Lai, M.M.C. (1995). Interactions between the cytoplasmic proteins and the intergenic (promoter) sequence of mouse hepatitis virus RNA: correlation with the amounts of subgenomic mRNA transcribed. J. Virol. *69*, 1637–1644.

Coronavirus RNA Synthesis: Transcription

4

Luis Enjuanes, Isabel Sola, Sonia Zuñiga, and J.L. Moreno

Abstract

Coronavirus replication involves both viral and cellular proteins, and takes place in the cytoplasm in a membrane-protected microenvironment. Coronaviruses may control the cell machinery by locating some of their proteins in the host cell nucleus, and their replication and transcription possibly require cross-talk between the 5′ and 3′ ends. Coronavirus transcription involves a discontinuous RNA synthesis (template switch) during the extension of a negative copy of the subgenomic (sg) mRNAs, a process that is controlled by transcription-regulating sequences (TRSs) preceding each gene. The requirement for base pairing during transcription has been formally demonstrated in arteriviruses and coronaviruses. The template switch requires base pairing between the nascent minus RNA strand and the leader TRS with minimum free energy (ΔG). Coronavirus N proteins have RNA chaperone activity that may help to overcome the energy barrier threshold associated with template switch. The potential role in transcription of selected proteins will be addressed.

Cell compartment distribution of coronavirus RNA synthesis

Coronavirus replication employs complex mechanisms that involve both viral and cellular proteins. Replication complexes of all plus strand eukaryotic RNA viruses are associated with intracellular membranes (Lee *et al.*, 2001), and virus infection frequently leads to membrane proliferation or vesiculation. In addition, proliferation of membrane proteins is linked to lipid biosynthesis. Both phenomena apparently are important for the amplification of most viral genomes (Boguszewska-Chachulska and Haenni, 2005). Therefore, membrane association probably is important for viral RNA synthesis. The polymerase of polioviruses, togaviruses, arteriviruses, and flaviviruses is associated to double-membrane structures (Egger *et al.*, 2000; Kaariainen and Ahola, 2002; Mackenzie *et al.*, 1999; Moradpour *et al.*, 2003; Pedersen *et al.*, 1999; Schlegel *et al.*, 1996; Snijder *et al.*, 2001). Also, the nodavirus flock house virus (FHV) RNA-dependent RNA polymerase (RdRp) contains transmembrane domains, and the replication of its RNAs is associated with membranes (Miller *et al.*, 2001; Miller *et al.*, 2003). RNA synthesis of plant RNA viruses also occurs in cell membranes whose nature (ER-derived membranes, cytoplasmic inclusions, vacuolar membranes, invagination of chloroplast membranes, or multivesicular bodies derived from peroxisomes, vacuoles or mitochondria) depends on the virus (for a review see Boguszewska-Chachulska 2005).

Similarly to other positive strand RNA viruses, coronavirus genome replication takes place in the cytoplasm in a membrane-protected microenvironment that contains all the protein functions required for viral RNA synthesis. Electron microscopy studies of mouse hepatitis virus (MHV)-infected cells have shown that these structures consist of double-membrane vesicles (DMVs) (Gosert *et al.*, 2002) that are possibly generated by using cellular autophagy-related processes (Prentice *et al.*, 2004). The cellular origin of MHV DMVs is under debate. While some data support that DMVs are derived from late endosomal membranes (van der Meer *et al.*, 1999) other suggest that DMVs have their origin in the endoplasmic reticulum (Prentice *et al.*, 2004; Shi *et al.*, 1999). Structures similar to DMVs have been found in severe acute respiratory syndrome coronavirus (SARS-CoV) (Snijder *et al.*, 2006) and in the equine arteritis virus (EAV) infection (Pedersen *et al.*, 1999), suggesting that coronaviruses and arteriviruses have a common replication strategy. The viral replication complex associated with these membranes apparently includes cellular proteins and up to 16 non-structural viral proteins (nsps) most of them derived from the proteolytic processing of the replicase polyproteins pp1a and pp1ab (Snijder *et al.*, 2006).

Although coronavirus replication essentially takes place within the cytoplasm, coronaviruses may control the cell machinery by locating some of their proteins in the host cell nucleus. Up to date, two coronavirus proteins (N and 3b) (Hiscox *et al.*, 2001; Wurm *et al.*, 2001; Yuan *et al.*, 2005) and another two from arteriviruses (N and nsp1) (Rowland *et al.*, 1999; Tijms *et al.*, 2002) have been identified in the nucleolus of cells. The nucleolus has been implicated in many aspects of cell biology that include functions such as ribosomal rRNA synthesis and ribosome biogenesis, gene silencing, senescence and cell cycle regulation (Hiscox, 2002). Viruses interact with nucleolar antigens leading to their redistribution during infection (Hiscox, 2002). N protein from coronavirus genus α (transmissible gastroenteritis coronavirus, TGEV), β (MHV), and γ (infectious bronchitis virus, IBV) (Hiscox *et al.*, 2001; Wurm *et al.*, 2001), and also N protein from the arteriviruses porcine respiratory and reproductive syndrome virus (PRRSV) and EAV (Rowland *et al.*, 1999; Tijms *et al.*, 2002) localize within the nucleolus. This may be a common feature among all nidovirus N proteins influencing host cell proliferation (Wurm *et al.*, 2001). However, the association of N protein with the nucleus may be cell dependent as TGEV N protein has been identified in the nucleus of transfected LLC-PK1 and Vero cells (Wurm *et al.*, 2001), but not in infected ST cells (Calvo *et al.*, 2005).

Coronavirus Genome Replication

Coronavirus replication and transcription possibly require recognition of RNA genome 5′ and 3′ ends by viral and cellular proteins. Like all other positive-stranded RNA viruses, coronavirus genome replication is mediated through the synthesis of a negative-strand RNA, which in turn is the template for the synthesis of progeny virus genomes. Mapping studies with MHV defective-interfering (DI) RNAs have indicated that 470 nt from the 5′ end and 436 nt from the 3′ end are required from DI RNA replication (Kim *et al.*, 1993; Lin and Lai, 1993). Both ends of the genome are necessary for positive-strand synthesis, whereas only the last 55 nt and the poly(A) tail at the 3′ end are required for the synthesis of MHV negative strand (Lin *et al.*, 1994). Since during the synthesis of the positive-strand RNA, the 3′ end of the genome is the last region reached by the viral polymerase, the replication signal at the 3′ end of the genome may interact with signals at the 5′ end to exert its effect on RNA synthesis. Based on this information, it has been postulated that the 5′ and 3′ ends of the genome interact during RNA replication. Similar observations have been made for the regulation of MHV subgenomic (sg) mRNA transcription (Lin *et al.*, 1996).

Although there is no apparent sequence complementarity between the 5′ and 3′ ends of coronavirus genome, a direct interaction between both ends could be possible, as it has been predicted for MHV and TGEV RNA genomes in protein-free media using computer programs (Holt *et al.*, 2004; Sgro *et al.*, 2004). However, there is experimental evidence for MHV (Huang and Lai, 2001) and TGEV (C. Galán, F. Almazan and L. Enjuanes, unpublished) supporting the idea that the cross-talk between the 5′ and 3′ ends is mediated by proteins through RNA-protein and protein-protein interactions. Heterogeneous nuclear ribonucleoprotein (hnRNP) A1 has been identified as a major protein species binding to the 3′ end of MHV RNA, while polypyrimidine tract-binding protein (PTB) binds to the 5′ end. Furthermore, hnRNP A1 and PTB bind to the complementary strands at the 5′ and 3′ ends of MHV RNA and together mediate the formation *in vitro* of a ribonucleoprotein (RNP) complex involving the 5′ and 3′ ends of MHV RNA (Huang and Lai, 2001; Li *et al.*, 1999; Li *et al.*, 1997) (Figure 4.1).

Coronavirus N protein forms a ribonucleoprotein complex with genomic RNA. In addition to its role in virus assembly (Escors *et al.*, 2001; Risco *et al.*, 1998), N protein probably also has a prominent role in coronavirus replication and transcription as it influences many viral and cellular processes. In fact, at early times after infection, coronavirus N protein colocalizes with the replication complex in DMVs, the site of coronavirus RNA synthesis (Bost *et al.*, 2001). Furthermore, the presence of N protein enhances the rescue of infectious virus from cDNA clones generated from different coronaviruses, such as IBV (Casais *et al.*, 2001), human coronavirus (HCoV)-229E (Schelle *et al.*, 2005), and TGEV using RNA *in vitro* transcripts (Yount *et al.*, 2000).

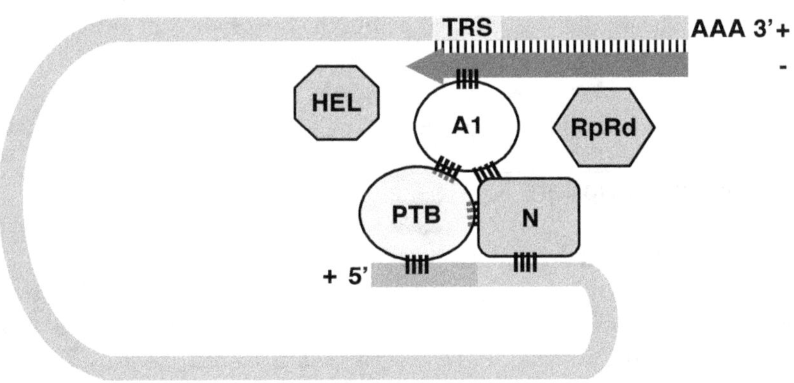

Figure 4.1 Potential formation of transcription complex. A postulated scheme on the interaction of viral and cellular components among themselves, and with the viral RNA, taken place in coronavirus transcription is shown. The medium grey and dark grey bars represent the genome and the nascent minus RNA strand. Light grey box, transcription regulating sequence (TRS). A1, chaperone protein hnRNAP A1; PTB, polypyrimidine tract binding protein, N nucleoprotein; RpRd, RNA-dependent RNA polymerase; HEL, helicase.

The requirement of N protein for virus replication and transcription has been under debate. Certain observations suggest that N protein plays a role in coronavirus RNA synthesis (Baric *et al.*, 1988; Compton *et al.*, 1987; Kim and Makino, 1995; Laude and Masters, 1995; Nelson *et al.*, 2000; Stohlman *et al.*, 1988), while others using either coronavirus (Thiel *et al.*, 2001) or arterivirus systems (Molenkamp *et al.*, 2000) claim that N protein is not essential. In the case of arteriviruses, a replicon constructed by reverse genetics was amplified in the absence of N protein. It would be interesting to determine whether the addition in trans of N protein would increase arterivirus replicon amplification. Two groups have shown that only background levels of coronavirus RNA synthesis are produced in the absence of N protein (Almazan *et al.*, 2004; Schelle *et al.*, 2005). In these systems, the presence either in cis or in trans of N protein is required for efficient coronavirus RNA synthesis (Almazan *et al.*, 2004). A quantitative analysis of TGEV replicon activity showed an increase of more than 100-fold when N protein was provided in cis, and more than 1000-fold when N protein was provided in trans. Whether the effect of N protein is at the level of replication, transcription or both, remains to be determined.

Despite the variable size of N protein from different coronaviruses, it presents a conserved pattern of secondary structural elements with a three-domain organization (Parker and Masters, 1990). N protein domains I and III are the most unstructured and divergent between coronaviruses, while domain II is more conserved and includes highly conserved alpha helices and a serine-rich domain (Britton, 1991; Calvo *et al.*, 2005). Several active N protein domains have been mapped, such as an RNA binding domain (Laude and Masters, 1995; Nelson *et al.*, 2000), the oligomerization domain (He *et al.*, 2004a; Surjit *et al.*, 2004), and the M protein binding domain (He *et al.*, 2004b; Hurst *et al.*, 2005), also involved in the N protein oligomerization. N protein activity should be a consequence of its self-interaction and interaction with other viral and cellular proteins and with virus and host cell nucleic acids. The phosphorylation state has been proposed to regulate N protein functions and to cause conformational changes in N protein structure (Chen *et al.*, 2005; Mohandas and Dales, 1991). TGEV (Calvo *et al.*, 2005) and IBV (Chen *et al.*, 2005) phosphoserine residues have been mapped within the coronavirus N protein primary and secondary structures. SARS-CoV N protein phosphorylation has also been demonstrated, but the precise amino acid residues involved have not been mapped (Surjit *et al.*, 2005). N protein phosphorylation could affect RNA binding activity and its biological function.

Coronavirus transcription

Introduction: coronavirus transcription

Coronavirus transcription, and in general transcription in the *Nidovirales* order, is an RNA-dependent RNA synthesis including a discontinuous step during the production of sg mRNAs (Lai and Cavanagh, 1997; Sawicki and Sawicki, 1998). This transcription process ultimately generates a nested set of sg mRNAs that are 5′- and 3′-coterminal with the virus genome. Three mechanisms of transcription have been described in the positive strand RNA viruses. In one of them, there is an *internal initiation* of the sg mRNAs using as template a negative copy of the viral genome, and was initially described for Brome mosaic virus (Miller *et al.*, 1985). The second model, *premature termination* of minus strand synthesis, takes place when the polymerase reaches a transcription regulating sequence

(TRS) in the genomic RNA (White, 2002). The third model shares with the second the premature termination of the RNA synthesis during the extension of the negative strand and, in addition, it has the distinctive peculiarity of being a *discontinuous process* by which a copy of the leader is added to the nascent RNA chain. This negative RNA of subgenomic length is used to generate sg mRNAs. Historically, the internal initiation model was proposed as the mechanism operating in coronavirus transcription (Baric *et al.*, 1983; Lai *et al.*, 1984; Spaan *et al.*, 1983). Later on, the seminal observation by D. Brian's group (Sethna *et al.*, 1989) with the identification of sg RNAs of negative polarity having the same length as the sg mRNAs, stimulated the proposal of the third model, discontinuous extension during the synthesis of the negative strand (Sawicki and Sawicki, 1998). Since then, this model has received strong support based on biochemical evidence using the MHV (Baric and Yount, 2000; Sawicki *et al.*, 2001), or genetic evidence using either the arterivirus model (den Boon *et al.*, 1996; Pasternak *et al.*, 2001; van Marle *et al.*, 1999a; van Marle *et al.*, 1995), or the coronavirus system (Sola *et al.*, 2005; Zuñiga *et al.*, 2004).

Working model for coronavirus transcription

The common 5′-terminal leader sequence, that in TGEV has 93 nucleotides (nt), is fused to the 5′ end of the mRNA coding sequence (body) by a discontinuous transcription mechanism. Sequences preceding each gene represent signals for the synthesis of sg mRNAs. These are the TRSs that include a conserved core sequence (CS), that in TGEV is 5′-CUAAAC-3′, identical for all genes (CS-B), and the 5′ and 3′ flanking sequences (5′ TRS and 3′ TRS, respectively) that regulate transcription (Alonso *et al.*, 2002; Sola *et al.*, 2005). As this CS sequence is also found at the 3′ end of the leader sequence (CS-L), it could basepair with the nascent minus strand complementary to each CS-B (cCS-B). In fact, the requirement for base pairing during transcription has been formally demonstrated in arteriviruses (Pasternak *et al.*, 2001; van Marle *et al.*, 1999a) and coronaviruses (Zuñiga *et al.*, 2004) by experiments in which base pairing between CS-L and the complement of CS-B was engineered in infectious genomic cDNAs. The data obtained are compatible with a transcription model including three steps (Figure 4.2) (i) formation of transcription initiation 5′–3′ complexes in the genomic RNA, (ii) slow down or stop and base pairing scanning of the nascent minus RNA strand by the TRS-L, and (iii) template switch to the leader region, during synthesis of the negative strand, to complete the minus sg RNA. This template switch takes place after copying the CS sequence and can be predicted *in silico* based on high base pairing score between the nascent minus RNA strand and the TRS-L, and minimum free energy (ΔG) of the duplex formation (Sola *et al.*, 2005; Zuñiga *et al.*, 2004). The synthesis of sg mRNAs proceeds only when the CS is located in an optimal sequence context (Alonso *et al.*, 2002; Ozdarendeli *et al.*, 2001; Sola *et al.*, 2005; Zuñiga *et al.*, 2004). It has been shown that base pairing between leader and body beyond the CS is a determinant regulatory factor in coronavirus transcription (Sola *et al.*, 2005). In TRS mutants with increasing complementarity between TRS-L and cTRS-B, a tendency to reach a plateau in ΔG values was observed, suggesting that a more precise definition of the TRS limits might be proposed, consisting of the central CS and around four nucleotides flanking 5′ and 3′ the CS (Sola *et al.*, 2005).

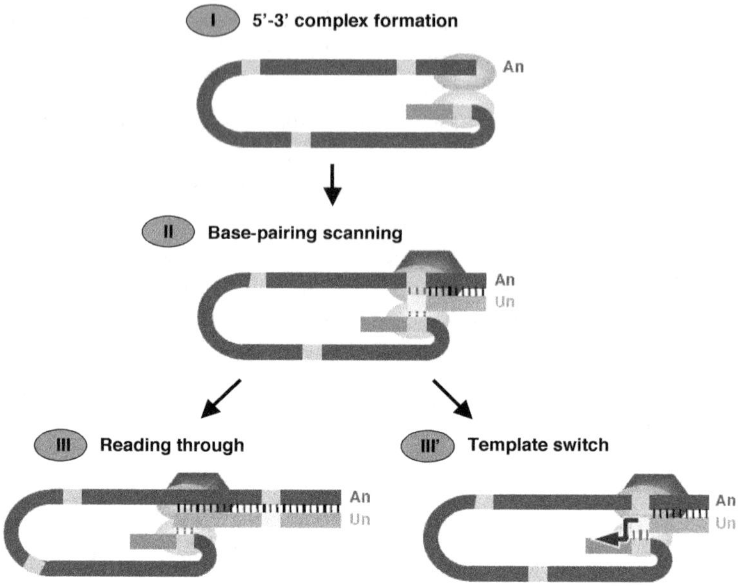

Figure 4.2 Three-step working model of coronavirus transcription. (I) 5′-3′ complex formation step. Proteins binding the 5′ and 3′ end TGEV sequences are represented by ellipsoids. Leader sequence is indicated with medium grey bar, CS sequences are indicated with a lighter grey bar. An, poly(A) tail. (II) Base pairing scanning step. Minus strand RNA is in a lighter grey compared with positive strand RNA. The transcription complex is represented by the hexagon. Vertical dotted bars represent the base pairing scanning by the TRS-L sequence in the transcription process. Vertical solid bars indicate complementarity between the genomic (gRNA) and the nascent minus strand. Un, poly(U) tail. (III) The synthesis of the negative strand can continue to make a longer sgRNA (III), or a template switch step can take place (III′) as indicated in the text. The thick arrow indicates the switch in the template made by the transcription complex to complete the synthesis of (-) sgRNA.

Formation of transcription pre-complexes

According to the proposed model of coronavirus transcription (Sola *et al.*, 2005; Zuñiga *et al.*, 2004), the first step could be the formation of a complex between the 5′ and 3′ genome ends. This step is required to bring into proximity the leader region and the newly synthesized cTRS-B minus RNA, facilitating the base pairing scanning between the leader TRS and the cTRS-B. Protein-RNA and protein-protein interactions probably mediate genome RNA ends interaction in transcription. PTB, hnRNPA1, and coronavirus N protein could help the formation and stabilization of 5′–3′ end complexes as described above (Figure 4.1).

The formation of a circular genomic structure by interaction of the 5′ and 3′ ends may be a general mechanism of initiating replication in positive-stranded RNA viruses, such as coronaviruses (Huang and Lai, 2001; Lai, 1998), polioviruses (Herold and Andino, 2001) and the pestivirus bovine viral diarrhea virus (BVDV) (Isken *et al.*, 2003). Replication of RNA virus genomes requires binding of the replicase complex to both 5′ and 3′ ends. A circularized ribonucleoprotein complex, consisting of RNA-protein-protein-RNA interactions is required within polioviruses to initiate minus strand RNA synthesis. In this viral system, an RNP complex formed around the 5′ cloverleaf RNA structure interacts with the poly(A) binding protein (PABP) bound to the 3′ poly(A) tail, circularizing viral RNA (Herold and Andino, 2001). In poliovirus infections, the switch between replication and translation is mediated by differential ternary ribonucleoprotein complexes. Binding of the cellular poly(rC) binding protein (PCBP) to the 5′ cloverleaf RNA structure enhances viral translation. In contrast, binding of the protease-polymerase precursor (3CD) promotes the synthesis of negative stranded RNA (Gamarnik and Andino, 1997; Parsley *et al.*, 1997). Circularization of BVDV genome is also mediated by the binding of three proteins of the NFAR group (NF90/NFAR-1, NF45, and RNA helicase A [RHA]) with the 5′ and 3′ UTRs (Isken *et al.*, 2003). Mutation of specific RNA 3′ end sites reduced both protein binding and RNA replication. Available data suggest that NFAR proteins mediate circularization of the viral RNA leading to a switch from translation to replication.

In coronaviruses, initiation of negative RNA synthesis from the genomic 3′ end could be a common process for replication and transcription. Switch from one function to another might be controlled by the binding of viral or cellular factors to replication or transcription complexes. It was described that vesicular stomatitis virus (VSV) replicase complex contains N protein, absent in transcriptase complexes. On the other hand, several cellular factors were only present in the transcriptase complex (Qanungo *et al.*, 2004). Viral proteins such as nsp2 and nsp3 modulate sg mRNA synthesis in alphaviruses (Miller and Koev, 2000), while proteins nsp1 and nsp10 influence RNA transcription in the EAV arterivirus (Tijms *et al.*, 2001; van Marle *et al.*, 1999b).

Predicted steps required for template switch during sg RNA synthesis

Within the template switch, different steps could be differentiated: (i) slow down or stop of RNA synthesis; (ii) template switch itself; (iii) reassociation of the nascent minus RNA strand with the leader TRS; and (iv) elongation of the minus strand using the leader as template.

RdRp pause and stop signals

A fundamental aspect of the discontinuous transcription model is the identification of RNA signals responsible for RdRp slow down or stop before template switch. These signals must be located around body TRS sequences preceding each transcriptional unit in the viral genome. The termination of nascent strand synthesis might be mediated by primary or secondary RNA structures located either in the template or in the nascent strand, as described for other prokaryotic (Wilson and von Hippel, 1995) or eukaryotic (Reeder and Lang, 1997) RNA polymerases. Alternatively, specific regulatory proteins might interact with these structures promoting the pause or stop of viral RdRp. Since the processivity of the RdRp complex is unknown, both alternatives, pause or stop of RdRp, are possible in coronavirus discontinuous transcription. At the postulated pausing site, the 3′ end of the nascent minus RNA strand would include the cCS-B sequence, complementary to the CS-L. A three RNA strand intermediate including template RNA, nascent minus RNA, and leader TRS RNA could be formed before template switch takes place. Then, the transcription complex could evaluate the base pairing between the nascent minus RNA strand and the leader TRS. If the ΔG associated to the formation of this duplex has reached a minimum threshold, template switching would take place. Otherwise, transcription complex would continue the synthesis of a longer minus sg RNA. Since the leader-body junction in coronavirus sg RNAs takes place at the CS, it can be postulated that this hexanucleotide, within the TRS, together with the immediate flanking sequences, act as the signal promoting the dissociation of the nascent minus RNA chain from the genomic RNA template, in order to stimulate a strand transfer to the leader region. This model would explain the results obtained in the TGEV (Sola *et al.*, 2005; Zuñiga *et al.*, 2004), and the bovine coronavirus (BCoV) showing that sequences downstream the CS exert stronger influence on the template-switching decision (Ozdarendeli *et al.*, 2001). In fact, both the CS and the TRS 3′-flanking sequences combine their complementarity to decide whether template switch would take place. Most likely, the additional complementarity provided within the 3′ TRS-B flank contributes to reach the ΔG threshold, leading to a template-switch at the CS site with a higher frequency. In contrast, if complementarity with the leader TRS is only extended towards the 5′ TRS flanking sequences, when the transcription complex arrives at the CS, the overall complementarity between nascent minus RNA and TRS-L will be mostly dependent on the six CS nucleotides. Therefore, strand transfer at this position will be a less frequent event, reducing but not excluding a role for the 5′ TRS flanking sequences in template switch.

TRS RNA secondary structure also could be involved in stopping of the transcription complex and dissociation of the minus RNA from the first template. Nidovirus transcription, particularly in the case of torovirus non-discontinuous transcription (Smits *et al.*, 2005), resembles the premature transcription termination mechanism of tombosviruses (Lin and White, 2004; White, 2002; White and Nagy, 2004). In these viruses, it has been proposed that RNA secondary structures formed in the genome, upstream to the transcription initiation sites, could facilitate transcription termination (White and Nagy, 2004). The presence of transcription attenuating structures located upstream of body TRSs has been recently proposed (Curtis *et al.*, 2004). Prediction of RNA structural motifs at the 5′ flanking sequences of the TRSs preceding TGEV genes has shown that a stem–loop is present not far from the CS in all cases. Furthermore, the predictions have also shown stem–loop structures at the 5′ TRSs associated with non-canonical CS sequences that, in some cases, generate sg RNAs (Zuñiga *et al.*, 2004). This kind of structure could be a stalling element for the transcription complex, similarly to what has been described for bacteria intrinsic terminators (Yarnell and Roberts, 1999).

The transcription complex most likely includes a helicase activity that could be involved in the dissociation of the nascent minus RNA chain from the template RNA. In contrast, the possibility of the removal of secondary structures during RNA synthesis by the helicase seems unlikely, as nidovirus helicases, encoded by the viral nsp13 (coronavirus) or nsp10 (arterivirus) proteins (Gorbalenya *et al.*, 1989; Ziebuhr, 2005), have a 5′-to-3′ polarity (Ivanov *et al.*, 2004; Seybert *et al.*, 2000a; Seybert *et al.*, 2000b). Additional information is required to determine whether 5′-TRS secondary structures are formed *in vivo* and their importance in RdRp stalling during coronavirus transcription.

Template switch

Coronavirus transcription resembles a copy-choice similarity-assisted recombination process (Pasternak 2001), being the leader TRS the acceptor RNA and the body TRS the donor RNA. A copy-choice mechanism implies that the transcription complex leaves the first template (the donor) and continues synthesis on a second template (the acceptor) (Kim and Kao, 2001). In similarity-assisted recombination, sequence similarity between RNAs influences the frequency and site of the recombination event, but additional RNA determinants, generally only present in one of the RNAs, are essential for efficient recombination (Nagy and Simon, 1997). In turnip crinkle virus (TCV), efficient recombination needs RNA sequence similarity and a hairpin structure present only in the acceptor RNA (Nagy *et al.*, 1998). In arteriviruses, it has been described that the leader TRS hairpin (LTH) stem–loop structure, present at the 5′ end of the genomic RNA, is necessary for efficient sgRNA synthesis (van den Born *et al.*, 2005). Results obtained with coronaviruses also indicate the importance of the LTH structure (Zuñiga *et al.*, 2004).

Template switch during coronavirus transcription could be interpreted as a transition between two states. In the first one, a duplex between the nascent minus RNA strand and the genomic positive RNA used as template is formed. In the second one, the nascent RNA strand is paired with the leader TRS. RNA chaperones could be involved in template switch by decreasing the energy required for the transition from the first to the second duplex. RNA chaperones are non-specific nucleic acid binding proteins that rescue RNAs trapped in unproductive folding states (Cristofari and Darlix, 2002; Herschlag, 1995; Lorsch, 2002; Schroeder *et al.*, 2004). A series of *in vitro* assays have been used to determine the RNA chaperone activity of a protein: (i) enhancement of ribozyme cleavage, (ii) allowance of rapid and accurate RNA annealing, and (iii) facilitation of RNA strand transfer and exchange (Cristofari and Darlix, 2002). From these activities, the first one has the strongest chaperone requirements (Rajkowitsch *et al.*, 2005). RNA chaperone activity cannot be predicted based on domain conservation. Nevertheless, RNA chaperones are proteins with the highest frequency of long intrinsically disordered regions (Ivanyi-Nagy *et al.*, 2005).

The importance of coronavirus N protein in RNA synthesis (Almazan *et al.*, 2004), its characteristics as non-specific RNA binding protein, and its intrinsic disorder, led us to investigate the potential nucleic acid chaperone activity of TGEV N protein. This protein enhanced RNA ribozyme self-cleavage (Figure 4.3) and nucleic acid annealing similarly as hnRNP A1, a cellular RNA chaperone, does (Zuñiga *et al.*, 2006). In addition, SARS-CoV N protein also displayed RNA chaperone activity (Zuñiga *et al.*, 2006). Therefore, nucleic acid chaperone activity could be a common property of all coronavirus N proteins.

Previous to the identification of coronavirus N proteins as RNA chaperones (Zuñiga *et al.*, 2006), three other viral RNA chaperones were already described, all of them being nucleocapsid proteins of RNA viruses: retrovirus (Bertrand and Rossi, 1994; Tsuchihashi and Brown, 1994), hepatitis delta virus (Huang *et al.*, 2003; Huang and Wu, 1998), and hepatitis C virus (Cristofari *et al.*, 2004). This RNA chaperone list is quickly growing with the recent inclusion of two other proteins, nucleocapsid from hantavirus (Mir and Panganiban, 2006), and 3AB from poliovirus (DeStefano and Titilope, 2006). The identification of chaperone activity in coronaviruses, and its potential involvement in template switch during coronavirus transcription links thermodynamic and molecular aspects of coronavirus transcription (Zuñiga *et al.*, 2006).

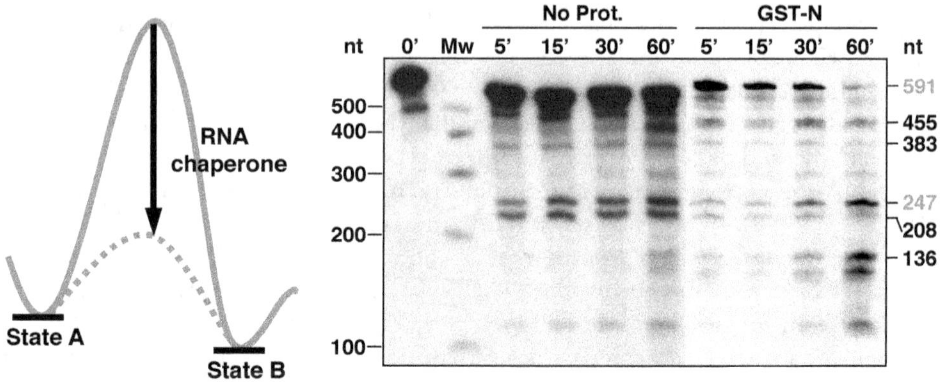

Figure 4.3 RNA chaperone activity of TGEV N protein. Left panel illustrates the action of an RNA chaperone decreasing the activation energy needed for a transition between two states. Right panel, hammerhead ribozyme self-cleavage. A time-course in cleavage conditions is presented in the absence or the presence of N protein. Molecular weight markers are indicated on the left of the gel. Cleavage products are indicated in the right of the gel, uncleaved and completely cleaved products are indicated in grey.

Secondary RNA structures play an important role in nidovirales transcription

Nidovirus genome replication and sg RNA transcription initiate at the 3′ end of the genomic RNA (gRNA). Therefore, it is expected that this region would contain at least part of the sequences or structures involved in the regulation of these processes. Among the sequences from the 3′ UTR needed for sg RNA synthesis, a predicted RNA pseudoknot structure, highly conserved in different coronaviruses, is required. This structure was also important for replication (Brian and Baric, 2005; Lin *et al.*, 1996; Williams *et al.*, 1999). At the 3′ UTR of the viral genome, in addition to the pseudoknot, a potential stem–loop is also present, although only one of these structures is present at a given time. A molecular switch between these structures has been described that could affect transcription (Goebel *et al.*, 2004), but the role of this switch in coronavirus RNA synthesis remains to be elucidated.

There is a well-established requirement of 5′-end sequences and RNA structures for viral replication (Brian and Baric, 2005). Previous publications using the MHV system showed that the leader RNA of MHV contain an enhancer-like element for sg mRNA synthesis (Wang and Zhang, 2000). In addition, according to the current model of coronavirus transcription (Pasternak *et al.*, 2001; Sawicki and Sawicki, 1998; Sola *et al.*, 2005; Zuñiga *et al.*, 2004), there is a requirement for the leader TRS, that should be in an adequate structural context to serve as the acceptor for the nascent minus RNA strand transfer (Figure 4.4). In fact, a stem–loop structure (LTH) was predicted for several nidoviruses and demonstrated for arterivirus EAV (van den Born *et al.*, 2004). This kind of structure generally exposes the leader conserved CS in the loop in different coronaviruses facilitating the interaction with the complementary sequence cCS-B (Figure 4.5). The LTH structure of EAV is required for efficient sg mRNA synthesis (van den Born *et al.*, 2005), as shown by using an infectious cDNA clone in which the stem loop and flanking sequences were duplicated and moved 300 nucleotides downstream of its original position in the genome

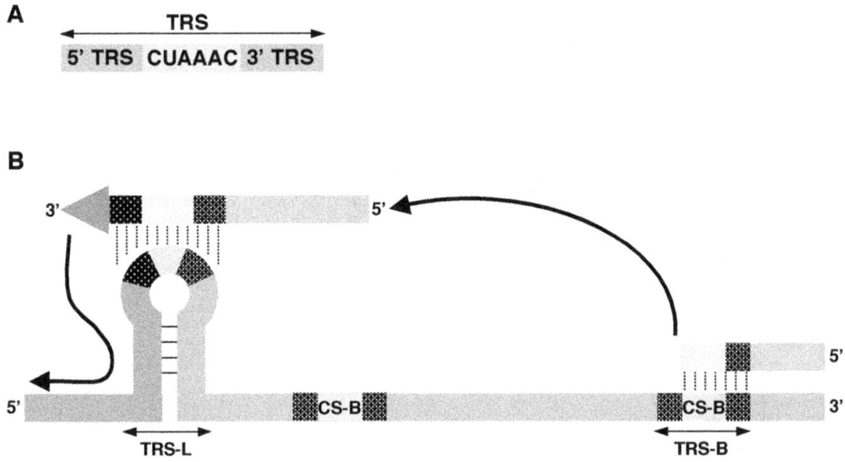

Figure 4.4 TRS and LTH in the context of full-length genome. (A) Structure of transcription regulating sequences (TRS) in coronaviruses. The conserved core sequence (CS), 5′-CUAAAC-3′ in TGEV, identical for all genes (CS-B) and also present at the 3′ end of the leader, is flanked by 5′ and 3′ flanking sequences (5′ TRS and 3′ TRS, respectively). (B) The leader TRS (TRS-L) in nidovirales is probably located in a hairpin loop structure, as supported by phylogenetic analysis, structure probing and site-directed mutagenesis (van den Born *et al.*, 2004; van den Born *et al.*, 2005). Body TRS sequences (TRS-B) would act as slow down or stop signals for RdRp. The 3′ end of the nascent minus RNA strand would include the cCS-B sequence, complementary to the CS-L. The presence of the CS-L in a hairpin structure in the acceptor strand may be necessary for strand transfer associated to discontinuous transcription. Following the template switch to the TRS-L, the RdRp completes the synthesis of minus sg RNA by adding the complement of the leader.

(van den Born *et al.*, 2004; van den Born *et al.*, 2005). Mutagenesis experiments within the duplicated LTH domain have shown that the LTH and its immediate flanking sequences form an independent functional entity that is essential for efficient sg mRNA synthesis.

Early studies on coronavirus transcription using DI RNAs from MHV suggested the formation of RNA secondary TRS structures preceding each gene (TRS-B) important for the generation of sg RNAs (An and Makino, 1998). These structures would include the CS and both upstream and downstream flanking sequences. It has been postulated that active TRS-B sequences tend to be in single-stranded regions, and inactive TRS-Bs generally are in predicted base paired domains. Nevertheless, (Ozdarendeli *et al.*, 2001; Pasternak *et al.*, 2000) most of the experimental data indicate that secondary RNA structure probably is not the primary determinant in the choice of a TRS for leader to body fusion (Alonso *et al.*, 2002; Ozdarendeli *et al.*, 2001). In addition, no conserved secondary structure has been found among the TRS-Bs of the same viral isolate (TGEV) (Figure 4.6), suggesting that primary TRS-B structure is more relevant in transcription than its secondary structure. This observation has been recently demonstrated for TGEV, as the activation of a previously inactive TRS-B was reached just by increasing its complementarity with the TRS-L (Sola *et al.*, 2005).

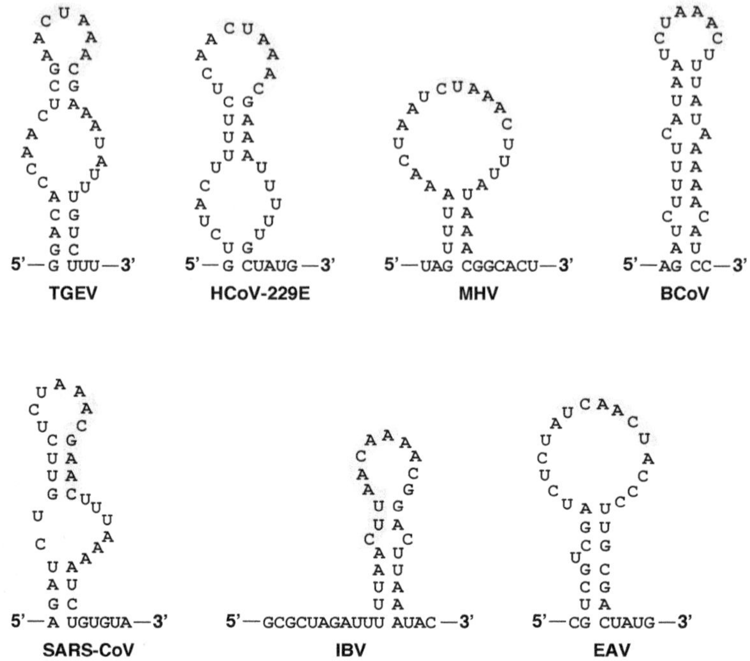

Figure 4.5 Predicted structure of TRS-L in different coronaviruses. The LTH structures obtained for different coronaviruses are indicated in the figure, compared with the LTH predicted for EAV arterivirus. Predictions were made with m-fold 3.2 (Mathews *et al.*, 1999; Zuker, 2003) using a 36 nt sequence centered in the CS-L (grey box). Sequences used were from TGEV (GeneBank accession number AJ271965), HCoV-229E (AF304460), MHV strain A59 (AF029248), BCoV (AF391541), SARS-CoV Urbani strain (AJ278741), IBV Beaudette strain (M95169) and EAV arterivirus (X53459).

Alternative transcription models

One alternative to the formation of 5′-end-3′-end complexes mediated by protein-RNA and protein-protein interactions, which locate the TRS-L in close proximity to sequences at the 3′ end of genomic RNA, is the formation of a collection of complexes between the TRS-L and each of the TRS-Bs present along the coronavirus genome. Each coronavirus RNA genome would be forming a different complex at a given time. These complexes should be pre-formed based on sequence composition and protein-RNA interactions. The presence of each of these complexes could be a requirement for the synthesis of the sg mRNAs. These complexes could act as a slow down or stop signal to promote the detachment of the nascent minus RNA chain together with the transcriptase complex. Nevertheless, the relative amount of the sg mRNAs would be determined by the ΔG of the TRS-L with cTRS-B duplex formation.

Two variant transcription models could be postulated depending on whether transcription complex detachment or base pairing of the nascent minus RNA strand with the leader TRS takes place first during transcription. Both transcription models postulate that in arteriviruses and coronaviruses the discontinuous synthesis of sg RNAs takes place during minus strand synthesis (Hussain *et al.*, 2005; Pasternak *et al.*, 2001; Zuñiga *et al.*,

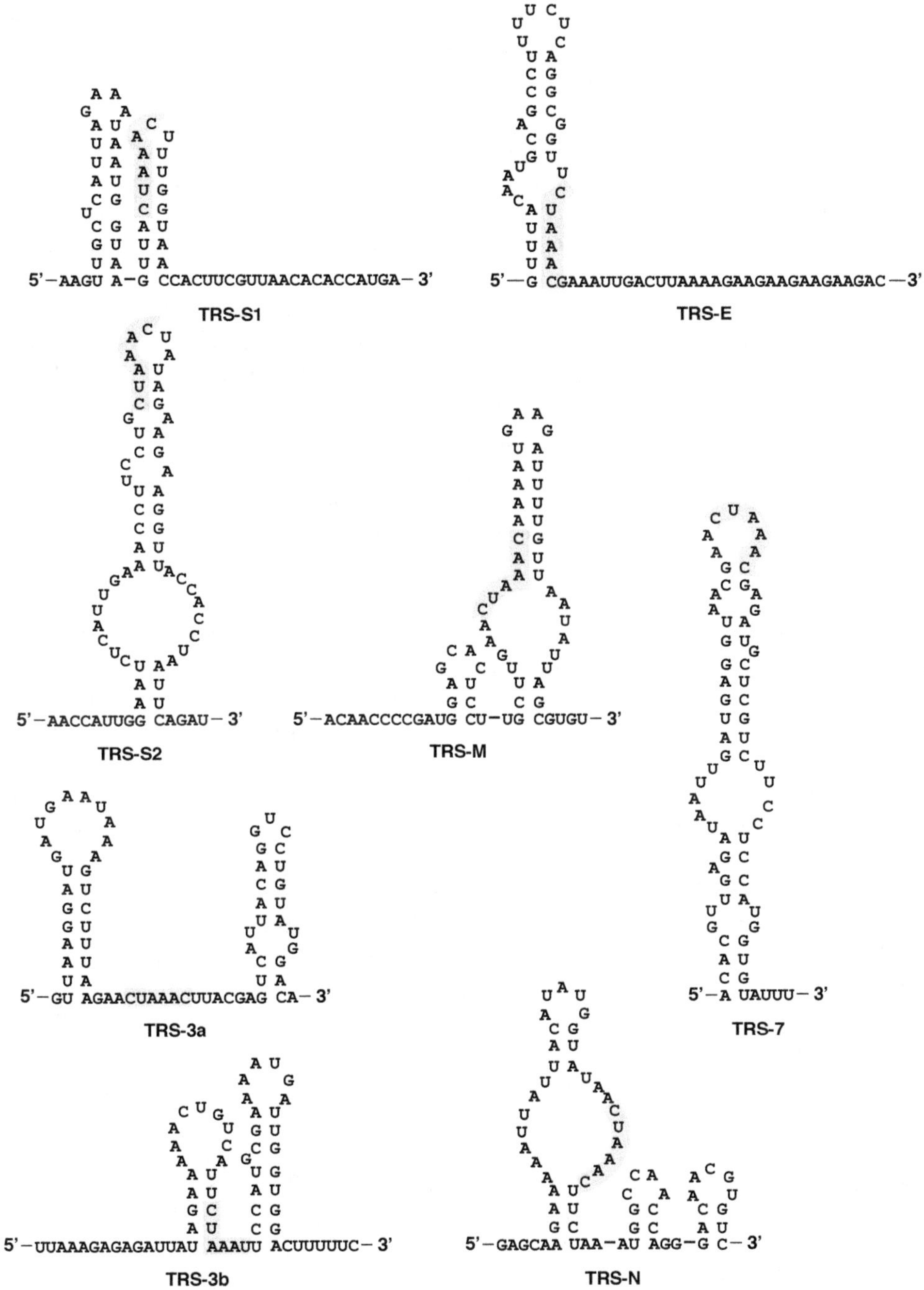

Figure 4.6 Predicted structure of TRS-Bs in TGEV. The predicted structures for TRS-Bs from TGEV are presented in the figure. Predictions were made with m-fold 3.2 (Mathews *et al.*, 1999; Zuker, 2003), using a sequence of 66 nt centered in the CS-B (grey box).

2004). Also, a common feature of both models, could be that the low stability of the top of the LTH stem, characterized by a large number of A-U base pairs and the presence of a bulge below the LTH loop could promote the unwinding of the LTH structure required for its use as template (van den Born *et al.*, 2004).

In the arterivirus working model, it is postulated that detachment of the nascent minus RNA strand takes place prior to the reassociation with the LTH. In contrast, in the TGEV coronavirus model (Sola *et al.*, 2005; Zuñiga *et al.*, 2004), a variant mechanism has been postulated. According to this model, during template switch, the first step would be the formation of an intermediate involving three RNA strands: the genome template, the nascent minus strand and, the leader TRS. The physical proximity between the nascent minus RNA strand and leader TRS would allow the evaluation of the potential base pairing between TRS-L and cTRS-B. If the ΔG associated to the TRS-L and cTRS-B duplex formation reaches a minimum threshold, template switching from the genome to the leader template would take place. Additional experimental evidence is needed to determine which of these working models might be correct. Identification of the correct model would require the engineering of virus mutants accumulating transcription intermediates that should be sufficiently stable to allow their detection.

Relative value of factors affecting transcription

Multiple factors seem to regulate the transcription process (Alonso *et al.*, 2002; Pasternak *et al.*, 2001; Pasternak *et al.*, 2003). One of them, probably the most important one, is the *complementarity* between TRS-L and cTRS-B (Sola *et al.*, 2005; Zuñiga *et al.*, 2004). Previous studies with coronavirus DI RNAs also support the concept that base pairing within the transcription-regulating sequences is part of the signal promoting the template switch event (Hiscox *et al.*, 1995; Hsue and Masters, 1999; Makino *et al.*, 1991; Ozdarendeli *et al.*, 2001; van Marle *et al.*, 1995). In fact, an *in silico* analysis method used to quantify the identity between the genomic RNA and the TRS-L (Huang and Miller, 1991), as an estimation of base pairing between TRS-L and cTRS-B during viral transcription, resulted reliable for the prediction of most sg mRNAs synthesized in TGEV, HCoV-229E and BCoV, including those derived from canonical and non-canonical CSs (Zuñiga *et al.*, 2004). Similarly, in the arterivirus EAV, the requirement for base pairing during transcription has been formally demonstrated (Pasternak *et al.*, 2001; Pasternak *et al.*, 2003; van Marle *et al.*, 1999a).

In the arteriviruses, it has been proposed that discontinuous RNA synthesis also seems to depend on the *primary sequence of the body TRS*, that is, on the chemistry of each nucleotide within the CS (Pasternak *et al.*, 2001). From an extensive site-directed co-variation mutagenesis study of the leader CS and RNA7 body CS, it was observed that the effects of single CS mutations were mostly base-specific, i.e. different nucleotide substitutions at the same position affected sg mRNA synthesis to different extent. In some cases, body CS mutations appear to be dominant over the duplex-restoring effects in the double mutants.

In contrast to these results, co-variation analysis of leader-body CSs in the TGEV coronavirus system showed that synthesis of sg mRNAs was partially or completely restored when leader-body duplex was restored in double mutants. In general, sg RNA synthesis could be explained by the stability (expressed as ΔG) of the duplex between CS-L and cCS-B. Differences in sg mRNA synthesis associated to mutations in the first and last CS positions could be interpreted in terms of stabilizing the formation of the duplex between the exposed CS-L and the cCS-B (Zuñiga *et al.*, 2004). Mutational analysis of the body CS in the MHV coronavirus, using DI RNAs, also support the idea that leader-body CS complementarity is the most critical factor for transcription (Joo and Makino, 1992; van der Most *et al.*, 1994).

An additional factor, the *secondary structure of body TRS* regions might regulate transcription activity as proposed for arteriviruses (Pasternak *et al.*, 2000). However, there is no conclusive data confirming the relevance of TRS-B secondary structure in regulating transcription efficiency. Furthermore, in the bovine coronavirus, it was shown that secondary structure surrounding a TRS motif might not be a primary determinant of the activity of that TRS (Ozdarendeli *et al.*, 2001). *The RNA structure of the nascent minus RNA* might influence pausing or strand transfer as shown for termination of transcription by stable hairpin structures in *E. coli* RNA polymerase (Wilson and von Hippel, 1995).

Viral and host components involved in protein-RNA and protein-protein recognition probably are also decisive in transcription (Lai, 1998). Differences in affinity of such factors for body TRSs might regulate transcription in coronaviruses. The body TRS might be recognized by viral or cellular proteins mediating the pause of minus RNA strand synthesis or the subsequent strand transfer to the TRS-L template. A similar mechanism has been described in the DNA-dependent RNA-polymerase I termination system, in which specific proteins bind to termination sequences (Reeder and Lang, 1997).

The *proximity of body TRS to the 3′ end of the genome* might also regulate sg RNA synthesis. This is the case for all viruses that produce multiple sg mRNAs (Miller and Koev, 2000) and, in general, for coronaviruses and arteriviruses. During infection, RNA species produced from more 3′-proximal TRSs are usually more abundant than the more distal ones in accordance with the proposed role of the body TRS as an attenuator of minus-strand synthesis (de Vries *et al.*, 1990; Lai and Cavanagh, 1997; Penzes *et al.*, 2001). However, similar results were observed for brome mosaic virus, in which sg mRNA synthesis is not mediated by an attenuation mechanism (Miller *et al.*, 1985). Both results might be explained by differences in the time required to synthesize sg mRNAs of different sizes. In the EAV arterivirus system, it was not possible to discriminate the influence in sg mRNA abundance between the relative order of body TRSs in the genome and the size differences between the sg mRNAs (Pasternak *et al.*, 2004). Furthermore, in TGEV generally the ΔG of the duplex formation between TRS-L and cTRS-B is more favorable for more 3′-proximal TRSs, making it very difficult to dissect the effects of ΔG of the duplex formation and TRS proximity to the 3′ end on sg RNA synthesis.

The precise relevance in transcription of the different factors considered above, i.e. sequence complementarity, TRS nucleotide chemical nature, TRS-B secondary structure, protein-RNA interaction, and proximity to the 3′ end has not been fully established. Nevertheless, in both, the TGEV and the EAV systems, complementarity between TRS-L and cTRS-B seems to be the most relevant factor.

Transcription enhancement in coronavirus

The ΔG of duplex formation between TGEV TRS-L and cTRS-B is determinant in coronavirus transcription as described above, but other factors must be involved. In fact, in this system a good correlation has been observed between the levels of sg mRNAs and their ΔG values for duplex formation with TRS-L. A clear exception to this postulate, the atypically high production of TGEV sg mRNA N, has been identified. N gene sg mRNA is the most abundant one synthesized during infection, although the ΔG for the duplex between its cTRS-B and TRS-L has one of the lowest values. In other coronaviruses, such as MHV, sg mRNA N is also the most abundant viral sg mRNA. In MHV, N gene is the most 3′ end gene in the genome and, therefore, a positional effect in sg mRNA N transcription cannot be excluded. In contrast, in the TGEV genome, gene 7, located downstream of the N gene, is the most 3′ end gene. Nevertheless, mRNA 7 is less abundant than mRNA N, suggesting that transcriptional activation of mRNA N is modulated by an additional mechanism.

In principle, several factors might explain the highest levels of sg mRNA N in TGEV: (i) an increased stability of mRNA N. Studies performed by metabolic ^{33}P-labeling of sg mRNAs showed that mRNA N stability is similar to that of the other viral sg mRNAs during TGEV infection, making unlikely a role for stability of sg mRNA in this process; and (ii) the presence of RNA *cis*-acting elements into the coding or transcriptional regulating sequences of N gene. This postulate would explain why the up-regulation of mRNA N levels exclusively affects to this mRNA. The sequences involved in the transcriptional activation of mRNA N could be similar to the enhancer-like activity previously described for other positive-stranded RNA viruses, like brome mosaic virus (Ranjith-Kumar *et al.*, 2003) and turnip crinkle virus (Nagy *et al.*, 1999). A variant of this model is the production of *trans*-acting factors, such as cellular and viral proteins, that specifically interact with genome RNA sequences regulating mRNA production. A similar protein-RNA interaction has been postulated for tombosvirus sg mRNA transcription (White and Nagy, 2004). In this virus, the protein encoded by the sg mRNA controls the transcriptional process by a feedback-type mechanism. It cannot be excluded that several of the proposed factors are simultaneously operating. Analysis of the elements regulating transcriptional activation of mRNA N in TGEV is in progress.

Proofreading systems and transcription

Bioinformatics studies on the conserved domains in the C-terminal region of coronavirus pp1ab showed the presence of new RNA processing enzymes, that are only found in the *Nidovirales* and include putative 3′-to-5′ exoribonuclease (ExoN, or nsp14), endoribonuclease (NendoU, or nsp15), and 2′-O-ribose methyltransferase (2′-O-MT, or nsp16) (Snijder *et al.*, 2003; von Grotthuss *et al.*, 2003; Yan *et al.*, 2003). All three domains are conserved in coronaviruses, toroviruses and roniviruses, but only the NendoU domain is also conserved in arteriviruses (nsp11). NendoU is considered a major genetic marker that discriminates nidoviruses from all other RNA virus families. Recent studies have confirmed the *in vitro* endonuclease activity of NendoU in the SARS-CoV, HCoV-229E, MHV and IBV, as well as a critical function in HCoV-229E RNA synthesis (Bhardwaj *et al.*, 2004; Ivanov *et al.*, 2004a). Similarly, it has also been demonstrated that the arterivirus NendoU is involved in viral RNA synthesis and in additional steps of the virus life cycle (Posthuma *et al.*, 2006). According to its conservation and specificity in the order Nidovirales, the NendoU domain could be involved in a nidovirus-specific, conserved function such as synthesis of sg mRNAs. It has been suggested that NendoU may play a role by specifically cutting double stranded RNA (transcriptive intermediates) generated during the synthesis of the nascent RNA of negative polarity. Coronavirus NendoU nuclease has a strong preference to cleave at GU(U) sequences in double stranded RNA substrates (Ivanov *et al.*, 2004a). The GU(U) sequence at the 3′ terminus of nascent minus-strand RNAs, which corresponds to conserved AAC nucleotides in the core of the coronavirus gene TRS elements, might be the substrate of this activity. Data from our laboratory, in which around 90 different sg RNAs generated during the mutagenesis of a TGEV CS were analyzed (Zuñiga *et al.*, 2004), and also from other laboratory (Schelle *et al.*, 2005), support the functional relevance of the AAC sequence in transcription, but further studies are required to provide a direct link between transcription and enzymes such as NendoU.

Homology of the coronavirus ExoN activity with some cellular DNA-processing activities (like exonuclease I and the exonuclease domain of DNA polymerases) has led to speculate a link of this enzymatic activity with RNA proofreading, repair, or recombination. Proofreading and repair are unknown activities in RNA viruses and could represent a particular adaptation of coronavirus extremely large genomes to biosynthesis and maintenance (Snijder *et al.*, 2003). This would be a unique feature among RNA viruses, characterized by a high error frequency of their RdRp and an apparent absence of proofreading systems. Thus, those replicase subunits that are unique in coronaviruses and are not represented in other positive stranded RNA viruses might be candidates for a proofreading mechanism.

Transcription mechanism and nidovirales evolution

The evolution of nidovirales transcription mechanism does not seem to be associated to the evolution of their genome size (Gorbalenya *et al.*, 2006). In fact, discontinuous sg mRNA synthesis takes place in arteri- and coronaviruses but not in roniviruses, while in toroviruses only sg mRNA 2 is produced by discontinuous synthesis, but not the rest of their sg mRNAs. The closteroviruses, with an intermediate size between that of arteriviruses and toro- and coronaviruses and a nested set of sg mRNAs, do not employ discontinuous transcription and have no TRS-like common elements in the region that directs sg mRNA synthesis (Dolja *et al.*, 2006; Karasev *et al.*, 1997; Kreuze *et al.*, 2002; Peremyslov and Dolja, 2002). Although the transcription model in closteroviruses has not been fully defined, the description of complementary minus strands of the sg RNAs in closterovirus-infected cells (Dolja *et al.*, 1990; Hilf *et al.*, 1995) is compatible with the nidovirus-like, minus-strand termination model (Dolja *et al.*, 2006). In addition, the discontinuous transcription observed in arteri- and coronaviruses could be considered a variant of the similarity-assisted template switch recombination that has been observed with small genomes and relatively simple replicase genes (Gorbalenya *et al.*, 2006; Lai, 1992). Therefore, the evolution of the transcription mechanism is not directly linked to the expansion of the genome.

The direction of the evolution in the nidovirales and related closteroviruses is not known. Nevertheless, it could be speculated that RNA viruses expand rather shrink their genomes (Gorbalenya *et al.*, 2006). Toroviruses could have represented a divergence point in the evolution of transcription mechanism, leading to the roniviruses in one direction, and the coronaviruses in the other one. Closteroviruses represent an evolutionary branch differentiated from that within the nidovirales (Dolja *et al.*, 2006).

Termination of the RNA minus strand seems related in the closteroviruses and toroviruses to the presence of small hairpin structures (Dolja *et al.*, 2006; van Vliet *et al.*, 2002), but the precise mechanism involved is not known. In fact, many questions remain to be addressed within the nidovirales transcription, and probably more in the case of no leader addition to the sg mRNA. Is the transcription stop signal in toroviruses and roniviruses conserved within the different genes of a virus? Does the 5′ end of the genome influence the synthesis of sg mRNAs in these viruses? Is there a leader TRS in the ronivirales? How divergent are the two different transcription models required within the nidovirales, depending on whether their sg mRNA production is based on continuous or discontinuous RNA synthesis?

The evolution from the arteriviruses to the large sized coronaviruses and roniviruses may have been accompanied by the acquisition of the ExoN domain (Snijder *et al.*, 2003) and a gradual increase of 14 kb in genome size. The ExoN domain may have been acquired to improve the low fidelity of the RdRp during RNA replication through its 3′ to 5′ exonuclease activity (Minskaia *et al.*, 2006).

Viral and cellular proteins involved in transcription

The identification of host factors participating in viral genome replication and expression has been based on specific binding assays, such as binding of factors to replicase or to selected regions of the genome, or on two functional assays, such as identification of host cell mutants affecting viral RNA amplification and expression, or inhibition of replication or transcription by small interfering RNAs (siRNAs), reducing expression of candidate proteins. One of the largest difficulties in the identification of factors affecting replication or transcription is to distinguish between direct and indirect effects. In addition, frequently, factors affecting RNA synthesis also affect translation (Boguszewska-Chachulska and Haenni, 2005). Binding to replicase alone does not guarantee involvement in RNA synthesis. Therefore, it becomes essential to study the role of factors in functional assays, such as *in vivo* or *in vitro* genome replication or transcription.

According to the working model of transcription proposed by our laboratory (Figure 4.2), the first transcription step would be the formation of a complex in which the leader TRS and the 3′ end of the genome interact with a transcription complex formed by viral and cellular proteins. Candidate proteins potentially forming the transcription complex have been reported by several laboratories. On the viral side, essential proteins in transcription should be the RNA-dependent RNA-polymerase (RdRp), the helicase (Hel), and possibly N protein (see above). In addition, other viral replicase proteins probably contribute to the regulation of transcription and are most likely involved in the formation of this complex. In fact, MHV RdRp (nsp12) interacts with other replicase-encoded products such as nsp5 (3CLpro), nsp8, and nsp9 (Brockway *et al.*, 2003). Biochemical studies have shown that SARS-CoV nsp9 is a single-stranded RNA binding protein (Egloff *et al.*, 2004; Sutton *et al.*, 2004). In addition, MHV nsp10 has been implicated in minus-strand RNA synthesis (Siddell *et al.*, 2001).

EAV nsp1 has been involved in arterivirus transcription (Tijms *et al.*, 2001; van Dinten *et al.*, 1997) and interacts with cellular transcription activating protein p100 (Tijms and Snijder, 2003). It has been speculated that this interaction might be important for viral sg RNA synthesis, either directly or by recruiting a p100 binding protein to the viral RdRp complex. Both nsp1 and Hel contain potential zinc-binding domains that could mediate RNA-protein interactions (Tijms *et al.*, 2001; van Dinten *et al.*, 2000). In fact, mutagenesis in the nsp1 and Hel zinc-binding domains affects transcription regulation.

In addition to viral, cell proteins most likely play a role in coronavirus transcription regulation. It has been demonstrated that regulatory regions of coronavirus genomic RNA interact with at least three proteins: hnRNP A1 (Li *et al.*, 1997; Shi *et al.*, 2000), PTB (Huang and Lai, 1999; Li *et al.*, 1999), and viral N protein (Baric *et al.*, 1988; Stohlman *et al.*, 1988). These proteins may act as mediators between the leader TRS and the transcription complex at the body TRS. In fact, binding between hnRNP A1 and PTB (Huang and Lai, 2001), hnRNP A1 and N protein (Wang and Zhang, 1999), and PTB and N protein (Choi *et al.*, 2002) have been documented. Furthermore, there are data directly involving hnRNP A1 and PTB in coronavirus transcription. hnRNP A1 has been identified as a major protein species binding to nt 90–170 (strongly) and nt 260–350 (weakly) from the 3′ end of MHV RNA. These binding sites are complementary to the sites on the negative-strand RNA that bind another cellular protein, PTB. Furthermore, hnRNP A1 and PTB also bind to the complementary strands at the 5′ end of MHV RNA and together mediate the formation *in vitro* of a ribonucleoprotein (RNP) complex involving the 5′ and 3′ ends of MHV RNA (Huang and Lai, 2001; Li *et al.*, 1999; Li *et al.*, 1997). The functional relevance of hnRNP A1 in MHV replication has been established by experiments showing that over-expression of hnRNP A1 facilitates MHV replication, whereas dominant-negative mutants of hnRNP A1 reduce replication (Shi *et al.*, 2000). Furthermore, mutations of the hnRNP A1 and PTB binding regions also impaired transcription of subgenomic RNA (Huang and Lai, 1999), suggesting a functional role in replication and transcription for the interaction between hnRNP A1 and PTB.

PTB binds to a sequence domain of four UCUAA repeats in the 5′ leader of the MHV genome and to the complementary of the 3′ UTR in the minus strand. Binding to these sites is required for transcription (Huang and Lai, 1999; Zhang and Lai, 1995). To further elucidate the role of PTB in MHV RNA synthesis, DI RNAs harboring different reporter genes were transfected into stable cell lines over-expressing full-length or truncated PTB (Choi *et al.*, 2002). Overexpression of PTB had a dominant-negative effect on translation or transcription. Transfection of premade RNAs into these cell lines suggested that PTB effect was on transcription and not on translation.

Mitochondrial aconitase specifically binds to a plus MHV 3′ RNA region harboring the poly(A) tail, leading to an increase in the production of infectious virus early in infection, when the cultures are supplemented with iron. Mitochondrial aconitase is involved in the citric acid cycle; it is a post-tanscriptional regulator and possesses iron-regulatory functions (Nanda and Leibowitz, 2001). Binding of mitochondrial aconitase to the 3′ UTR of the viral genome might increase the stability of the mRNA, thereby favoring translation and the production of virus particles at early times after infection.

The La autoantigen is a ubiquitous phosphoprotein and an RNA binding protein. La autoantigen has been shown to bind to TGEV coronavirus 5′ and 3′ ends. Similarly, La autoantigen interacts with the UTRs of several RNA viruses (Gutierrez-Escolano *et al.*, 2000; Gutierrez-Escolano *et al.*, 2003), suggesting that La protein might have a role in transcription. Nevertheless, a cautious interpretation of these results must be made, as La autoantigen can bind RNA unspecifically.

The cellular factors involved in transcription described above, probably represent just the iceberg tip of a collection of host factors that participate in a complex transcription regulatory network. The large amount of work that has already been initiated in different laboratories will be required to decipher relevant nodes of this network.

References

Almazan, F., Galan, C., and Enjuanes, L. (2004). The nucleoprotein is required for efficient coronavirus genome replication. J. Virol. *78*, 12683–12688.

Alonso, S., Izeta, A., Sola, I., and Enjuanes, L. (2002). Transcription regulatory sequences and mRNA expression levels in the coronavirus transmissible gastroenteritis virus. J. Virol. *76*, 1293–1308.

An, S., and Makino, S. (1998). Characterizations of coronavirus *cis*-acting RNA elements and the transcription step affecting its transcription efficiency. Virology *243*, 198–207.

Baric, R.S., Nelson, G.W., Fleming, J.O., Deans, R.J., Keck, J.G., Casteel, N., and Stohlman, S.A. (1988). Interactions between coronavirus nucleocapsid protein and viral RNAs: implications for viral transcription. J. Virol. *62*, 4280–4287.

Baric, R.S., Stohlman, S.A., and Lai, M.M.C. (1983). Characterization of replicative intermediate RNA of mouse hepatitis virus: presence of leader RNA sequences on nascent chains. J. Virol. *48*, 633–640.

Baric, R.S., and Yount, B. (2000). Subgenomic negative-strand RNA function during mouse hepatitis virus infection. J. Virol. *74*, 4039–4046.

Bertrand, E.L., and Rossi, J.J. (1994). Facilitation of hammerhead ribozyme catalysis by the nucleocapsid protein of HIV-1 and the heterogeneous nuclear ribonucleoprotein A1. EMBO J. *13*, 2904–2912.

Bhardwaj, K., Guarino, L., and Kao, C.C. (2004). The severe acute respiratory syndrome coronavirus Nsp15 protein is an endoribonuclease that prefers manganese as a cofactor. J. Virol. *78*, 12218–11224.

Boguszewska-Chachulska, A.M., and Haenni, A.L. (2005). RNA viruses redirect host factors to better amplify their genome. Adv. Virus Res. *65*, 29–61.

Bost, A.G., Prentice, E., and Denison, M.R. (2001). Mouse hepatitis virus replicase protein complexes are translocated to sites of M protein accumulation in the ERGIC at late times of infection. Virology *285*, 21–29.

Brian, D.A., and Baric, R.S. (2005). Coronavirus genome structure and replication. In Coronavirus replication and reverse genetics, L. Enjuanes, ed. (Berlin, Springer), pp. 1–30.

Britton, P. (1991). Coronavirus motif. Nature *353*, 394.

Brockway, S.M., Clay, C.T., Lu, X.T., and Denison, M.R. (2003). Characterization of the expression, intracellular localization, and replication complex association of the putative mouse hepatitis virus RNA-dependent RNA polymerase. J. Virol. *77*, 10515–10527.

Calvo, E., Escors, D., Lopez, J.A., Gonzalez, J.M., Alvarez, A., Arza, E., and Enjuanes, L. (2005). Phosphorylation and subcellular localization of transmissible gastroenteritis virus nucleocapsid protein in infected cells. J. Gen. Virol. *86*, 2255–2267.

Casais, R., Thiel, V., Siddell, S.G., Cavanagh, D., and Britton, P. (2001). Reverse genetics system for the avian coronavirus infectious bronchitis virus. J. Virol. *75*, 12359–12369.

Chen, H., Gill, A., Dove, B.K., Emmett, S.R., Kemp, C.F., Ritchie, M.A., Dee, M., and Hiscox, J.A. (2005). Mass spectroscopic characterization of the coronavirus infectious bronchitis virus nucleoprotein and elucidation of the role of phosphorylation in RNA binding by using surface plasmon resonance. J. Virol. *79*, 1164–1179.

Choi, K.S., Huang, P., and Lai, M.M. (2002). Polypyrimidine-tract-binding protein affects transcription but not translation of mouse hepatitis virus RNA. Virology *303*, 58–68.

Compton, S.R., Rogers, D.B., Holmes, K.V., Fertsch, D., Remenick, J., and Mc. Gowan, J.J. (1987). In vitro replication of mouse hepatitis virus strain A59. J. Virol. *69*, 2313–2321.

Cristofari, G., and Darlix, J.L. (2002). The ubiquitous nature of RNA chaperone proteins. Prog. Nucleic Acid Res. Mol. Biol. *72*, 223–268.

Cristofari, G., Ivanyi-Nagy, R., Gabus, C., Boulant, S., Lavergne, J.P., Penin, F., and Darlix, J.L. (2004). The hepatitis C virus Core protein is a potent nucleic acid chaperone that directs dimerization of the viral (+) strand RNA in vitro. Nucleic Acids Res. *32*, 2623–2631.

Curtis, K.M., Yount, B., Sims, A.C., and Baric, R.S. (2004). Reverse genetic analysis of the transcription regulatory sequence of the coronavirus transmissible gastroenteritis virus. J. Virol. *78*, 6061–6066.

de Vries, A.A., Chirnside, E.D., Bredenbeek, P.J., Gravestein, L.A., Horzinek, M.C., and Spaan, W.J. (1990). All subgenomic mRNAs of equine arteritis virus contain a common leader sequence. Nucleic Acids Res. *18*, 3241–3247.

den Boon, J.A., Kleijnen, M.F., Spaan, W.J.M., and Snijder, E.J. (1996). Equine arteritis virus subgenomic mRNA synthesis: analysis of leader-body junctions and replicative-form RNAs. J. Virol. *70*, 4291–4298.

DeStefano, J.J., and Titilope, O. (2006). Poliovirus protein 3AB displays nucleic acid chaperone and helix-destabilizing activities. J. Virol. *80*, 1662–1671.

Dolja, V.V., Karasev, A.V., and Agranovsky, A.A. (1990). Organization of the beet yellows closterovirus genome. In New aspects of positive-strand RNA viruses, M.A. Brinton, and F.X. Heinz, eds. (Washington, DC, ASM Press), pp. 31–35.

Dolja, V.V., Kreuze, J.F., and Valkonen, J.P. (2006). Comparative and functional genomics of closteroviruses. Virus Res. *117*, 38–51.

Egger, D., Teterina, N., Ehrenfeld, E., and Bienz, K. (2000). Formation of the poliovirus replication complex requires coupled viral translation, vesicle production, and viral RNA synthesis. J. Virol. *74*, 6570–6580.

Egloff, M.P., Ferron, F., Campanacci, V., Longhi, S., Rancurel, C., Dutartre, H., Snijder, E.J., Gorbalenya, A.E., Cambillau, C., and Canard, B. (2004). The severe acute respiratory syndrome-coronavirus replicative protein nsp9 is a single-stranded RNA-binding subunit unique in the RNA virus world. Proc. Natl. Acad. Sci. USA *101*, 3792–3796.

Escors, D., Camafeita, E., Ortego, J., Laude, H., and Enjuanes, L. (2001). Organization of two transmissible gastroenteritis coronavirus membrane protein topologies within the virion and core. J. Virol. *75*, 12228–12240.

Gamarnik, A.V., and Andino, R. (1997). Two functional complexes formed by KH domain containing proteins with the 5′ noncoding region of poliovirus RNA. RNA *3*, 882–892.

Goebel, S.J., Hsue, B., Dombrowski, T.F., and Masters, P.S. (2004). Characterization of the RNA components of a putative molecular switch in the 3′ untranslated region of the murine coronavirus genome. J. Virol. *78*, 669–682.

Gorbalenya, A.E., Enjuanes, L., Ziebuhr, J., and Snijder, E.J. (2006). Nidovirales: Evolving the largest RNA virus genome. Virus Res. *117*, 17–37.

Gorbalenya, A.E., Koonin, E.V., Donchenko, A.P., and Blinov, V.M. (1989). Coronavirus genome: prediction of putative functional domains in the non-structural polyprotein by comparative amino acid sequence analysis. Nucleic Acids Res. *17*, 4847–4861.

Gosert, R., Kanjanahaluethai, A., Egger, D., Bienz, K., and Baker, S.C. (2002). RNA replication of mouse hepatitis virus takes place at double-membrane vesicles. J. Virol. *76*, 3697–3708.

Gutierrez-Escolano, A.L., Brito, Z.U., del Angel, R.M., and Jiang, X. (2000). Interaction of cellular proteins with the 5′ end of Norwalk virus genomic RNA. J. Virol. *74*, 8558–8562.

Gutierrez-Escolano, A.L., Vazquez-Ochoa, M., Escobar-Herrera, J., and Hernandez-Acosta, J. (2003). La, PTB, and PAB proteins bind to the 3(′) untranslated region of Norwalk virus genomic RNA. Biochem. Biophys. Res. Commun. *311*, 759–766.

He, R., Dobie, F., Ballantine, M., Leeson, A., Li, Y., Bastien, N., Cutts, T., Andonov, A., Cao, J., Booth, T.F., *et al.* (2004a). Analysis of multimerization of the SARS coronavirus nucleocapsid protein. Biochem. Biophys. Res. Commun. *316*, 476–483.

He, R., Leeson, A., Ballantine, M., Andonov, A., Baker, L., Dobie, F., Li, Y., Bastien, N., Feldmann, H., Strocher, U., *et al.* (2004b). Characterization of protein-protein interactions between the nucleocapsid protein and membrane protein of the SARS coronavirus. Virus Res. *105*, 121–125.

Herold, J., and Andino, R. (2001). Poliovirus RNA replication requires genome circularization through a protein-protein bridge. Mol. Cell *7*, 581–591.

Herschlag, D. (1995). RNA chaperones and the RNA folding problem. J. Biol. Chem. *270*, 20871–20874.

Hilf, M.E., Karasev, A.V., Pappu, H.R., Gumpf, D.J., Niblett, C.L., and Garnsey, S.M. (1995). Characterization of citrus tristeza virus subgenomic RNAs in infected tissue. Virology *208*, 576–582.

Hiscox, J.A. (2002). The nucleolus—a gateway to viral infection? Arch. Virol. *147*, 1077–1089.

Hiscox, J.A., Mawditt, K.L., Cavanagh, D., and Britton, P. (1995). Investigation of the control of coronavirus subgenomic mRNA transcription by using T7-generated negative-sense RNA transcripts. J. Virol. *69*, 6219–6227.

Hiscox, J.A., Wurm, T., Wilson, L., Britton, P., Cavanagh, D., and Brooks, G. (2001). The coronavirus infectious bronchitis virus nucleoprotein localizes to the nucleolus. J. Virol. *75*, 506–512.

Holt, J., Sgro, J.Y., Zuker, M., and Palmenberg, A. (2004). Computer folding of full-length viral genomes: a new toolkit for automated analysis of RNAs longer than 10000 bases. Seventh International Symposium on Positive Strand RNA Viruses San Francisco, California, USA.

Hsue, B., and Masters, P.S. (1999). Insertion of a new transcriptional unit into the genome of mouse hepatitis virus. J. Virol. *73*, 6128–6135.

Huang, P., and Lai, M.M.C. (1999). Polypyrimidine tract-binding protein binds to the complementary strand of the mouse hepatitis virus 3′ untranslated region, thereby altering RNA conformation. J. Virol. *73*, 9110–9116.

Huang, P., and Lai, M.M.C. (2001). Heterogeneous nuclear ribonucleoprotein A1 binds to the 3′-untranslated region and mediates potential 5′-3′-end cross talks of mouse hepatitis virus RNA. J. Virol. *75*, 5009–5017.

Huang, X., and Miller, W. (1991). A time-efficient, linear-space local similarity algorithm. Adv. Appl. Math. *12*, 337–357.

Huang, Z.S., Su, W.H., Wang, J.L., and Wu, H.N. (2003). Selective strand annealing and selective strand exchange promoted by the N-terminal domain of hepatitis delta antigen. J. Biol. Chem. *278*, 5685–5693.

Huang, Z.S., and Wu, H.N. (1998). Identification and characterization of the RNA chaperone activity of hepatitis delta antigen peptides. J. Biol. Chem. *273*, 26455–26461.

Hurst, K.R., Kuo, L., Koetzner, C.A., Ye, R., Hsue, B., and Masters, P.S. (2005). A major determinant for membrane protein interaction localizes to the carboxy-terminal domain of the mouse coronavirus nucleocapsid protein. J. Virol. *79*, 13285–13297.

Hussain, S., Pan, J., Chen, Y., Yang, Y., Xu, J., Peng, Y., Wu, Y., Li, Z., Zhu, Y., Tien, P., and Guo, D. (2005). Identification of novel subgenomic RNAs and noncanonical transcription initiation signals of severe acute respiratory syndrome coronavirus. J. Virol. *79*, 5288–5295.

Isken, O., Grassmann, C.W., Sarisky, R.T., Kann, M., Zhang, S., Grosse, F., Kao, P. N., and Behrens, S.-E. (2003). Members of the NF90/NFAR protein group are involved in the life cycle of a positive-strand RNA virus. EMBO J. *22*, 5655–5665.

Ivanov, K.A., Hertzig, T., Rozanov, M., Bayer, S., Thiel, V., Gorbalenya, A.E., and Ziebuhr, J. (2004a). Major genetic marker of nidoviruses encodes a replicative endoribonuclease. Proc. Natl. Acad. Sci. USA *101*, 12694–12699.

Ivanov, K.A., Thiel, V., Dobbe, J.C., van der Meer, Y., Snijder, E.J., and Ziebuhr, J. (2004). Multiple enzymatic activities associated with severe acute respiratory syndrome coronavirus helicase. J. Virol. *78*, 5619–5632.

Ivanyi-Nagy, R., Davidovic, L., Khandjian, E.W., and Darlix, J.L. (2005). Disordered RNA chaperone proteins: from functions to disease. Cell. Mol. Life Sci. *62*, 1409–1417.

Joo, M., and Makino, S. (1992). Mutagenic analysis of the coronavirus intergenic consensus sequence. J. Virol. *66*, 6330–6337.

Kaariainen, L., and Ahola, T. (2002). Functions of alphavirus nonstructural proteins in RNA replication. Prog. Nucleic Acid Res. Mol. Biol. *71*, 187–222.

Karasev, A.V., Hilf, M.E., Garnsey, S.M., and Dawson, W.O. (1997). Transcriptional strategy of closteroviruses: mapping the 5′ termini of the citrus tristeza virus subgenomic RNAs. J. Virol. *71*, 6233–6236.

Kim, M.J., and Kao, C. (2001). Factors regulating template swith *in vitro* by viral RNA-dependent RNA polymerases: implications for RNA-RNA recombination. Proc. Natl. Acad. Sci. USA *98*, 4972–4977.

Kim, Y., Jeong, Y., and Makino, S. (1993). Analysis of *cis*-acting sequences essential for coronavirus defective interfering RNA replication. Virology *197*, 53–63.

Kim, Y.-N., and Makino, S. (1995). Characterization of a murine coronavirus defective interfering RNA internal *cis*-acting replication signal. J. Virol. *69*, 4963–4971.

Kreuze, J.F., Savenkov, E.I., and Valkonen, J.P. (2002). Complete genome sequence and analyses of the subgenomic RNAs of sweet potato chlorotic stunt virus reveal several new features for the genus Crinivirus. J. Virol. *76*, 9260–9270.

Lai, M.M.C. (1992). RNA recombination in animal and plant viruses. Microbiol. Rev. *56*, 61–79.

Lai, M.M.C. (1998). Cellular factors in the transcription and replication of viral RNA genomes: a parallel to DNA-dependent RNA transcription. Virology *244*, 1–12.

Lai, M.M.C., Baric, R.S., Brayton, P.R., and Stohlman, S.A. (1984). Characterization of leader RNA sequences on the virion and mRNAs of mouse hepatitis virus, a cytoplasmic RNA virus. Proc. Natl. Acad. Sci. USA *81*, 3626–3630.

Lai, M.M.C., and Cavanagh, D. (1997). The molecular biology of coronaviruses. Adv. Virus Res. *48*, 1–100.

Laude, H., and Masters, P.S. (1995). The coronavirus nucleocapsid protein. In The Coronaviridae, S.G. Siddell, ed. (New York, Plenum press), pp. 141–158.

Lee, W.M., Ishikawa, M., and Ahlquist, P. (2001). Mutation of host delta9 fatty acid desaturase inhibits brome mosaic virus RNA replication between template recognition and RNA synthesis. J. Virol. *75*, 2097–2106.

Li, H.P., Huang, P., Park, S., and Lai, M.M.C. (1999). Polypyrimidine tract-binding protein binds to the leader RNA of mouse hepatitis virus and serves as a regulator of viral transcription. J. Virol. *73*, 772–777.

Li, H.P., Zhang, X., Duncan, R., Comai, L., and Lai, M.M.C. (1997). Heterogeneous nuclear ribonucleoprotein A1 binds to the transcription-regulatory region of mouse hepatitis virus RNA. Proc. Natl. Acad. Sci. USA *94*, 9544–9549.

Lin, H.X., and White, K.A. (2004). A complex network of RNA-RNA interactions controls subgenomic mRNA transcription in a tombusvirus. EMBO J. *23*, 3365–3374.

Lin, Y.J., and Lai, M.M.C. (1993). Deletion mapping of a mouse hepatitis virus defective interfering RNA reveals the requirement of an internal and discontinuous sequence for replication. J. Virol. *67*, 6110–6118.

Lin, Y.J., Liao, C.L., and Lai, M.M.C. (1994). Identification of the cis-acting signal for minus-strand RNA synthesis of a murine coronavirus: implications for the role of minus-strand RNA in RNA replication and transcription. J. Virol. *68*, 8131–8140.

Lin, Y.J., Zhang, X., Wu, R.C., and Lai, M.M.C. (1996). The 3′ untranslated region of coronavirus RNA is required for subgenomic mRNA transcription from a defective interfering RNA. J. Virol. *70*, 7236–7240.

Lorsch, J.R. (2002). RNA chaperones exist and DEAD box proteins get a life. Cell *109*, 797–800.

Mackenzie, J.M., Jones, M.K., and Westaway, E.G. (1999). Markers for trans-Golgi membranes and the intermediate compartment localize to induced membranes with distinct replication functions in flavivirus-infected cells. J. Virol. *73*, 9555–9567.

Makino, S., Joo, M., and Makino, J.K. (1991). A system for study of coronavirus messenger RNA synthesis: a regulated, expressed subgenomic defective interfering RNA results from intergenic site insertion. J. Virol. *65*, 6031–6041.

Mathews, D.H., Sabina, J., Zuker, M., and Turner, D.H. (1999). Expanded sequence dependence of thermodynamic parameters improves prediction of RNA secondary structure. J. Mol. Biol. *288*, 911–940.

Miller, D.J., Schwartz, M.D., and Ahlquist, P. (2001). Flock house virus RNA replicates on outer mitochondrial membranes in Drosophila cells. J. Virol. *75*, 11664–11676.

Miller, D.J., Schwartz, M.D., Dye, B.T., and Ahlquist, P. (2003). Engineered retargeting of viral RNA replication complexes to an alternative intracellular membrane. J. Virol. *77*, 12193–121202.

Miller, W.A., Dreher, T.W., and Hall, T.C. (1985). Synthesis of brome mosaic virus subgenomic RNA in vitro by internal initiation on (-)-sense genomic RNA. Nature *313*, 68–70.

Miller, W.A., and Koev, G. (2000). Synthesis of subgenomic RNAs by positive-strand RNA virus. Virology *273*, 1–8.

Minskaia, E., Hertzig, T., Gorbalenya, A.E., Campanacci, V., Cambillau, C., Canard, B., and Ziebuhr, J. (2006). Discovery of an RNA virus 3′→5′ exoribonuclease that is critically involved in coronavirus RNA synthesis. Proc. Natl. Acad. Sci. USA *103*, 5108–5113.

Mir, M.A., and Panganiban, A.T. (2006). The bunyavirus nucleocapsid protein is an RNA chaperone: possible roles in viral RNA panhandle formation and genome replication. RNA *12*, 272–282.

Mohandas, D.V., and Dales, S. (1991). Endosomal association of a protein phosphatase with high dephosphorylating activity against a coronavirus nucleocapsid protein. FEBS Lett. *282*, 419–424.

Molenkamp, R., van Tol, H., Rozier, B.C., van der Meer, Y., Spaan, W.J., and Snijder, E.J. (2000). The arterivirus replicase is the only viral protein required for genome replication and subgenomic mRNA transcription. J. Gen. Virol. *81*, 2491–2496.

Moradpour, D., Gosert, R., Egger, D., Penin, F., Blum, H.E., and Bienz, K. (2003). Membrane association of hepatitis C virus nonstructural proteins and identification of the membrane alteration that harbors the viral replication complex. Antiviral Res. *60*, 103–109.

Nagy, P.D., Pogany, J., and Simon, A.E. (1999). RNA elements required for RNA recombination function as replication enhancers in vitro and in vivo in a plus-strand RNA virus. EMBO J. *18*, 5653–5665.

Nagy, P.D., and Simon, A.E. (1997). New insights into the mechanisms of RNA recombination. Virology *235*, 1–9.

Nagy, P.D., Zhang, C., and Simon, A.E. (1998). Dissecting RNA recombination in vitro: role of RNA sequences and the viral replicase. EMBO J. *17*, 2392–2403.

Nanda, S.K., and Leibowitz, J.L. (2001). Mitochondrial aconitase binds to the 3′ untranslated region of the mouse hepatitis virus genome. J. Virol. *75*, 3352–3362.

Nelson, G.W., Stohlman, S.A., and Tahara, S.M. (2000). High affinity interaction between nucleocapsid protein and leader/intergenic sequence of mouse hepatitis virus RNA. J. Gen. Virol. *81*, 181–188.

Ozdarendeli, A., Ku, S., Rochat, S., Williams, G.D., Senanayake, S.D., and Brian, D.A. (2001). Downstream sequences influence the choice between a naturally occurring noncanonical and closely positioned upstream canonical heptameric fusion motif during bovine coronavirus subgenomic mRNA synthesis. J. Virol. *75*, 7362–7374.

Parker, M.M., and Masters, P.S. (1990). Sequence comparison of the N genes of five strains of the coronavirus mouse hepatitis virus suggests a three domain structure for the nucleocapsid protein. Virology *179*, 463–468.

Parsley, T.B., Towner, J.S., Blyn, L.B., Ehrenfeld, E., and Semler, B.L. (1997). Poly (rC) binding protein 2 forms a ternary complex with the 5′-terminal sequences of poliovirus RNA and the viral 3CD proteinase. RNA *3*, 1124–1134.

Pasternak, A.O., Gultyaev, A.P., Spaan, W.J., and Snijder, E.J. (2000). Genetic manipulation of arterivirus alternative mRNA leader-body junction sites reveals tight regulation of structural protein expression. J. Virol. *74*, 11642–11653.

Pasternak, A.O., Spaan, W.J., and Snijder, E.J. (2004). Regulation of relative abundance of arterivirus subgenomic mRNAs. J. Virol. *78*, 8102–8113.

Pasternak, A.O., van den Born, E., Spaan, W.J.M., and Snijder, E.J. (2001). Sequence requirements for RNA strand transfer during nidovirus discontinuous subgenomic RNA synthesis. EMBO J. *20*, 7220–7228.

Pasternak, A.O., van den Born, E., Spaan, W.J.M., and Snijder, E.J. (2003). The stability of the duplex between sense and antisense transcription-regulating sequences is a crucial factor in arterivirus subgenomic mRNA synthesis. J. Virol. *77*, 1175–1183.

Pedersen, K.W., van der Meer, Y., Roos, N., and Snijder, E.J. (1999). Open reading frame 1a-encoded subunits of the arterivirus replicase induce endoplasmic reticulum-derived double-membrane vesicles which carry the viral replication complex. J. Virol. *73*, 2016–2026.

Penzes, Z., González, J.M., Calvo, E., Izeta, A., Smerdou, C., Mendez, A., Sánchez, C.M., Sola, I., Almazán, F., and Enjuanes, L. (2001). Complete genome sequence of transmissible gastroenteritis coronavirus PUR46-MAD clone and evolution of the Purdue virus cluster. Virus Genes *23*, 105–118.

Peremyslov, V.V., and Dolja, V.V. (2002). Identification of the subgenomic mRNAs that encode 6-kDa movement protein and Hsp70 homolog of Beet yellows virus. Virology *295*, 299–306.

Posthuma, C.C., Nedialkova, D.D., Zevenhoven-Dobbe, J.C., Blokhuis, J.H., Gorbalenya, A.E., and Snijder, E.J. (2006). Site-directed mutagenesis of the Nidovirus replicative endoribonuclease NendoU exerts pleiotropic effects on the arterivirus life cycle. J. Virol. *80*, 1653–1661.

Prentice, E., Jerome, W.G., Yoshimori, T., Mizushima, N., and Denison, M.R. (2004). Coronavirus replication complex formation utilizes components of cellular autophagy. J. Biol. Chem. *279*, 10136–10141.

Qanungo, K.R., Shaji, D., Mathur, M., and Banerjee, A.K. (2004). Two RNA polymerase complexes from vesicular stomatitis virus-infected cells that carry out transcription and replication of genome RNA. Proc. Natl. Acad. Sci. USA *101*, 5952–5957.

Rajkowitsch, L., Semrad, K., Mayer, O., and Schroeder, R. (2005). Assays for the RNA chaperone activity of proteins. Biochem. Soc. Trans. *33*, 450–456.

Ranjith-Kumar, C.T., Zhang, X., and Kao, C.C. (2003). Enhancer-like activity of a brome mosaic virus RNA promoter. J. Virol. *77*, 1830–1839.

Reeder, R.H., and Lang, W.H. (1997). Terminating transcription in eukaryotes: lessons learned from RNA polymerase I. Trends Biochem. Sci. *22*, 473–477.

Risco, C., Muntión, M., Enjuanes, L., and Carrascosa, J.L. (1998). Two types of virus-related particles are found during transmissible gastroenteritis virus morphogenesis. J. Virol. *72*, 4022–4031.

Rowland, R.R., Kervin, R., Kuckleburg, C., Sperlich, A., and Benfield, D.A. (1999). The localization of porcine reproductive and respiratory syndrome virus nucleocapsid protein to the nucleolus of infected cells and identification of a potential nucleolar localization signal sequence. Virus Res. *64*, 1–12.

Sawicki, D.L., Wang, T., and Sawicki, S.G. (2001). The RNA structures engaged in replication and transcription of the A59 strain of mouse hepatitis virus. J. Gen. Virol. *82*, 386–396.

Sawicki, S.G., and Sawicki, D.L. (1998). A new model for coronavirus transcription. Adv. Exp. Med. Biol. *440*, 215–219.

Schelle, B., Karl, N., Ludewig, B., Siddell, S.G., and Thiel, V. (2005). Selective replication of coronavirus genomes that express nucleocapsid protein. J. Virol. *79*, 6620–6630.

Schlegel, A., Giddings, T.H., Jr., Ladinsky, M.S., and Kirkegaard, K. (1996). Cellular origin and ultrastructure of membranes induced during poliovirus infection. J. Virol. *70*, 6576–6588.

Schroeder, R., Barta, A., and Semrad, K. (2004). Strategies for RNA folding and assembly. Nat. Rev. Mol. Cell Biol. *5*, 908–919.

Sethna, P.B., Hung, S.-L., and Brian, D.A. (1989). Coronavirus subgenomic minus-strand RNAs and the potential for mRNA replicons. Proc. Natl. Acad. Sci. USA *86*, 5626–5630.

Seybert, A., Hegyi, A., Siddell, S.G., and Ziebuhr, J. (2000a). The human coronavirus 229E superfamily 1 helicase has RNA and DNA duplex-unwinding activities with 5′-to-3′ polarity. RNA *6*, 1056–1068.

Seybert, A., van Dinten, L.C., Snijder, E.J., and Ziebuhr, J. (2000b). Biochemical characterization of the equine arteritis virus helicase suggests a close functional relationship between arterivirus and coronavirus helicases. J. Virol. *74*, 9586–9593.

Sgro, J.Y., Holt, J., Zuker, M., and Palmenberg, A. (2004). RNA folding of the complete SARS and MHV coronavirus genomes. Seventh International Symposium on Positive Strand RNA Viruses San Francisco, California, USA.

Shi, S.T., Huang, P., Li, H.P., and Lai, M.M.C. (2000). Heterogeneous nuclear ribonucleoprotein A1 regulates RNA synthesis of a cytoplasmic virus. EMBO J. *19*, 4701–4711.

Shi, S.T., Schiller, J.J., Kanjanahaluethai, A., Baker, S.C., Oh, J.W., and Lai, M.M.C. (1999). Colocalization and membrane association of murine hepatitis virus gene 1 produts and de novo-synthesized viral RNA in infected cells. J. Virol. *73*, 5957–5969.

Siddell, S.G., Sawicki, D., Meyer, Y., Thiel, V., and Sawicki, S. (2001). Identification of the mutations responsible for the phenotype of three MHV RNA-negative ts mutants. Adv. Exp. Med. Biol. *494*, 453–458.

Smits, S.L., van Vliet, A.L., Segeren, K., el Azzouzi, H., van Essen, M., and de Groot, R.J. (2005). Torovirus non-discontinuous transcription: mutational analysis of a subgenomic mRNA promoter. J. Virol. *79*, 8275–8281.

Snijder, E.J., van der Meer, Y., Zevenhoven-Dobbe, J., Onderwater, J.J., van der Meulen, J., Koerten, H.K., and Mommaas, A.M. (2006). Ultrastructure and origin of membrane vesicles associated with the severe acute respiratory syndrome coronavirus replication complex. J. Virol. *80*, 5927–5940.

Snijder, E.J., Bredenbeek, P.J., Dobbe, J.C., Thiel, V., Ziebuhr, J., Poon, L.L.M., Guan, Y., Rozanov, M., Spaan, W.J.M., and Gorbalenya, A.E. (2003). Unique and conserved features of genome and proteome of SARS-coronavirus, and early split-off from the coronavirus group 2 lineage. J. Mol. Biol. *331*, 991–1004.

Snijder, E.J., van Tol, H., Roos, N., and Pedersen, K.W. (2001). Non-structural proteins 2 and 3 interact to modify host cell membranes during the formation of the arterivirus replication complex. J. Gen. Virol. *82*, 985–994.

Sola, I., Moreno, J. L., Zuñiga, S., Alonso, S., and Enjuanes, L. (2005). Role of nucleotides immediately flanking the transcription-regulating sequence core in coronavirus subgenomic mRNA synthesis. J. Virol. *79*, 2506–2516.

Spaan, W.H., Delius, H., Skinner, M., Armstrong, J., Rottier, P., Smeekens, S., van der Zeijst, B.A.M., and Siddell, S.G. (1983). Coronavirus mRNA synthesis involves fusion of non-contiguous sequences. EMBO J. *2*, 1839–1844.

Stohlman, S.A., Baric, R.S., Nelson, G.N., Soe, L.H., Welter, L.M., and Deans, R.J. (1988). Specific interaction between coronavirus leader RNA and nucleocapsid protein. J. Virol. *62*, 4288–4295.

Surjit, M., Kumar, R., Mishra, R.N., Reddy, M.K., Chow, V.T., and Lal, S.K. (2005). The severe acute respiratory syndrome coronavirus nucleocapsid protein is phosphorylated and localizes in the cytoplasm by 14-3-3-mediated translocation. J. Virol. *79*, 11476–11486.

Surjit, M., Liu, B., Kumar, P., Chow, V.T., and Lal, S.K. (2004). The nucleocapsid protein of the SARS coronavirus is capable of self-association through a C-terminal 209 amino acid interaction domain. Biochem. Biophys. Res. Commun. *317*, 1030–1036.

Sutton, G., Fry, E., Carter, L., Sainsbury, S., Walter, T., Nettleship, J., Berrow, N., Owens, R., Gilbert, R., Davidson, A., *et al.* (2004). The nsp9 replicase protein of SARS-coronavirus, structure and functional insights. Structure (Camb) *12*, 341–353.

Thiel, V., Herold, J., Schelle, B., and Siddell, S.G. (2001). Viral replicase gene products suffice for coronavirus discontinuous transcription. J. Virol. *75*, 6676–6681.

Tijms, M.A., and Snijder, E.J. (2003). Equine arteritis virus non-structural protein 1, an essential factor for viral subgenomic mRNA synthesis, interacts with the cellular transcription co-factor p100. J. Gen. Virol. *84*, 2317–2322.

Tijms, M.A., van der Meer, Y., and Snijder, E.J. (2002). Nuclear localization of non-structural protein 1 and nucleocapsid protein of equine arteritis virus. J. Gen. Virol. *83*, 795–800.

Tijms, M.A., van Dinten, L.C., Gorbalenya, A.E., and Snijder, E.J. (2001). A zinc finger-containing papain-like protease couples subgenomic mRNA synthesis to genome translation in a positive-stranded RNA virus. Proc. Natl. Acad. Sci. USA *98*, 1889–1894.

Tsuchihashi, Z., and Brown, P.O. (1994). DNA strand exchange and selective DNA annealing promoted by the Human Immunodeficiency Virus Type I nucleocapsid protein. J. Virol. *68*, 5863–5870.

van den Born, E., Gultyaev, A.P., and Snijder, E.J. (2004). Secondary structure and function of the 5′-proximal region of the equine arteritis virus RNA genome. RNA *10*, 424–437.

van den Born, E., Posthuma, C.C., Gultyaev, A.P., and Snijder, E.J. (2005). Discontinuous subgenomic RNA synthesis in arteriviruses is guided by an RNA hairpin structure located in the genomic leader region. J. Virol. *79*, 6312–6324.

van der Meer, Y., Snijder, E.J., Dobbe, J.C., Schleich, S., Denison, M.R., Spaan, W.J.M., and Krijnse Locker, J. (1999). Localization of mouse hepatitis virus nonstructural proteins and RNA synthesis indicates a role for late endosomes in viral replication. J. Virol. *73*, 7641–7657.

van der Most, R.G., De Groot, R.J., and Spaan, W.J.M. (1994). Subgenomic RNA synthesis directed by a synthetic defective interfering RNA of mouse hepatitis virus: a study of coronavirus transcription initiation. J. Virol. *68*, 3656–3666.

van Dinten, L.C., den Boon, J.A., Wassenaar, A.L.M., Spaan, W.J.M., and Snijder, E.J. (1997). An infectious arterivirus cDNA clone: identification of a replicase point mutation that abolishes discontinuous mRNA transcription. Proc. Natl. Acad. Sci. USA *94*, 991–996.

van Dinten, L.C., van Tol, H., Gorbalenya, A.E., and Snijder, E.J. (2000). The predicted metal-binding region of the arterivirus helicase protein is involved in subgenomic mRNA synthesis, genome replication, and virion biogenesis. J. Virol. *74*, 5213–5223.

van Marle, G., Dobbe, J.C., Gultyaev, A.P., Luytjes, W., Spaan, W.J.M., and Snijder, E.J. (1999a). Arterivirus discontinuous mRNA transcription is guided by base pairing between sense and antisense transcription-regulating sequences. Proc. Natl. Acad. Sci. USA *96*, 12056–12061.

van Marle, G., Luytjes, W., Van der Most, R.G., van der Straaten, T., and Spaan, W.J.M. (1995). Regulation of Coronavirus mRNA transcription. J. Virol. *69*, 7851–7856.

van Marle, G., van Dinten, L.C., Spaan, W.J., Luytjes, W., and Snijder, E.J. (1999b). Characterization of an equine arteritis virus replicase mutant defective in subgenomic mRNA synthesis. J. Virol. *73*, 5274–5281.

van Vliet, A.L.W., Smits, S.L., Rottier, P.J.M., and de Groot, R.J. (2002). Discontinuous and non-discontinuous subgenomic RNA transcription in a nidovirus. EMBO J. *21*, 6571–6580.

von Grotthuss, M., Wyrwicz, L.S., and Rychlewski, L. (2003). mRNA cap-1 methyltransferase in the SARS genome. Cell *113*, 701–702.

Wang, Y., and Zhang, X. (1999). The nucleocapsid protein of coronavirus mouse hepatitis virus interacts with the cellular heterogeneous nuclear ribonucleoprotein A1 in vitro and in vivo. Virology *265*, 96–109.

Wang, Y., and Zhang, X. (2000). The leader RNA of coronavirus mouse hepatitis virus contains an enhancer-like element for subgenomic mRNA transcription. J. Virol. *74*, 10571–10580.

White, K.A. (2002). The premature termination model: a possible third mechanism for subgenomic mRNA transcription in (+)-strand RNA viruses. Virology *304*, 147–154.

White, K.A., and Nagy, P.D. (2004). Advances in the molecular biology of tombusviruses: gene expression, genome replication, and recombination. Prog. Nucleic Acid Res. Mol. Biol. *78*, 187–226.

Williams, G.D., Chang, R.-Y., and Brian, D.A. (1999). A phylogenetically conserved hairpin-type 3′ untranslated region pseudoknot functions in coronavirus RNA replication. J. Virol. *73*, 8349–8355.

Wilson, K.S., and von Hippel, P.H. (1995). Transcription termination at intrinsic terminators: the role of the RNA hairpin. Proc. Natl. Acad. Sci. USA *92*, 8793–8797.

Wurm, T., Chen, H., Hodgson, T., Britton, P., Brooks, G., and Hiscox, J.A. (2001). Localization to the nucleolus is a common feature of coronavirus nucleoproteins, and the protein may disrupt host cell division. J. Virol. *75*, 9345–9356.

Yan, L., Velikanov, M., Flook, P., Zheng, W., Szalma, S., and Kahn, S. (2003). Assessment of putative protein targets derived from the SARS genome. FEBS Lett. *554*, 257–263.

Yarnell, W.S., and Roberts, J.W. (1999). Mechanism of intrinsic transcription termination and antitermination. Science *284*, 611–615.

Yount, B., Curtis, K.M., and Baric, R.S. (2000). Strategy for systematic assembly of large RNA and DNA genomes: the transmissible gastroenteritis virus model. J. Virol. *74*, 10600–10611.

Yuan, X., Yao, Z., Shan, Y., Chen, B., Yang, Z., Wu, J., Zhao, Z., Chen, J., and Cong, Y. (2005). Nucleolar localization of non-structural protein 3b, a protein specifically encoded by the severe acute respiratory syndrome coronavirus. Virus Res. *114*, 70–79.

Zhang, X., and Lai, M.M.C. (1995). Interactions between the cytoplasmic proteins and the intergenic (promoter) sequence of mouse hepatitis virus RNA: correlation with the amounts of subgenomic mRNA transcribed. J. Virol. *69*, 1637–1644.

Ziebuhr, J. (2005). The coronavirus replicase. In Coronavirus replication and reverse genetics, L. Enjuanes, ed. (Springer), pp. 57–94.

Zuker, M. (2003). Mfold web server for nucleic acid folding and hybridization prediction. Nucleic Acids Res. *31*, 3406–3415.

Zuñiga, S., Sola, I., Alonso, S., and Enjuanes, L. (2004). Sequence motifs involved in the regulation of discontinuous coronavirus subgenomic RNA synthesis. J. Virol. *78*, 980–994.

Zuñiga, S., Sola, I., Moreno, J. L., Alonso, S., and Enjuanes, L. (2006). Regulation of coronavirus transcription: viral and cellular proteins interacting with transcription-regulating sequences. In The Nidovirus: Towards control of SARS and other nidovirus diseases, S. Perlman, and K. Holmes, eds. (New York, Springer), pp. 31–35.

Reverse Genetic Analysis of Coronavirus Replication

5

Volker Thiel

Abstract

Coronavirus reverse genetics has greatly contributed to our current understanding of coronavirus molecular biology and diseases. Starting in the early 1990s reverse genetic techniques have been applied to the modification of the 3′-third of coronavirus genomes by targeted recombination. In 2000, reverse genetic systems based on full-length cDNAs made the entire coronavirus genome amenable to mutagenesis. Coronavirus reverse genetic techniques have been applied to the analysis of all aspects of coronavirus infections. These include studies on the tropism, pathogenesis, virus entry, virion formation and RNA synthesis of coronaviruses. In addition to the generation of recombinant coronaviruses, reverse genetic approaches have also been used for the development of coronavirus replicon RNAs and novel vectors designed for heterologous gene expression.

Introduction

The extreme large size of coronavirus genomes and the instability of specific coronavirus cDNA sequences in bacterial cloning systems have been the main obstacles to generate recombinant coronaviruses by traditional reverse genetic techniques. Consequently, current reverse genetic systems for coronaviruses are exclusively based on non-traditional approaches. In the early 1990s, Masters and colleagues reported a method termed "targeted RNA recombination" as a convenient system to manipulate coronavirus genomes (Koetzner *et al.*, 1992; Masters, 1999; Masters and Rottier, 2005). This system is based on the intrinsic property of the coronavirus replication machinery to recombine RNA molecules, and in this case, the genome of a temperature-sensitive mouse hepatitis virus (MHV) mutant has been recombined with a donor RNA (containing a mutation of choice) provided in trans. Initially, recombinant coronaviruses generated by targeted RNA recombination have been selected against the temperature sensitive parental virus. However, a more sophisticated and stringent method based on a switch in species specificity has recently been established by the groups of Masters and Rottier (Kuo *et al.*, 2000; Masters and Rottier, 2005). Targeted RNA recombination has been successfully used for the manipulation of the 3′ third of coronavirus genomes, including all structural genes, accessory genes and 3′ untranslated region (Masters and Rottier, 2005), and it has been applied to two other coronaviruses, feline infectious peritonitis virus (FIPV) (Haijema *et al.*, 2003) and transmissible gastroenteritis virus (TGEV) (Sanchez *et al.*, 1999).

For technical reasons, targeted RNA recombination has not yet been extended to the modification of the 5′ two-thirds of the genome. Thus, the reverse genetic manipulation of the replicase gene and 5′ untranslated region had to await the development of systems based on the construction of full-length coronavirus cDNAs. Eventually, in 2000, Enjuanes and colleagues reported the first cDNA-based reverse genetic system (Almazan *et al.*, 2000). Recombinant TGEV has been recovered from a full-length cDNA cloned in a bacterial artificial chromosome (BAC). Soon after, two other groups also succeeded in the recovery of recombinant coronaviruses from cloned cDNA. First, Baric and colleagues reported a system based on the systematic assembly of a full-length TGEV cDNA by *in vitro* ligation using DNA fragments derived from six adjoining cDNA subclones spanning the entire TGEV genome (Yount *et al.*, 2000). Finally, an alternative to the cDNA cloning in bacterial systems has been reported by Siddell and colleagues, who used vaccinia virus as an eukaryotic cloning vector to stably insert a full-length cDNA of the human coronavirus 229E (HCoV-229E) genome (Thiel *et al.*, 2001a). Although different cloning vectors and protocols for the recovery of recombinant coronaviruses have been established (Figure 5.1), full-length cDNA-based systems have in common that they enable the modification of all regions within coronavirus genomes, including the replicase gene and 5′ untranslated region.

The list of coronaviruses that can now be studied by reverse genetic techniques is constantly expanding and currently includes a number of human (e.g. HCoV-229E, HCoV-OC43, SARS-CoV) and animal coronaviruses (e.g. mouse hepatitis virus (MHV), TGEV, avian infectious bronchitis virus (IBV), FIPV) (Almazan *et al.*, 2006; Almazan *et al.*, 2000; Baric and Sims, 2005; Casais *et al.*, 2001; Coley *et al.*, 2005; Enjuanes *et al.*, 2005; Fang *et al.*, 2006; Gonzalez *et al.*, 2002; Haijema *et al.*, 2003; Koetzner *et al.*, 1992; Kuo *et al.*, 2000; Lai, 2000; Masters, 1999; Masters and Rottier, 2005; St-Jean *et al.*, 2006; Thiel *et al.*, 2001a; Thiel and Siddell, 2005; Yount *et al.*, 2000; Yount *et al.*, 2003; Yount *et al.*, 2002). Reverse genetic analyses of these viruses have greatly contributed to extending our knowledge on virtually all aspects of coronavirus infections, including tropism, pathogenesis, packaging, and mechanisms and requirements for viral RNA synthesis. This knowledge and the ability to modify coronavirus genomes at will, has also paved the way for the development of replicon RNAs and novel coronavirus vectors and vaccines.

Reverse genetic analysis of recombinant coronaviruses

Requirements for coronavirus RNA synthesis

Coronavirus RNA synthesis involves the replication of viral genomes and transcription of subgenomic mRNAs (Siddell *et al.*, 2005). Both processes take place within the host cell cytoplasm, most likely at ER-derived membrane vesicles that are induced upon infection (Bi *et al.*, 1998; Gosert *et al.*, 2002; Prentice *et al.*, 2004a; Shi *et al.*, 1999; Snijder *et al.*, 2006; van der Meer *et al.*, 1999). The enzymatic activities required for coronavirus replication and transcription are encoded by the replicase gene, that is comprised of two large open reading frames (ORFs), ORF1a and ORF1b, located at the 5′ two-thirds of the genome (the coronavirus replicase is reviewed in Chapter 2 of this volume by Ziebuhr and Snijder). Translation of the genomic RNA gives rise to primary translation products, termed polyproteins pp1a and pp1ab, that are co- and post-translationally processed by

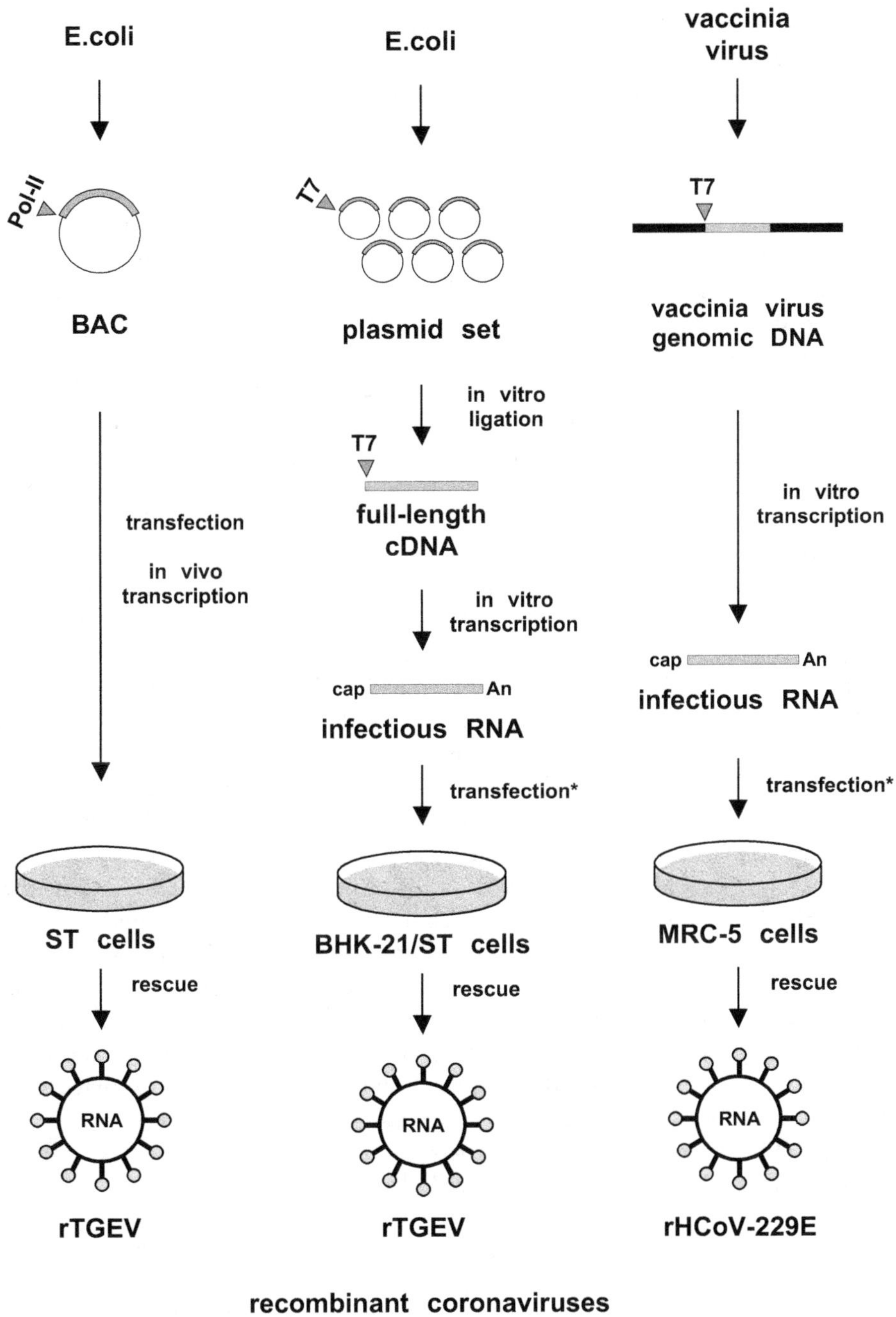

Figure 5.1 Full-length cDNA-based coronavirus reverse genetic systems. The rescue procedures of recombinant coronaviruses are illustrated according to their first descriptions for full-length cDNAs cloned in a bacterial artificial chromosome (BAC, left panel) (Almazan *et al.*, 2000), vaccinia virus (right panel) (Thiel *et al.*, 2001a), or for a set of plasmid cDNA together encompassing an entire coronavirus genome (middle panel) (Yount *et al.*, 2000).

virus-encoded proteinases to smaller subunits forming the replicase/transcriptase complex (Ziebuhr, 2005; Ziebuhr *et al.*, 2000). This complex is believed to interact with *cis*-acting elements in order to replicate and transcribe viral RNAs (Brian and Baric, 2005). Although reverse genetic analyses have also been used to study coronavirus *cis*-acting elements, they are not discussed here, but can be found in Chapters 3 and 4.

Reverse genetic analysis of ORF1b-encoded replicase gene products

The complexity of the coronavirus replicase/transcriptase is best illustrated by the fact that the replicase gene-encoded polyproteins pp1a and pp1ab are proteolytically processes to 15–16 processing end products (termed non-structural proteins (nsps) 1–16) and various processing intermediates. The discovery of a coronavirus causing severe acute respiratory syndrome (SARS-CoV) has significantly accelerated the research aimed at elucidating replicase protein functions (Gorbalenya *et al.*, 2006; Snijder *et al.*, 2003; Ziebuhr, 2005). Thus, enzymatic activities and functions for a number of nsps have been predicted and some of these have already been experimentally verified. Specifically, the ORF1b encodes a number of enzymes that are associated with replicase key functions (Gorbalenya *et al.*, 1989; Snijder *et al.*, 2003), namely the RNA-dependent RNA polymerase (RdRp, nsp12) (Cheng *et al.*, 2005) and the NTPase/helicase (HEL, nsp13) (Ivanov and Ziebuhr, 2004; Seybert *et al.*, 2000; Thiel *et al.*, 2003a), followed by RNA processing enzymes comprising a 3′-to-5′ exonuclease (ExoN, nsp14) (Minskaia *et al.*, 2006; Snijder *et al.*, 2003), an uridylate-specific endoribonuclease (NendoU, nsp15) (Bhardwaj *et al.*, 2004; Bhardwaj *et al.*, 2006; Ivanov *et al.*, 2004; Snijder *et al.*, 2003) and a S-adenosylmethionine-dependent ribose 2′-O-methyltransferase (O-MT, nsp16) (Snijder *et al.*, 2003). It is tempting to speculate that this enzymatic repertoire may be the prerequisite that enable coronaviruses to perform their unique type of discontinuous transcription and/or to stably maintain the extraordinary size of their genome (Gorbalenya *et al.*, 2006). Reverse genetic studies indeed revealed that the ExoN and NendoU are critically involved in coronavirus RNA synthesis (Ivanov *et al.*, 2004; Minskaia *et al.*, 2006; Ziebuhr, 2005). HCoV-229E mutant genomes that lack either the ExoN or the NendoU activity are not able to initiate a productive infection cycle and RNA synthesis is markedly reduced or not detectable, respectively. Although these enzymes have been characterized biochemically to some extend, future studies are needed to elucidate when, where and how these activities are required during coronavirus RNA synthesis. However, it should be kept in mind that at least some coronavirus nsps may be multifunctional proteins. It has been shown that a single point mutation in nsp14 (ExoN) of MHV is attenuating in mice, whereas virus growth kinetics or RNA synthesis were not affected in tissue culture (Sperry *et al.*, 2005).

Reverse genetic analysis of ORF1a-encoded replicase gene products

Similar to ORF1b-encoded replicase gene products, a number of domains and functions have been proposed for ORF1a-encoded nsps (Gorbalenya *et al.*, 1989; Snijder *et al.*, 2003; Ziebuhr *et al.*, 2001). Of particular importance are the viral proteases that are engaged in the extensive proteolytic processing of replicase polyproteins pp1a and pp1ab. Their functions and putative cleavage sites within pp1a and pp1ab have been predicted almost two decades ago (Gorbalenya *et al.*, 1989; Lee *et al.*, 1991). Particularly, the main proteinase (M^{pro}, also called 3C-like proteinase; nsp5) has been studied extensively and crystal structures of a number of M^{pro}s from different coronaviruses have been determined (Anand *et al.*, 2002; Anand *et al.*, 2003; Chen *et al.*, 2005; Thiel *et al.*, 2003a; Yang *et al.*, 2005; Yang *et al.*, 2003; Ziebuhr, 2005; Ziebuhr *et al.*, 1995; Ziebuhr *et al.*, 2000) (see also Chapter 2). All coronaviruses encode one M^{pro} and this enzyme is considered indispensable for coronavirus replication as it is responsible for the proteolytic release of nsps 5–16 from replicase polyprotein precursors.

In addition to the M^{pro}, coronaviruses encode one or two papain-like proteinases ($PL1^{pro}$ and $PL2^{pro}$, also called accessory proteinases), that reside in nsp3 and are responsible for the proteolytic release of nsp1–4 (with the nsp4 C-terminus being cleaved by the Mpro) (Ziebuhr, 2005; Ziebuhr *et al.*, 2000). Interestingly, there are differences in the PL^{pro}-mediated proteolytic processing pathways among the various groups and subgroups of coronaviruses. Thus, for example, coronaviruses of (sub)groups 2b and 3 specify only a single papain-like protease activity. Their PL^{pro} cleavage sites have three strictly conserved residues at their P4–P1 positions (Thiel *et al.*, 2003a) and their PL^{pro} enzymes correspond to $PL2^{pro}$ enzymes found in other coronaviruses (Thiel *et al.*, 2003b; Ziebuhr *et al.*, 2001). Noteworthy, IBV still contains sequences that are homologous to coronavirus $PL1^{pro}$ enzymes, however, this $PL1^{pro}$ remnant is considered proteolytically inactive because some sequences essential for proteolytic activity are missing (Ziebuhr *et al.*, 2001). Like all currently known group 3 coronaviruses, IBV does not encode a counterpart of nsp1, indicating that $PL1^{pro}$ activity may be dispensable because the main substrate, the nsp1|nsp2 site, may have been lost during evolution (Lim *et al.*, 2000; Ziebuhr *et al.*, 2001). In contrast, a nsp1 is encoded by the SARS-CoV replicase gene despite the presence of only one PL^{pro} activity. Thus, the SARS-CoV PL^{pro} can efficiently process the entire nsp1–4 region (Harcourt *et al.*, 2004; Prentice *et al.*, 2004b; Snijder *et al.*, 2003; Thiel *et al.*, 2003a). In group 1 and 2a coronaviruses, which encode two PL^{pro}s, the regulation of the proteolytic processing pathways seems to be more complex. For example, in mouse hepatitis virus (MHV), the prototype of coronavirus group 2a, $PL1^{pro}$ cleaves the nsp1|nsp2 and nsp2|nsp3 sites, while $PL2^{pro}$ cleaves the nsp3|nsp4 site (Baker *et al.*, 1989; Bonilla *et al.*, 1995; Bonilla *et al.*, 1997; Dong and Baker, 1994; Graham and Denison, 2006; Kanjanahaluethai and Baker, 2000; Teng *et al.*, 1999). Accordingly, the MHV $PL1^{pro}$ and $PL2^{pro}$ cleavage sites resemble each other only remotely in that both of them have small amino acid residues at their P2 and/or P1 positions. In human coronavirus 229E (HCoV-229E), a group 1 coronavirus, $PL1^{pro}$ and $PL2^{pro}$ have overlapping substrate specificities, as both can cleave at the nsp2|nsp3 site (though, $PL2^{pro}$ cleaves this site more efficient *in vitro*). Similarly, our recent analyses indicate that both HCoV-229E PL^{pro}s can also cleave the nsp1|nsp2 site, however here, $PL1^{pro}$ appears to cleave more efficiently (Ziebuhr *et al.*, 2007). In view of these results the question arises why most coronaviruses possess two PL^{pro} activities and if both activities are actually required. Using our reverse genetic system for HCoV-229E we could recently show that $PL1^{pro}$, but not $PL2^{pro}$, is dispensable for virus replication. This indicates that the proteolytic activity of $PL1^{pro}$ is not strictly required for HCoV-229E RNA synthesis and virus production. Nevertheless, $PL1^{pro}$-deficient viruses that carry an $PL1^{pro}$ active site mutation rapidly revert to the wild-type sequence within three passages in tissue culture, indicating that $PL1^{pro}$ activity improves viral fitness (Ziebuhr *et al.*, 2007). These results corroborate reverse genetic analyses of nsp1|nsp2 cleavage site mutants of MHV and TGEV performed in the laboratories of Denison, Baric and Enjuanes (Denison *et al.*, 2004; Galan *et al.*, 2005). Mutations that abolish nsp1|nsp2 cleavage are tolerated, but viral growth in tissue culture is significantly reduced in both virus systems. Like the HCoV-229E PL1^{pro} mutant, a TGEV nsp1|nsp2 cleavage site mutant rapidly reverts to the wild-type sequence upon passaging in tissue culture, highlighting the close relationship of HCoV-229E and TGEV that are both group 1 coronaviruses (Galan *et al.*, 2005; Thiel and Siddell, 2005) (Ziebuhr *et al.*, 2007). In the case of MHV, a group 2a coronavirus, nsp1|nsp2 cleavage site mutations that abolish nsp1|nsp2 cleavage appear to be stable in MHV mutant genomes, as no reversion to the wild-type sequence has been observed even after 10 passages in tissue culture (Denison *et al.*, 2004). Thus, there might be different constrains governing the regulation of PL^{pro}-mediated processing events among different phylogenetic lineages of coronaviruses. The SARS-CoV PL^{pro} has recently been studied biochemically by several groups and the crystal structure of the catalytic core has been resolved (Barretto *et al.*, 2005; Han *et al.*, 2005; Harcourt *et al.*, 2004; Lindner *et al.*, 2005; Ratia *et al.*, 2006; Sulea *et al.*, 2005; Sulea *et al.*, 2006; Thiel *et al.*, 2003a). These analyses revealed that SARS-CoV PL^{pro} displays biochemical and structural similarities to deubiquitinating enzymes (Barretto *et al.*, 2005; Lindner *et al.*, 2005; Ratia *et al.*, 2006; Sulea *et al.*, 2005; Sulea *et al.*, 2006). Although it remains to be determined how a deubiquitinating activity can be placed in relation to the viral replication cycle, these findings indicate that virus-host interactions may represent an additional driving force in the evolution of coronaviral PL^{pro}s.

In addition to proteinase and cleavage site mutants within the nsp1–4 region, there are now first reverse genetic studies targeting nsp1, nsp2 and nsp3 directly. Brockway and Denison studied MHV mutants containing deletions or point mutations in nsp1 (Brockway and Denison, 2005). This study revealed that the carboxyterminal half of MHV nsp1 (K124–L241) is dispensable for virus replication *in vitro*, but deletion of the entire nsp1 or the aminoterminal half of nsp1 did not result in viable viruses. It remains to be determined if the aminoterminal half of nsp1 is required for MHV replication or if essential 5′-*cis*-acting elements located within the nsp1-coding region have been deleted. Single point mutations in the aminoterminal half of nsp1 resulted in a panel of recombinant viruses of which some displayed altered replication and RNA synthesis. The function of nsp1 in viral RNA synthesis is still unknown, but recently the SARS-CoV nsp1 has been implicated in virus-host cell interaction as a factor promoting degradation of cellular mRNAs (Kamitani *et al.*, 2006). For coronavirus nsp2 there is yet no function proposed. Reverse genetic studies on MHV and SARS-CoV nsp2 indicate that nsp2 is dispensable for viral replication in both viruses (Graham *et al.*, 2005). Remarkably, the entire nsp2-coding region could be deleted resulting in mutant viruses that displayed only modest growth and RNA synthesis defects and no detectable effect on polyprotein processing. Although MHV nsp1 and nsp2 have been shown to co-localize with replicase complexes, similar studies on nsp 1 and nsp2 of other coronaviruses are required to determine if these processing products may display the same cellular localization pattern. It should be noted that coronavirus replicase genes differ most in the nsp1 and nsp2-encoding region with very limited similarities of nsp 1 and nsp2 even amongst members of the same coronavirus group. The Denison laboratory has now provided the first genetic evidence that nsp1 (at least in part) and nsp2 may not be directly involved in viral RNA synthesis (Brockway and Denison, 2005; Graham *et al.*,

2005), indicating that these protein may have roles in the fine regulation of viral RNA synthesis and/or in virus-host interactions and pathogenesis.

Finally, nsp3, the largest replicase polyprotein processing product with a molecular mass of approximately 200 kDa, has also been analyzed using reverse genetics. Comparative sequence analyses revealed a (multi)domain organization with five conserved domains that are encoded in the order (N- to C-terminal): Ac (acidic domain), $PL1^{pro}$, ADRP (ADP-ribose-1″-monophosphatase, also called X domain), $PL2^{pro}$ and Y (containing several *trans*-membrane domains) (Ziebuhr *et al.*, 2001). Although there are variations concerning nsp3-encoded domains in IBV and SARS-CoV, for instance SARS-CoV nsp3 is lacking a $PL1^{pro}$ domain, but instead has acquired a domain with unknown function, termed SARS-CoV unique domain (SUD) (Snijder *et al.*, 2003), the Ac, ADRP, $PL2^{pro}$ and Y domains are conserved in all coronaviruses. Recently, the SARS-CoV ADRP crystal structure has been elucidated and the SARS-CoV and HCoV-229E ADRP domains were shown to display indeed a ADP-ribose-1″-monophosphatase (Appr-1″-p) processing activity (Egloff *et al.*, 2006; Putics *et al.*, 2005; Saikatendu *et al.*, 2005). Interestingly, mutations in the active site of the HCoV-229E ADRP that abolish the Appr-1″-p processing activity did not significantly affect mutant virus growth and RNA synthesis (Putics *et al.*, 2005). Thus, it appears that the replicase gene-encoded ADRP activity is not required for viral replication in tissue culture. Further studies involving the analysis of viral replication in different cell types and *in vivo* experiments are required to determine if nsps 1–3 may function in virus-host interactions that possibly impact on virus pathogenicity.

Reverse genetic studies targeting the nsp4–11 region of ORF1a have not yet been reported, although it can be expected that such studies are under way. This region is processed by the M^{pro} and displays a high degree of homology between different coronaviruses suggesting that these nsps may be critically involved in coronavirus RNA synthesis. Biochemical and structural data on nsps 7–10 indicate that these proteins are nucleic acid-binding proteins (Egloff *et al.*, 2004; Imbert *et al.*, 2006; Joseph *et al.*, 2006; Matthes *et al.*, 2006; Su *et al.*, 2006; Sutton *et al.*, 2004; Zhai *et al.*, 2005). Interestingly, the SARS-CoV nsp8 has been shown to possess a RNA-dependent RNA polymerase activity that allows for the synthesis of small RNAs *in vitro* (Imbert *et al.*, 2006). It has been proposed that nsp8 may act as a primase that produces small RNA primers utilized by the "main" RdRp (nsp12). Although it is not clear at the moment if nsps 7–10 recognize specific coronaviral RNA sequences or structures, it will be interesting to learn whether they are engaged in interactions with coronavirus *cis*-acting RNA elements that govern genome replication and/or transcription.

Efficient coronavirus genome replication requires nucleocapsid protein expression

Reverse genetic analyses of replicase gene products revealed that a number of nsps are indeed critically involved in coronavirus RNA synthesis. In contrast, genes encoded downstream the replicase gene (accessory and structural genes) were considered to be not involved in coronavirus RNA synthesis. However, there is one notable exception, the nucleocapsid (N) protein. The coronavirus N protein is known as a structural protein that is associated with viral genomic RNA within the virion and it has been described as an RNA-binding protein (Masters, 2006; Parker and Masters, 1990; Zuniga *et al.*, 2006).

After the establishment of reverse genetic systems based on cloned cDNA it has been reported that the recovery of recombinant coronaviruses is greatly facilitated by N protein expression (Almazan *et al.*, 2004; Baric and Sims, 2005; Casais *et al.*, 2001; Coley *et al.*, 2005; Schelle *et al.*, 2005; Thiel *et al.*, 2001a; Thiel *et al.*, 2001b; Thiel *et al.*, 2003b; Thiel and Siddell, 2005; Yount *et al.*, 2000; Yount *et al.*, 2003; Yount *et al.*, 2002). Similarly, the heterologous gene expression by coronavirus vectors is markedly increased under conditions of sustained N protein expression (Almazan *et al.*, 2006; Almazan *et al.*, 2004; Baric and Sims, 2005; Curtis *et al.*, 2002; Enjuanes *et al.*, 2005; Hertzig *et al.*, 2004; Schelle *et al.*, 2005; Thiel *et al.*, 2003b; Thiel and Siddell, 2005). Notably, we have observed that coronavirus transcription in the context of HCoV-229E vectors can take place in the absence of N protein (Thiel *et al.*, 2001b; Thiel *et al.*, 2003b). However, efficient full-length vector replication is dependent on N protein expression in *cis* or in *trans* (Schelle *et al.*, 2005; Thiel *et al.*, 2003b). This finding corroborates a study of the Enjuanes laboratory that described a similar set of experiments using TGEV replicons (Almazan *et al.*, 2004). To further confirm this hypothesis, we also analyzed full-length HCoV-229E genomes for their ability to replicate in correlation with N protein expression (Schelle *et al.*, 2005). By modifying the transcription regulatory sequence (TRS) of the N protein mRNA in the HCoV-229E genome to include a stretch of three random nucleotides, we could demonstrate selective replication of genomes that encode a functional TRS, and thus are able to mediate N protein expression. Taken together, reverse genetic studies provide conclusive evidence that N protein expression is required for efficient genome replication.

Reverse genetic analyses of the structural gene region

The 3′-terminal third of coronavirus genomes encodes mainly structural proteins and is referred to as the structural gene region. There are four structural proteins that can be found in all coronaviruses: the spike glycoprotein S, the envelope protein E, the membrane protein M and the nucleocapsid protein N (Masters, 2006; Siddell *et al.*, 2005). S, E and M proteins are embedded in the viral envelope that surrounds the viral ribonucleocapsid comprised of N proteins and the viral RNA genome. Some group 2a coronaviruses encode an additional structural protein, the hemagglutinin-esterase (HE) which is also embedded into the viral envelope (Masters, 2006; Siddell *et al.*, 2005). Interspersed between the structural genes coronaviruses also encode so-called accessory genes (formerly called group-specific genes). The number and location of accessory genes varies within coronavirus genomes (Masters, 2006; Siddell *et al.*, 2005).

Analysis of recombinant coronaviruses encoding rearranged structural genes

All coronaviruses have a canonical organization of structural genes in the order 5′-S-E-M-N-3′. The significance of this remarkable conservation has been challenged by de Haan and colleagues (de Haan *et al.*, 2002b) by producing a number of recombinant viruses encoding rearranged structural genes (5′-S-M-E-N-3′, 5′-S-N-E-M-3′, 5′-M-S-E-N-3′ and 5′-E-M-S-N-3′). Surprisingly, all of these viruses were found to be viable in tissue culture. One recombinant virus (5′-E-M-S-N-3′) was also shown to replicate *in vivo*, although to reduced levels compared to wild-type virus. The results of this study highlighted the extraordinary plasticity of coronavirus genomes and, in addition, lead to an improvement of targeted recombination (Goebel *et al.*, 2004; Masters and Rottier, 2005). Thus, by using a recombinant virus with rearranged order of structural genes as parental virus for targeted recombination background progeny recombinants generated by double crossovers could virtually be eliminated.

Reverse genetic analyses of coronavirus structural proteins

Reverse genetic studies targeting the structural genes of coronaviruses were pioneered by the generation and analysis of recombinant viruses generated by targeted recombination. Studies in this area involved the analysis of interactions of structural proteins during particle formation, functional aspects of structural proteins in receptor binding, virus-cell fusion, tropism and pathogenesis (Masters and Rottier, 2005).

Using the initial version of targeted recombination, the N protein has been analyzed resulting in the mapping of thermolability of MHV virions to a point mutation and deletion, respectively (Koetzner *et al.*, 1992; Masters and Rottier, 2005). Along with the improvement of the system more severely impaired mutants have been rescued. Thus, it was possible to generate a M gene mutant virus, MΔ2, containing a 2-amino-acid carboxy-terminal truncation of the M protein (Kuo and Masters, 2002). Initially, MΔ2 formed tiny plaques but, upon passage, gave raise to a number of revertants. Interestingly, some second site revertants contained compensatory mutations within the N gene, suggesting a structural interaction between M and N proteins. To further corroborate the interaction of MHV N and M carboxy-termini, similar experiments were done by constructing clustered charged-to-alanine N gene mutants, and indeed, some second-site revertants contained mutations within the M gene that could compensate for the original defect in N protein (Hurst *et al.*, 2005).

A clustered charged-to-alanine mutagenesis approach has also been used to study MHV E protein function (Fischer *et al.*, 1998). Two resulting mutants revealed a temperature-sensitive phenotype and electron microscopic analysis of virions derived from the more impaired mutant showed an aberrant morphology. This finding together with earlier reports on the requirements of MHV E protein for virus-like particle (VLP) formation (Bos *et al.*, 1996; Vennema *et al.*, 1996) solidified a pivotal role of E protein in coronavirus assembly. Reverse genetic studies of TGEV revealed indeed an indispensable role of E for TGEV replication in tissue culture, as deduced from recombinant viral vectors lacking E that were dependent on E protein expression in trans (Curtis *et al.*, 2002; Ortego *et al.*, 2002). Therefore, these vectors were described as replication-competent and propagation-deficient. Surprisingly, however, recombinant viruses lacking the E gene were viable in the MHV system, although the mutant virus produced tiny plaques and replicated to low titers (Kuo and Masters, 2003). Thus, for MHV the E protein greatly enhances virion formation but is not essential. Future studies are required to elucidate if the apparent differences between the requirement of E protein in particle formation of TGEV (group 1 coronavirus) and MHV (group 2a coronavirus) are related to basic (group-specific?) differences of E protein function or reflect the relative robustness of MHV growth and plaque formation in tissue culture.

The most intensively studied coronavirus structural protein is the S protein (see Chapter 1). This reflects its important roles in receptor binding, membrane fusion activity, tropism and pathogenesis. Using a VLP system Godeke and colleagues identified that the carboxyterminally 64 residues comprising the endodomain and transmembrane domain of the MHV spike protein confer packaging into MHV VLPs (Godeke *et al.*, 2000). This knowledge has then been used to generate recombinant viruses encoding a chimeric spike protein comprised of the carboxyterminal 64 residues of MHV fused to the FIPV spike ectodomain (Kuo *et al.*, 2000). This virus, designated fMHV, lost the ability to infect murine cells but could now infect feline cells. Beside the demonstration that the MHV/FIPV chimeric spike protein conferred a switch in species tropism, the fMHV virus is now used in a modified protocol of the targeted recombination-based reverse genetic system that allows for the rescue of mutants that are severely impaired in their growth in tissue culture (Masters and Rottier, 2005). Similar conclusions on the role of spike as the major determinant of tropism can be drawn from experiments using TGEV and IBV (Hodgson *et al.*, 2006; Hodgson *et al.*, 2004; Sanchez *et al.*, 1999; Tarpey *et al.*, 2006). In both systems the tropism correlated with the spike protein expressed by the recombinant viruses. Interestingly, a recombinant IBV mutant, based on the apathogenic Beaudette strain but containing the spike gene of the virulent M41 strain (BeauR-M41(S)), was still apathogenic in chicken despite a changed cell tropism. Therefore, other factors than the spike protein most likely contribute to the pathogenic phenotype of the IBV M41 strain. It should be noted that the recombinant BeauR-M41(S) virus induced protection against IBV M41 infections indicating that a spike gene exchange represents a promising approach for IBV vaccine development (Hodgson *et al.*, 2006; Hodgson *et al.*, 2004; Tarpey *et al.*, 2006). The role of coronavirus spike proteins as the major determinant of viral tropism and pathogenesis has been analyzed in more detail, mainly by the Weiss laboratory, in the MHV system. There is a large number of MHV strains available that differ for example in their ability to efficiently replicate in the liver or the CNS. Thus, in a series of experiments it has been shown that the spike proteins of MHV-2, a MHV strain that replicates to high titers in the liver, is sufficient to induce a severe hepatitis with extensive hepatocellular necrosis when expressed in the MHV-A59 genetic background (Das Sarma *et al.*, 2000; Navas *et al.*, 2001). Similarly, the expression of the spike protein of MHV-4, a highly neurovirulent isolate of the MHV-JHM strain, confer a high neurovirulent phenotype when expressed in the MHV-A59 genetic background (Phillips *et al.*, 1999). In addition, this chimeric virus, like the parental MHV-4, replicated only poorly in the liver and induced few if any pathological changes (Navas *et al.*, 2001). However, it should be noted that the MHV-A59 spike protein in the MHV-JHM genetic background did not reproduce the MHV-A59 hepatotropic phenotype, indicating that other factors, as observed in the IBV system, can play a (minor) role in coronavirus tropism and pathogenesis (Iacono *et al.*, 2006; MacNamara *et al.*, 2005; Navas and Weiss, 2003). Finally, it should be mentioned that reverse genetic analyses of coronavirus spike proteins also addressed functional aspects of coronavirus spike proteins, such as virus–cell and cell–cell fusion activity, the impact of amino acid substitutions in the spike heptad repeat regions or S1-S2 cleavage site and the analyses of amino acid substitutions resulting in changes in virus tropism, virulence or replication *in vivo* (see Chapter 1).

The fifth structural protein of coronaviruses is the hemagglutinin esterase (HE). This glycoprotein in expressed only by a subset of closely related group 2a coronaviruses, such as bovine coronavirus (BCoV), human coronaviruses OC43 (HCoV-OC43) and HKU1 (HCoV-HKU1) and some strains of MHV (Brian *et al.*, 1995; Woo *et al.*, 2005).

Coronavirus HE proteins have been described as sialic acid-binding, type I envelope glycoproteins with sialate-O-actetylesterase activity (Smits *et al.*, 2005). Accordingly, coronavirus HE glycoprotein function has been implicated in initial virion-cell attachment by its sialic acid binding activity and/or efficient release of coronavirus virions from infected cells due to its sialate-O-actylesterase activity. Interestingly, in many laboratory strains of MHV the HE gene expression was lost upon passaging in tissue culture, indicating that HE expression is dispensable *in vitro* (Luytjes *et al.*, 1988; Shieh *et al.*, 1989; Yokomori *et al.*, 1991). Even *in vivo*, HE expression is apparently not required (La Monica *et al.*, 1991; Yokomori *et al.*, 1993), although it has been speculated that HE expression by some MHV strains may be associated with a (more) neurovirulent phenotype in mice (Yokomori *et al.*, 1995). Notably, phylogenetic analyses indicate that the HE gene in rodent coronavirus field isolates is strictly maintained, supporting the idea that HE expression may contribute to viral fitness during natural infections (Smits *et al.*, 2005). To shed more light on the role of HE *in vitro* and *in vivo*, de Groot and colleagues constructed a set of recombinant viruses based on MHV-A59 expressing either the wide-type HE protein, an enzymatic inactive HE protein, or no HE protein at all (Kazi *et al.*, 2005; Lissenberg *et al.*, 2005). Although there were no gross differences detectable in single step growth curves *in vitro*, HE expression was rapidly lost during serial passaging of HE-encoding viruses due to mutations that precluded the incorporation of HE into virions. The observed reduced viral fitness of HE-expressing viruses *in vitro* was not due to the O-acetylesterase activity, since compared to viruses that do not express HE a similarly reduced fitness of HE-expressing viruses was observed in competition–propagation experiments irrespectively if the expressed HE is enzymatically active or inactive. The analysis of the same set of recombinant viruses *in vivo* did not reveal any significant differences in the tropism, pathogenicity or virus spread in mice. However, when a mirror set of recombinants encoding the spike protein of MHV-JHM was analyzed an enhanced neurovirulence was observed when recombinant viruses expressed HE in combination with MHV-JHM-derived spike (Kazi *et al.*, 2005). Noteworthy, the enhanced neurovirulence had been observed regardless of whether the HE retained esterase activity or not. These data further support the idea that, although there is an apparent selection pressure against HE expression in tissue culture, HE has a biological role to play *in vivo*.

Reverse genetic analyses of coronavirus accessory proteins

The coronavirus transcription mechanism enable the production of multiple subgenomic (sg) viral mRNAs (Pasternak *et al.*, 2006; Sawicki *et al.*, 2006), mainly encoding the structural proteins that are incorporated into the virion. The number of sg mRNAs produced by a particular coronavirus usually exceeds the number of encoded structural proteins and, consequently, coronaviruses are able to express additional, so-called—accessory—genes. The number of accessory genes differ between the coronavirus groups but some concordance, also regarding the positions where these genes are encoded within the genome, can be observed within groups or subgroups of coronaviruses. Therefore, the ability to transcribe one or more sg mRNA(s) encoding accessory proteins might be one mechanism that enables coronaviruses to express new genetic material, that may have been acquired by recombination. The functions of coronavirus accessory proteins are largely unknown. For one accessory protein, namely the NS2 protein encoded on mRNA2 of MHV and some other group 2a coronaviruses, a cyclic phosphodiesterase (CPD) activity has been predicted on the basis of homology to a cellular protein that is involved in pre-tRNA splicing (Snijder *et al.*, 2003). Although the enzymatic activity of coronavirus CPD has not yet been experimentally verified, there is evidence to believe that this protein (or activity) may play a role *in vivo* rather than *in vitro*. A single amino acid substitution (Leu94Pro) in the MHV CPD resulted in reduced virulence in mice whereas no detectable differences of viral growth were encountered in tissue culture (Sperry *et al.*, 2005). In general, reverse genetic analyses have generated a clear picture that accessory proteins are not required for viral replication in tissue culture but there is increasing evidence that their deletion usually results in reduced virulence in the natural host (Casais *et al.*, 2005; de Haan *et al.*, 2002a; Fischer *et al.*, 1997; Haijema *et al.*, 2004; Hodgson *et al.*, 2006; Masters, 2006; Ortego *et al.*, 2003). Interestingly, a direct contribution of an accessory protein to an enhancement of virulence has been demonstrated for the SARS-CoV ORF6 protein when inserted into the genome of an attenuated MHV strain (Pewe *et al.*, 2005). Here reverse genetics has been used to produce a gain-of-function mutant and it will be interesting to learn how the SARS-CoV ORF6 protein may function to increase virus virulence *in vivo*.

Coronavirus replicon RNAs: towards the generation of coronavirus-based vaccine vectors

The extraordinary large genome and the unique transcription strategy make coronaviruses promising candidates for the development of multigene expression vectors (Bredenbeek and Rice, 1992; Enjuanes *et al.*, 2001). The observation that accessory genes are dispensable for virus replication allows for a replacement of these genes by heterologous genes. In principle, coronavirus vectors have room for the insertion of large heterologous genes and the expression of multiple genes is feasible (Bredenbeek and Rice, 1992; Enjuanes *et al.*, 2001; Thiel *et al.*, 2003b). Consequently, coronavirus reverse genetic systems are currently being used to assess the efficacy of coronavirus-based expression vectors for the production of high amounts of heterologous proteins or as immunogenic vectors in the context of vaccine development and immunotherapy. In addition, reverse genetic techniques have also been used to introduce reporter genes into coronavirus genomes or replicon RNAs in order to facilitate the analysis of coronavirus replication.

Coronavirus replicon RNAs

Replicon RNAs are defined as autonomous replicating RNAs and, consequently, replicon RNAs must encode all replicative viral proteins and *cis*-acting elements required for RNA replication. Thus, in order to construct coronavirus replicons it is important to identify viral genes and sequence elements that are indispensable for efficient coronavirus replication. By analogy to other positive stranded RNA viruses it had been assumed that the replicase gene and *cis*-acting elements at the 5′ and 3′ genomic termini suffice for efficient coronavirus replication and transcription. However, as described above, reverse genetic analyses revealed a role of coronavirus N proteins for efficient coronavirus replication (Almazan *et al.*, 2004; Schelle *et al.*, 2005). Therefore, the basic units of a coronaviral replicon are (i) the replicase gene, (ii) the 5′ end 3′ genomic termini (containing *cis*-acting elements required for replication and transcription) and (iii) the nucleocapsid gene.

Coronavirus replicons are currently being used as a non-infectious systems to analyze coronavirus replication and transcription (Almazan *et al.*, 2006; Chen *et al.*, 2005; Hertzig *et al.*, 2004). These systems are particularly useful if the corresponding virus grows poorly in tissue culture or if the pathogenicity of the virus is a matter of concern. In order to facilitate the analysis of replicon amplification, reporter genes have been inserted as individual transcriptional units into coronavirus replicons. The general principle of expressing heterologous genes in coronavirus systems is to make use of the coronavirus transcription mechanism by inserting a transcriptional cassette, comprised of a transcription regulatory sequence (TRS) located upstream of the gene of interest, into a coronavirus genome or replicon RNA (Figure 5.2) (Thiel and Siddell, 2005). These heterologous transcription units are encoded downstream the replicase gene in the 3′-third of the genome or replicon RNA. It is, however, also possible to express heterologous genes by modifying the replicase gene. This has been demonstrated for a HCoV-229E-based replicon RNA where a selectable marker, the neomycin resistance (neo) gene, has been inserted between nsp1 and nsp2 (Figure 5.2c) (Hertzig *et al.*, 2004). Since the carboxy-terminus of nsp1 and the amino-terminus of nsp2 are released by proteolytic processing in the wild-type context, a "2A-like" autoprocessing peptide has been used to liberate a slightly modified nsp1 carboxy-terminus. The neo gene was cloned downstream of the 2A-like element followed by an internal ribosomal entry site (IRES) that drives the translation of nsp2-nsp16. This "selection cassette," comprised of the 2A-like element, the neo gene and the IRES element, has been used to generate stable replicon-containing cell lines using conventional G418 selection after replicon RNA transfection into eukaryotic cells. An attractive application for these non-infectious replicon cell lines is the identification and evaluation of coronavirus replicase inhibitors (Chen *et al.*, 2005; Hertzig *et al.*, 2004).

Generation of coronavirus-based vector RNAs

The unique characteristics of coronavirus transcription have also been used for the development of coronavirus-based vector RNAs. Soon after reverse genetic systems became available, replication-competent but propagation-deficient coronavirus vectors have been generated (Curtis *et al.*, 2002; Ortego *et al.*, 2002). The rationale of this concept is to generate coronavirus replicons by deleting one or more structural genes and, in addition, to provide a system for the production of virus-like particles (VLPs) by expressing the missing structural gene(s) in trans in packaging cells. Proof of principle for this strategy has been demonstrated by two groups using recombinant TGEV lacking the E gene (Curtis *et al.*, 2002; Ortego *et al.*, 2002). Packaging of ΔE vectors to VLPs was done by expression of E via host cell transcription or alphaviral expression systems. Both groups reported that VLP production of TGEV-based vectors lacking the E gene are dependent on E protein expression in trans, indicating that E is indispensable for TGEV virus propagation. Although for MHV it has been reported that E is not absolutely required for virus propagation (see above; (Kuo and Masters, 2003)), ΔE vectors are still considered safe vectors, since recombinant MHV lacking the E gene did only replicate to extremely low titers. Furthermore, neither in the MHV system, nor in the TGEV system reconstitution of parental viruses, that might have acquired the E gene by recombination within the packaging cells, have been reported. In order to further improve the safety of such vectors it should be possible to virtually exclude reconstitution of wild-type viruses by using vec-

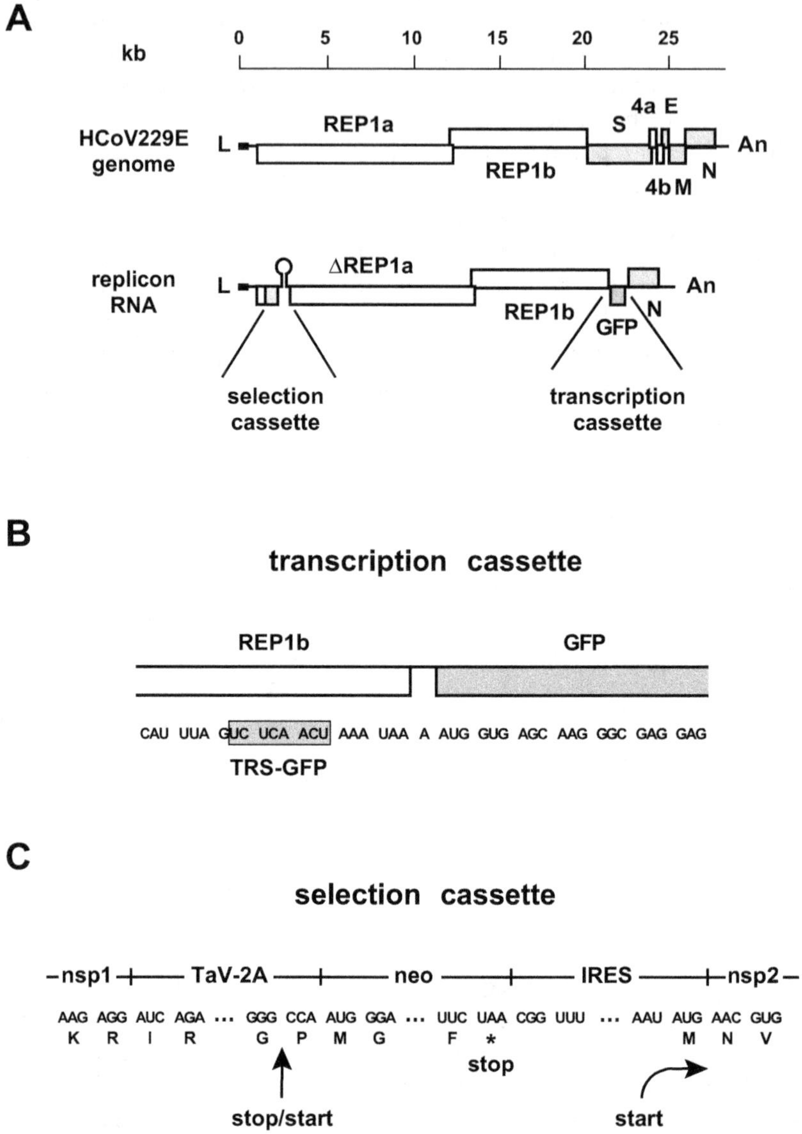

Figure 5.2 Structure of HCoV-229E-based replicon RNA. A. The structure of a HCoV-229E-based replicon RNA (Hertzig *et al.*, 2004) is shown together with the structure of the parental HCoV-229E genomic RNA. ORFs are shown as boxes. Regions that are modified in the replicon RNA are depicted as selection cassette and transcription cassette. L, leader sequence; An, polyA-tail. B. Illustration of the transcription cassette encoded by the HCoV-229E-based replicon RNA. TRS, transcription regulatory sequence. C. Illustration of the selection cassette encoded by the HCoV-229E replicon. TaV-2A, sequence of a 2A-like autoprocessing peptide derived from *Thosea asigna* virus. This element is proposed to mediate a translational termination and re-initiation at the carboxyterminal glycine/proline residues (arrow) of the TaV-2A peptide (Donnelly *et al.*, 2001a; Donnelly *et al.*, 2001b), resulting in an nsp1 that is extended at its carboxyterminus by the short TaV-2A sequence (19 residues) and an neomycin resistance protein that is aminoterminally extended by an additional proline residue. The expression of nsp2-nsp16 is mediated by the EMCV IRES element.

tors with rearranged structural genes (de Haan *et al.*, 2002b) or by using a strategy that "rewires" the coronavirus transcription circuit (Yount *et al.*, 2006). The latter approach has been successfully demonstrated for SARS-CoV (Yount *et al.*, 2006). Recombinant viruses have been produced that contained changed TRS elements, resulting in an incompatibility of recombinant and wild-type TRS elements. It should also be noted that coronavirus-based vectors usually lack all accessory genes, a strategy that further increases vector safety (de Haan *et al.*, 2002a; Haijema *et al.*, 2004). The concept of coronavirus-based vectors was put forward to the expression of multiple heterologous genes in the HCoV-229E system. A multigene vector was constructed to encode three heterologous genes, namely the chloramphenicol-actyltransferase (CAT) gene, the firefly luciferase (LUC) gene and the GFP gene (Thiel *et al.*, 2003b). Vector-mediated expression of CAT, LUC and GFP was demonstrated on the RNA level (i.e. sequencing of specific leader-body junctions of reporter gene mRNAs) and on the protein level using appropriate reporter protein assays. Although all structural genes were deleted, this vector RNA could be packaged to VLPs by the use of a helper virus providing the structural protein in trans. It should be noted, however, that due to the deletion of all structural proteins VLP titers were low under these conditions and a substantial amount of helper virus was present in VLP stocks.

An important consideration for viral vaccine vectors is their potential for efficient delivery of their genetic material to specific target cells. For example, targeting of viral vaccine vectors to professional antigen-presenting cells, such as dendritic cells (DCs), is highly desirable in order to optimize vaccine efficacy. It is important to note that the HCoV 229E receptor, human aminopeptidase N (hAPN or CD13), is expressed at high levels on human DCs. This implies that HCoV 229E-based VLPs could be used to efficiently (receptor-mediated uptake) transduce these cells. Using a HCoV-229E-based multigene vector is has been demonstrated that VLPs can be used to transduce immature and mature human DCs and that heterologous gene expression can be achieved in these cells (Thiel *et al.*, 2003b). Similarly, the receptor of MHV, carcinoembryonic antigen-related cell adhesion molecule (CEACAM) 1, is expressed on murine DCs and it has been shown that MHV can infect murine DCs (Cervantes-Barragan *et al.*, 2007; Turner *et al.*, 2004; Zhou and Perlman, 2006). Noteworthy, the concept of coronavirus-based vectors has already been adapted to the murine system. MHV-based vector RNAs lacking the structural genes E and M have been produced and successfully packaged in E and M-expressing packaging cell lines (Eriksson *et al.*, 2006). Most importantly, VLPs containing MHV-based vector RNAs were able to transduce murine DCs and to mediate GFP expression in murine DCs. The murine model will allow a thorough evaluation of coronavirus vectors due to the availability of well established immunological techniques and well characterized inbred and transgenic mice. Therefore, recombinant MHV vectors in the context of a murine model may guide the development of coronavirus vaccine vectors in general, and may pave the way for the generation of coronavirus-based vaccines in humans (Eriksson *et al.*, 2006; Thiel and Siddell, 2005).

Perspectives

Coronavirus reverse genetics has been applied to the generation of recombinant coronaviruses and vector RNAs and has considerably contributed to a better understanding of general principles governing the molecular biology of coronaviruses. Nevertheless it should be kept in mind that reverse genetics has its limitations. For example, it is difficult to study viral gene functions that are essential for coronavirus RNA synthesis. There are already a number of reports describing the introduction of lethal mutations in studies that aimed to investigate coronavirus replicase gene functions. To overcome these limitations, individual gene functions can be studied by bioinformatic, biochemical or structural means. Moreover, classical "forward" genetics represents an attractive tool to analyze essential viral gene functions. The power of classical genetics has recently been demonstrated through the detailed analysis of temperature-sensitive MHV mutants that are unable to synthesize viral RNA when the infection is performed at the non-permissive temperature (Sawicki *et al.*, 2005). Such mutants are likely to be defective in some aspect(s) of the coronavirus RNA synthesis, and since they are conditionally lethal they are experimentally accessible. Therefore, the knowledge deduced from bioinformatic, biochemical, structural and classical genetic studies can be put to great advantage to exploit the full potential of coronavirus reverse genetics.

Acknowledgments

The coronavirus research at the Kantonal Hospital in St. Gallen was supported by grants from the Swiss National Science Foundation, the National Institutes of Health, USA, the Gebert-Rüf Foundation, Switzerland, the Swiss Cancer League, and the European Commission in the context of the activities of the Euro-Asian SARS-DTV Network (SP22-CT-2004-511064).

References

Almazan, F., Dediego, M.L., Galan, C., Escors, D., Alvarez, E., Ortego, J., Sola, I., Zuniga, S., Alonso, S., Moreno, J.L., *et al.* (2006). Construction of a severe acute respiratory syndrome coronavirus infectious cDNA clone and a replicon to study coronavirus RNA synthesis. J. Virol. *80*, 10900–10906.

Almazan, F., Galan, C., and Enjuanes, L. (2004). The nucleoprotein is required for efficient coronavirus genome replication. J. Virol. *78*, 12683–12688.

Almazan, F., Gonzalez, J.M., Penzes, Z., Izeta, A., Calvo, E., Plana-Duran, J., and Enjuanes, L. (2000). Engineering the largest RNA virus genome as an infectious bacterial artificial chromosome. Proc. Natl. Acad. Sci. USA *97*, 5516–5521.

Anand, K., Palm, G.J., Mesters, J.R., Siddell, S.G., Ziebuhr, J., and Hilgenfeld, R. (2002). Structure of coronavirus main proteinase reveals combination of a chymotrypsin fold with an extra alpha-helical domain. EMBO J. *21*, 3213–3224.

Anand, K., Ziebuhr, J., Wadhwani, P., Mesters, J.R., and Hilgenfeld, R. (2003). Coronavirus main proteinase (3CLpro) structure: basis for design of anti-SARS drugs. Science *300*, 1763–1767.

Baker, S.C., Shieh, C.K., Soe, L.H., Chang, M.F., Vannier, D.M., and Lai, M.M. (1989). Identification of a domain required for autoproteolytic cleavage of murine coronavirus gene A polyprotein. J. Virol. *63*, 3693–3699.

Baric, R.S., and Sims, A.C. (2005). Development of mouse hepatitis virus and SARS-CoV infectious cDNA constructs. Curr. Top. Microbiol. Immunol. *287*, 229–252.

Barretto, N., Jukneliene, D., Ratia, K., Chen, Z., Mesecar, A.D., and Baker, S.C. (2005). The papain-like protease of severe acute respiratory syndrome coronavirus has deubiquitinating activity. J. Virol. *79*, 15189–15198.

Bhardwaj, K., Guarino, L., and Kao, C.C. (2004). The severe acute respiratory syndrome coronavirus Nsp15 protein is an endoribonuclease that prefers manganese as a cofactor. J. Virol. *78*, 12218–12224.

Bhardwaj, K., Sun, J., Holzenburg, A., Guarino, L.A., and Kao, C.C. (2006). RNA recognition and cleavage by the SARS coronavirus endoribonuclease. J. Mol. Biol. *361*, 243–256.

Bi, W., Pinon, J.D., Hughes, S., Bonilla, P.J., Holmes, K.V., Weiss, S.R., and Leibowitz, J.L. (1998). Localization of mouse hepatitis virus open reading frame 1A derived proteins. J. Neurovirol. *4*, 594–605.

Bonilla, P.J., Hughes, S.A., Pinon, J.D., and Weiss, S.R. (1995). Characterization of the leader papain-like proteinase of MHV-A59: identification of a new in vitro cleavage site. Virology *209*, 489–497.

Bonilla, P.J., Hughes, S.A., and Weiss, S.R. (1997). Characterization of a second cleavage site and demonstration of activity in trans by the papain-like proteinase of the murine coronavirus mouse hepatitis virus strain A59. J. Virol. *71*, 900–909.

Bos, E.C., Luytjes, W., van der Meulen, H.V., Koerten, H.K., and Spaan, W.J. (1996). The production of recombinant infectious DI-particles of a murine coronavirus in the absence of helper virus. Virology *218*, 52–60.

Bredenbeek, P., and Rice, C. (1992). Animal RNA virus expression systems. Semin. Virol. *3*, 297–310.

Brian, D.A., and Baric, R.S. (2005). Coronavirus genome structure and replication. Curr. Top. Microbiol. Immunol. *287*, 1–30.

Brian, D.A., Hogue, B.G., and Kienzle, T.E. (1995). The coronavirus hemagglutinin esterase glycoprotein. In The Coronaviridae, S.G. Siddell, ed. (New York, Plenum Press), pp. 165–179.

Brockway, S.M., and Denison, M.R. (2005). Mutagenesis of the murine hepatitis virus nsp1-coding region identifies residues important for protein processing, viral RNA synthesis, and viral replication. Virology *340*, 209–223.

Casais, R., Davies, M., Cavanagh, D., and Britton, P. (2005). Gene 5 of the avian coronavirus infectious bronchitis virus is not essential for replication. J. Virol. *79*, 8065–8078.

Casais, R., Thiel, V., Siddell, S.G., Cavanagh, D., and Britton, P. (2001). Reverse genetics system for the avian coronavirus infectious bronchitis virus. J. Virol. *75*, 12359–12369.

Cervantes-Barragan, L., Zust, R., Weber, F., Spiegel, M., Lang, K.S., Akira, S., Thiel, V., and Ludewig, B. (2006). Control of coronavirus infection through plasmacytoid dendritic cell-derived type I interferon. Blood, prepublished online September 19, 2006; DOI 10.1182/blood-2006–05–023770.

Chen, L., Gui, C., Luo, X., Yang, Q., Gunther, S., Scandella, E., Drosten, C., Bai, D., He, X., Ludewig, B., *et al.* (2005). Cinanserin is an inhibitor of the 3C-like proteinase of severe acute respiratory syndrome coronavirus and strongly reduces virus replication in vitro. J. Virol. *79*, 7095–7103.

Cheng, A., Zhang, W., Xie, Y., Jiang, W., Arnold, E., Sarafianos, S.G., and Ding, J. (2005). Expression, purification, and characterization of SARS coronavirus RNA polymerase. Virology *335*, 165–176.

Coley, S.E., Lavi, E., Sawicki, S.G., Fu, L., Schelle, B., Karl, N., Siddell, S.G., and Thiel, V. (2005). Recombinant mouse hepatitis virus strain A59 from cloned, full-length cDNA replicates to high titers in vitro and is fully pathogenic *in vivo*. J. Virol. *79*, 3097–3106.

Curtis, K.M., Yount, B., and Baric, R.S. (2002). Heterologous gene expression from transmissible gastroenteritis virus replicon particles. J. Virol. *76*, 1422–1434.

Das Sarma, J., Fu, L., Tsai, J.C., Weiss, S.R., and Lavi, E. (2000). Demyelination determinants map to the spike glycoprotein gene of coronavirus mouse hepatitis virus. J. Virol. *74*, 9206–9213.

de Haan, C.A., Masters, P.S., Shen, X., Weiss, S., and Rottier, P.J. (2002a). The group-specific murine coronavirus genes are not essential, but their deletion, by reverse genetics, is attenuating in the natural host. Virology *296*, 177–189.

de Haan, C.A., Volders, H., Koetzner, C.A., Masters, P.S., and Rottier, P.J. (2002b). Coronaviruses maintain viability despite dramatic rearrangements of the strictly conserved genome organization. J. Virol. *76*, 12491–12502.

Denison, M.R., Yount, B., Brockway, S.M., Graham, R.L., Sims, A.C., Lu, X., and Baric, R.S. (2004). Cleavage between replicase proteins p28 and p65 of mouse hepatitis virus is not required for virus replication. J. Virol. *78*, 5957–5965.

Dong, S., and Baker, S.C. (1994). Determinants of the p28 cleavage site recognized by the first papain-like cysteine proteinase of murine coronavirus. Virology *204*, 541–549.

Donnelly, M.L., Hughes, L.E., Luke, G., Mendoza, H., ten Dam, E., Gani, D., and Ryan, M.D. (2001a). The "cleavage" activities of foot-and-mouth disease virus 2A site-directed mutants and naturally occurring "2A-like" sequences. J. Gen. Virol. *82*, 1027–1041.

Donnelly, M.L., Luke, G., Mehrotra, A., Li, X., Hughes, L.E., Gani, D., and Ryan, M.D. (2001b). Analysis of the aphthovirus 2A/2B polyprotein "cleavage" mechanism indicates not a proteolytic reaction, but a novel translational effect: a putative ribosomal "skip." J. Gen. Virol. *82*, 1013–1025.

Egloff, M.P., Ferron, F., Campanacci, V., Longhi, S., Rancurel, C., Dutartre, H., Snijder, E.J., Gorbalenya, A.E., Cambillau, C., and Canard, B. (2004). The severe acute respiratory syndrome-coronavirus replicative protein nsp9 is a single-stranded RNA-binding subunit unique in the RNA virus world. Proc. Natl. Acad. Sci. USA *101*, 3792–3796.

Egloff, M.P., Malet, H., Putics, A., Heinonen, M., Dutartre, H., Frangeul, A., Gruez, A., Campanacci, V., Cambillau, C., Ziebuhr, J., *et al.* (2006). Structural and functional basis for ADP-ribose and poly(ADP-ribose) binding by viral macro domains. J. Virol. *80*, 8493–8502.

Enjuanes, L., Sola, I., Almazan, F., Ortego, J., Izeta, A., Gonzalez, J.M., Alonso, S., Sanchez, J.M., Escors, D., Calvo, E., *et al.* (2001). Coronavirus derived expression systems. J. Biotechnol. *88*, 183–204.

Enjuanes, L., Sola, I., Alonso, S., Escors, D., and Zuniga, S. (2005). Coronavirus reverse genetics and development of vectors for gene expression. Curr. Top. Microbiol. Immunol. *287*, 161–197.

Eriksson, K.K., Makia, D., Maier, R., Cervantes, L., Ludewig, B., and Thiel, V. (2006). Efficient transduction of dendritic cells using coronavirus-based vectors. Adv. Exp. Med. Biol. *581*, 203–206.

Fang, S., Chen, B., Tay, F.P., Ng, B.S., and Liu, D.X. (2006). An arginine-to-proline mutation in a domain with undefined functions within the helicase protein (Nsp13) is lethal to the coronavirus infectious bronchitis virus in cultured cells. Virology, doi:10.1016/j.virol.2006.08.020.

Fischer, F., Peng, D., Hingley, S.T., Weiss, S.R., and Masters, P.S. (1997). The internal open reading frame within the nucleocapsid gene of mouse hepatitis virus encodes a structural protein that is not essential for viral replication. J. Virol. *71*, 996–1003.

Fischer, F., Stegen, C.F., Masters, P.S., and Samsonoff, W.A. (1998). Analysis of constructed E gene mutants of mouse hepatitis virus confirms a pivotal role for E protein in coronavirus assembly. J. Virol. *72*, 7885–7894.

Galan, C., Enjuanes, L., and Almazan, F. (2005). A point mutation within the replicase gene differentially affects coronavirus genome versus minigenome replication. J. Virol. *79*, 15016–15026.

Godeke, G.J., de Haan, C.A., Rossen, J.W., Vennema, H., and Rottier, P.J. (2000). Assembly of spikes into coronavirus particles is mediated by the carboxy-terminal domain of the spike protein. J. Virol. *74*, 1566–1571.

Goebel, S.J., Hsue, B., Dombrowski, T.F., and Masters, P.S. (2004). Characterization of the RNA components of a putative molecular switch in the 3′ untranslated region of the murine coronavirus genome. J. Virol. *78*, 669–682.

Gonzalez, J.M., Penzes, Z., Almazan, F., Calvo, E., and Enjuanes, L. (2002). Stabilization of a full-length infectious cDNA clone of transmissible gastroenteritis coronavirus by insertion of an intron. J. Virol. *76*, 4655–4661.

Gorbalenya, A.E., Enjuanes, L., Ziebuhr, J., and Snijder, E.J. (2006). Nidovirales: evolving the largest RNA virus genome. Virus Res. *117*, 17–37.

Gorbalenya, A.E., Koonin, E.V., Donchenko, A.P., and Blinov, V.M. (1989). Coronavirus genome: prediction of putative functional domains in the non-structural polyprotein by comparative amino acid sequence analysis. Nucleic Acids Res. *17*, 4847–4861.

Gosert, R., Kanjanahaluethai, A., Egger, D., Bienz, K., and Baker, S.C. (2002). RNA replication of mouse hepatitis virus takes place at double-membrane vesicles. J. Virol. *76*, 3697–3708.

Graham, R.L., and Denison, M.R. (2006). Replication of murine hepatitis virus is regulated by papain-like proteinase 1 processing of nonstructural proteins 1, 2, and 3. J. Virol. *80*, 11610–11620.

Graham, R.L., Sims, A.C., Brockway, S.M., Baric, R.S., and Denison, M.R. (2005). The nsp2 replicase proteins of murine hepatitis virus and severe acute respiratory syndrome coronavirus are dispensable for viral replication. J. Virol. *79*, 13399–13411.

Haijema, B.J., Volders, H., and Rottier, P.J. (2003). Switching species tropism: an effective way to manipulate the feline coronavirus genome. J. Virol. *77*, 4528–4538.

Haijema, B.J., Volders, H., and Rottier, P.J. (2004). Live, attenuated coronavirus vaccines through the directed deletion of group-specific genes provide protection against feline infectious peritonitis. J. Virol. *78*, 3863–3871.

Han, Y.S., Chang, G.G., Juo, C.G., Lee, H.J., Yeh, S.H., Hsu, J.T., and Chen, X. (2005). Papain-like protease 2 (PLP2) from severe acute respiratory syndrome coronavirus (SARS-CoV): expression, purification, characterization, and inhibition. Biochemistry *44*, 10349–10359.

Harcourt, B.H., Jukneliene, D., Kanjanahaluethai, A., Bechill, J., Severson, K.M., Smith, C.M., Rota, P.A., and Baker, S.C. (2004). Identification of severe acute respiratory syndrome coronavirus replicase products and characterization of papain-like protease activity. J. Virol. *78*, 13600–13612.

Hertzig, T., Scandella, E., Schelle, B., Ziebuhr, J., Siddell, S.G., Ludewig, B., and Thiel, V. (2004). Rapid identification of coronavirus replicase inhibitors using a selectable replicon RNA. J. Gen. Virol. *85*, 1717–1725.

Hodgson, T., Britton, P., and Cavanagh, D. (2006). Neither the RNA nor the proteins of open reading frames 3a and 3b of the coronavirus infectious bronchitis virus are essential for replication. J. Virol. *80*, 296–305.

Hodgson, T., Casais, R., Dove, B., Britton, P., and Cavanagh, D. (2004). Recombinant infectious bronchitis coronavirus Beaudette with the spike protein gene of the pathogenic M41 strain remains attenuated but induces protective immunity. J. Virol. *78*, 13804–13811.

Hurst, K.R., Kuo, L., Koetzner, C.A., Ye, R., Hsue, B., and Masters, P.S. (2005). A major determinant for membrane protein interaction localizes to the carboxy-terminal domain of the mouse coronavirus nucleocapsid protein. J. Virol. *79*, 13285–13297.

Iacono, K.T., Kazi, L., and Weiss, S.R. (2006). Both spike and background genes contribute to murine coronavirus neurovirulence. J. Virol. *80*, 6834–6843.

Imbert, I., Guillemot, J.C., Bourhis, J.M., Bussetta, C., Coutard, B., Egloff, M.P., Ferron, F., Gorbalenya, A.E., and Canard, B. (2006). A second, non-canonical RNA-dependent RNA polymerase in SARS Coronavirus. EMBO J. *25*, 4933–4942.

Ivanov, K.A., Hertzig, T., Rozanov, M., Bayer, S., Thiel, V., Gorbalenya, A.E., and Ziebuhr, J. (2004). Major genetic marker of nidoviruses encodes a replicative endoribonuclease. Proc. Natl. Acad. Sci. USA *101*, 12694–12699.

Ivanov, K.A., and Ziebuhr, J. (2004). Human coronavirus 229E nonstructural protein 13: characterization of duplex-unwinding, nucleoside triphosphatase, and RNA 5′-triphosphatase activities. J. Virol. *78*, 7833–7838.

Joseph, J.S., Saikatendu, K.S., Subramanian, V., Neuman, B.W., Brooun, A., Griffith, M., Moy, K., Yadav, M.K., Velasquez, J., Buchmeier, M.J., *et al.* (2006). Crystal structure of nonstructural protein 10 from the severe acute respiratory syndrome coronavirus reveals a novel fold with two zinc-binding motifs. J. Virol. *80*, 7894–7901.

Kamitani, W., Narayanan, K., Huang, C., Lokugamage, K., Ikegami, T., Ito, N., Kubo, H., and Makino, S. (2006). Severe acute respiratory syndrome coronavirus nsp1 protein suppresses host gene expression by promoting host mRNA degradation. Proc. Natl. Acad. Sci. USA *103*, 12885–12890.

Kanjanahaluethai, A., and Baker, S.C. (2000). Identification of mouse hepatitis virus papain-like proteinase 2 activity. J. Virol. *74*, 7911–7921.

Kazi, L., Lissenberg, A., Watson, R., de Groot, R.J., and Weiss, S.R. (2005). Expression of hemagglutinin esterase protein from recombinant mouse hepatitis virus enhances neurovirulence. J. Virol. *79*, 15064–15073.

Koetzner, C.A., Parker, M.M., Ricard, C.S., Sturman, L.S., and Masters, P.S. (1992). Repair and mutagenesis of the genome of a deletion mutant of the coronavirus mouse hepatitis virus by targeted RNA recombination. J. Virol. *66*, 1841–1848.

Kuo, L., Godeke, G.J., Raamsman, M.J., Masters, P.S., and Rottier, P.J. (2000). Retargeting of coronavirus by substitution of the spike glycoprotein ectodomain: crossing the host cell species barrier. J. Virol. *74*, 1393–1406.

Kuo, L., and Masters, P.S. (2002). Genetic evidence for a structural interaction between the carboxy termini of the membrane and nucleocapsid proteins of mouse hepatitis virus. J. Virol. *76*, 4987–4999.

Kuo, L., and Masters, P.S. (2003). The small envelope protein E is not essential for murine coronavirus replication. J. Virol. *77*, 4597–4608.

La Monica, N., Banner, L.R., Morris, V.L., and Lai, M.M. (1991). Localization of extensive deletions in the structural genes of two neurotropic variants of murine coronavirus JHM. Virology *182*, 883–888.

Lai, M.M. (2000). The making of infectious viral RNA: No size limit in sight. Proc. Natl. Acad. Sci. USA *97*, 5025–5027.

Lee, H.J., Shieh, C.K., Gorbalenya, A.E., Koonin, E.V., La Monica, N., Tuler, J., Bagdzhadzhyan, A., and Lai, M.M. (1991). The complete sequence (22 kilobases) of murine coronavirus gene 1 encoding the putative proteases and RNA polymerase. Virology *180*, 567–582.

Lim, K.P., Ng, L.F., and Liu, D.X. (2000). Identification of a novel cleavage activity of the first papain-like proteinase domain encoded by open reading frame 1a of the coronavirus Avian infectious bronchitis virus and characterization of the cleavage products. J. Virol. *74*, 1674–1685.

Lindner, H.A., Fotouhi-Ardakani, N., Lytvyn, V., Lachance, P., Sulea, T., and Menard, R. (2005). The papain-like protease from the severe acute respiratory syndrome coronavirus is a deubiquitinating enzyme. J. Virol. *79*, 15199–15208.

Lissenberg, A., Vrolijk, M.M., van Vliet, A.L., Langereis, M.A., de Groot-Mijnes, J.D., Rottier, P.J., and de Groot, R.J. (2005). Luxury at a cost? Recombinant mouse hepatitis viruses expressing the accessory hemagglutinin esterase protein display reduced fitness in vitro. J. Virol. *79*, 15054–15063.

Luytjes, W., Bredenbeek, P.J., Noten, A.F., Horzinek, M.C., and Spaan, W.J. (1988). Sequence of mouse hepatitis virus A59 mRNA 2: indications for RNA recombination between coronaviruses and influenza C virus. Virology *166*, 415–422.

MacNamara, K.C., Chua, M.M., Phillips, J.J., and Weiss, S.R. (2005). Contributions of the viral genetic background and a single amino acid substitution in an immunodominant CD8+ T-cell epitope to murine coronavirus neurovirulence. J. Virol. *79*, 9108–9118.

Masters, P.S. (1999). Reverse genetics of the largest RNA viruses. Adv. Virus Res. *53*, 245–264.

Masters, P.S. (2006). The molecular biology of coronaviruses. Adv. Virus Res. *66*, 193–292.

Masters, P.S., and Rottier, P.J. (2005). Coronavirus reverse genetics by targeted RNA recombination. Curr. Top. Microbiol. Immunol. *287*, 133–159.

Matthes, N., Mesters, J.R., Coutard, B., Canard, B., Snijder, E.J., Moll, R., and Hilgenfeld, R. (2006). The non-structural protein Nsp10 of mouse hepatitis virus binds zinc ions and nucleic acids. FEBS Lett. *580*, 4143–4149.

Minskaia, E., Hertzig, T., Gorbalenya, A.E., Campanacci, V., Cambillau, C., Canard, B., and Ziebuhr, J. (2006). Discovery of an RNA virus 3′→5′ exoribonuclease that is critically involved in coronavirus RNA synthesis. Proc. Natl. Acad. Sci. USA *103*, 5108–5113.

Navas, S., Seo, S.H., Chua, M.M., Sarma, J.D., Lavi, E., Hingley, S.T., and Weiss, S.R. (2001). Murine coronavirus spike protein determines the ability of the virus to replicate in the liver and cause hepatitis. J. Virol. *75*, 2452–2457.

Navas, S., and Weiss, S.R. (2003). Murine coronavirus-induced hepatitis: JHM genetic background eliminates A59 spike-determined hepatotropism. J. Virol. *77*, 4972–4978.

Ortego, J., Escors, D., Laude, H., and Enjuanes, L. (2002). Generation of a replication-competent, propagation-deficient virus vector based on the transmissible gastroenteritis coronavirus genome. J. Virol. *76*, 11518–11529.

Ortego, J., Sola, I., Almazan, F., Ceriani, J.E., Riquelme, C., Balasch, M., Plana, J., and Enjuanes, L. (2003). Transmissible gastroenteritis coronavirus gene 7 is not essential but influences in vivo virus replication and virulence. Virology *308*, 13–22.

Parker, M.M., and Masters, P.S. (1990). Sequence comparison of the N genes of five strains of the coronavirus mouse hepatitis virus suggests a three domain structure for the nucleocapsid protein. Virology *179*, 463–468.

Pasternak, A.O., Spaan, W.J., and Snijder, E.J. (2006). Nidovirus transcription: how to make sense...? J. Gen. Virol. *87*, 1403–1421.

Pewe, L., Zhou, H., Netland, J., Tangudu, C., Olivares, H., Shi, L., Look, D., Gallagher, T., and Perlman, S. (2005). A severe acute respiratory syndrome-associated coronavirus-specific protein enhances virulence of an attenuated murine coronavirus. J. Virol. *79*, 11335–11342.

Phillips, J.J., Chua, M.M., Lavi, E., and Weiss, S.R. (1999). Pathogenesis of chimeric MHV4/MHV-A59 recombinant viruses: the murine coronavirus spike protein is a major determinant of neurovirulence. J. Virol. *73*, 7752–7760.

Prentice, E., Jerome, W.G., Yoshimori, T., Mizushima, N., and Denison, M.R. (2004a). Coronavirus replication complex formation utilizes components of cellular autophagy. J. Biol. Chem. *279*, 10136–10141.

Prentice, E., McAuliffe, J., Lu, X., Subbarao, K., and Denison, M.R. (2004b). Identification and characterization of severe acute respiratory syndrome coronavirus replicase proteins. J. Virol. *78*, 9977–9986.

Putics, A., Filipowicz, W., Hall, J., Gorbalenya, A.E., and Ziebuhr, J. (2005). ADP-ribose-1″-monophosphatase: a conserved coronavirus enzyme that is dispensable for viral replication in tissue culture. J. Virol. *79*, 12721–12731.

Ratia, K., Saikatendu, K.S., Santarsiero, B.D., Barretto, N., Baker, S.C., Stevens, R.C., and Mesecar, A.D. (2006). Severe acute respiratory syndrome coronavirus papain-like protease: structure of a viral deubiquitinating enzyme. Proc. Natl. Acad. Sci. USA *103*, 5717–5722.

Saikatendu, K.S., Joseph, J.S., Subramanian, V., Clayton, T., Griffith, M., Moy, K., Velasquez, J., Neuman, B.W., Buchmeier, M.J., Stevens, R.C., and Kuhn, P. (2005). Structural basis of severe acute respiratory syndrome coronavirus ADP-ribose-1″-phosphate dephosphorylation by a conserved domain of nsP3. Structure (Camb) *13*, 1665–1675.

Sanchez, C.M., Izeta, A., Sanchez-Morgado, J.M., Alonso, S., Sola, I., Balasch, M., Plana-Duran, J., and Enjuanes, L. (1999). Targeted recombination demonstrates that the spike gene of transmissible gastroenteritis coronavirus is a determinant of its enteric tropism and virulence. J. Virol. *73*, 7607–7618.

Sawicki, S.G., Sawicki, D.L., and Siddell, S.G. (2006). A contemporary view of coronavirus transcription. J. Virol., doi:10.1128/JVI.01358-06.

Sawicki, S.G., Sawicki, D.L., Younker, D., Meyer, Y., Thiel, V., Stokes, H., and Siddell, S.G. (2005). Functional and genetic analysis of coronavirus replicase-transcriptase proteins. PLoS Pathog. *1*, e39.

Schelle, B., Karl, N., Ludewig, B., Siddell, S.G., and Thiel, V. (2005). Selective replication of coronavirus genomes that express nucleocapsid protein. J. Virol. *79*, 6620–6630.

Seybert, A., Hegyi, A., Siddell, S.G., and Ziebuhr, J. (2000). The human coronavirus 229E superfamily 1 helicase has RNA and DNA duplex-unwinding activities with 5′-to-3′ polarity. RNA *6*, 1056–1068.

Shi, S.T., Schiller, J.J., Kanjanahaluethai, A., Baker, S.C., Oh, J.W., and Lai, M.M. (1999). Colocalization and membrane association of murine hepatitis virus gene 1 products and De novo-synthesized viral RNA in infected cells. J. Virol. *73*, 5957–5969.

Shieh, C.K., Lee, H.J., Yokomori, K., La Monica, N., Makino, S., and Lai, M.M. (1989). Identification of a new transcriptional initiation site and the corresponding functional gene 2b in the murine coronavirus RNA genome. J. Virol. *63*, 3729–3736.

Siddell, S.G., Ziebuhr, J., and Snijder, E.J. (2005). Coronaviruses, toroviruses, and arteriviruses. In Topley and Wilson's microbiology and microbial infections, B. W. J. Mahy, and V. ter Meulen, eds. (London, Hodder Arnold), pp. 823–856.

Smits, S.L., Gerwig, G.J., van Vliet, A.L., Lissenberg, A., Briza, P., Kamerling, J.P., Vlasak, R., and de Groot, R.J. (2005). Nidovirus sialate-O-acetylesterases: evolution and substrate specificity of coronaviral and toroviral receptor-destroying enzymes. J. Biol. Chem. *280*, 6933–6941.

Snijder, E.J., Bredenbeek, P.J., Dobbe, J.C., Thiel, V., Ziebuhr, J., Poon, L.L., Guan, Y., Rozanov, M., Spaan, W.J., and Gorbalenya, A.E. (2003). Unique and conserved features of genome and proteome of SARS-coronavirus, an early split-off from the coronavirus group 2 lineage. J. Mol. Biol. *331*, 991–1004.

Snijder, E.J., van der Meer, Y., Zevenhoven-Dobbe, J., Onderwater, J.J., van der Meulen, J., Koerten, H.K., and Mommaas, A.M. (2006). Ultrastructure and origin of membrane vesicles associated with the severe acute respiratory syndrome coronavirus replication complex. J. Virol. *80*, 5927–5940.

Sperry, S.M., Kazi, L., Graham, R.L., Baric, R.S., Weiss, S.R., and Denison, M.R. (2005). Single-amino-acid substitutions in open reading frame (ORF) 1b-nsp14 and ORF 2a proteins of the coronavirus mouse hepatitis virus are attenuating in mice. J. Virol. *79*, 3391–3400.

St-Jean, J.R., Desforges, M., Almazan, F., Jacomy, H., Enjuanes, L., and Talbot, P.J. (2006). Recovery of a neurovirulent human coronavirus OC43 from an infectious cDNA clone. J. Virol. *80*, 3670–3674.

Su, D., Lou, Z., Sun, F., Zhai, Y., Yang, H., Zhang, R., Joachimiak, A., Zhang, X.C., Bartlam, M., and Rao, Z. (2006). Dodecamer structure of severe acute respiratory syndrome coronavirus nonstructural protein nsp10. J. Virol. *80*, 7902–7908.

Sulea, T., Lindner, H.A., Purisima, E.O., and Menard, R. (2005). Deubiquitination, a new function of the severe acute respiratory syndrome coronavirus papain-like protease? J. Virol. *79*, 4550–4551.

Sulea, T., Lindner, H.A., Purisima, E.O., and Menard, R. (2006). Binding site-based classification of coronaviral papain-like proteases. Proteins *62*, 760–775.

Sutton, G., Fry, E., Carter, L., Sainsbury, S., Walter, T., Nettleship, J., Berrow, N., Owens, R., Gilbert, R., Davidson, A., *et al.* (2004). The nsp9 replicase protein of SARS-coronavirus, structure and functional insights. Structure *12*, 341–353.

Tarpey, I., Orbell, S.J., Britton, P., Casais, R., Hodgson, T., Lin, F., Hogan, E., and Cavanagh, D. (2006). Safety and efficacy of an infectious bronchitis virus used for chicken embryo vaccination. Vaccine *24*, 6830–6838.

Teng, H., Pinon, J.D., and Weiss, S.R. (1999). Expression of murine coronavirus recombinant papain-like proteinase: efficient cleavage is dependent on the lengths of both the substrate and the proteinase polypeptides. J. Virol. *73*, 2658–2666.

Thiel, V., Herold, J., Schelle, B., and Siddell, S.G. (2001a). Infectious RNA transcribed in vitro from a cDNA copy of the human coronavirus genome cloned in vaccinia virus. J. Gen. Virol. *82*, 1273–1281.

Thiel, V., Herold, J., Schelle, B., and Siddell, S.G. (2001b). Viral replicase gene products suffice for coronavirus discontinuous transcription. J. Virol. *75*, 6676–6681.

Thiel, V., Ivanov, K.A., Putics, A., Hertzig, T., Schelle, B., Bayer, S., Weissbrich, B., Snijder, E.J., Rabenau, H., Doerr, H.W., *et al.* (2003a). Mechanisms and enzymes involved in SARS coronavirus genome expression. J. Gen. Virol. *84*, 2305–2315.

Thiel, V., Karl, N., Schelle, B., Disterer, P., Klagge, I., and Siddell, S.G. (2003b). Multigene RNA vector based on coronavirus transcription. J. Virol. *77*, 9790–9798.

Thiel, V., and Siddell, S.G. (2005). Reverse genetics of coronaviruses using vaccinia virus vectors. Curr. Top. Microbiol. Immunol. *287*, 199–227.

Turner, B.C., Hemmila, E.M., Beauchemin, N., and Holmes, K.V. (2004). Receptor-dependent coronavirus infection of dendritic cells. J. Virol. *78*, 5486–5490.

van der Meer, Y., Snijder, E.J., Dobbe, J.C., Schleich, S., Denison, M.R., Spaan, W.J., and Locker, J.K. (1999). Localization of mouse hepatitis virus nonstructural proteins and RNA synthesis indicates a role for late endosomes in viral replication. J. Virol. *73*, 7641–7657.

Vennema, H., Godeke, G.J., Rossen, J.W., Voorhout, W.F., Horzinek, M.C., Opstelten, D.J., and Rottier, P.J. (1996). Nucleocapsid-independent assembly of coronavirus-like particles by co-expression of viral envelope protein genes. EMBO J *15*, 2020–2028.

Woo, P.C., Lau, S.K., Chu, C.M., Chan, K.H., Tsoi, H.W., Huang, Y., Wong, B.H., Poon, R.W., Cai, J.J., Luk, W.K., *et al.* (2005). Characterization and complete genome sequence of a novel coronavirus, coronavirus HKU1, from patients with pneumonia. J. Virol. *79*, 884–895.

Yang, H., Xie, W., Xue, X., Yang, K., Ma, J., Liang, W., Zhao, Q., Zhou, Z., Pei, D., Ziebuhr, J., *et al.* (2005). Design of wide-spectrum inhibitors targeting coronavirus main proteases. PLoS Biol. *3*, e324.

Yang, H., Yang, M., Ding, Y., Liu, Y., Lou, Z., Zhou, Z., Sun, L., Mo, L., Ye, S., Pang, H., *et al.* (2003). The crystal structures of severe acute respiratory syndrome virus main protease and its complex with an inhibitor. Proc. Natl. Acad. Sci. USA *100*, 13190–13195.

Yokomori, K., Asanaka, M., Stohlman, S.A., Makino, S., Shubin, R.A., Gilmore, W., Weiner, L.P., Wang, F.I., and Lai, M.M. (1995). Neuropathogenicity of mouse hepatitis virus JHM isolates differing in hemagglutinin-esterase protein expression. J. Neurovirol. *1*, 330–339.

Yokomori, K., Banner, L.R., and Lai, M.M. (1991). Heterogeneity of gene expression of the hemagglutinin-esterase (HE) protein of murine coronaviruses. Virology *183*, 647–657.

Yokomori, K., Stohlman, S.A., and Lai, M.M. (1993). The detection and characterization of multiple hemagglutinin-esterase (HE)-defective viruses in the mouse brain during subacute demyelination induced by mouse hepatitis virus. Virology *192*, 170–178.

Yount, B., Curtis, K.M., and Baric, R.S. (2000). Strategy for systematic assembly of large RNA and DNA genomes: transmissible gastroenteritis virus model. J. Virol. *74*, 10600–10611.

Yount, B., Curtis, K.M., Fritz, E.A., Hensley, L.E., Jahrling, P.B., Prentice, E., Denison, M.R., Geisbert, T.W., and Baric, R.S. (2003). Reverse genetics with a full-length infectious cDNA of severe acute respiratory syndrome coronavirus. Proc. Natl. Acad. Sci. USA *100*, 12995–13000.

Yount, B., Denison, M.R., Weiss, S.R., and Baric, R.S. (2002). Systematic assembly of a full-length infectious cDNA of mouse hepatitis virus strain A59. J. Virol. *76*, 11065–11078.

Yount, B., Roberts, R.S., Lindesmith, L., and Baric, R.S. (2006). Rewiring the severe acute respiratory syndrome coronavirus (SARS-CoV) transcription circuit: engineering a recombination-resistant genome. Proc. Natl. Acad. Sci. USA *103*, 12546–12551.

Zhai, Y., Sun, F., Li, X., Pang, H., Xu, X., Bartlam, M., and Rao, Z. (2005). Insights into SARS-CoV transcription and replication from the structure of the nsp7-nsp8 hexadecamer. Nat. Struct. Mol. Biol. *12*, 980–986.

Zhou, H., and Perlman, S. (2006). Preferential infection of mature dendritic cells by mouse hepatitis virus strain JHM. J. Virol. *80*, 2506–2514.

Ziebuhr, J. (2005). The coronavirus replicase. Curr. Top. Microbiol. Immunol. *287*, 57–94.

Ziebuhr, J., Schelle, B., Karl, N., Minskaia, E., Bayer, S., Siddell, S.G., Gorbalenya, A.E., and Thiel, V. (2007) Human coronavirus 229E papain-like proteases have overlapping specificities but distinct functions in viral replication. J. Virol. *81*, 3922–3932.

Ziebuhr, J., Herold, J., and Siddell, S.G. (1995). Characterization of a human coronavirus (strain 229E) 3C-like proteinase activity. J. Virol. *69*, 4331–4338.

Ziebuhr, J., Snijder, E.J., and Gorbalenya, A.E. (2000). Virus-encoded proteinases and proteolytic processing in the Nidovirales. J. Gen. Virol. *81*, 853–879.

Ziebuhr, J., Thiel, V., and Gorbalenya, A.E. (2001). The autocatalytic release of a putative RNA virus transcription factor from its polyprotein precursor involves two paralogous papain-like proteases that cleave the same peptide bond. J. Biol. Chem. *276*, 33220–33232.

Zuniga, S., Sola, I., Moreno, J.L., Sabella, P., Plana-Duran, J., and Enjuanes, L. (2006). Coronavirus nucleocapsid protein is an RNA chaperone. Virology, doi:10.1016/j.virol.2006.07.046.

Coronavirus Genome Packaging

6

Krishna Narayanan and Shinji Makino

Abstract

The assembly of infectious coronavirus particles requires the selection of viral genomic RNA from a cellular pool that contains an abundant excess of non-viral and viral RNAs. Among the seven to ten specific viral mRNAs synthesized in virus-infected cells, only the full length genomic RNA is packaged efficiently into coronavirus particles. Studies have revealed *cis*-acting elements and *trans*-acting viral factors involved in coronavirus genome encapsidation and packaging. Understanding the molecular mechanisms of genome selection and packaging is critical for developing antiviral strategies and viral expression vectors based on the coronavirus genome. In this review, we have compiled these collective findings to provide some insights into the initial stages of infectious coronavirus particle assembly.

RNA packaging

To replicate itself, the virus must first introduce its genome successfully into a susceptible host. In infected cells, during the viral replication cycle, successive replication and transcription of the viral genome generates multitudes of genomic and subgenomic RNAs. As RNA accumulates and when sufficient viral structural proteins are produced, the assembly of progeny virus initiates through complex interactions between the proteins and the nucleic acid. A critical step in the production of an infectious progeny virus is to package the viral genome into virus particles. This step involves encapsidation of viral genomic nucleic acid with viral nucleocapsid protein (N) to form a complex of ribonucleoprotein (RNP) also called the nucleocapsid structure, which becomes incorporated or packaged into new virus particles. The viral nucleocapsid protein condenses the relatively large RNA molecule so that it fits into the relatively small virus particle; this important function involves the neutralization of the negatively charged RNA molecule by the basic residues (lysines and arginines) in the nucleocapsid protein. In the case of enveloped viruses, the assembly of infectious virus particles requires the nucleocapsid structure to interact with viral envelope glycoproteins and/or the cellular membrane (Simons and Garoff, 1980). Occasional packaging of non-genomic viral RNAs and cellular RNAs results in non-infectious viruses. The intracellular viral genomic RNA must be able to compete for the limited viral and host assembly factors to form infectious viruses. A highly specific mechanism must exist for the selection of genomic RNA from the pool of viral subgenomic RNAs, antigenomic RNAs, host-cell mRNAs, rRNAs and tRNAs. Several RNA viruses have devised elaborate strategies to selectively incorporate the intracellular genomic RNA into virus particles. A specific *cis*-acting RNA element, called the packaging signal, has been identified for several RNA viruses (Adam and Miller, 1988; Banks *et al.*, 1999; Beck *et al.*, 1997; Berkowitz *et al.*, 1995; Cologna and Hogue, 2000; Dalton *et al.*, 2001; Escors *et al.*, 2003; Fosmire *et al.*, 1992; Fujimura *et al.*, 1990; Hsieh *et al.*, 2005; Luytjes *et al.*, 1989; Pattnaik *et al.*, 1995; Qu and Morris, 1997; Whelan and Wertz, 1999). The packaging signal is defined as the site(s) in the viral genome that guides the selective incorporation of viral genomic RNA into the virus particle. One of the features of a packaging signal is that the insertion of this sequence into a non-viral RNA directs the packaging of a heterologous RNA into virus particles. These packaging-signal RNA sequence elements form specific secondary structures, which are recognized by *trans*-acting factor(s) like viral proteins to direct the packaging of RNA into virus particles. In most viruses, the interaction of the viral nucleocapsid protein with the packaging sequence is considered to be the first step in the encapsidation of the genomic RNA and is assumed to ensure the specificity of RNA packaging (Draper, 1995; Geigenmuller and Linial, 1996; Owen and Kuhn, 1996; Weiss *et al.*, 1989). In some viruses, an additional viral factor(s) is required to selectively incorporate the intracellular nucleocapsid into virus particles. The subject of this review is to provide insights into the current ideas about the mechanism of coronavirus genome packaging.

Coronavirus genome selection

Coronaviruses, belonging to the family Coronaviridae, have a broad host range and cause a number of clinically relevant and economically important diseases in humans and animals. Several strains of coronaviruses are associated with upper respiratory, gastrointestinal and neurological infections, including the recent outbreak of the severe respiratory disease in humans called severe acute respiratory syndrome (SARS) caused by SARS coronavirus (SCoV) (Drosten *et al.*, 2003; Ksiazek *et al.*, 2003; Murray *et al.*, 1992; Peiris *et al.*, 2003; Wege *et al.*, 1982; Weiss and Navas-Martin, 2005). Some representative members of the family include infectious bronchitis virus (IBV), mouse hepatitis virus (MHV), bovine coronavirus (BCoV), porcine transmissible gastroenteritis virus (TGEV), feline infectious peritonitis virus (FIPV) and turkey coronavirus (TCV).

Coronaviruses are spherical, enveloped RNA viruses with a positive-sense, single-stranded RNA genome of 27–32 kilobases (kb) (Lai and Cavanagh, 1997). Coronaviruses possess the largest genome among RNA viruses (Boursnell and Brown, 1984; Lee *et al.*, 1991). The viral genomic RNA is capped and polyadenylated and functions as an mRNA (Lai and Stohlman, 1978; Lai and Stohlman, 1981). In infected cells, seven to ten specific mRNAs, ranging in size from 1.7 kilobases to genomic length, are synthesized. These virus-specific mRNAs have a 3′ coterminal nested set structure and a common leader sequence at the 5′-end (Lai *et al.*, 1984; Spaan *et al.*, 1983). Hence, all the mRNAs are structurally polycistronic, except for the smallest mRNA. However, with certain exceptions in which the internal open reading frames (ORFs) are translated, usually only the 5′-most ORF of each mRNA is translated; thus the coronavirus mRNAs are functionally monocistronic (Lai and Cavanagh, 1997). For some coronaviruses, like MHV, only the genome-length RNA is efficiently packaged into the virus particle. In MHV, the subgenomic RNAs are not incorporated into the virus particles (Lai and Stohlman, 1978; Makino *et al.*, 1988; Makino *et al.*, 1990). TGEV, BCoV and IBV are reported to contain subgenomic RNAs in an amount that is about 5–200-fold lower than the genomic RNA (Cavanagh *et al.*, 1993; Hofmann *et al.*, 1990; Sethna *et al.*, 1991; Sethna *et al.*, 1989). All these studies point towards the presence of a specific packaging signal in coronavirus genomic RNA that drives its efficient incorporation into virus particles. The first specific RNA packaging signal was precisely identified in MHV (Fosmire *et al.*, 1992; Makino *et al.*, 1990; van der Most *et al.*, 1991). Nearly a decade later several subsequent studies have led to the identification of the packaging signal in BCoV, IBV and TGEV (Cologna and Hogue, 2000; Dalton *et al.*, 2001; Escors *et al.*, 2003). More recently, the location of the packaging signal in SCoV genome was also identified (Hsieh *et al.*, 2005).

MHV packaging signal

Of all the coronavirus packaging signals, the nature and function of MHV's packaging signal is the most extensively characterized. The region of the MHV genome that is important for packaging was identified by in-depth analysis of defective interfering (DI) RNA. Analysis of a variety of naturally occurring DI RNAs, after passage of MHV in tissue culture at a high multiplicity of infection, showed that some DI RNAs are efficiently packaged into the virus particle while others are packaged inefficiently (Fosmire *et al.*, 1992; Makino *et al.*, 1990; van der Most *et al.*, 1991). Because the size of inefficiently packaged DI RNAs are both smaller and larger than the efficiently packaged DI RNAs, the role of the size of the RNAs in efficiency of packaging was ruled out (Fosmire *et al.*, 1992; Makino *et al.*, 1990; van der Most *et al.*, 1991). The alternative possibility was the presence of a *cis*-acting packaging signal in the efficiently packaged DI RNAs. Because the 5′ and 3′-end sequences of these subgenomic RNAs and DI RNAs are identical to the efficiently packaged genomic RNA, the possible localization of the packaging signal was predicted to be within the downstream coding region of the 5′-most gene (gene 1). Sequence analysis of efficiently packaged DI RNAs further mapped the packaging signal to a putative region located within ORF1b of the genome in a domain that is not present in the viral subgenomic RNAs (Makino *et al.*, 1990; van der Most *et al.*, 1991). Studies involving deletion analysis of MHV DI RNAs narrowed the precise location of the MHV packaging signal to a 190-nt sequence located between nucleotides 20244 to 20433 from the 5′-end of genomic RNA (Fosmire *et al.*, 1992) (Figure 6.1). Site-directed mutagenesis of this 190-nt sequence showed that a 69-nt stem–loop structure within this 190-nt region is important for the biological function of the packaging signal (Fosmire *et al.*, 1992). Mutations that disrupt the stem–loop structure abrogate the packaging of DI RNA into virus particles. More importantly, this 69-nt region can function as an autonomous packaging signal and is sufficient to drive the incorporation of subgenomic DI RNA and heterologous non-viral RNA into MHV particles (Bos *et al.*, 1997; Woo *et al.*, 1997) showing that the specificity of RNA packaging in MHV is determined by the 69-nt packaging signal. Some past studies, however, have indicated that though specificity of packaging could be determined by the 69-nt packaging signal, the efficiency of packaging could involve additional packaging enhancement signals and/or factors (Bos *et al.*, 1997). In those studies, the packaging efficiency was analyzed in serially passaged virus samples that could have allowed the amplification of even poorly packaged replicating RNAs. A later study partially addressed this issue by directly comparing the packaging efficiency of different DI RNAs into MHV particles (Narayanan and Makino, 2001); it showed that relative to the 69-nt region the

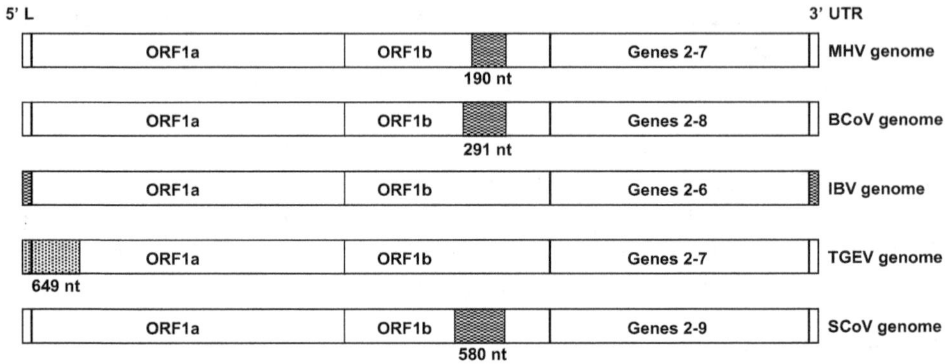

Figure 6.1 Schematic representation of the location of coronavirus packaging signals. The positions of the packaging signals are indicated with shaded boxes. Numbers below the shaded boxes represent the approximate nucleotide length of the packaging signals. The length of the packaging signal in IBV has not been determined as yet. L, leader sequence; UTR, untranslated region; ORF, open reading frame. The figure is not drawn to scale.

190-nt packaging signal provided a higher packaging efficiency to DI RNAs (Narayanan and Makino, 2001). This was the first study to demonstrate the role of the size of the packaging signal in determining packaging efficiency of RNAs into coronavirus particles (Narayanan and Makino, 2001). Additional sequences 5′ to the packaging signal did not alter the packaging efficiency of DI RNAs (Narayanan and Makino, 2001). While DI RNAs are known to require a helper virus for their replication in infected cells, it is not clear whether MHV DI RNAs are copackaged with the viral genomic RNA or packaged separately into virus particles. Genome selection could be a competitive process between genomic RNA and DI RNAs regulating the packaging efficiencies of these RNAs. Additional studies are still required to rule out the existence of packaging enhancement signals in the MHV genome.

Some studies have suggested that replication could play a role in determining the efficiency of MHV RNA packaging (Bos *et al.*, 1997). The packaging of non-replicating MHV subgenomic DI RNA has been shown to be six fold less efficient than the replicating genomic DI RNA (Bos *et al.*, 1997). The incorporation of non-replicating, heterologous RNAs into MHV particles was suggested to be inefficient (Woo *et al.*, 1997). Although replication is not required for RNA to be packaged into MHV particles, folding of RNA into secondary structures as it replicates can easily be imagined to facilitate presentation at the cellular budding compartment, thereby increasing that RNA's incorporation into MHV particles.

BCoV packaging signal

BCoV and MHV are closely related group 2 coronaviruses. An RNA packaging signal similar to that of MHV is found in the ORF1b region of BCoV (Cologna and Hogue, 2000) (Figure 6.1). Insertion of a 291-nt region from ORF1b of BCoV into a non-coronavirus RNA was sufficient to drive its incorporation into BCoV particles (Cologna and Hogue, 2000). The BCoV packaging signal also allowed the incorporation of the heterologous RNA into MHV particles (Cologna and Hogue, 2000). This region was shown to exhibit

72% sequence homology to the MHV packaging signal, which could possibly explain the functional interchangeability of the signal (Cologna and Hogue, 2000). This also suggests that related coronaviruses might use similar protein- and RNA-recognition elements for RNA packaging. The role of this packaging signal in BCoV DI RNA packaging could not be determined because insertion of this sequence into DI RNA affected its replication significantly (Cologna and Hogue, 2000). Note that several lines of evidence have suggested that RNA packaging in BCoV is different from that of MHV. Unlike MHV, abundant levels of subgenomic mRNAs are detected in BCoV particles (Chang and Brian, 1996; Cologna and Hogue, 2000; Hofmann *et al.*, 1990). A BCoV DI RNA that does not have any ORF1b region is also packaged into virus particles (Chang and Brian, 1996; Cologna and Hogue, 2000). This BCoV DI RNA contains the 5′-terminal 498 nt, the N ORF, and the 3′ end of the genome suggesting that an additional region of the BCoV genome could also serve as the packaging signal for BCoV genome and DI RNA (Cologna and Hogue, 2000). Further studies are required to address the contribution of the BCoV packaging signal in regulating BCoV genome incorporation into virus particles.

IBV packaging signal

Deletion analysis of IBV DI RNAs showed that the *cis*-acting sequences in the 5′ untranslated regions (UTR) and/or 3′ UTR of the IBV genome are required for rescue of DI RNAs (Dalton *et al.*, 2001) (Figure 6.1). Rescue of DI RNAs indicated the replication and packaging of RNA into passaged IBV virus particles. No specific part of ORF1b region is required for efficient packaging of IBV DI RNAs. This is in contrast to MHV where an internal region in ORF1b serves as the packaging signal. Whether or not the identified sequences in IBV could mediate the packaging of heterologous RNAs into IBV particles is not yet known. Furthermore, some DI RNAs containing the 5′- and 3′-terminal sequences that replicated well could not be rescued suggesting that the RNA motif involved in packaging of IBV genome requires more delineation.

TGEV packaging signal

The TGEV packaging signal is located within the first 649 nt of the 5′ end of the genome (Escors *et al.*, 2003) (Figure 6.1). Most of this domain is not present in TGEV mRNAs. Using highly immunopurified virions and quantitative RT-PCR, this study convincingly demonstrated that the detection of subgenomic mRNAs in TGEV particles is a function of the degree of purity of the virus preparation (Escors *et al.*, 2003). It was shown that only the packaging signal containing TGEV genome and minigenome (DI RNA) are incorporated by an active process independent of the purity of the virus preparation while the mRNAs are non-specifically associated with the virus particle (Escors *et al.*, 2003). Like IBV (group 3 coronavirus), TGEV (group 1 coronavirus) also does not require regions of ORF1b for RNA packaging. Of note is that the first 649 nt from the TGEV genome is also required for minigenome replication suggesting that the RNA packaging in TGEV could be coupled to genome replication. An interesting observation was that the minigenome packaging efficiency was about 30-fold higher than that of the TGEV genome (Escors *et al.*, 2003). This data suggested that the recognition of the same signal in two different RNAs by the packaging machinery is a competing process leading to the interference with genome packaging. Additional work is required to determine whether the domains for

RNA packaging and replication can be mapped to different regions. Also, it remains to be determined whether the TGEV packaging signal can function as an autonomous sequence in guiding the incorporation of heterologous non-replicating RNA into virus particles.

SCoV packaging signal

A recent study identified the region in the SCoV genome responsible for packaging of heterologous reporter RNA into SCoV virus-like particles (VLPs) (Hsieh *et al.*, 2005). The SCoV VLPs are produced by coexpression of viral M and E proteins (Hsieh *et al.*, 2005; Mortola and Roy, 2004). A 580-nt region from nucleotides 19715 to 20294 in the SCoV genome was identified as the SCoV packaging signal (Hsieh *et al.*, 2005). Like MHV, this domain mapped to the ORF1b region. Bioinformatic analysis showed that this region contains a putative core signal of 63-nt (PS_{core}) that forms a stem–loop structure and this structure is conserved in the 580-nt packaging signal. Insertion of this region into green fluorescent protein (GFP) reporter mRNA confers packaging of this RNA into SCoV VLPs. RNA packaging was measured indirectly by tracking the production of GFP upon infection of cells with SCoV VLPs; this study also showed that the viral N protein is required for the packaging of reporter RNA into SCoV VLPs. A careful and direct analysis of RNA packaging, by measuring RNA levels in VLP, is required to calculate the packaging efficiency conferred by the SCoV packaging signal. Additionally, the role of the 580-nt region in RNA packaging into SCoV particles needs to be demonstrated.

Trans-acting factors involved in selective coronavirus genome packaging

Genome incorporation in RNA viruses is a coordinated process that involves the recognition of the *cis*-acting packaging signal by *trans*-acting factors like viral and host proteins. Data suggests that genome recognition can be achieved by different mechanisms. The viral nucleocapsid protein can selectively bind the cognate packaging signal in the genome and form a specific RNP structure that is recognized by the viral envelope proteins. In this case, the selectivity is determined entirely by the specific interaction of the nucleocapsid protein with the packaging signal in the genome. Another mechanism involves the role of additional viral factors in discriminating between the different RNP complexes in infected cells. In this case, the different viral RNAs form RNP complexes, yet the genomic RNP complex is selected by the viral packaging machinery. Both types of packaging lead to the selective enrichment of genomic RNPs in assembled virions.

The selective packaging mechanism of coronavirus genomic RNPs into virus particles has been extensively studied in MHV. In MHV, N protein wraps the viral genomic RNA to form a helical nucleocapsid structure (Escors *et al.*, 2001; Sturman *et al.*, 1980). The N protein of MHV is a basic phosphoprotein with a molecular weight of approximately 55 kDa (Armstrong *et al.*, 1983; Parker and Masters, 1990); it is phosphorylated at serine residues (Stohlman and Lai, 1979). RNA binding motifs, like arginine-rich motifs (ARM) or zinc fingers, are not found in N protein (Burd and Dreyfuss, 1994; Draper, 1995; Holmes and Behnke, 1981).

Several studies have examined the RNA binding specificity of MHV N protein to determine whether N protein might discriminate between different viral RNAs and provide the selectivity in MHV RNA packaging. N protein specifically interacts, *in vitro*, with the leader region and the packaging signal within the MHV genome (Cologna *et al.*, 2000; Molenkamp and Spaan, 1997; Stohlman *et al.*, 1988). N protein also binds to RNAs of non-viral origin (Cologna *et al.*, 2000; Masters, 1992; Robbins *et al.*, 1986). Most significantly, N protein associates with all MHV RNAs and non-MHV RNAs in infected cells (Cologna *et al.*, 2000; Masters, 1992; Robbins *et al.*, 1986). All this evidence ruled out the role of the specific interaction of N protein with RNA elements in the selective recognition of MHV genome for packaging.

In view of the fact that all the MHV RNAs form intracellular RNP complexes with N protein, how is the intracellular genomic RNP complex selected for incorporation into virus particle? Several studies have shown that the viral transmembrane envelope M protein distinguishes between different intracellular viral RNP complexes and determines the specificity and efficiency of incorporation of viral genomic RNPs into budding MHV particles (Narayanan *et al.*, 2000; Narayanan and Makino, 2001). Several lines of evidence strongly support this conclusion. The M protein specifically interacts with the nucleocapsid in MHV-infected cells and in virions (Narayanan *et al.*, 2000; Sturman *et al.*, 1980). MHV M protein preferentially associates with genomic RNA and N protein in infected cells (Narayanan *et al.*, 2000). M protein interacts selectively with RNAs containing the MHV packaging signal and more importantly, only those RNAs that interact selectively with M protein are efficiently packaged into MHV particles (Narayanan and Makino, 2001). This study showed a strong correlation between the efficiency of interaction of M protein with RNA and the packaging efficiency of RNA into virus particles. These findings are surprising because M protein is a transmembrane viral envelope protein with no known RNA binding motifs. The question is how does the M protein specifically recognize genomic RNA and how does it selectively incorporate the intracellular genomic RNP complex into virus particles? Is the recognition mediated by M protein-N protein interaction or is there a direct interaction between M protein and the packaging signal? Some studies have suggested the presence of M protein-N protein interaction, independent of viral RNA, in the M protein-nucleocapsid complex (Narayanan *et al.*, 2000). In a recent study, genetic evidence showed that the carboxy terminus of M protein binds to N protein in MHV (Kuo and Masters, 2002). However, a direct M protein-N protein interaction has not been demonstrated as yet.

Indirect evidence for an interaction between M protein and viral genomic RNA was provided in a seminal study; Renografin density gradient centrifugation of NP-40-solubilized MHV particles results in cosedimentation of MHV M protein with genomic RNA (Sturman *et al.*, 1980). Stronger evidence for a direct recognition of RNA packaging signal by M protein followed in a study that showed that MHV M protein interacts selectively with non-coronavirus RNAs carrying the MHV packaging signal in the absence of N protein and other virus structural and non-structural proteins (Narayanan *et al.*, 2003). VLPs, produced by the coexpression of MHV M and E proteins, have been used to dissect the *trans*-acting determinants of RNA packaging into MHV particles in the absence of other virus non-structural proteins (Bos *et al.*, 1996; Narayanan *et al.*, 2003). These VLPs are morphologically similar to infectious virus particles and are produced in the absence of viral RNP complex (Bos *et al.*, 1996; Vennema *et al.*, 1996). Remarkably, the non-viral RNAs carrying the packaging signal are specifically packaged into MHV VLPs in the absence of N protein and other virus non-structural proteins (Narayanan *et al.*,

2003). Collectively, these studies showed that in MHV, the envelope M protein determines the specificity and selectivity of RNA packaging into virus particles. These findings highlight a novel mechanism of viral genome packaging in RNA viruses in which a viral envelope protein selectively recognizes the *cis*-acting packaging signal independently of the nucleocapsid protein and guides the packaging of specific RNA species into virus particles without the nucleocapsid protein. A direct interaction between M protein with packaging signal remains to be demonstrated and future studies in this direction will provide some valuable insight into the mechanism of RNA packaging in coronaviruses.

The collective data suggest that MHV genome packaging might proceed through the following mechanism. First, MHV N protein binds to the intracellular form of genomic RNA during nascent synthesis or as soon as the genomic RNA is synthesized on intracellular membranes (Bost *et al.*, 2001; Gosert *et al.*, 2002). Subsequently, the envelope M protein, which accumulates in the intermediate compartment between the endoplasmic reticulum (ER) and the Golgi complex (the MHV budding site) (Klumperman *et al.*, 1994; Locker *et al.*, 1995; Tooze *et al.*, 1984), binds to the packaging signal present in the genomic RNA; this binding determines the selective genomic RNA packaging and excludes the packaging of MHV subgenomic mRNAs lacking the packaging signal. The abundance of M protein and the proximity of M to N protein that is wrapped on the genomic RNA facilitates the association of N protein with M protein, perhaps by virtue of the N-RNA complex being more condensed (Escors *et al.*, 2001; Narayanan *et al.*, 2000). Finally, the M protein-genomic RNP complex assembles into virus particles in association with E protein.

A few studies have examined the *trans*-acting factors involved in the packaging of RNA in other coronaviruses (Kuo and Masters, 2002). In TGEV, the virion M protein specifically binds to the nucleocapsid through a 16-amino acid region located within the carboxy terminus of M protein suggesting that this domain could be responsible for nucleocapsid incorporation into TGEV particles (Escors *et al.*, 2001). Evidence suggests that the mechanism of RNA packaging in SCoV is different from MHV, because SCoV N protein is required for RNA packaging into SCoV VLPs; this study, however, did not directly analyze RNA packaging into VLPs; cells were infected with SCoV VLPs carrying reporter RNA, and RNA packaging was indirectly measured from reporter gene expression. Further studies are required to confirm the role of N protein in SCoV RNA packaging.

Concluding remarks

In summary, the location of the packaging signal differs among the coronaviruses examined so far. For group 1 and group 3 coronaviruses like TGEV and IBV, the packaging signal is located at the genome ends whereas for group 2 coronaviruses like MHV and BCoV, the packaging signal is present in an internal region within ORF1b. The packaging signal in group 2b SCoV, (Snijder *et al.*, 2003), is also located within the ORF1b region. The presence of additional RNA motifs working in conjunction with the major packaging signal to regulate RNA packaging in coronaviruses cannot be ruled out. Future studies making the most of reverse genetics systems to look at RNA packaging in the context of full-length genomes are warranted. Results obtained with MHV have answered some key questions related to the mechanism of selective genome packaging in coronaviruses.

It remains to be demonstrated whether the packaging machinery in other coronaviruses employ a similar mechanism for genomic RNA enrichment in virus particles. An attractive area of investigation is the role of host proteins in coronavirus genome recognition and packaging. Host proteins have been shown to bind to the coronavirus genome (Huang and Lai, 2001; Lau *et al.*, 2000; Li *et al.*, 1999; Li *et al.*, 1997; Shi *et al.*, 2000; Shi *et al.*, 2001; Shi and Lai, 2005; Shi *et al.*, 2003; Spagnolo and Hogue, 2000; Zhang *et al.*, 1999). Some interesting questions regarding the initial stages of coronavirus assembly still remain to be addressed. How does the genome reach the site of virus assembly? What role does replication play in regulating the site and efficiency of RNA packaging? Some answers to these questions would allow us to better understand genome packaging during the early stages of coronavirus assembly and direct us towards designing better coronavirus-based gene delivery systems and expression vectors.

References

Adam, M.A., and Miller, A.D. (1988). Identification of a signal in a murine retrovirus that is sufficient for packaging of nonretroviral RNA into virions. J. Virol. *62*, 3802–3806.

Armstrong, J., Smeekens, S., and Rottier, P. (1983). Sequence of the nucleocapsid gene from murine coronavirus MHV-A59. Nucleic Acids Res. *11*, 883–891.

Banks, J.D., Kealoha, B.O., and Linial, M.L. (1999). An Mpsi-containing heterologous RNA, but not env mRNA, is efficiently packaged into avian retroviral particles. J. Virol. *73*, 8926–8933.

Beck, J., Bartos, H., and Nassal, M. (1997). Experimental confirmation of a hepatitis B virus (HBV) epsilon-like bulge-and-loop structure in avian HBV RNA encapsidation signals. Virology *227*, 500–504.

Berkowitz, R.D., Hammarskjold, M.L., Helga-Maria, C., Rekosh, D., and Goff, S.P. (1995). 5′ regions of HIV-1 RNAs are not sufficient for encapsidation: implications for the HIV-1 packaging signal. Virology *212*, 718–723.

Bos, E.C., Dobbe, J.C., Luytjes, W., and Spaan, W.J. (1997). A subgenomic mRNA transcript of the coronavirus mouse hepatitis virus strain A59 defective interfering (DI) RNA is packaged when it contains the DI packaging signal. J. Virol. *71*, 5684–5687.

Bos, E.C., Luytjes, W., van der Meulen, H.V., Koerten, H.K., and Spaan, W.J. (1996). The production of recombinant infectious DI-particles of a murine coronavirus in the absence of helper virus. Virology *218*, 52–60.

Bost, A.G., Prentice, E., and Denison, M.R. (2001). Mouse hepatitis virus replicase protein complexes are translocated to sites of M protein accumulation in the ERGIC at late times of infection. Virology *285*, 21–29.

Boursnell, M.E., and Brown, T.D. (1984). Sequencing of coronavirus IBV genomic RNA: a 195-base open reading frame encoded by mRNA B. Gene *29*, 87–92.

Burd, C.G., and Dreyfuss, G. (1994). Conserved structures and diversity of functions of RNA-binding proteins. Science *265*, 615–621.

Cavanagh, D., Shaw, K., and Zhao, X. (1993). Analysis of messenger RNA within virions of IBV. Adv. Exp. Med. Biol. *342*, 123–128.

Chang, R.Y., and Brian, D.A. (1996). cis Requirement for N-specific protein sequence in bovine coronavirus defective interfering RNA replication. J. Virol. *70*, 2201–2207.

Cologna, R., and Hogue, B.G. (2000). Identification of a bovine coronavirus packaging signal. J. Virol. *74*, 580–583.

Cologna, R., Spagnolo, J.F., and Hogue, B.G. (2000). Identification of nucleocapsid binding sites within coronavirus- defective genomes. Virology *277*, 235–249.

Dalton, K., Casais, R., Shaw, K., Stirrups, K., Evans, S., Britton, P., Brown, T.D., and Cavanagh, D. (2001). cis-acting sequences required for coronavirus infectious bronchitis virus defective-RNA replication and packaging. J. Virol. *75*, 125–133.

Draper, D.E. (1995). Protein-RNA recognition. Annu. Rev. Biochem. *64*, 593–620.

Drosten, C., Gunther, S., Preiser, W., van der Werf, S., Brodt, H.R., Becker, S., Rabenau, H., Panning, M., Kolesnikova, L., Fouchier, R.A., *et al.* (2003). Identification of a novel coronavirus in patients with severe acute respiratory syndrome. N. Engl. J. Med. *348*, 1967–1976.

Escors, D., Izeta, A., Capiscol, C., and Enjuanes, L. (2003). Transmissible gastroenteritis coronavirus packaging signal is located at the 5′ end of the virus genome. J. Virol. *77*, 7890–7902.

Escors, D., Ortego, J., Laude, H., and Enjuanes, L. (2001). The membrane M protein carboxy terminus binds to transmissible gastroenteritis coronavirus core and contributes to core stability. J. Virol. *75*, 1312–1324.

Fosmire, J.A., Hwang, K., and Makino, S. (1992). Identification and characterization of a coronavirus packaging signal. J. Virol. *66*, 3522–3530.

Fujimura, T., Esteban, R., Esteban, L.M., and Wickner, R.B. (1990). Portable encapsidation signal of the L-A double-stranded RNA virus of S. cerevisiae. Cell *62*, 819–828.

Geigenmuller, U., and Linial, M.L. (1996). Specific binding of human immunodeficiency virus type 1 (HIV-1) Gag- derived proteins to a 5′ HIV-1 genomic RNA sequence. J. Virol. *70*, 667–671.

Gosert, R., Kanjanahaluethai, A., Egger, D., Bienz, K., and Baker, S.C. (2002). RNA replication of mouse hepatitis virus takes place at double-membrane vesicles. J. Virol. *76*, 3697–3708.

Hofmann, M.A., Sethna, P.B., and Brian, D.A. (1990). Bovine coronavirus mRNA replication continues throughout persistent infection in cell culture. J. Virol. *64*, 4108–4114.

Holmes, K.V., and Behnke, J.N. (1981). Evolution of a coronavirus during persistent infection in vitro. Adv. Exp. Med. Biol. *142*, 287–299.

Hsieh, P.K., Chang, S.C., Huang, C.C., Lee, T.T., Hsiao, C.W., Kou, Y.H., Chen, I.Y., Chang, C.K., Huang, T.H., and Chang, M.F. (2005). Assembly of severe acute respiratory syndrome coronavirus RNA packaging signal into virus-like particles is nucleocapsid dependent. J. Virol. *79*, 13848–13855.

Huang, P., and Lai, M.M. (2001). Heterogeneous nuclear ribonucleoprotein A1 binds to the 3′-untranslated region and mediates potential 5′-3′-end cross talks of mouse hepatitis virus RNA. J. Virol. *75*, 5009–5017.

Klumperman, J., Locker, J.K., Meijer, A., Horzinek, M.C., Geuze, H.J., and Rottier, P.J. (1994). Coronavirus M proteins accumulate in the Golgi complex beyond the site of virion budding. J. Virol. *68*, 6523–6534.

Ksiazek, T.G., Erdman, D., Goldsmith, C.S., Zaki, S.R., Peret, T., Emery, S., Tong, S., Urbani, C., Comer, J.A., Lim, W., *et al.* (2003). A novel coronavirus associated with severe acute respiratory syndrome. N. Engl. J. Med. *348*, 1953–1966.

Kuo, L., and Masters, P.S. (2002). Genetic Evidence for a Structural Interaction between the Carboxy Termini of the Membrane and Nucleocapsid Proteins of Mouse Hepatitis Virus. J. Virol. *76*, 4987–4999.

Lai, M.M., Baric, R.S., Brayton, P.R., and Stohlman, S.A. (1984). Characterization of leader RNA sequences on the virion and mRNAs of mouse hepatitis virus, a cytoplasmic RNA virus. Proc. Natl. Acad. Sci. USA *81*, 3626–3630.

Lai, M.M., and Cavanagh, D. (1997). The molecular biology of coronaviruses. Adv. Virus Res. *48*, 1–100.

Lai, M.M., and Stohlman, S.A. (1978). RNA of mouse hepatitis virus. J. Virol. *26*, 236–242.

Lai, M.M., and Stohlman, S.A. (1981). Comparative analysis of RNA genomes of mouse hepatitis viruses. J. Virol. *38*, 661–670.

Lau, J.S., Baumeister, P., Kim, E., Roy, B., Hsieh, T.Y., Lai, M., and Lee, A.S. (2000). Heterogeneous nuclear ribonucleoproteins as regulators of gene expression through interactions with the human thymidine kinase promoter. J. Cell. Biochem. *79*, 395–406.

Lee, H.J., Shieh, C.K., Gorbalenya, A.E., Koonin, E.V., La Monica, N., Tuler, J., Bagdzhadzhyan, A., and Lai, M.M. (1991). The complete sequence (22 kilobases) of murine coronavirus gene 1 encoding the putative proteases and RNA polymerase. Virology *180*, 567–582.

Li, H.P., Huang, P., Park, S., and Lai, M.M. (1999). Polypyrimidine tract-binding protein binds to the leader RNA of mouse hepatitis virus and serves as a regulator of viral transcription. J. Virol. *73*, 772–777.

Li, H.P., Zhang, X., Duncan, R., Comai, L., and Lai, M.M. (1997). Heterogeneous nuclear ribonucleoprotein A1 binds to the transcription-regulatory region of mouse hepatitis virus RNA. Proc. Natl. Acad. Sci. USA *94*, 9544–9549.

Locker, J.K., Opstelten, D.J., Ericsson, M., Horzinek, M.C., and Rottier, P.J. (1995). Oligomerization of a trans-Golgi/trans-Golgi network retained protein occurs in the Golgi complex and may be part of its retention. J. Biol. Chem. *270*, 8815–8821.

Luytjes, W., Krystal, M., Enami, M., Pavin, J.D., and Palese, P. (1989). Amplification, expression, and packaging of foreign gene by influenza virus. Cell *59*, 1107–1113.

Makino, S., Shieh, C.K., Keck, J.G., and Lai, M.M. (1988). Defective-interfering particles of murine coronavirus: mechanism of synthesis of defective viral RNAs. Virology *163*, 104–111.

Makino, S., Yokomori, K., and Lai, M.M. (1990). Analysis of efficiently packaged defective interfering RNAs of murine coronavirus: localization of a possible RNA-packaging signal. J. Virol. *64*, 6045–6053.

Masters, P.S. (1992). Localization of an RNA-binding domain in the nucleocapsid protein of the coronavirus mouse hepatitis virus. Arch. Virol. *125*, 141–160.

Molenkamp, R., and Spaan, W.J. (1997). Identification of a specific interaction between the coronavirus mouse hepatitis virus A59 nucleocapsid protein and packaging signal. Virology *239*, 78–86.

Mortola, E., and Roy, P. (2004). Efficient assembly and release of SARS coronavirus-like particles by a heterologous expression system. FEBS Lett. *576*, 174–178.

Murray, R.S., Cai, G.Y., Hoel, K., Zhang, J.Y., Soike, K.F., and Cabirac, G.F. (1992). Coronavirus infects and causes demyelination in primate central nervous system. Virology *188*, 274–284.

Narayanan, K., Chen, C.J., Maeda, J., and Makino, S. (2003). Nucleocapsid-independent specific viral RNA packaging via viral envelope protein and viral RNA signal. J. Virol. *77*, 2922–2927.

Narayanan, K., Maeda, A., Maeda, J., and Makino, S. (2000). Characterization of the coronavirus M protein and nucleocapsid interaction in infected cells. J. Virol. *74*, 8127–8134.

Narayanan, K., and Makino, S. (2001). Cooperation of an RNA packaging signal and a viral envelope protein in coronavirus RNA packaging. J. Virol. *75*, 9059–9067.

Owen, K.E., and Kuhn, R.J. (1996). Identification of a region in the Sindbis virus nucleocapsid protein that is involved in specificity of RNA encapsidation. J. Virol. *70*, 2757–2763.

Parker, M.M., and Masters, P.S. (1990). Sequence comparison of the N genes of five strains of the coronavirus mouse hepatitis virus suggests a three domain structure for the nucleocapsid protein. Virology *179*, 463–468.

Pattnaik, A.K., Ball, L.A., LeGrone, A., and Wertz, G.W. (1995). The termini of VSV DI particle RNAs are sufficient to signal RNA encapsidation, replication, and budding to generate infectious particles. Virology *206*, 760–764.

Peiris, J.S., Lai, S.T., Poon, L.L., Guan, Y., Yam, L.Y., Lim, W., Nicholls, J., Yee, W.K., Yan, W.W., Cheung, M.T., *et al.* (2003). Coronavirus as a possible cause of severe acute respiratory syndrome. Lancet *361*, 1319–1325.

Qu, F., and Morris, T.J. (1997). Encapsidation of turnip crinkle virus is defined by a specific packaging signal and RNA size. J. Virol. *71*, 1428–1435.

Robbins, S.G., Frana, M.F., McGowan, J.J., Boyle, J.F., and Holmes, K.V. (1986). RNA-binding proteins of coronavirus MHV: detection of monomeric and multimeric N protein with an RNA overlay-protein blot assay. Virology *150*, 402–410.

Sethna, P.B., Hofmann, M.A., and Brian, D.A. (1991). Minus-strand copies of replicating coronavirus mRNAs contain antileaders. J. Virol. *65*, 320–325.

Sethna, P.B., Hung, S.L., and Brian, D.A. (1989). Coronavirus subgenomic minus-strand RNAs and the potential for mRNA replicons. Proc. Natl. Acad. Sci. USA *86*, 5626–5630.

Shi, S.T., Huang, P., Li, H.P., and Lai, M.M. (2000). Heterogeneous nuclear ribonucleoprotein A1 regulates RNA synthesis of a cytoplasmic virus. EMBO J. *19*, 4701–4711.

Shi, S.T., Huang, P., Li, H.P., and Lai, M.M. (2001). Regulation of mouse hepatitis virus RNA synthesis by heterogeneous nuclear ribonucleoprotein A1. Adv. Exp. Med. Biol. *494*, 429–436.

Shi, S.T., and Lai, M.M. (2005). Viral and cellular proteins involved in coronavirus replication. Curr. Top. Microbiol. Immunol. *287*, 95–131.

Shi, S.T., Yu, G.Y., and Lai, M.M. (2003). Multiple type A/B heterogeneous nuclear ribonucleoproteins (hnRNPs) can replace hnRNP A1 in mouse hepatitis virus RNA synthesis. J. Virol. *77*, 10584–10593.

Simons, K., and Garoff, H. (1980). The budding mechanisms of enveloped animal viruses. J. Gen. Virol. *50*, 1–21.

Snijder, E.J., Bredenbeek, P.J., Dobbe, J.C., Thiel, V., Ziebuhr, J., Poon, L.L., Guan, Y., Rozanov, M., Spaan, W.J., and Gorbalenya, A.E. (2003). Unique and conserved features of genome and proteome of SARS-coronavirus, an early split-off from the coronavirus group 2 lineage. J. Mol. Biol. *331*, 991–1004.

Spaan, W., Delius, H., Skinner, M., Armstrong, J., Rottier, P., Smeekens, S., van der Zeijst, B.A., and Siddell, S.G. (1983). Coronavirus mRNA synthesis involves fusion of non-contiguous sequences. EMBO J. *2*, 1839–1844.

Spagnolo, J.F., and Hogue, B.G. (2000). Host protein interactions with the 3′ end of bovine coronavirus RNA and the requirement of the poly(A) tail for coronavirus defective genome replication. J. Virol. *74*, 5053–5065.

Stohlman, S.A., Baric, R.S., Nelson, G.N., Soe, L.H., Welter, L.M., and Deans, R.J. (1988). Specific interaction between coronavirus leader RNA and nucleocapsid protein. J. Virol. *62*, 4288–4295.

Stohlman, S.A., and Lai, M.M. (1979). Phosphoproteins of murine hepatitis viruses. J. Virol. *32*, 672–675.

Sturman, L.S., Holmes, K.V., and Behnke, J. (1980). Isolation of coronavirus envelope glycoproteins and interaction with the viral nucleocapsid. J. Virol. *33*, 449–462.

Tooze, J., Tooze, S., and Warren, G. (1984). Replication of coronavirus MHV-A59 in sac-cells: determination of the first site of budding of progeny virions. Eur. J. Cell. Biol. *33*, 281–293.

van der Most, R.G., Bredenbeek, P.J., and Spaan, W.J. (1991). A domain at the 3′ end of the polymerase gene is essential for encapsidation of coronavirus defective interfering RNAs. J. Virol. *65*, 3219–3226.

Vennema, H., Godeke, G. J., Rossen, J.W., Voorhout, W.F., Horzinek, M.C., Opstelten, D.J., and Rottier, P.J. (1996). Nucleocapsid-independent assembly of coronavirus-like particles by co-expression of viral envelope protein genes. EMBO J. *15*, 2020–2028.

Wege, H., Siddell, S., and ter Meulen, V. (1982). The biology and pathogenesis of coronaviruses. Curr. Top. Microbiol. Immunol. *99*, 165–200.

Weiss, B., Nitschko, H., Ghattas, I., Wright, R., and Schlesinger, S. (1989). Evidence for specificity in the encapsidation of Sindbis virus RNAs. J. Virol. *63*, 5310–5318.

Weiss, S.R., and Navas-Martin, S. (2005). Coronavirus pathogenesis and the emerging pathogen severe acute respiratory syndrome coronavirus. Microbiol. Mol. Biol. Rev. *69*, 635–664.

Whelan, S.P., and Wertz, G.W. (1999). The 5′ terminal trailer region of vesicular stomatitis virus contains a position-dependent cis-acting signal for assembly of RNA into infectious particles. J. Virol. *73*, 307–315.

Woo, K., Joo, M., Narayanan, K., Kim, K.H., and Makino, S. (1997). Murine coronavirus packaging signal confers packaging to nonviral RNA. J. Virol. *71*, 824–827.

Zhang, X., Li, H.P., Xue, W., and Lai, M.M. (1999). Formation of a ribonucleoprotein complex of mouse hepatitis virus involving heterogeneous nuclear ribonucleoprotein A1 and transcription- regulatory elements of viral RNA. Virology *264*, 115–124.

Molecular Evolution of Group 2 Coronaviruses

7

Leen Vijgen, Els Keyaerts, and Marc Van Ranst

Abstract

Coronaviruses are well-equipped to adapt rapidly to changing ecological niches based on two major forces that drive their viral evolution: mutation and recombination. Based on the available nucleotide and amino acid sequence information, the evolutionary relationships among coronaviruses can be inferred by using molecular phylogenetic approaches. The calculation of an evolutionary rate provides an indication of the rate at which mutations become fixed in the coronavirus population. The divergence of ancestral strains of existing coronaviruses with a different host-specificity can be dated back in history, thereby estimating the time in which potential interspecies transmission events occurred. A similar cross-species jumping led to the emergence of the SARS coronavirus, that has been proposed to be an early split-off from group 2 coronaviruses. Within group 2 a subdivision into murine hepatitis virus-related and bovine coronavirus-related coronaviruses has been suggested. The remarkably high genetic similarity among the bovine coronavirus-related coronaviruses indicates a relatively recent common evolutionary history. In this chapter, we discuss the molecular evolution of these closely related group 2 coronaviruses.

Introduction

Since the renewed interest in the field of the *Coronaviridae* following the SARS outbreak, the virus family has been expanded with several novel members infecting both humans and animals. The SARS-coronavirus (SARS-CoV) is most closely related to group 2 CoVs, and is considered as an early split-off from the lineage containing all other group 2 CoVs known at present (Snijder *et al.*, 2003; Gorbalenya *et al.*, 2004). Several studies have been performed in order to gain insights in the evolutionary events that are responsible for the genetic and antigenic relatedness observed among members of the same coronavirus group (Song *et al.*, 2005; Vijgen *et al.*, 2005b; Sanchez *et al.*, 1992). Virus evolution can be defined as a process of changes in the genetic make up of a virus population, which is mainly controlled by the forces of genetic drift and natural selection. A traditional approach to investigate the evolutionary relationships among organisms was by comparing their morphological characteristics. The increasing availability of molecular information, such as nucleotide and amino acid sequence data, which reflect the past actions of genetic variation within a population, stimulated the use of molecular approaches for reconstruction of the evolutionary relationships among groups of organisms. An interesting aspect of evolutionary research is the ability to estimate the time-scale of evolution. A useful tool to investigate the time of evolutionary events is provided by the molecular clock hypothesis, which assumes that genes evolve at a roughly constant rate, and therefore implicates that the amount of nucleotide or amino acid differences between two homologous sequences is proportional to the time that elapsed since their evolutionary separation (Bromham and Penny, 2003). However, the rate-constancy assumption is frequently violated for sequence data sets, and different approaches have been developed to estimate evolutionary rates and divergence times under a relaxed molecular clock hypothesis (Bromham and Penny, 2003). For fast evolving populations such as RNA viruses, a statistically significant number of genetic differences can be observed in sequences sampled at different points in time, and the molecular clock can be calibrated using the temporal information in the serially sampled sequences (Drummond *et al.*, 2005).

Genetic analysis of group 2 coronaviruses

The most extensively studied group 2 coronavirus by far is murine hepatitis virus (MHV), subsequently followed by bovine coronavirus (BCoV). MHV represents a group of serologically and genetically related but distinct strains, causing a wide spectrum of clinical syndromes in mice, ranging from subclinical infection to enteritis, hepatitis, encephalitis, (chronic) demyelinating diseases, and death. More than 25 strains have currently been reported, and most studies have been performed using the prototype strains MHV-A59, MHV-JHM, MHV-1, -2, and -3, and MHV-S. A similar heterogeneity is present among BCoV strains, which are associated with either enteric or respiratory disease. MHV and BCoV can be considered as representatives of the two distinct evolutionary lineages of group 2 coronaviruses. Combined with differences in serological reactivity profiles and in substrate specificity of the hemagglutinin-esterase glycoprotein, a subdivision into MHV-like and BCoV-like group 2 coronaviruses has been suggested (Wurzer *et al.*, 2002).

Most of the group 2 coronaviruses identified thus far show close relatedness to BCoV, and include human coronavirus OC43 (HCoV-OC43), porcine hemagglutinating encephalo-myelitis virus (PHEV), canine respiratory coronavirus (CRCoV), and equine coronavirus (ECoV). Rat sialodacryoadenitis virus (SDAV) on the other hand, is found to be more closely related to MHV. Phylogenetic analysis revealed that the recently identified human coronavirus HKU1 (HCoV-HKU1) is a novel member of group 2 coronaviruses, forming a distinct branch within this group (Figure 7.1) (Woo *et al.*, 2005). To date, complete genomic sequence data of BCoV-related coronaviruses are available for several BCoV and HCoV-OC43 strains, and recently also for a PHEV-VW572 isolate. However, nucleotide sequence data of other BCoV-related group 2 coronaviruses, like CRCoV and ECoV, remain limited to partial genome fragments. Although this partial sequence information can be considered valuable for a preliminary classification as group 2 coronaviruses, the knowledge of full-length genome sequence data is essential for a more accurate evolutionary analysis of the BCoV subcluster. Also, determining the complete genome sequence of these coronavirus prototype strains has the advantage of revealing possible anomalies in the genomic organization and common genomic features of these closely related viruses, which can be of functional and/or evolutionary importance.

The complete genomic characterization of HCoV-OC43, one of the common cold human coronaviruses known since 1967, confirmed the remarkable close genetic relationship between HCoV-OC43 and BCoV, not only in their structural protein genes as described earlier (Kamahora *et al.*, 1989; Mounir and Talbot, 1992; Zhang *et al.*, 1992), but throughout their complete genomes (Vijgen *et al.*, 2005b; St-Jean *et al.*, 2004). Phylogenetic analysis of coronavirus ORF1b amino acid sequences shows an even higher divergence between different MHV strains or between different infectious bronchitis virus (IBV) strains than between HCoV-OC43 and BCoV, implicating a fairly recent common ancestry of this coronavirus pair (Figure 7.1). Although genetically very alike, a major difference between HCoV-OC43 and BCoV genomes is present in the region located between the spike and the ns12.9 gene (Vijgen *et al.*, 2005b; Mounir and Talbot, 1993). In this region, a 290 nucleotide (nt) deletion in HCoV-OC43 leads to the absence of BCoV non-structural protein genes ns4.9 and ns4.8, suggesting that these are not essential for coronavirus replication.

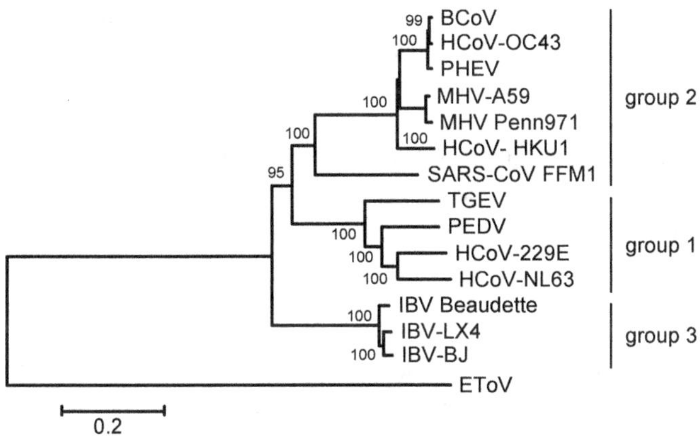

Figure 7.1 Phylogenetic analysis of coronavirus ORF1b amino acid sequences. Group 1: human coronavirus 229E (HCoV-229E, GenBank accession number AF304460); human coronavirus NL63 (HCoV-NL63, AY567487); porcine epidemic diarrhea virus strain CV777 (PEDV, AF353511); porcine transmissible gastroenteritis virus Purdue strain (TGEV, AJ271965). Group 2: human coronavirus OC43 (HCoV-OC43, AY391777); bovine coronavirus Mebus strain (BCoV, U00735); porcine hemagglutinating encephalomyelitis virus VW572 (PHEV-VW572, DQ011855); murine hepatitis virus Penn 97-1 strain (MHV-Penn97-1, AF208066); murine hepatitis virus A59 (MHV-A59, NC_001846); human coronavirus HKU1 (HCoV-HKU1, NC_006577). Group 3: avian infectious bronchitis virus Beaudette strain (IBV-Beaudette, M95169), avian infectious bronchitis virus BJ strain (IBV-BJ, AY319651) and avian infectious bronchitis virus LX4 strain (IBV-LX4, AY338732). SARS coronavirus Frankfurt-1 strain (SARS-CoV, AY291315). Outgroup: equine Berne torovirus (EToV, X52374). All columns containing gaps were removed. The resulting alignment included 2144 amino acid characters (1199 being parsimony-informative), and contained the meld of the following PHEV-VW572 pp1ab protein fragments: amino acids 4493–4505, 4509–4528, 4530–4537, 4550–4560, 4563–4595, 4617–4657, 4680–4849, 4853–4882, 4884–4507, 4512–4578, 4586–5155, 5159–5170, 5173–5187, 5192–5216, 5219–5266, 5278–5382, 5388–5501, 5503–5547, 5549–5627, 5635–5685, 5687–5709, 5711–5718, 5724–5773, 5785–5876, 5906–5935, 5937–5942, 5954–5966, 6000–6077, 6081–6106, 6108–6131, 6135–6240, 6245–6268, 6271–6350, 6355–6419, 6424–6472, 6490–6529, 6693–6709, 6720–6798, 6806–6834, 6836–6852, 6854–6886, 6888–6900, 6906–6933, 6946–7029, 7034–7074, 7077–7089. The frequency of occurrence of particular bifurcations (percentage of 10000 bootstrap replicate calculations) is indicated at the nodes. Bootstrap values over 75% are shown.

Analysis of the complete genome sequence of circulating HCoV-OC43 strains provided similar conclusions regarding the close relatedness to BCoV (Vijgen *et al.*, 2005a; St-Jean *et al.*, 2004). Contrary to the HCoV-OC43 ATCC and Paris strain, the Belgian circulating strains have a spike protein proteolytic cleavage motif identical to that of BCoV (N-Arg-Arg-Xxx-Arg-Arg-C) (Abraham *et al.*, 1990). In most MHV strains, in BCoV and also in the group 3 coronavirus IBV, the spike protein is cleaved during or after virus maturation by a cellular protease to yield two subunits, S1 and S2, which remain non-covalently associated in the viral spikes (Binns *et al.*, 1985; Frana *et al.*, 1985; Abraham *et al.*, 1990). Proteolytic cleavage of the coronavirus spike protein occurs adjacent to a sequence of basic amino acids on the carboxyterminus of S1 (Luytjes *et al.*, 1987). In paramyxoviruses (Glickman *et al.*, 1988), many retroviruses (Perez and Hunter, 1987) and influenza A viruses (Bosch *et al.*, 1981), all of which contain envelope proteins that are post-translationally cleaved, analogous stretches of basic amino acids are present adjacent to the cleavage site. In the HCoV-OC43 ATCC strain, the last amino acid at the carboxyterminus of S1 is a glycine, an uncharged instead of a basic amino acid, which might account for the observation that its spike protein remains uncleaved (Kunkel and Herrler, 1993). The presence of an intact spike protein cleavage motif in the HCoV-OC43 circulating strains might therefore have potential functional consequences. In MHV and BCoV, the cleavage process may enhance their cell fusion activity and their viral infectivity (Gombold *et al.*, 1993; Sturman *et al.*, 1985). Nevertheless, spike proteins of group 1 coronaviruses are not cleaved, but these viruses can still induce cell fusion (de Groot *et al.*, 1989).

Spike gene analysis of contemporary Belgian HCoV-OC43 strains provided evidence for a relatively high genetic diversity among circulating HCoV-OC43 strains (Vijgen *et al.*, 2005a). Whether the HCoV-OC43 spike gene variability would be correlated with possible differences in virulence, tissue tropism or pathogenesis of the variants, as described for other coronaviruses, could not be confirmed (Ballesteros *et al.*, 1997; Leparc-Goffart *et al.*, 1997). In a study by Vabret and colleagues reporting on an outbreak of HCoV-OC43 respiratory infections in Normandy, France, partial M gene sequence data of 20 HCoV-OC43 clinical isolates were analyzed, revealing several nucleotide variations in the M gene fragments, leading to a phylogenetic clustering in different groups (Vabret *et al.*, 2003). As all these HCoV-OC43 isolates were sampled in February and March 2001, this study also provided evidence for a relatively high HCoV-OC43 genetic diversity, even among strains circulating in one area during the same short period of time. However, more sequence data, also from isolates from other geographical areas are required to investigate more thoroughly the extent of variation among geographically and chronologically distinct HCoV-OC43 strains.

Another group 2 coronavirus, PHEV, which causes encephalomyelitis and vomiting and wasting disease in young pigs, shows a very close genetic and antigenic relatedness to BCoV and HCoV-OC43 (Vautherot and Laporte, 1983; Vieler *et al.*, 1996). The complete genomic sequence of a PHEV-VW572 isolate was determined and a significant truncation of the group 2-specific ns2 gene was found in PHEV, which resulted from a 211-nt deletion, compared to BCoV and HCoV-OC43 (Vijgen *et al.*, 2006). Although group 2-specific genes are not essential for viral growth, it was shown that deletion of the MHV group 2-specific non-structural genes led to a significant *in vivo* attenuation of the virus (de Haan *et al.*, 2002). This approach showed to be valuable for the development of a live attenuated coronavirus vaccine against feline infectious peritonitis, and is promising for future coronavirus vaccine development (Haijema *et al.*, 2004). Whether the carboxyterminal truncation of PHEV ns2 has led to an alteration in its biological function remains to be elucidated. In the recently discovered HCoV-HKU1 an 800-nt deletion was shown to be present between ORF1b and the HE gene, leading to the complete absence of the ns2 gene (Woo *et al.*, 2005). This finding also demonstrates that ns2 is not essential for the replication of group 2 coronaviruses.

The availability of complete genome sequence data of BCoV, HCoV-OC43, and PHEV, allows a thorough comparative genomic analysis of these closely related group 2 coronaviruses, which might yield clues to reconstruct their evolutionary history. As has been described above for HCoV-OC43, a similar nucleotide deletion in the region between the spike and ns12.9 gene, in comparison to BCoV, is present in PHEV leading to the absence of BCoV ns4.8 and a truncation of the ns4.9 ORF. When examining this region in other group 2 coronaviruses, we observed a similar deletion in this region of the ECoV genome, although much smaller than in HCoV-OC43 and PHEV. On the other hand, in MHV and in SDAV, this region is approximately 100 nt longer than in BCoV, and harbors one large ORF, coding for a non-structural protein of 17.8 kDa in MHV, and of 15 kDa in SDAV. Analysis of this region in HCoV-HKU1 showed an interesting finding. HCoV-HKU1 is suggested to be a group 2 coronavirus with the highest genetic similarity to MHV, although phylogenetic analysis demonstrates that HCoV-HKU1 does not belong to the MHV-subcluster but forms a separate branch within the group 2 coronavirus cluster (Woo *et al.*, 2005). With only 40 nt, the length of the genomic region between the HCoV-HKU1 spike gene and the ns12.9 gene is the smallest observed for all group 2 coronaviruses characterized thus far. More interestingly, when analyzing this 40-nt region more intensively, a 33-nt stretch identical to the 3′ end of the leader sequence of HCoV-HKU1 is found. The occurrence of a fragment of the leader sequence within the genomic RNA, additional to its presence at the 5′ end of the genome, is a rather rare coronavirus event, that at present has only been described in HCoV-OC43 and recently also in HCoV-NL63 (Mounir and Talbot, 1993; Pyrc *et al.*, 2004; Vijgen *et al.*, 2005b). In HCoV-OC43, a 33-nt sequence containing 32 nt identical to the HCoV-OC43 leader sequence is present upstream of the ns12.9 ORF. A similar phenomenon cannot be seen in BCoV, PHEV, or any other group 2 coronavirus characterized at present. In HCoV-NL63, a 13-nt fragment of the HCoV-NL63 leader sequence was found upstream of the E gene. As a possible explication for this finding, a potential compensation mechanism for the absence of an optimal transcription regulating sequence (TRS) core was suggested (Pyrc *et al.*, 2004). TRS elements, also referred to as intergenic sequences (IGS), play a role in the coronavirus transcription process, in which they act as transcription termination sequences in order to allow a polymerase jumping for production of subgenomic RNAs. The core TRS sequence in HCoV-NL63 was defined as AACUAAA, and this consensus sequence was found upstream of every gene, except for the E gene, which was preceded by a sub-optimal TRS core, AACUAUA (Pyrc *et al.*, 2004). In the HCoV-OC43 genome, the core TRS sequence was predicted to be UC(U/C)AAAC, but it was not always conserved throughout the genome. For example, the TRS of the ns12.9 ORF was UCUUAAG (St-Jean *et al.*, 2004). Whether the occurrence of a fragment of the leader sequence within the genomes of HCoV-OC43 and HCoV-HKU1 can also be considered as a compensation mechanism for the presence of a sub-optimal TRS remains unknown. Moreover, this theory seems rather unlikely for HCoV-HKU1 since it contains an UCUAAAC motif at the 3′ end of the 33 nucleotide stretch, which might act as an optimal TRS for the downstream ns12.9 ORF. Another hypothesis, suggested by Mounir and Talbot, speculated about a potential recombination event in which the loss of the ns4.9 and ns4.8 sequences in HCoV-OC43 stimulated the development of a mechanism for conservation of the ns12.9 ORF (Mounir and Talbot, 1993).

Analysis of the genetic similarity between the complete genomes of HCoV-OC43, BCoV, and PHEV shows the close relatedness between HCoV-OC43 and BCoV in all genomic regions (Figure 7.2). In more than two-third of the PHEV genome, a similar genetic distance to the HCoV-OC43 genome, as found for BCoV, is demonstrated except for the PHEV genome region containing the HE and spike gene, which is remarkably divergent from the corresponding regions in the HCoV-OC43 and BCoV genomes (Vijgen

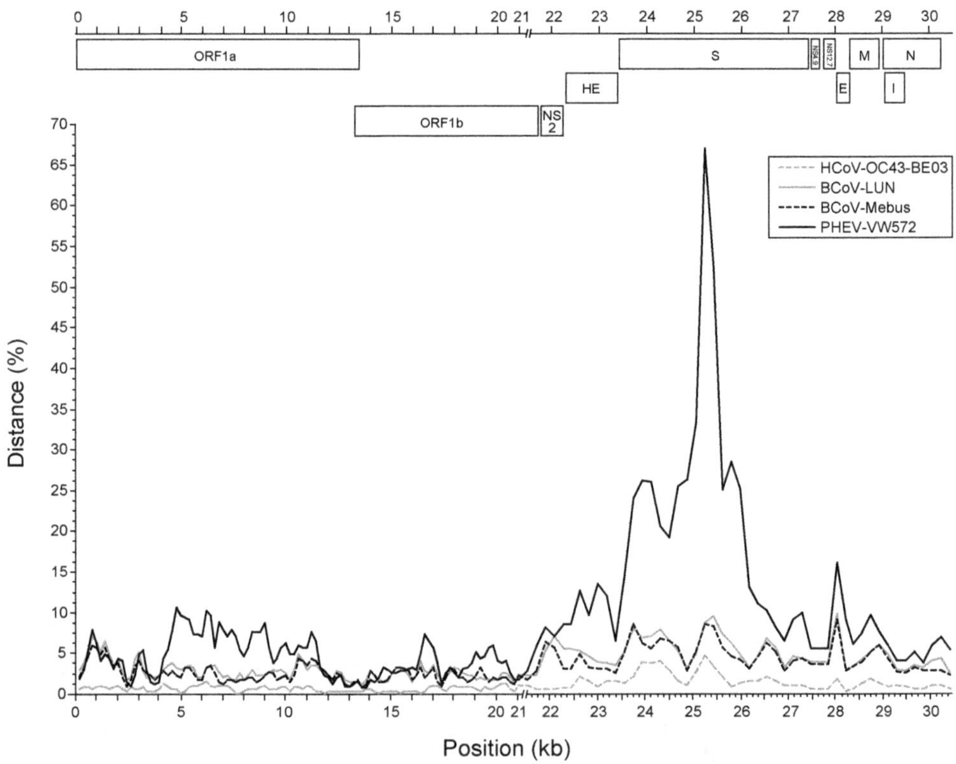

Figure 7.2 Linear representation of the PHEV-VW572 genome (GenBank accession number DQ011855) and SimPlot analysis of complete genome sequence data of PHEV-VW572, BCoV-LUN (AF391542), BCoV-Mebus (U00735), HCoV-OC43 BE03 (AY903459), and HCoV-OC43 ATCC (American Type Culture Collection, AY391777). Each point plotted is the percentage genetic distance within a sliding window of 400 nt wide, centered on the position plotted, with a step size of 200 nt. Each curve represents a comparison of the sequence data of PHEV-VW572, the BCoV strains, and HCoV-OC43 BE03 to the reference sequence data of the ATCC HCoV-OC43 strain.

et al., 2006). The variability in PHEV in this region was shown to be mainly caused by synonymous substitutions making the possibility of positive selection less likely, unless the synonymous substitutions would have been selected for their role in the secondary RNA structure of this genomic region. Another explanation for the remarkably higher genetic distance in this genomic region could be a recombination event between PHEV and another coronavirus, which has not been sampled yet. However, this would not explain why synonymous substitutions were mainly responsible for the higher genetic distance, and therefore we cannot provide conclusive evidence for these speculations.

At present, the complete genome sequences of two known BCoV-related group 2 coronaviruses, ECoV and CRCoV remain to be determined. In particular, the genomic characterization of CRCoV would be interesting in order to investigate the role of CRCoV in the evolutionary history of HCoV-OC43 and BCoV. CRCoV was recently identified as a cause of upper and lower respiratory tract infections in dogs, and in contrast to the group 1 canine enteric coronavirus (CCoV), CRCoV was assigned to group 2 coronaviruses (Erles *et al.*, 2003). To date, only the spike gene and fragments of the HE and the polymerase genes of CRCoV have been sequenced, and these partial sequence data show a remarkably high similarity to the corresponding BCoV sequences (Erles *et al.*, 2003). For the full-length CRCoV spike gene, a nucleotide identity percentage of 97% was found with the BCoV spike gene, while only 93% identity was shown with the HCoV-OC43 spike gene. In our study, analyzing the close genetic relationship between HCoV-OC43 and BCoV (Vijgen *et al.*, 2005b), we found in contrast to the 97% identity between the BCoV and CRCoV spike genes, only 93.5% similarity between the BCoV and HCoV-OC43 spike genes. Based on these findings we can speculate about an even closer genetic and possible evolutionary relationship between BCoV and CRCoV than between BCoV and HCoV-OC43. However, complete genome sequence data of CRCoV are required to allow a complete comparative genetic analysis of all ORFs, and a close investigation of the region between the spike and ns12.9 gene, in which the presence or absence of BCoV ns4.9 and ns4.8 ORFs could be an indication for the level of genetic relatedness to BCoV. The second known BCoV-related coronavirus that still needs to be completely genomically characterized, ECoV, was isolated from a diarrheic foal, implicating an etiologic role of coronaviruses in enteric diseases in foals and adult horses (Guy *et al.*, 2000). To date, only sequence data of the 3′ last 8 kb of the genome, comprising the genes coding for structural proteins, are available for ECoV. Phylogenetic analysis of these sequence data shows a close relationship between ECoV and the BCoV-related group 2 coronaviruses, although ECoV forms a separate branch within the subcluster, suggesting a slightly higher genetic distance to BCoV than observed for the other BCoV-related coronaviruses.

Estimation of the rate of coronavirus evolution

When studying the evolution of RNA viruses, one has to deal with their enormous potential to generate genetic diversity through their high mutation rates, short replication times, and often immense population sizes. The error-prone nature of RNA virus replication can be explained by a lack of a 3′-to-5′ proofreading activity in the RNA dependent RNA polymerase (RdRp), allowing mutations to occur at a high frequency throughout their genomes. Mutation rates of RNA viruses are the highest of any group of organisms, and are estimated at approximately one mutation per genome, per replication cycle (Drake and

Holland, 1999; Moya *et al.*, 2004). As a consequence, RNA viruses are known to exist as a dynamic distribution of quasispecies that are closely related, non-identical genomes centered around one master sequence (Domingo *et al.*, 1998). In function of their mutation rates and replication rates, which can vary substantially between RNA viruses, the evolutionary rates of RNA viruses, including coronaviruses, are estimated to be in the range of 10^{-4} to 10^{-3} nucleotide substitutions per site per year, although some variation exists (Moya *et al.*, 2004). These values are several orders of magnitude larger than those encountered during replication of DNA viruses, and many orders of magnitude higher than those of cellular DNA (Drake, 1991). Although high mutation rates provide RNA viruses with a useful tool to increase their level of genetic variability and thereby their ability to adapt quickly to changing environments, they also present a certain adaptive constraint, by limiting the genome size of RNA viruses, as is explained in the concept of the error threshold (Eigen, 1987). According to this theory, only a maximum error rate is tolerable for a genome of a particular size. Beyond this threshold, genomes would be over-burdened with deleterious mutations leading to dramatic fitness losses and eventual extinction. Given the high mutation rates of RNA viruses, genome sizes must be small to prevent the accumulation of a lethal number of deleterious mutations, and most RNA viruses therefore evolve permanently close to error catastrophe (Crotty *et al.*, 2001). The relationship between genome size and mutation rate is of particular importance for coronaviruses, as they have the largest genomes among all RNA viruses, which might therefore lead to lowering of their mutation rates or to a balancing of coronavirus evolution near the threshold of error catastrophe. Another potential impact on coronavirus mutation rates was recently hypothesized, as a 3′-to-5′ exonuclease domain was found in the replicase pp1ab polyprotein of coronaviruses, for which a potential RNA proofreading mechanism might be speculated (Snijder *et al.*, 2003). However, if this domain plays a functional role in coronavirus replication, and whether this role would involve a possible proofreading or repair activity, cannot yet be confirmed.

Although often confused in literature, a distinction between the terms of mutation rates and substitution rates has to be made. While mutation rates represent the rate at which mutational errors are incorporated into a genome during replication, substitution rates are a property of the viral population as a whole and represent the rate at which new mutations spread and become fixed in a population as a result of random genetic drift or natural selection (Drummond *et al.*, 2003). In MHV, a high mutation rate of 2.4 (range of 1.5–7.4) mutations per genome round of replication was found (Baric *et al.*, 1995). Recently, we estimated the substitution rate, also referred to as the evolutionary rate, of the BCoV/HCoV-OC43 pair at $3.5–4.3 \times 10^{-4}$ nucleotide substitutions per site per year (Vijgen *et al.*, 2005b; Vijgen *et al.*, 2005c; Vijgen *et al.*, 2005a). In these analyses, coronaviruses were assumed to evolve according to a molecular clock, with a constant rate of evolution over time. A relaxed molecular clock analysis, that allows rates to change among lineages, was furthermore used to estimate the evolutionary rate of BCoV, HCoV-OC43, and the closely related PHEV, based on two sequence data sets. For the spike gene region, a higher evolutionary rate (6×10^{-4} nucleotide substitutions per site per year) was estimated, while the estimate for the nucleocapsid gene region was consistent with the previously calculated evolutionary rate of BCoV/HCoV-OC43 (Vijgen *et al.*, 2006). These evolutionary rates of BCoV-related group 2 coronaviruses compare well to those estimated for other coronaviruses. For the SARS-coronavirus (SARS-CoV), an evolutionary rate ranging from 4.0×10^{-4} to 2.4×10^{-3} nucleotide substitutions per site per year was estimated, depending on the sequence data used, and on the number of strains under analysis (Salemi *et al.*, 2004; Zhao *et al.*, 2004). An evolutionary rate of 7×10^{-4} nucleotide substitutions per site per year was calculated for transmissible gastroenteritis coronavirus (TGEV), a group 1 coronavirus, by using spike gene sequence data (Sanchez *et al.*, 1992). These rates all fall in the range of evolutionary rates reported for most other RNA viruses (Domingo and Holland, 1988; Jenkins *et al.*, 2002). Coronaviruses evolve slower than the classical fast evolving RNA viruses such as human immunodeficiency virus type 1 (HIV-1) and influenza A virus, which have been reported to have substitution rates in excess of 1×10^{-3} nucleotide substitutions per site per year (Jenkins *et al.*, 2002).

While the impact of the accumulation of point mutations is reflected in the estimated coronavirus evolutionary rates, the influence of another major force that drives coronavirus evolution, recombination, is not incorporated in these calculations. However, based on phylogenetic analysis, the occurrence of recombination events in the evolutionary history of coronaviruses can be suggested when a phylogenetic incongruence, in which different gene regions produce different tree topologies, is demonstrated. Coronaviruses undergo homologous RNA recombination at a high frequency, and this process is believed to play a role in the emergence of novel coronavirus types (Lai and Cavanagh, 1997). The occurrence of recombination events is not only beneficial for increasing the level of genetic diversity, it may also serve as a correction mechanism counteracting the accumulation of deleterious mutations, a process known as Muller's ratchet. However, in RNA viruses an opposite hypothesis has also been suggested, in which recombination is more likely to create combinations of deleterious mutations than to prevent them. In doing so, recombination would reduce the error threshold so that dramatic fitness reductions would occur at lower mutation rates (Boerlijst *et al.*, 1996). However, no experimental data supporting these hypotheses have yet been obtained. The mechanism for generating coronavirus recombinants is inherent to the process of coronavirus replication, in which polymerase jumping occurs during subgenomic RNA synthesis. Coronavirus recombination is assumed to be mediated by a "copy choice mechanism," similar to RNA recombination in poliovirus, implicating a template switching during RdRp jumping (Kirkegaard and Baltimore, 1986; Herrewegh *et al.*, 1998). Template switching occurs randomly, and in case of MHV, the frequency of recombination in the entire genome was estimated to be around 25% (Baric *et al.*, 1990). To date, natural recombination has been described between group 1 feline and canine enteric coronaviruses, between strains of MHV, and between strains of IBV (Herrewegh *et al.*, 1998; Makino *et al.*, 1986; Kottier *et al.*, 1995). A high potential for recombination among group 2 coronaviruses has recently been suggested, and in an experimental system recombination between BCoV and HCoV-OC43 was demonstrated (Wu *et al.*, 2003). Possible recombination crossover sites have also been found in the SARS-CoV genome suggesting a possible recombinant history of the SARS-CoV, in which a recombination event between a mammalian coronavirus assigned to group 1 or 2, and an avian coronavirus belonging to group 3 was proposed (Stavrinides and Guttman, 2004; Rest and Mindell, 2003).

Another factor that increases genetic diversity in coronavirus genomes is the occurrence of nucleotide insertions and deletions, together called "indels" since only circumstantial evidence can clarify whether such a difference between two genomes is the result of an insertion in one sequence or a deletion in the other. Insertion or deletion mutations with biological importance generally occur within the coronavirus spike gene, leading to alterations in e.g. tissue tropism and immune sensitivity (Rasschaert *et al.*, 1990; Gallagher *et al.*, 1990). Similarities in insertion and deletion patterns might also yield important information on the evolutionary history of coronaviruses.

Evolutionary history of the BCoV/HCoV-OC43/PHEV subcluster of group 2 coronaviruses: a story of interspecies transmission events

The close genetic relatedness among BCoV, HCoV-OC43 and PHEV, three coronaviruses with a different host-specificity, which show in certain genome regions an even higher genetic similarity than is observed between some MHV strains, suggests that these coronaviruses diverged fairly recently. Reconstruction of the evolutionary history of these coronaviruses, demonstrated a prior divergence of an ancestral strain of PHEV from the ancestor of BCoV and HCoV-OC43 (Vijgen *et al.*, 2006). This event in coronavirus evolution could be dated back to the end of the 19th century (1878, 95% highest posterior density interval: 1747–1954) based on nucleocapsid gene sequence data of several strains of BCoV, HCoV-OC43 and PHEV. When spike gene sequence data were used in this analysis, this event was dated approximately 100 years earlier, which is the result of a significantly higher genetic distance of PHEV to BCoV and HCoV-OC43 in the spike gene region than in other genome regions. An explanation for this incongruence could not yet be provided, but one of the hypotheses might be a recombination event in the spike gene region between an ancestral strain of PHEV and another hitherto unknown coronavirus. However, this hypothesis would not explain why an excess of synonymous substitutions is responsible for the high genetic distance in the PHEV spike gene region. Whether the most recent common ancestor of PHEV, BCoV and HCoV-OC43 was a virus replicating in a porcine, bovine or human host, or in all three species, or even in another species could not be inferred from the present data, but we can speculate that interspecies transmission events have occurred prior to the emergence of PHEV, BCoV, and HCoV-OC43. Coronavirus host-switching events have been reported before, such as for BCoV that has the potential to infect other species, including humans (Zhang *et al.*, 1994). The isolation of the SARS-CoV from masked palm civets also indicated that this coronavirus was enzootic in an animal species before suddenly emerging as a virulent human pathogen. Phylogenetic analysis of SARS-CoV sequences from palm civets and humans during the second, smaller, SARS outbreak in 2003–2004, suggested that this new SARS episode resulted from a novel viral transmission from animals to humans, which occurred independently from the previous interspecies transmission event(s) that led to the initial SARS outbreak (Song *et al.*, 2005). The SARS-CoV is therefore likely to easily cross the species barrier, and its fast evolution in humans as well as in civets suggests an adaptation process in these species, hypothesizing that the SARS-CoV originally resides in another, yet unknown, animal species. Another potential reservoir for the SARS-CoV was suggested when a SARS-CoV-related coronavirus was identified in bats (Lau *et al.*, 2005; Li *et al.*, 2005). Such viral promiscuity reported for coronaviruses could account for the incongruence observed between coronavirus and host phylogenies, in which a monophyly of the host species does not correspond to a monophyly of the coronaviruses isolated from these species (Rest and Mindell, 2003).

The divergence of BCoV and HCoV-OC43 strains could be dated back to around 1890 (Figure 7.3) (Vijgen *et al.*, 2005b). Whether the interspecies transmission event, which is inextricably associated with the divergence of these genetically similar viruses, took place in a bovine-to-human or in a human-to-bovine direction is not known. However, we suggested that the occurrence of a 290 nucleotide deletion, corresponding to the absence of BCoV ns4.9 and ns4.8, in HCoV-OC43 in reference to the BCoV genome, supports the speculation of a bovine-to-human direction of the transmission, as this additional sequence fragment is also present in MHV and SDAV. We hypothesized that at the end of the 19th century, an ancestral BCoV strain crossed the species barrier and became able to infect humans, resulting in the emergence of a new human coronavirus type, a scenario similar to the origin of the SARS outbreak. However, this hypothesis is very speculative, and no conclusive evidence for the bovine-to-human direction of the interspecies transmission event is available.

Interestingly, the most recent common ancestors of the PHEV strains, the BCoV strains, and the HCoV-OC43 strains individually, were all estimated to have existed in a recent past i.e. only 50 to 60 years ago, approximately 10 to 20 years before their first isolation (Vijgen *et al.*, 2006). The isolation areas of the PHEV, BCoV and HCoV-OC43 strains used in these analyses, were distributed across the North-American and European continents, indicating that these three coronaviruses might have spread in their natural hosts over a large geographical region in a relatively short period of time. Another explana-

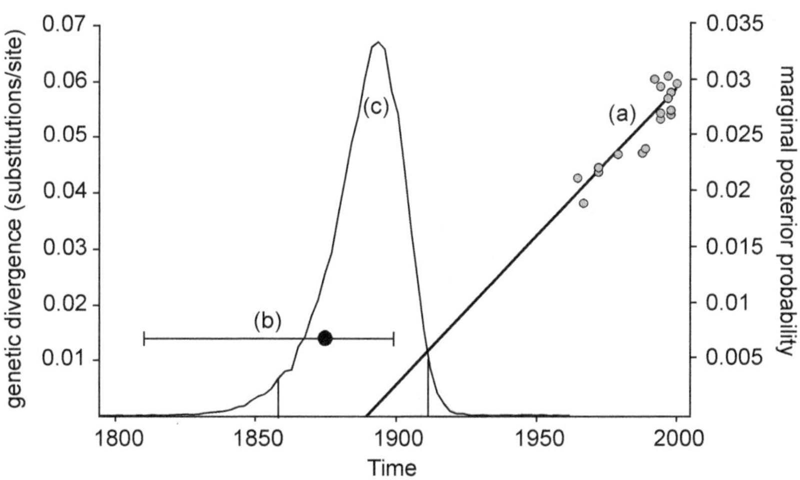

Figure 7.3 Results of the evolutionary rate and timing of the most recent common ancestor (MRCA) analyses for the HCoV-OC43/BCoV pair. (a) Linear regression of root-to-tip divergence (Y-axis) versus sampling time (X-axis). The point at which the regression line crosses the time-axis indicates the time to the most recent common ancestor (TMRCA, 1891). (b) Maximum likelihood estimate (1873) with 95% confidence intervals (1815–1899) for the TMRCA. (c) Marginal posterior probability (secondary Y-axis) for the TMRCA obtained using the Bayesian coalescent approach. The vertical bars in the distribution represent the 95% highest posterior density interval. Isolations dates of HCoV-OC43 and BCoV strains are indicated by grey dots.

tion for this observation could be that for some species the most recent common ancestor only relates to the coronaviruses that are currently circulating. Continual extinction events might have replaced earlier lineages in these species (Holmes, 2001). However, in the corresponding phylogenetic tree, the species-specific lineages were not connected by particularly long internal branches.

Conclusions and perspective

When studying the evolutionary history of the BCoV-related group 2 coronaviruses, the reconstruction of the evolutionary relationships among BCoV, HCoV-OC43, and PHEV might represent only a small part of the story. The limited sequence data that are available for CRCoV suggest an even closer genetic and possible evolutionary relationship between BCoV and CRCoV than has been demonstrated for BCoV and HCoV-OC43. A complete genome characterization of this novel canine coronavirus with respiratory tropism is however required to perform a more intensive evolutionary analysis. Certain genome characteristics like genome regions with a potential recombinant origin or particular nucleotide insertion and deletion patterns, can provide important information regarding the evolutionary history of these viruses. Furthermore, the recent discoveries of novel human and animal coronaviruses suggest that, when looked for, coronaviruses might be present in each animal species. Moreover, several species harbor a set of genetically distinct coronaviruses, as has been demonstrated for e.g. pigs (from which the group 1 coronaviruses TGEV and PDEV, and the group 2 PHEV can be isolated), dogs (harboring the group 1 CCoV and the group 2 CRCoV), and humans (for which five types of human coronaviruses have been described, belonging to group 1 and group 2, as well as to an early split-off from group 2). It is therefore possible that several, yet unknown, animal and human coronaviruses might also have played a role in the evolutionary history of the BCoV-related group 2 coronaviruses.

Acknowledgments

We would like to thank all the colleagues of the laboratory of Clinical and Epidemiological Virology, Department of Microbiology and Immunology, Rega Institute for Medical Research, University of Leuven, Belgium for helpful comments and discussion. This work was supported by a postdoctoral fellowship of the Research Fund K.U.Leuven to Leen Vijgen.

References

Abraham, S., Kienzle, T.E., Lapps, W., and Brian, D.A. (1990). Deduced sequence of the bovine coronavirus spike protein and identification of the internal proteolytic cleavage site. Virology *176*, 296–301.

Ballesteros, M.L., Sanchez, C.M., and Enjuanes, L. (1997). Two amino acid changes at the N-terminus of transmissible gastroenteritis coronavirus spike protein result in the loss of enteric tropism. Virology *227*, 378–388.

Baric, R.S., Fu, K., Chen, W., and Yount, B. (1995). High recombination and mutation rates in mouse hepatitis virus suggest that coronaviruses may be potentially important emerging viruses. Adv. Exp. Med. Biol. *380*, 571–576.

Baric, R.S., Fu, K., Schaad, M.C., and Stohlman, S.A. (1990). Establishing a genetic recombination map for murine coronavirus strain A59 complementation groups. Virology *177*, 646–656.

Binns, M.M., Boursnell, M.E., Cavanagh, D., Pappin, D.J., and Brown, T.D. (1985). Cloning and sequencing of the gene encoding the spike protein of the coronavirus IBV. J. Gen. Virol. *66*, 719–726.

Boerlijst, M.C., Bonhoeffer, S., and Nowak, M.A. (1996). Viral quasi-species and recombination. Proc. R. Soc. Lond. B *263*, 1577–1584.

Bosch, F.X., Garten, W., Klenk, H.D., and Rott, R. (1981). Proteolytic cleavage of influenza virus hemagglutinins: primary structure of the connecting peptide between HA1 and HA2 determines proteolytic cleavability and pathogenicity of Avian influenza viruses. Virology *113*, 725–735.

Bromham, L., and Penny, D. (2003). The modern molecular clock. Nat. Rev. Genet. *4*, 216–224.

Crotty, S., Cameron, C.E., and Andino, R. (2001). RNA virus error catastrophe: direct molecular test by using ribavirin. Proc. Natl. Acad. Sci. USA *98*, 6895–6900.

de Groot, R.J., Van Leen, R.W., Dalderup, M.J., Vennema, H., Horzinek, M.C., and Spaan, W.J. (1989). Stably expressed FIPV peplomer protein induces cell fusion and elicits neutralizing antibodies in mice. Virology *171*, 493–502.

de Haan, C.A., Masters, P.S., Shen, X., Weiss, S., and Rottier, P.J. (2002). The group-specific murine coronavirus genes are not essential, but their deletion, by reverse genetics, is attenuating in the natural host. Virology *296*, 177–189.

Domingo, E., Baranowski, E., Ruiz-Jarabo, C.M., Martin-Hernandez, A.M., Saiz, J.C., and Escarmis, C. (1998). Quasispecies structure and persistence of RNA viruses. Emerg. Infect. Dis. *4*, 521–527.

Domingo,E., and Holland,J.J. (1988). High error rates, population equilibrium and evolution of RNA replication systems. In RNA genetics, E.Domingo, J.J.Holland, and P.Ahlquist, eds. (Boca Raton, FL: CRC Press), pp. 3–36.

Drake, J.W. (1991). A constant rate of spontaneous mutation in DNA-based microbes. Proc. Natl. Acad. Sci. USA. *88*, 7160–7164.

Drake, J.W., and Holland, J.J. (1999). Mutation rates among RNA viruses. Proc. Natl. Acad. Sci. USA. *96*, 13910–13913.

Drummond, A., Pybus, O.G., and Rambaut, A. (2003). Inference of viral evolutionary rates from molecular sequences. Adv. Parasitol. *54*, 331–358.

Drummond, A.J., Pybus, O.G., Rambaut, A., Forsberg, R., and Rodrigo, A.G. (2005) Measurably evolving populations. Trends in Ecology and Evolution *18*, 481–488.

Eigen, M. (1987). New concepts for dealing with the evolution of nucleic acids. Cold Spring Harb. Symp. Quant. Biol. *52*, 307–320.

Erles, K., Toomey, C., Brooks, H.W., and Brownlie, J. (2003). Detection of a group 2 coronavirus in dogs with canine infectious respiratory disease. Virology *310*, 216–223.

Frana, M.F., Behnke, J.N., Sturman, L.S., and Holmes, K.V. (1985). Proteolytic cleavage of the E2 glycoprotein of murine coronavirus: host-dependent differences in proteolytic cleavage and cell fusion. J. Virol. *56*, 912–920.

Gallagher, T.M., Parker, S.E., and Buchmeier, M.J. (1990). Neutralization-resistant variants of a neurotropic coronavirus are generated by deletions within the amino-terminal half of the spike glycoprotein. J. Virol. *64*, 731–741.

Glickman, R.L., Syddall, R.J., Iorio, R.M., Sheehan, J.P., and Bratt, M.A. (1988). Quantitative basic residue requirements in the cleavage-activation site of the fusion glycoprotein as a determinant of virulence for Newcastle disease virus. J. Virol. *62*, 354–356.

Gombold, J.L., Hingley, S.T., and Weiss, S.R. (1993). Fusion-defective mutants of mouse hepatitis virus A59 contain a mutation in the spike protein cleavage signal. J. Virol. *67*, 4504–4512.

Gorbalenya, A.E., Snijder, E.J., and Spaan, W.J. (2004). Severe acute respiratory syndrome coronavirus phylogeny: toward consensus. J. Virol. *78*, 7863–7866.

Guy, J.S., Breslin, J.J., Breuhaus, B., Vivrette, S., and Smith, L.G. (2000). Characterization of a coronavirus isolated from a diarrheic foal. J. Clin. Microbiol. *38*, 4523–4526.

Haijema, B.J., Volders, H., and Rottier, P.J. (2004). Live, attenuated coronavirus vaccines through the directed deletion of group-specific genes provide protection against feline infectious peritonitis. J. Virol. *78*, 3863–3871.

Herrewegh, A.A., Smeenk, I., Horzinek, M.C., Rottier, P.J., and de Groot, R.J. (1998). Feline coronavirus type II strains 79–1683 and 79–1146 originate from a double recombination between feline coronavirus type I and canine coronavirus. J. Virol. *72*, 4508–4514.

Holmes, E.E. (2001). Estimating risks in declining populations with poor data. Proc. Natl. Acad. Sci. USA. *98*, 5072–5077.

Jenkins, G.M., Rambaut, A., Pybus, O.G., and Holmes, E.C. (2002). Rates of molecular evolution in RNA viruses: a quantitative phylogenetic analysis. J. Mol. Evol. *54*, 156–165.

Kamahora, T., Soe, L.H., and Lai, M.M. (1989). Sequence analysis of nucleocapsid gene and leader RNA of human coronavirus OC43. Virus Res. *12*, 1–9.

Kirkegaard, K., and Baltimore, D. (1986). The mechanism of RNA recombination in poliovirus. Cell *47*, 433–443.

Kottier, S.A., Cavanagh, D., and Britton, P. (1995). Experimental evidence of recombination in coronavirus infectious bronchitis virus. Virology *213*, 569–580.

Kunkel, F., and Herrler, G. (1993). Structural and functional analysis of the surface protein of human coronavirus OC43. Virology *195*, 195–202.

Lai, M.M., and Cavanagh, D. (1997). The molecular biology of coronaviruses. Adv. Virus Res. *48*, 1–100.

Lau, S.K., Woo, P.C., Li, K.S., Huang, Y., Tsoi, H.W., Wong, B.H., Wong, S.S., Leung, S.Y., Chan, K.H., and Yuen, K.Y. (2005). Severe acute respiratory syndrome coronavirus-like virus in Chinese horseshoe bats. Proc. Natl. Acad. Sci. USA. *102*, 14040–14045.

Leparc-Goffart, I., Hingley, S.T., Chua, M.M., Jiang, X., Lavi, E., and Weiss, S.R. (1997). Altered pathogenesis of a mutant of the murine coronavirus MHV-A59 is associated with a Q159L amino acid substitution in the spike protein. Virology *239*, 1–10.

Li, W., Shi, Z., Yu, M., Ren, W., Smith, C., Epstein, J.H., Wang, H., Crameri, G., Hu, Z., Zhang, H., Zhang, J., McEachern, J., Field, H., Daszak, P., Eaton, B.T., Zhang, S., and Wang, L.F. (2005). Bats are natural reservoirs of SARS-like coronaviruses. Science *310*, 676–679.

Luytjes, W., Sturman, L.S., Bredenbeek, P.J., Charite, J., van der Zeijst, B.A., Horzinek, M.C., and Spaan, W.J. (1987). Primary structure of the glycoprotein E2 of coronavirus MHV-A59 and identification of the trypsin cleavage site. Virology *161*, 479–487.

Makino, S., Keck, J.G., Stohlman, S.A., and Lai, M.M. (1986). High-frequency RNA recombination of murine coronaviruses. J. Virol. *57*, 729–737.

Mounir, S., and Talbot, P.J. (1992). Sequence analysis of the membrane protein gene of human coronavirus OC43 and evidence for O-glycosylation. J. Gen. Virol. *73*, 2731–2736.

Mounir, S., and Talbot, P.J. (1993). Human coronavirus OC43 RNA 4 lacks two open reading frames located downstream of the S gene of bovine coronavirus. Virology *192*, 355–360.

Moya, A., Holmes, E.C., and Gonzalez-Candelas, F. (2004). The population genetics and evolutionary epidemiology of RNA viruses. Nat. Rev. Microbiol. *2*, 279–288.

Perez, L.G., and Hunter, E. (1987). Mutations within the proteolytic cleavage site of the Rous sarcoma virus glycoprotein that block processing to gp85 and gp37. J. Virol. *61*, 1609–1614.

Pyrc, K., Jebbink, M.F., Berkhout, B., and van der Hoek, L. (2004). Genome structure and transcriptional regulation of human coronavirus NL63. Virol. J. *1*, 7.

Rasschaert, D., Duarte, M., and Laude, H. (1990). Porcine respiratory coronavirus differs from transmissible gastroenteritis virus by a few genomic deletions. J. Gen. Virol. *71*, 2599–2607.

Rest, J.S., and Mindell, D.P. (2003). SARS associated coronavirus has a recombinant polymerase and coronaviruses have a history of host-shifting. Infect. Genet. Evol. *3*, 219–225.

Salemi, M., Fitch, W.M., Ciccozzi, M., Ruiz-Alvarez, M.J., Rezza, G., and Lewis, M.J. (2004). Severe acute respiratory syndrome coronavirus sequence characteristics and evolutionary rate estimate from maximum likelihood analysis. J. Virol. *78*, 1602–1603.

Sanchez, C.M., Gebauer, F., Sune, C., Mendez, A., Dopazo, J., and Enjuanes, L. (1992). Genetic evolution and tropism of transmissible gastroenteritis coronaviruses. Virology *190*, 92–105.

Snijder, E.J., Bredenbeek, P.J., Dobbe, J.C., Thiel, V., Ziebuhr, J., Poon, L.L., Guan, Y., Rozanov, M., Spaan, W.J., and Gorbalenya, A.E. (2003). Unique and conserved features of genome and proteome of SARS-coronavirus, an early split-off from the coronavirus group 2 lineage. J. Mol. Biol. *331*, 991–1004.

Song, H.D., Tu, C.C., Zhang, G.W., Wang, S.Y., Zheng, K., Lei, L.C., Chen, Q.X., Gao, Y.W., Zhou, H.Q., Xiang, H., Zheng, H.J., Chern, S.W., Cheng, F., Pan, C.M., Xuan, H., Chen, S.J., Luo, H.M., Zhou, D.H., Liu, Y.F., He, J.F., Qin, P.Z., Li, L.H., Ren, Y.Q., Liang, W.J., Yu, Y.D., Anderson, L., Wang, M., Xu, R.H., Wu, X.W., Zheng, H.Y., Chen, J.D., Liang, G., Gao, Y., Liao, M., Fang, L., Jiang, L.Y., Li, H., Chen, F., Di, B., He, L.J., Lin, J.Y., Tong, S., Kong, X., Du, L., Hao, P., Tang, H., Bernini, A., Yu, X.J., Spiga, O., Guo, Z.M., Pan, H.Y., He, W.Z., Manuguerra, J.C., Fontanet, A., Danchin, A., Niccolai, N., Li, Y.X., Wu, C.I., and Zhao, G.P. (2005). Cross-host evolution of severe acute respiratory syndrome coronavirus in palm civet and human. Proc. Natl. Acad. Sci. USA *102*, 2430–2435.

St-Jean, J.R., Jacomy, H., Desforges, M., Vabret, A., Freymuth, F., and Talbot, P.J. (2004). Human respiratory coronavirus OC43: genetic stability and neuroinvasion. J. Virol. *78*, 8824–8834.

Stavrinides, J., and Guttman, D.S. (2004). Mosaic evolution of the severe acute respiratory syndrome coronavirus. J. Virol. *78*, 76–82.

Sturman, L.S., Ricard, C.S., and Holmes, K.V. (1985). Proteolytic cleavage of the E2 glycoprotein of murine coronavirus: activation of cell-fusing activity of virions by trypsin and separation of two different 90K cleavage fragments. J. Virol. *56*, 904–911.

Vabret, A., Mourez, T., Gouarin, S., Petitjean, J., and Freymuth, F. (2003). An outbreak of coronavirus OC43 respiratory infection in Normandy, France. Clin. Infect. Dis. *36*, 985–989.

Vautherot, J.F., and Laporte, J. (1983). Utilization of monoclonal antibodies for antigenic characterization of coronaviruses. Ann. Rech. Vet. *14*, 437–444.

Vieler, E., Schlapp, T., and Herbst, W. (1996). The region between the M and S genes of porcine haemagglutinating encephalomyelitis virus is highly similar to human coronavirus OC43. J. Gen. Virol. *77*, 1443–1447.

Vijgen, L., Keyaerts, E., Lemey, P., Maes, P., Van Reeth, K., Nauwynck, H., Pensaert, M., and Van Ranst, M. (2006). Evolutionary history of the closely related group 2 coronaviruses: porcine hemagglutinating encephalomyelitis virus, bovine coronavirus, and human coronavirus OC43. J. Virol. *80*, 7270-7274.

Vijgen, L., Keyaerts, E., Lemey, P., Moës, E., Li, S., Vandamme, A.M., and Van Ranst, M. (2005a). Circulation of genetically distinct contemporary human coronavirus OC43 strains. Virology *337*, 85–92.

Vijgen, L., Keyaerts, E., Moes, E., Thoelen, I., Wollants, E., Lemey, P., Vandamme, A.M., and Van Ranst, M. (2005b). Complete genomic sequence of human coronavirus OC43: molecular clock analysis suggests a relatively recent zoonotic coronavirus transmission event. J. Virol. *79*, 1595–1604.

Vijgen, L., Lemey, P., Keyaerts, E., and Van Ranst, M. (2005c). Genetic variability of human respiratory coronavirus OC43. J. Virol. *79*, 3223–3224.

Woo, P.C., Lau, S.K., Chu, C.M., Chan, K.H., Tsoi, H.W., Huang, Y., Wong, B.H., Poon, R.W., Cai, J.J., Luk, W.K., Poon, L.L., Wong, S.S., Guan, Y., Peiris, J.S., and Yuen, K.Y. (2005). Characterization and complete genome sequence of a novel coronavirus, coronavirus HKU1, from patients with pneumonia. J. Virol. *79*, 884–895.

Wu, H.Y., Guy, J.S., Yoo, D., Vlasak, R., Urbach, E., and Brian, D.A. (2003). Common RNA replication signals exist among group 2 coronaviruses: evidence for in vivo recombination between animal and human coronavirus molecules. Virology *315*, 174–183.

Wurzer, W.J., Obojes, K., and Vlasak, R. (2002). The sialate-4-O-acetylesterases of coronaviruses related to mouse hepatitis virus: a proposal to reorganize group 2 Coronaviridae. J. Gen. Virol. *83*, 395–402.

Zhang, X.M., Herbst, W., Kousoulas, K.G., and Storz, J. (1994). Biological and genetic characterization of a hemagglutinating coronavirus isolated from a diarrhoeic child. J. Med. Virol. *44*, 152–161.

Zhang, X.M., Kousoulas, K.G., and Storz, J. (1992). The hemagglutinin/esterase gene of human coronavirus strain OC43: phylogenetic relationships to bovine and murine coronaviruses and influenza C virus. Virology *186*, 318–323.

Zhao, Z., Li, H., Wu, X., Zhong, Y., Zhang, K., Zhang, Y.P., Boerwinkle, E., and Fu, Y.X. (2004). Moderate mutation rate in the SARS coronavirus genome and its implications. BMC. Evol. Biol. *4*, 21.

Part II

Avian Coronavirus Diseases and Infectious Bronchitis Virus Vaccine Development

8

Paul Britton and Dave Cavanagh

Abstract

Infectious bronchitis virus (IBV), a Group 3 coronavirus, is responsible for economic losses and welfare problems with chickens (the domestic fowl) globally. In addition to all respiratory epithelial surfaces it replicates throughout the alimentary tract, kidneys and gonads. Despite live and inactivated vaccines, it continues to be a major problem due to extensive antigenic variation; there is poor cross-protection. Susceptibility to the outcome of IBV infection is genetically determined.

Coronaviruses that are genetically and antigenically similar to IBV cause enteric disease in turkeys, and respiratory and kidney disease in pheasants. IBV-like viruses have recently been reported in several other gallinaceous (fowl-like, order Galliformes) birds—peafowl, partridge, guinea fowl—and in a duck (teal); some of these isolates may be IBVs that have crossed into other species. Group 3 coronaviruses with additional open reading frames have been detected in graylag goose, pigeon and mallard duck. A group 2 coronavirus has been reportedly isolated from a Manx shearwater (*Puffinus puffinus*), and a coronavirus of an undetermined Group has been isolated from a parrot.

Inactivated vaccines are ineffective unless preceded by live attenuated vaccines to prime the protective immune response. The S1 spike protein subunit is necessary and sufficient to induce protective immunity. Differences in as few as 5% of the amino acids in S1 can decrease cross-protection. Genetic manipulation of the IBV genome is under way for the rational attenuation of IBV, and for the development of IB vaccines that can be applied *in ovo*.

Introduction

During the course of a year there are approximately 40×10^9 chickens (the domestic fowl, *Gallus gallus domesticus*). The majority of modern meat-type chickens (broilers) are raised until only five to six weeks of age. Consequently there are many crops within a year. Poultry is an increasingly important source of meat protein. Globally there has been a 16% increase in poultry meat production since 1995. In the developing countries the corresponding increase is a staggering 77% increase (Anon, 2006). A government-sponsored report published in 2005 revealed that the number one cause of infectious disease-related economic loss in the UK poultry industry was infectious bronchitis (IB) (http://www.defra.gov.uk/science/Project_Data/DocumentLibrary/ZZ0102/ZZ0102_1215_FRP.

doc). The causative agent, IB virus (IBV), was the first coronavirus to be discovered, in the 1930s in the USA. Despite the application of live and killed IB vaccines it has continued to be of major economic importance in the poultry industry. Why is that? Reasons include: the ubiquity of IBV, effectively global; the intensive nature of poultry production, which is conducive to disease spread and exacerbation of clinical outcome; IBV spreads easily; the huge numbers of juvenile chickens kept in confined spaces; the short-lived nature of immunity to IB in the respiratory tract; and the co-existence of numerous (dozens, maybe hundreds) of serotypes, cross-protection being poor. Serotype and the induction of protective immunity are features of the surface spike (S) protein.

An interesting feature of IBV, which is not revealed by its name, is that it can replicate in a great many tissues: essentially at every epithelial surface, not only of the respiratory tract or, indeed, of the alimentary tract, but also in epithelial cells of other organs e.g. the kidney, oviduct and Harderian gland (a lymphoid gland near the eye). Consequently, infection by IBV can manifest itself clinically in many ways.

Another feature of IBV (coronavirus Group 3), one which has emerged in recent years, is that its host range may be greater than previously thought (Cavanagh, 2005). In addition there are coronaviruses of other species e.g. turkey and pheasant, which, though considered to be separate coronavirus species, are genetically very similar to IBV. Most recently Group 3 coronaviruses have been detected in wild geese, ducks and pigeons (Jonassen *et al.*, 2005). Experimental infection of turkeys with bovine coronavirus (BCoV; Group 2) resulted not only in virus replication but also disease (Ismail *et al.*, 2001). This raises the prospect that birds may be infected with coronaviruses other than those of Group 3.

Imperfect control of IB has been achieved by the application of vaccines (Cavanagh, 2003). Unfortunately inactivated IBV has poor efficacy unless the chickens have previously been primed by vaccination with a live virus vaccine. Expression of the S1 subunit of IBV from a fowl adenovirus vector induced good protective immunity after a single live application. This and similar findings, combined with manipulation of the IBV genome, offer hope for the development of a new generation of rationally attenuated vaccine strains with selected S protein genes.

These are the topics that are expanded upon in this chapter.

Diseases caused by infectious bronchitis virus

As suggested by its name, IBV causes respiratory disease, though its replication is not limited to the respiratory tract. IBV causes deciliation of the ciliated epithelia of the nose and trachea (Cavanagh and Naqi, 2003; Chen *et al.*, 1996; Hodgson *et al.*, 2004). Chickens, especially those of only a few days or weeks of age, exhibit nasal discharge, snicking (similar to sneezing), râles (a vibration, perhaps caused by air passing through airways that are partially blocked by mucus), watery eyes and lethargy (Cavanagh and Naqi, 2003; Dhinaker Raj and Jones, 1997). Small areas of pneumonia may be observed in the lungs (Dhinaker Raj and Jones, 1997). Notwithstanding this, and the fact that titers of virus in the lungs can be similar to those in the nose and trachea, IB is not considered to cause pneumonia.

Alimentary tract tissues are susceptible to IBV, although infection of these does not usually manifest itself clinically. Nephritis is not uncommon in a proportion of naturally IBV-infected meat-type birds, environmental factors, including protein content of feed and temperature changes, playing a role. Some IBV strains are intrinsically nephropathogenic i.e. they reproducibly cause nephritis when inoculated experimentally into specific pathogen free (SPF) chickens (Cook *et al.*, 2001; Lambrechts *et al.*, 1993; Li and Yang, 2001; Pensaert and Lambrechts, 1994). The oviduct is also susceptible to IBV, which may contribute to diminished egg production. Recently it has been shown to replicate in the gonads of male chickens (Boltz *et al.*, 2004).

The hardest hit from a welfare point of view are chicks of only a few days or weeks of age. Some may die directly from the viral infection but a greater number die following secondary bacterial infection (Matthijs *et al.*, 2003). Most economic loss, though, is from reduced growth. Juvenile and mature birds suffer less from IBV infection although the economic consequences of infection in egg-laying stock can be great, as egg production drops precipitously and usually does not rise back to normal in the flock as a whole.

Tissue tropism of IBV

IBV strains as a group productively infect a large range of epithelial surfaces, literally from the top to the bottom of the chicken. Isolates differ in their extent of replication in the non-respiratory tissues, and some produce clinical disease in non-respiratory tissues, most notably kidney and proventriculus.

IBV initially infects the upper respiratory tract. Titers of live virus are maximal in the nose and trachea by three days post inoculation (by eye-drop and intranasally) and remain so for two to five days further, depending on the strain (Ambali and Jones, 1990; Hofstad and Yoder, 1966); Table 2 in Cavanagh (2003)). Thereafter the titer usually falls to below detectable levels, although small amounts of virus were still detectable at 14 days in some studies (Ambali and Jones, 1990). The virus is restricted to the ciliated and mucus-secreting cells (reviewed by Dhinaker Raj and Jones (1997)). Similar virus titers occur in the lungs and airsacs, sometimes peaking a little after those in nose and trachea.

Following the inoculation of SPF chickens experimentally, virus has been recovered from nose, Harderian gland (Toro *et al.*, 1996), a small lymphoid organ in the eye-socket, a major contributor to locally produced antibody for protecting oculonasal mucosae (reviewed by Dhinaker Raj and Jones (1997)), trachea, bronchus, lung, esophagus, proventriculus, duodenum, ileum, jejunum, bursa of Fabricius, ceca (cecal tonsils), kidney, rectum and cloaca (the common opening for release of eggs and feces) (Ambali and Jones, 1990; El Houadfi *et al.*, 1986; Hofstad and Yoder, 1966; Lucio and Fabricant, 1990; Yu *et al.*, 2001). Organ culture studies showed that whilst all strains studied by Bhattacharjee and Jones (1997) grew in proventriculus, bursa and kidney explants, some did not grow in cecal tonsil or rectal explants. The longest periods of isolation were in respect of the kidney and cecal tonsils, the latter being at the distal end of the alimentary tract in birds. Indeed, these organs can be the best place for isolation of IBV in the field, if the investigation is started after peak virus replication in the respiratory tract. Strains differ with respect to their relative titers in various tissues. Ambali and Jones (1990) showed that the Moroccan G strain of IBV produced titers in kidney that were similar to the high titers in trachea. In some other reports, cited above, titers in respiratory tissues were one to two orders of magnitude higher than elsewhere.

Beyond the respiratory tract, the most intensively studied tropism has been that for the kidney. Only a small proportion of IBV isolates exhibit high nephropathogenicity, causing up to 44% mortality within three weeks of intratracheal inoculation of unvaccinated birds (Liu and Kong, 2004; Pensaert and Lambrechts, 1994). Histopathological and immunochemical studies of the process showed that the MA-87 strain of virus attacked mainly the lower nephron down to the collecting duct epithelial cells (Chen *et al.*, 1996; Chen and Itakura, 1996). An ultrastructural investigation revealed that the virus replicated in all segments of tubules and ducts, but more frequently in the epithelial cells of the collecting ducts, collecting tubules, distal convoluted tubules and Henle's loops (Chen and Itakura, 1996). The extent of mortality is age-related (greatest in chicks; (Lambrechts *et al.*, 1993)) and varies according to the line of bird e.g. 18% and 44% of an egg layer-type bird and a meat-type bird, respectively (Pensaert and Lambrechts, 1994). These and other aspects of IBV have been reviewed by Cavanagh and Naqi (2003).

IBVs can be propagated readily in the allantoic cavity of embryonated chicken eggs, and in the chorioallantoic membrane of de-embryonated eggs, in which it can be radiolabeled (Cavanagh, 1981). The virus can also be adapted to good growth in primary chick kidney cells (Casais *et al.*, 2005; Hodgson *et al.*, 2006), and some strains have been adapted to Vero cells (Casais *et al.*, 2003; Coria and Ritchie, 1973).

Turkey coronaviral enteritis

Viruses that are genetically and antigenically closely related to IBV have been isolated from several avian species other than the domestic fowl (Cavanagh, 2005), the most well studied being that from turkeys.

First described in the USA in the 1940s, turkey coronaviral enteritis was initially called bluecomb disease or mud fever. In young poults infection with TCoV can be associated with mortality, which can be as high as 80% although is frequently much lower whilst still being of economic importance. A major contributor to mortality is secondary bacterial infection e.g. with enteropathogenic *E. coli* (Guy, 2000). In older birds it is debilitating, resulting in underperformance with regard to meat and egg production (Guy, 2000; Guy, 2003).

A coronavirus was shown thirty years ago as being the cause of bluecomb disease in the USA, where it has been most thoroughly studied (Guy, 2003). The virus has subsequently been detected in the UK (Cavanagh *et al.*, 2001; Culver *et al.*, 2006) and Brazil (L.Y.B. Villarreal, M.S. Assayag Jr., P.E. Brandão, J.L.V. Chacón, C.S.A Ferreira, N. Grando, M.R.S. Pequini, R. Soncini, M. Souza, and A.J.P. Ferreira, personal communication; Teixeira *et al.*, 2007). The virus replicates predominantly in enterocytes in the jejunum and ileum (Breslin *et al.*, 2000) and in the epithelium of the bursa of Fabricius (Guy *et al.*, 1997), in which gross lesions are manifest (reviewed by Guy (2003)). Destruction of the villous epithelium is probably responsible for the malabsorption, maldigestion and diarrhea that follow infection. No vaccine has been developed against TCoV. The disease has been reviewed by (Barnes *et al.*, 2000) in the context of other causes of enteric infections in turkeys.

Research in the USA, Europe and Brazil has shown that the coronaviruses from turkeys are genetically and antigenically closely related to IBV (Breslin *et al.*, 2000; Breslin *et al.*, 1999a; Breslin *et al.*, 1999b; Cavanagh *et al.*, 2001; Guy *et al.*, 1997; Ismail *et al.*, 2001; L.Y.B. Villarreal *et al.* personal communication; reviewed by Guy (2000)). Earlier reports, suggesting that TCoV was closely related to bovine coronavirus (Group 2), are erroneous (reviewed by Guy (2000)).

The sequences of genes 3 and 5 (equivalent to the accessory protein genes of IBV; (Casais *et al.*, 2005; Hodgson *et al.*, 2006) and the nucleocapsid protein (N) have been established for a number of TCoV isolates, including the strain (VR-911) in the American Type Culture Collection, demonstrating them to have ≥ 85% identity with those of IBV, which is high given that strains of IBV commonly differ from each other by this amount. The TCoVs N proteins have amino acid identities in the range 89–96% when compared with IBV strains (Figure 8.1).

The exception is the spike protein; recent research in the USA (Lin *et al.*, 2004) has revealed that there is only ~33% amino acid identity between the spike protein of the TCoVs and those of IBVs, a much lower identity than the lowest yet seen between IBV strains (61%; (Figure 8.2).

Unfortunately no cultured cells have been found to support the replication of TCoV. It is propagated in embryonated turkey eggs, inoculated by the amniotic sac route (Guy, 2003).

Pheasant coronavirus

Pheasants are infected by a coronavirus (PhCoV), sometimes associated with respiratory disease, and sometimes associated with kidney disease (Cavanagh *et al.*, 2002; Gough *et al.*, 1996; Lister *et al.*, 1985; Pennycott, 2000; Spackman and Cameron, 1983; Welchman Dde *et al.*, 2002). In the field mortalities, associated with the presence of PhCoV, have been high. For example, 15% of breeding pheasants died rapidly, the only clinical sign being sneezing. In another group of pheasants there were no respiratory signs but 45% had died by 10 weeks of age. Specific lesions that have been described are visceral gout and urolithiasis, with gross swelling and pallor of the kidneys. Moderately severe interstitial nephrosis was revealed by histological analysis.

The degree of genetic relatedness of PhCoV to IBV is the same as that between IBV and TCoV (Figures 8.1 and 8.2; Cavanagh *et al.*, 2002), except with regard to the S protein, in which PhCoV and IBV are more closely related to each other than to TCoV (Figure 8.2; Cavanagh *et al.*, 2002). The complete S gene of the ph/UK/606/95 strain of PhCoV has 79–83% identity with those of IBVs (Figure 8.2; our unpublished data).

The virus can replicate in chickens, as experimental inoculation resulted in antibody production, but without causing disease (Gough *et al.*, 1996). It can be propagated in the allantoic cavity of embryonated chicken eggs (Gough *et al.*, 1996; Lister *et al.*, 1985). Thus PhCoV and IBV have much in common, differing mostly in respect of the host in which they cause disease.

IBV replication in other avian species

As mentioned above, PhCoV can replicate asymptomatically in chickens (Gough *et al.*, 1996), likewise TCoV in chickens (Ismail *et al.*, 2003). Recent research in China suggests IBV can replicate in domestic peafowl (*Pavo cristatus*; Liu *et al.*, 2005), partridge (*Alectoris* sp.), guinea fowl (*Numida meleagris*) and teal (Anas sp. f duck; Liu *et al.*, 2005; reviewed by Cavanagh (2005)). These, like the domestic fowl, turkey and pheasant, are gallinaceous birds ("fowl-like," order Galliformes).

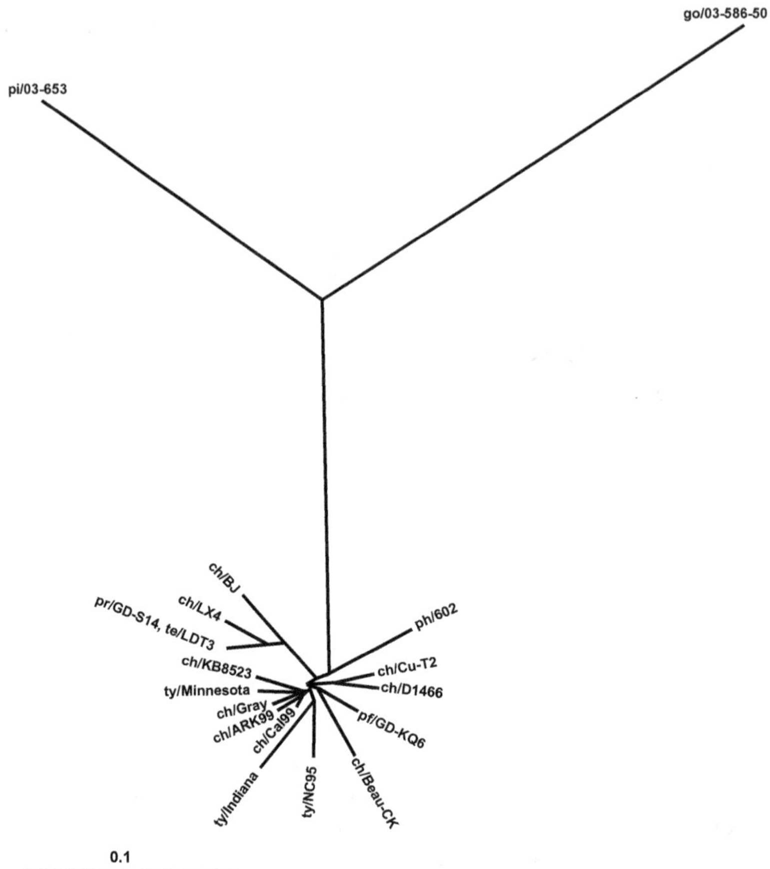

Figure 8.1 Phylogenetic analysis of the nucleocapsid (N) protein amino acid sequences of Group 3 coronaviruses isolated from various avian species; ch, chicken; ty, turkey; ph, pheasant; pr, partridge; pf, peafowl; te, teal; go, greylag goose; pi, pigeon. The sequences were aligned using ClustalX (Thompson *et al*., 1997) and the phylogenetic relationships of the aligned sequences were determined using the distance matrix and neighbor-joining methods of the PHYLIP package (Felsenstein, 1988). The scale bar represents the distances as numbers of substitutions per site, 0.1 nucleotide substitutions per site. The amino acid sequences used were from the following viruses deposited in GenBank, the accession numbers are given parentheses, the avian species from which the virus was isolated, the country of origin, isolate number and year of isolation are indicated in the names; ch/China/BJ/03 (AY319651), ch/China/LX4/02 (AY338732), ch/Japan/KB8523/88 (M21515), ch/USA/Gray/92 (S48137), ch/USA/Ark/99 (M85244), ch/USA/Cal/99 (AY514485), ch/USA/Beaudette/37 (AJ311317), ch/USA/Cu-T2/95 (U04805), ch/NL/D1466/78 (AF203006), ty/USA/Minnesota(ATCC VR-911)/69 (AF111996), ty/USA/Indiana/99 (AF111995), ty/USA/NC/95 (AF111997), ph/UK/602/95 (Unpublished), pr/China/GD-S14/03 (AY646283)" pf/China/GD-KQ6/03 (AY641576), te/China/LDT3/03 (AY702975).

Sequencing of the entire genome of the viruses isolated from these species revealed that they were IBV-like with respect to their genome organization and with the sequences of the genes encoded by their genes.

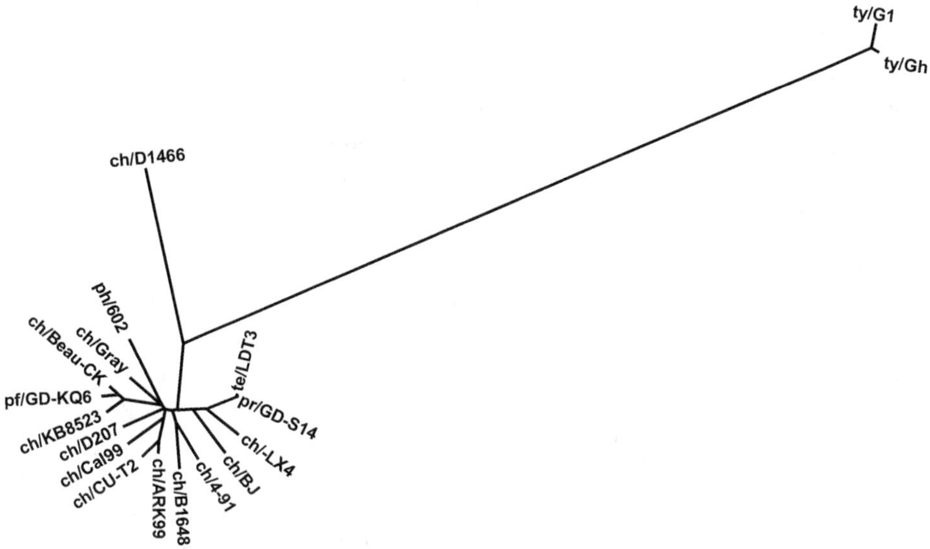

Figure 8.2 Phylogenetic analysis of the spike (S1 and S2) protein amino acid sequences of Group 3 coronaviruses isolated from various avian species; ch, chicken; ty, turkey; ph, pheasant; pr, partridge; pf, peafowl; te, teal. The sequences were aligned using ClustalX (Thompson *et al.*, 1997) and the phylogenetic relationships of the aligned sequences were determined using the distance matrix and neighbor-joining methods of the PHYLIP package (Felsenstein, 1988). The scale bar represents the distances as numbers of substitutions per site, 0.1 nucleotide substitutions per site. The amino acid sequences used were from the following viruses deposited in GenBank, the accession numbers are given parentheses, the avian species from which the virus was isolated, the country of origin, isolate number and year of isolation are indicated in the names; ch/UK/4/91 (Unpublished), ch/BE/B1648/84 (X87238), ch/NL/D207/78 (M21969 (S1) and X58003 (S2)), ch/China/BJ/03 (AY319651), ch/China/LX4/02 (AY189157), ch/Japan/KB8523/88 (M21515), ch/USA/Gray/92 (L14069 (S1) and AF394180 (S2)), ch/USA/Ark/99 (L10384), ch/USA/Cal/99 (AY514485), ch/USA/Beaudette/37 (AJ311317), ch/USA/Cu-T2/95 (U04739), ch/NL/D1466/78 (M21971 (S1) and X58001 (S2)), ty/USA/G1/03 (AAQ20922), ty/USA/Gh/03 (AAQ20921), ph/UK/602/95 (Unpublished), pr/China/GD-S14/03 (AY646283), pf/China/GD-KQ6/03 (AY641576), te/China/LDT3/03 (AY702975).

Details of the events which led to these sequences have only been documented in respect of the peafowl and teal reported by (Liu *et al.*, 2005). In both cases these two species were being maintained as domestic animals in the vicinity of chickens. The proteins of the peafowl virus (strain pf/China/LKQ3/03) had > 99% identity with the widely used IBV H120 live vaccine, making it almost certain that it was the H120 vaccine strain that had been recovered from the peafowl. The complete genome sequence of another coronavirus isolated from peafowl (pe/China/GD-KQ6/03) has been submitted to the databanks (J.D. Fu, M. Liao, T. Ren, G.H. Zhang, C.A. Xin and J.D. Chen, 2004: accession number AY641576). This had genes with ~95% identity with those of IBV strains (Figures 8.1 and 8.2).

The isolate from a teal (te/China/LDT3/03) had a spike protein that had ~90% identity with some known IBV strains, including a nephropathogenic one (Liu *et al.*, 2005; Figures 8.1 and 8.2). When this isolate was inoculated into chickens, it caused disease, including kidney involvement. This suggests that the teal isolate was actually an IBV strain that had spread to the teal from nearby chickens. Assuming that these findings did not result from cross-contamination in the laboratory, it would appear that IBV can replicate not only in the chicken and other gallinaceous birds, but also in teal, a non-gallincaceous bird. It is possible that the virus was a natural pathogen of the teal that was able to infect and cause disease in chickens.

Infection of guinea fowl with virus antigenically related to IBV had been reported earlier, in Brazil (Ito *et al.*, 1991). The guinea fowl had been suffering high mortality and low feed consumption, and enteritis was reported. When the isolate was inoculated intranasally into chickens and into guinea fowl, both species exhibited respiratory distress and aqueous feces. Whether the guinea fowl virus was a "genuine" guinea fowl coronavirus i.e. a separate species, or an IBV that had spread from nearby chickens to the guinea fowl, is not known. Nevertheless, the field observations and the laboratory infections indicate that the virus from the guinea fowl had a host range of more than one bird species.

Group 3 coronaviruses in wild birds

Jonassen *et al.* (2005) used a pan-coronavirus RT-PCR (corresponding to sequence within gene 1, potentially able to detect coronaviruses of all coronavirus groups) to look for coronaviruses in geese, pigeons and mallard duck. They detected coronaviruses in fecal samples or cloacal swabs of 40/163 graylag geese, 2/100 pigeons and 1/5 mallard ducks. Further RT-PCRs, for the 3′ end of the genome, revealed that these viruses were Group 3 coronaviruses. This study also revealed that these viruses were most definitely not IBV or like the coronaviruses isolated from gallinaceous birds. For example, there was a large insert in the hypervariable part of the 3′ untranslated region (UTR) of the genome of the viruses when compared to IBV (Figure 8.3). Moreover, the goose virus had two additional open reading frames (ORFs) downstream of the nucleocapsid protein gene (labeled 7 and 8 in Figure 8.3), whilst the pigeon virus had one of these ORFs (7 in Figure 8.3). Sequence of the corresponding part of the genome for the mallard duck virus was not available. Upstream of ORFs 7 and 8 and of the N protein gene was a putative transcription associated sequence (TAS) TTTAACAA, very similar to the TAS, CTTAACAA, of the other Group 3 coronaviruses. Consequently, one would expect that the pigeon and goose viruses would generate one and two, respectively, additional mRNAs, and the corresponding proteins.

Attempts to isolate live coronavirus from the geese, ducks and pigeons in the allantoic cavity of domestic fowl eggs were unsuccessful.

Non-Group 3 coronaviruses in birds

Avian species may not be limited to infection by coronaviruses of only Group 3. Nuttall and Harrap (1982) isolated a Group 2 coronavirus from the Manx shearwater (*Puffinus puffinus*), a bird that visits the shores of Britain in summer. It is not absolutely certain that the virus was from the birds. Virus was isolated following passage of shearwater material

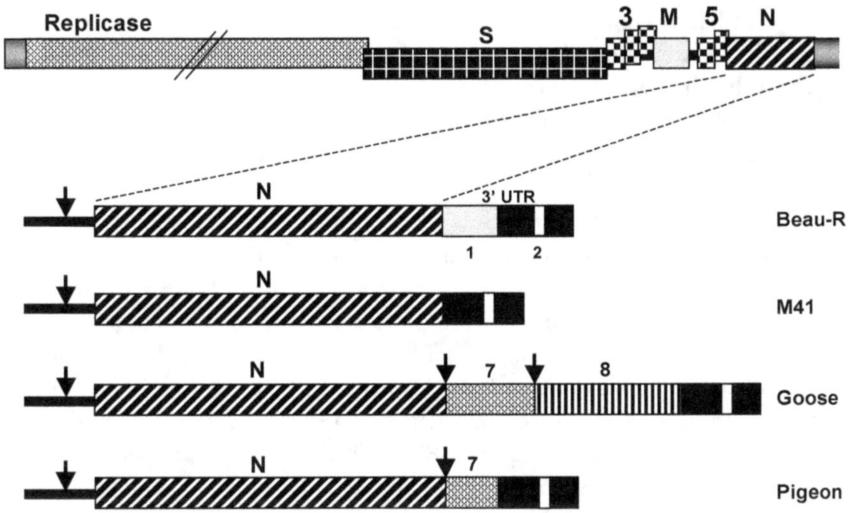

Figure 8.3 Comparison of the 3′ untranslated regions (UTRs) of IBV and the coronaviruses isolated from goose and pigeon. The top part of the diagram illustrates the entire IBV genome with the regions from the nucleocapsid gene of the IBV and the other coronaviruses expanded below. The IBV 3′ UTR is divided into two regions, indicated 1 and 2, in which region 1 is hypervariable between different coronaviruses and in some case, as illustrated for IBV M41, this region is deleted. Region 2 is highly conserved between different IBV strains and contains a sequence of 43 nt (indicated as a white boxed region) that forms a stem–loop structure, s2m, also found in the 3′ UTR of SARS-CoV and of many astroviruses and the picornavirus equine rhinovirus (Jonassen *et al.*, 1998). The sequence of region 2 of the 3′ UTRs from the goose and pigeon coronaviruses are very similar the corresponding IBV region 2 sequences including s2m (Jonassen *et al.*, 2005). However, as illustrated the hypervariable, region 1, of the goose and pigeon coronaviruses were found to be larger than their IBV counterparts and to potentially encode additional genes, labeled 7 and 8 for the coronavirus isolated from geese and 7 for the virus isolated from pigeons. The additional genes contain TASs upstream of the potential protein coding regions, indicating that additional genes sequences will probably result in the production of one or two additional subgenomic mRNAs, and corresponding proteins. The positions of the TASs upstream of the N, 7 and 8 genes are indicated with arrows.

in the brains of mice; it is possible that the virus is actually a murine coronavirus (a Group 2 coronavirus) that was present in the mice before inoculation.

Viruses that had coronavirus morphology were isolated from ticks (*Ixodes* spp.) collected from colonies of seabirds on the island of Runde, Norway (Traavik *et al.*, 1977). Extracts of the ticks were inoculated into the brains of suckling mice, which subsequently died. The virus replicated in BHK21/c13, BSC-1 and Vero cells, causing cytopathic effects. The virus agglutinated chicken erythrocytes. Sera collected, at a different location two years later, from the common puffin (*Fratercula arctica*), razor bill (*Alca torda*) and common murre (*Uria aalge*) resulted in lines of precipitation when diffused against virus antigen prepared from inoculated mouse brain or from inoculated BHK cells. This indicated that seabirds had been infected by a virus, related to the one under investigation, and is evidence against the possibility that the virus was simply a murine coronavirus recovered from the mice that had been inoculated. There was no evidence of mouse hepatitis in the mouse colony. Notwithstanding all of the above, there was no evidence, other than electron microscopy, that the virus was a coronavirus.

There is no doubt that chickens can be infected by a Group 2 coronavirus; when Ismail *et al.* (2001) inoculated bovine coronavirus (Group 2) into turkeys, it replicated, causing enteritis.

We have analyzed a coronavirus that was isolated from green-cheeked Amazon parrot (*Amazon viridigenalis Cassin*) by propagation in primary embryo liver cells derived from blue and yellow macaw (*Ara ararauna*) embryos (Gough *et al.*, 2006). A pan-coronavirus RT-PCR that amplified part of the RNA-dependent RNA polymerase domain in gene 1 gave a PCR product. Sequence analysis confirmed that the virus was a coronavirus, although the deduced sequence of 66 amino acids had only 66 to 74% amino acid identity with the corresponding sequence of coronaviruses in Groups 1, 2 and 3. Within these Groups the degree of amino acid identity is ≥ 85%, 100% and ≥ 89%, respectively. Several other oligonucleotide primer pairs that gave positive results with IBV, TCoV and PhCoV RNA (corresponding to genes 3, 5, N and the 3′ untranslated region) failed to do so with RNA from the parrot coronavirus.

Molecular basis of serotype variation

There are dozens, perhaps hundreds, of serotypes of IBV, as defined by virus neutralization tests (Cavanagh, 2001; Cook *et al.*, 1999; Darbyshire *et al.*, 1979; Farsang *et al.*, 2002; Ignjatovic and McWaters, 1991; Meulemans *et al.*, 2001).

The large spike glycoprotein (S) comprises a carboxy-terminal S2 subunit (approximately 625 amino acid residues), which anchors S in the virus envelope, and an amino-terminal S1 subunit (approximately 520 residues), believed to largely form the distal bulbous part of S (Cavanagh, 1995; Lai and Cavanagh, 1997).

It is the S1 subunit that induces virus-neutralizing (VN) antibody, demonstrated in various ways (Cavanagh *et al.*, 1986; Cavanagh *et al.*, 1988; Ignjatovic and Galli, 1994; Johnson *et al.*, 2003; Kant *et al.*, 1992; Koch *et al.*, 1990; Tomley *et al.*, 1987; Wang *et al.*, 2002a; Wang *et al.*, 2002b). Expression of S1 in chickens from a fowl adenovirus vector induced good protective immunity (Johnson *et al.*, 2003).

When comparing whole S protein sequences, most serotypes of IBV exhibit 80 to 90% identity amongst one another, although this drops to ~64% for some isolates e.g. D1466 in Figure 8.2.

Within S2, amino acid identity is usually ≥ 90%. It is S1 that exhibits the greatest variation. Most IBV serotypes differ from each other by 20 to 25% of S1 amino acids (Adzhar *et al.*, 1997; Kingham *et al.*, 2000), although e.g. D1466, some differ by up to 50% (Cavanagh *et al.*, 1997; Gelb *et al.*, 1997) and others by as little as 2–3% (Cavanagh *et al.*, 1992a). Generally speaking, the immunity induced by inoculation with one serotype protects poorly against infection by heterologous serotypes (Cavanagh and Naqi, 2003).

Monoclonal antibody resistant mutant investigations have revealed that many of the amino acids involved in the formation of VN epitopes are located within the first and third quarters of the linear S1 polypeptide (Cavanagh *et al.*, 1988; Kant *et al.*, 1992; Koch *et al.*, 1990). Sequence analysis of variants that are genetically very similar (> 95% amino acid identity in S1) has shown that most of the differences are within these two regions

(Adzhar *et al.*, 1997; Cavanagh *et al.*, 1992b). Moreover, some variants that differed by only 2 to 3% of S1 amino acids (10 to 15 residues) behaved as different serotypes (Cavanagh *et al.*, 1992b; Cook, 1984; Cook and Huggins, 1986).

Serotypes exhibit poor cross-protection

The degree of cross-protection induced by one serotype against another has been a cause for investigation over decades (Cavanagh *et al.*, 2005; Liu and Kong, 2004) (Gelb Jr *et al.*, 2005; Ladman *et al.*, 2006). Growth of challenge virus and observation of clinical changes has mostly been in regard to the respiratory tract (reviewed by Cavanagh, 2003).

In some experiments some heterologous challenge viruses broke through in 100% of the chickens i.e. 0% cross-protection (Hofstad, 1981; Marquardt *et al.*, 1982; Picault *et al.*, 1986). Other challenge viruses, known now to differ by 20% or so of amino acids in S1 from the virus used to initially infect the chickens, broke through immunity in 30 to 70% of cases (Hofstad, 1981; Rosenberger *et al.*, 1976). Collectively these and other reports show that cross-protection can range from very poor to moderate (poor for commercial purposes). In the field one might expect to get poorer cross-protection than observed in experiments under laboratory conditions.

Spray vaccination with an attenuated strain developed from a nephropathogenic IBV completely protected against mortality upon challenge four weeks later (Pensaert and Lambrechts, 1994). Vaccination with non-homologous, commercially available vaccines markedly, though not completely, reduced mortality. The homologous vaccine had prevented detectable growth of the challenge virus in the trachea but the heterologous vaccines had not done so.

Small differences in S1 sequence may adversely affect cross-protection

As noted above, some serotypes differ by only 2 to 3% of S1 amino acids. These viruses exhibited good, though not complete, cross-protection amongst each other (Cavanagh *et al.*, 1997). Slightly greater degrees of S1 difference can result in poor cross-protection.

Nix and colleagues (Nix *et al.*, 2000) inoculated chickens with several virulent isolates, of the same serotype, that had 94% or greater amino acid identity in S1. Some isolates cross-reacted in a VN test much less than others. Some of these isolates induced 100% protection, assessed by re-isolation of challenge virus. However, two other isolates induced cross-protection of only 55 to 58% of the chickens when the challenge virus was of lower serological identity.

Small (~5%) differences in S1 amino acids would appear to be sometimes responsible for poor cross-protection. Recently we (Casais *et al.*, 2003) substituted (swapped) the spike protein gene of our Beaudette infectious clone (receiver strain) with that of the M41 strain (donor strain). Both strains are of the Massachusetts serotype and have 95% S1 amino acid identity. The Beaudette and M41 strains are attenuated and pathogenic for chickens, respectively. When chickens were inoculated with Beaudette, M41 or spike-swapped recombinant IBV and challenged with M41, the Beaudette strain did induce some cross-protective immunity as assessed by reduced nasal discharge and lack of recovery of challenge virus (Hodgson *et al.*, 2004). However, the Beaudette strain induced almost no tracheal protection when assessed by tracheal ciliary activity and the clinical sign of snicking. In contrast, the recombinant IBV induced protection almost as good as the donor strain by these two and other criteria. This was not due to an increased virulence of the receiver strain, as the spike-swapped recombinant was as non-pathogenic as the Beaudette receiver strain, and both replicated poorly *in vivo*. As the receiver strain and recombinant virus had identical proteins except for the spike protein, it would appear that the poor immunity induced by the receiver strain against the donor strain was due to some of the 5% of amino acid differences (27 different residues) in S1.

Vaccination with live-attenuated vaccines

Broilers are usually vaccinated by spray on the day of hatch, in the hatchery (Cavanagh and Naqi, 2003). It is convenient to do it at that time, and the vaccine is delivered more efficiently than when the birds have been placed in their houses. Also, because IBV is endemic and ever-present in many poultry-rearing parts of the world; protection is needed as early as possible. As the modern broiler goes for processing at only five or six weeks of age, so vaccination has to be early.

Many studies have been undertaken during the last 30 years or so to examine the degree of protection induced by single inoculations of various serotypes of IBV (reviewed by Bijlenga *et al.*, 2004 and Cavanagh, 2003).

Duration of respiratory tract immunity following live-attenuated vaccination

Following a single live-attenuated virus vaccination under laboratory conditions there is a decline in the number of protected chickens from about six weeks after vaccination (Gough and Alexander, 1979; Darbyshire and Peters, 1984). In regions where a given serotype of IBV vaccine is consistently not giving acceptable protection against field challenge, the birds might be revaccinated, by spray, two to three weeks after the first application, with the homologous or a heterologous serotype IB vaccine. The latter approach sometimes gives protection against a broader range of serotypes (Cook *et al.*, 1999).

Vaccination with inactivated vaccines

For practical purposes, vaccination with inactivated IBV is ineffective without prior priming of the birds by live vaccine (Box *et al.*, 1980; Finney *et al.*, 1990; Gough *et al.*, 1977; Martins *et al.*, 1991; McDougall, 1969; Muneer *et al.*, 1987). Broilers live for too short a period—and killed vaccines are too expensive—for killed vaccines to be used in broilers. Instead they are used in breeding stock and commercial egg layers to give long-lasting immunity to the hen bird, to protect against drops in egg production resulting from IBV infection, and to maintain high levels of maternally derived antibody in embryos.

Comparatively high amounts of inactivated virus can induce protection in a proportion of chickens. Immunization of 11 chickens with a single dose of inactivated IBV strain M41 resulted in 36% to 45% (depending on the criterion used) developing respiratory tract protection (Cavanagh *et al.*, 1986). A dose of 100 μg was not more efficacious than 5 μg. In a second experiment, 29% (based on retention of ciliary activity in the trachea) to 76% (based on not isolating challenge virus) of 17 chickens were protected after a dose of 10 μg. In a third experiment 40% of 10 chickens were judged protected (retention of ciliary activity) following a single 10 μg dose. It is possible that a larger amount of virus was used in these three experiments than by previous workers, and that this was the reason for the partial success.

Ignjatovic and Galli (1994) gave four doses, each of 200 μg, of inactivated IBV; 40% and 20% of 10 birds had developed protection (assessed by non-reisolation of challenge virus) in the trachea and kidney, respectively, after three doses. These percentages did not increase after four doses. Song *et al.* (1998) immunized chickens up to three times with an inactivated nephropathogenic strain. Protection (assessed by non-reisolation of challenge virus) of kidney and trachea was 50% and 13%, respectively, of eight chickens, after two inoculations, rising to 88% and 50%, respectively, after three inoculations.

Genetic-based variation in susceptibility of chickens to IBV

IBV infection predisposes chickens to mortality when concurrently or subsequently infected with *Escherichia coli*. This has been observed in the field. Experimentally breeds of chicken demonstrated different degrees of mortality when inoculated with a mixture of IBV isolates alone, and in conjunction with a pool of *Escherichia coli* strains (Cook *et al.*, 1986; Smith *et al.*, 1985). A continuation of this work (Bumstead *et al.*, 1989) showed marked variation in mortality amongst several inbred lines of White Leghorn chickens. More detailed analysis of two of these lines, 15I (highly susceptible) and line C (relatively resistant), showed that the rate of virus production and titers in the respiratory tract were similar during the first four days after infection (Otsuki *et al.*, 1990). Thereafter, however, titers declined much more slowly in the susceptible line 15I, taking two-fold longer to decline to undetectable levels. In tracheal explants of the two lines there was no difference in the replication profile of the virus, showing that the difference in the birds themselves was not because of an intrinsic difference in the capacity of the trachea to support the replication of IBV. The underlying cause of the difference *in vivo* might be immunological. In these studies there was no evidence that the MHC influenced the outcome (Bumstead *et al.*, 1989). In contrast, a retrospective study by Bacon *et al.*, (2004) revealed that the mortality in chicks caused by a low-attenuation vaccine strain was influenced not only by the genetic line of the bird but suggested that it was also affected by the MHC(B) haplotype in chickens of three B congenic lines. Much more research is required in this area.

Genetic differences also affect the outcome of infection with nephropathogenic IBV (Pensaert and Lambrechts, 1994). Thus 0% and 2% of a specified pathogen free flock died within the periods 5–7 and 7–9 days, respectively, after tracheal inoculation of the B1648 strain. In contrast, 15% and 21% of commercial meat-type birds died in these periods.

Even within a given strain of bird there are variations in the responses of individual chickens to infection by IBV (reviewed by Cavanagh, 2003). Briefly, even when Rhode Island Red out bred chickens had been carefully inoculated with a high titer of a given strain of IBV, in three independent experiments, 10% of the birds remained non-protected against challenge with the homologous strain three weeks later (Cavanagh *et al.*, 1997). This has been observed in other studies (Cavanagh *et al.*, 1997; Cook *et al.*, 1999; Hofstad, 1981; Nix *et al.*, 2000; Parsons *et al.*, 1992; Picault *et al.*, 1986; Winterfield *et al.*, 1976). These results show that chickens (out-bred, though with restricted sets of parental breeding stocks) are not uniform in their response to IBV vaccination.

Presence and duration of antibody in serum, trachea and nasal secretions

The humoral immune response to IBV vaccination has mostly been investigated by measuring antibody levels in serum, using ELISA, VN or hemagglutination-inhibition (HAI) tests. Results have shown that IBV hemagglutinates very poorly until treated with enzyme preparations containing neuraminidases (Alexander *et al.*, 1983; Schultze *et al.*, 1992, and references therein). However, there have also been a few studies of IBV antibodies in nose and trachea. Many studies have shown that the presence or absence or the titer of serum antibody to IBV does not correlate with protection i.e. vaccinated chickens may be protected against respiratory disease IBV irrespective of the titer of serum antibody (e.g. Ignjatovic and Galli, 1994; Raggi and Lee, 1965). Antibody profiles following infection with IBV have been reviewed by Cavanagh (2003).

Nature of protective immune responses to IBV

Much effort has been expended over several decades measuring virus-neutralizing and hemagglutination-inhibiting antibodies, partly because these antibodies define serotype—and serotype differences are often reflective of poor cross-protection. But little is known about the immunological basis of protection against IBV.

Collisson and colleagues (Pei *et al.*, 2003; Seo and Collisson, 1997) have shown that cytotoxic T-cell (CTL) responses in chickens to IBV infection correlated with initial decreases in infection and clinical signs. CTL activity was major histocompatibility complex restricted, and lysis was mediated by CD8+ CD4- cells. Adoptive transfer of IBV-infection-induced alphabeta T cells bearing CD8 antigen protected chicks from challenge infection (Seo *et al.*, 2000); reviewed in Collisson *et al.* (2000). Earlier work has been reviewed by Dhinaker Raj and Jones (1997). Tissues from chickens infected with two field strains of IBV were assayed for interferon, which was readily detected in trachea and lung but at only low levels in plasma, kidney, liver and spleen (Otsuki *et al.*, 1987).

Attenuation of IBV pathogenicity for vaccine development

Commercial IB vaccines are developed by multiple passage of a field isolate in embryonated domestic fowl eggs, until the desired blend of non-pathogenicity and immunizing capacity has been achieved. Which mutations cause the attenuation of pathogenicity is not known. Replacement of the spike protein gene of the Beaudette strain (highly attenuated after ~300 passages in eggs) with that from the pathogenic M41 strain (Casais *et al.*, 2003) resulted in a recombinant virus that was still not pathogenic (Hodgson *et al.*, 2004). Thus proteins other than or in addition to the S protein must affect pathogenicity.

We observed that extensive passage in embryos (which attenuated pathogenicity, for development of a vaccine) of a field isolate resulted in selection of virus with an alanine residue at position 95 in the S1 protein, compared with a serine residue in the field virus (Cavanagh, 2005). Back-passage of the egg-adapted virus in chickens selected for virus with a serine i.e. like the original field isolate. Passage of the chicken-passaged virus in embryos again selected for virus with an alanine residue, characteristic of egg-passaged virus. Although passage of the attenuated virus in chickens selected for virus with an amino acid at position 95 that was characteristic of field virus, the virus remained attenuated i.e. this change was not associated with pathogenicity. Nevertheless, this was the first demonstration of host-specific variation of the IBV spike protein.

Induction of protection by sub-viral vaccines

It is the S1 subunit, rather than S2, that induces VN antibody (Cavanagh and Davis, 1986; Cavanagh *et al.*, 1988; Ignjatovic and Galli, 1994; Johnson *et al.*, 2003; Kant *et al.*, 1992; Koch *et al.*, 1990; Tomley *et al.*, 1987; Wang *et al.*, 2002a; Wang *et al.*, 2002b), and protective immunity (Johnson *et al.*, 2003; Song *et al.*, 1998); reviewed by Cavanagh (2003).

Ignjatovic and Galli (1994) dispersed the proteins of a nephropathogenic IBV using non-ionic detergent and purified S1, N and M by affinity chromatography using monoclonal antibodies. Chickens were inoculated on four occasions with one of the purified proteins (50 μg). Chickens that had been immunized with N or M proteins did not develop protective immunity (assessed by non-recovery of challenge virus) to either trachea or kidney following challenge. However, 71% and 86% of 10 chickens inoculated with S1 developed tracheal and kidney immunity, respectively, after four inoculations. The respective percentages were 70% and 10% after three inoculations.

The S1 protein has also been expressed in *Spodoptera frugiperda* cells from a recombinant *Autographa californica* baculovirus (Song *et al.*, 1998). When chickens were inoculated with these cells containing expressed S1, 43% and 0% of seven birds were protected (assessed by the inability to re-isolate the challenge nephropathogenic virus) in kidney and trachea, respectively. Protection rose to 50% and 25%, respectively, of eight chickens after three inoculations.

Induction of protection by vectored vaccines

The S1 protein has also been expressed *in situ* from fowlpox virus and adenovirus vectors. Three inoculations, in the wing web, of a recombinant fowlpox virus expressing S1 induced IBV-specific antibodies (Wang *et al.*, 2002b). After challenge with IBV, there was somewhat less recovery of challenge virus and less tracheal damage than in controls, and only mild clinical signs. Greater protection was achieved following a single oral application of a fowl adenovirus expressing S1 (Johnson *et al.*, 2003). Protection (assessed by the inability to re-isolate the challenge virus) was obtained in 90% and 100% of 10 to 13 chickens in two experiments.

Prospects for developing live IB vaccines by genetic manipulation

There is no question that the enormous, successful expansion of the poultry industry was due, in part, to the introduction of live and inactivated IB vaccines (Bijlenga *et al.*, 2004). That said, control is not as great as one would like, in large part because of the existing of poorly cross-protecting serotypes, discussed above. It is difficult to see how this situation can be circumvented; any new generation of IB vaccines will most probably have to continue to have S proteins of more than one serotype.

A major challenge to poultry vaccine producers is that of *in ovo* vaccination. That is, poultry producers would like to be able to vaccinate *in ovo*, at 18 days of age, three days prior to hatching. This permits vaccination in a very controlled, uniform way compared with vaccination by spray in the hatchery, or by drinking water in the poultry house. Unfortunately, even IB vaccines that are extremely mild in hatched chicks prevent hatching when inoculated *in ovo*. We have found that our molecularly cloned Beaudette strain of IBV does not prevent hatching. Moreover, our recombinant Beaudette with the S protein gene of the M41 strain induced immunity, when inoculated *in ovo*, to challenge with M41 (Tarpey *et al.*, 2006). Consequently there is the potential for the development of *in ovo* vaccines against IB.

Acknowledgments

The authors acknowledge the support of the following organizations: Department of the Environment, Food and Rural Affairs (Defra), Biotechnology and Biological Sciences Research Council (BBSRC), Intervet International, British Poultry Council, Merial Animal Health, British Egg Marketing Board Research and Education Trust.

References

Adzhar, A., Gough, R.E., Haydon, D., Shaw, K., Britton, P., and Cavanagh, D. (1997). Molecular analysis of the 793/B serotype of infectious bronchitis virus in Great Britain. Avian Pathol. 26, 625–640.

Alexander, D.J., Allan, W.H., Biggs, P.M., Bracewell, C.D., Darbyshire, J.H., Dawson, P.S., Harris, A.H., Jordan, F.T., MacPherson, I., McFerran, J.B., Randall, C.J., Stuart, J.C., Swarbrick, O., and Wilding, G.P. (1983). A standard technique for haemagglutination inhibition tests for antibodies to avian infectious bronchitis virus. Vet. Rec. 113, 64.

Ambali, A.G., and Jones, R.C. (1990). Early pathogenesis in chicks of infection with an enterotropic strain of infectious bronchitis virus. Avian Dis. 34, 809–817.

Anon (2006). Continued growth forecast for poultrymeat output and trade. Poultry Inter. 45, 4–6.

Bacon, L.D., Hunter, D.B., Zhang, H.M., Brand, K., and Etches, R. (2004). Retrospective evidence that the MHC (B haplotype) of chickens influences genetic resistance to attenuated infectious bronchitis vaccine strains in chickens. Avian Pathol. 33, 605–609.

Barnes, H.J., Guy, J.S., and Vaillancourt, J.-P. (2000). Poult enteritis complex. Rev. Sci. Tech. Off. Intern. Épiz. 19, 565–588.

Bhattacharjee, P.S., and Jones, R.C. (1997). Susceptibility of organ cultures from chicken tissues for strains of infectious bronchitis virus isolated from the intestine. Avian Pathol. 26, 553–563.

Bijlenga, G., Cook, J.K.A., Gelb, J., and de Wit, J.J. (2004). Development and use of the hours strain of avian infectious bronchitis virus from The Netherlands as a vaccine: a review. Avian Pathol. 33, 550–557.

Boltz, D.A., Nakai, M., and Bahra, J.M. (2004). Avian infectious bronchitis virus: a possible cause of reduced fertility in the rooster. Avian Dis. 48, 909–915.

Box, P.G., Beresford, A.W., and Roberts, B. (1980). Protection of laying hens against infectious bronchitis with inactivated emulsion vaccines. Vet. Rec. 106, 264–268.

Breslin, J.J., Smith, L.G., Barnes, H.J., and Guy, J.S. (2000). Comparison of virus isolation, immunohistochemistry, and reverse transcriptase-polymerase chain reaction procedures for detection of turkey coronavirus. Avian Dis. 44, 624–631.

Breslin, J.J., Smith, L.G., Fuller, F.J., and Guy, J.S. (1999a). Sequence analysis of the matrix/nucleocapsid gene region of turkey coronavirus. Intervirology 42, 22–29.

Breslin, J.J., Smith, L.G., Fuller, F.J., and Guy, J.S. (1999b). Sequence analysis of the turkey coronavirus nucleocapsid protein gene and 3′ untranslated region identifies the virus as a close relative of infectious bronchitis virus. Virus Res. 65, 187–193.

Bumstead, N., Huggins, M.B., and Cook, J.K.A. (1989). Genetic differences in susceptibility to a mixture of avian infectious bronchitis virus and Escherichia coli. Br. Poultry Sci. 30, 39–48.

Casais, R., Davies, M., Cavanagh, D., and Britton, P. (2005). Gene 5 of the avian coronavirus infectious bronchitis virus is not essential for replication. J. Virol. 79, 8065–8078.

Casais, R., Dove, B., Cavanagh, D., and Britton, P. (2003). Recombinant avian infectious bronchitis virus expressing a heterologous spike gene demonstrates that the spike protein is a determinant of cell tropism. J. Virol. *77*, 9084–9089.

Cavanagh, D. (1981). Structural polypeptides of coronavirus IBV. J. Gen. Virol. *53*, 93–103.

Cavanagh, D. (1995). The coronavirus surface glycoprotein, In The Coronaviridae, S.. G. Siddell, ed. (New York and London: Plenum Press), pp. 73–113.

Cavanagh, D. (2001). A nomenclature for avian coronavirus isolates and the question of species status. Avian Pathol. *30*, 109–115.

Cavanagh, D. (2003). Severe acute respiratory syndrome vaccine development: experiences of vaccination against avian infectious bronchitis coronavirus. Avian Pathology *32*, 567–582.

Cavanagh, D. (2005). Coronaviruses in poultry and other birds. Avian Pathol. *34*, 439–448.

Cavanagh, D., and Davis, P.J. (1986). Coronavirus IBV: removal of spike glycopolypeptide S1 by urea abolishes infectivity and haemagglutination but not attachment to cells. J. Gen. Virol. *67*, 1443–1448.

Cavanagh, D., Davis, P.J., and Cook, J.K.A. (1992a). Infectious-bronchitis virus—evidence for recombination within the Massachusetts serotype. Avian Pathol. *21*, 401–408.

Cavanagh, D., Davis, P.J., Cook, J.K.A., Li, D., Kant, A., and Koch, G. (1992b). Location of the amino-acid differences in the S1 spike glycoprotein subunit of closely related serotypes of infectious-bronchitis virus. Avian Pathol. *21*, 33–43.

Cavanagh, D., Davis, P.J., Darbyshire, J.H., and Peters, R.W. (1986). Coronavirus IBV: virus retaining spike glycopolypeptide S2 but not S1 is unable to induce virus-neutralizing or haemagglutination-inhibiting antibody, or induce chicken tracheal protection. J. Gen. Virol. *67*, 1435–1442.

Cavanagh, D., Davis, P.J., and Mockett, A.P. (1988). Amino acids within hypervariable region 1 of avian coronavirus IBV (Massachusetts serotype) spike glycoprotein are associated with neutralization epitopes. Virus Res. *11*, 141–150.

Cavanagh, D., Ellis, M.M., and Cook, J.K.A. (1997). Relationship between sequence variation in the S1 spike protein of infectious bronchitis virus and the extent of cross-protection in vivo. Avian Pathol. *26*, 63–74.

Cavanagh, D., Mawditt, K., Sharma, M., Drury, S.E., Ainsworth, H.L., Britton, P., and Gough, R.E. (2001). Detection of a coronavirus from turkey poults in Europe genetically related to infectious bronchitis virus of chickens. Avian Pathol. *30*, 355–368.

Cavanagh, D., Mawditt, K., Welchman, D.d.B., Britton, P., and Gough, R.E. (2002). Coronaviruses from pheasants (Phasianus colchicus) are genetically closely related to coronaviruses of domestic fowl (infectious bronchitis virus) and turkeys. Avian Pathol. *31*, 81–93.

Cavanagh, D., and Naqi, S. (2003). Infectious bronchitis, In Diseases of Poultry, Y.M. Saif, H.J. Barnes, J.R. Glisson, A.M. Fadly, L.R. McDougald, and D.E. Swayne, eds. (Ames, Iowa: Iowa State University Press), pp. 101–119.

Cavanagh, D., Picault, J.P., Gough, R., Hess, M., Mawditt, K., and Britton, P. (2005). Variation in the spike protein of the 793/B type of infectious bronchitis virus, in the field and during alternate passage in chickens and embryonated eggs. Avian Pathol. *34*, 20–25.

Chen, B.Y., Hosi, S., Nunoya, T., and Itakura, C. (1996). Histopathology and immunohistochemistry of renal lesions due to infectious bronchitis virus in chicks. Avian Pathol. *25*, 269–283.

Chen, B.Y., and Itakura, C. (1996). Cytopathology of chick renal epithelial cells experimentally infected with avian infectious bronchitis virus. Avian Pathol. *25*, 675–690.

Collisson, E. W., Pei, J., Dzielawa, J., and Seo, S.. H. (2000). Cytotoxic T lymphocytes are critical in the control of infectious bronchitis virus in poultry. Dev. Comp. Immunol. *24*, 187–200.

Cook, J.K., Chesher, J., Baxendale, W., Greenwood, N., Huggins, M.B., and Orbell, S.J. (2001). Protection of chickens against renal damage caused by a nephropathogenic infectious bronchitis virus. Avian Pathol. *30*, 423–426.

Cook, J.K., Smith, H.W., and Huggins, M.B. (1986). Infectious bronchitis immunity: its study in chickens experimentally infected with mixtures of infectious bronchitis virus and Escherichia coli. J. Gen. Virol. *67*, 1427–1434.

Cook, J.K.A. (1984). The classification of new serotypes of infectious bronchitis virus isolated from poultry flocks in Great Britain between 1981 and 1983. Avian Pathol. *13*, 733–741.

Cook, J.K.A., and Huggins, M.B. (1986). Newly isolated serotypes of infectious bronchitis virus: Their role in disease. Avian Pathol. *15*, 129–138.

Cook, J.K.A., Orbell, S.J., Woods, M.A., and Huggins, M.B. (1999). Breadth of protection of the respiratory tract provided by different live-attenuated infectious bronchitis vaccines against challenge with infectious bronchitis viruses of heterologous types. Avian Pathol. *28*, 471–479.

Coria, M.F., and Ritchie, A.E. (1973). Serial passage of 3 strains of avian infectious bronchitis virus in African green monkey kidney cells (VERO). Avian Dis. *17*, 697–704.

Culver, F., Dziva, F., Cavanagh, D., and Stevens, M.P. (2006). Poultry enteritis and mortality syndrome in turkeys in Great Britain. Vet. Rec. *159*, 209–210.

Darbyshire, J.H., and Peters, R.W. (1984). Sequential development of humoral immunity and assessement of protection in chickens following vaccination and challenge with avian infectious bronchitis virus. Res. Vet. Sci. *37*, 77–86.

Darbyshire, J.H., Rowell, J.G., Cook, J.K.A., and Peters, R.W. (1979). Taxonomic studies on strains of avian infectious bronchitis virus using neutralisation tests in tracheal organ cultures. Arch. Virol. *61*, 227–238.

Dhinaker Raj, G., and Jones, R.C. (1997). Infectious bronchitis virus: immunopathogenesis of infection in the chicken. Avian Pathol. *26*, 677–706.

El Houadfi, M., Jones, R.C., Cook, J.K.A., and Ambali, A.G. (1986). The isolation and characterisation of six avian infectious bronchitis viruses isolated in Morocco. Avian Pathol. *15*, 93–105.

Farsang, A., Ros, C., Renström, L.H.M., Baule, C., Soós, T., and Belák, S. (2002). Molecular epizootiology of infectious bronchitis virus in Sweden indicating the involvement of a vaccine strain. Avian Pathol. *31*, 229–236.

Felsenstein, J. (1988). Phylogenies from molecular sequences: inferences and reliability. Am. Rev. Genet. *22*, 512–565.

Finney, P.M., Box, P.G., and Holmes, H.C. (1990). Studies with a bivalent infectious bronchitis killed virus vaccine. Avian Pathol. *19*, 435–450.

Gelb, J., Keeler, C.L., Nix, W.A., Rosenberger, J.K., and Cloud, S.S. (1997). Antigenic and S-1 genomic characterization of the Delaware variant serotype of infectious bronchitis virus. Avian Dis. *41*, 661–669.

Gelb Jr, J., Weisman, Y., Ladman, B.S., and Meir, R. (2005). S1 gene characteristics and efficacy of vaccination against infectious bronchitis virus field isolates from the United States and Israel (1996 to 2000). Avian Pathol *34*, 194–203.

Gough, R.E., and Alexander, D.J. (1979). Comparison of duration of immunity in chickens infected with a live infectious bronchitis vaccine by three different routes. Res. Vet. Sci. *26*, 329–332.

Gough, R.E., Allan, W.H., and Nedelciu, D. (1977). Immune response to monovalent and bivalent Newcastle disease and infectious bronchitis inactivated vaccines. Avian Pathol. *6*, 131–142.

Gough, R.E., Cox, W.J., Winkler, C.E., Sharp, M.W., and Spackman, D. (1996). Isolation and identification of infectious bronchitis virus from pheasants. Vet. Rec. *138*, 208–209.

Gough, R.E., Drury, S.E., Culver, F., Britton, P., and Cavanagh, D. (2006). Isolation of a coronavirus from a green-cheeked Amazon parrot (Amazon viridigenalis Cassin). Avian Pathol. *35*, 122–126.

Guy, J.S. (2000). Turkey coronavirus is more closely related to avian infectious bronchitis virus than to mammalian coronaviruses: a review. Avian Pathol. *29*, 207–212.

Guy, J.S. (2003). Turkey coronavirus enteritis. In Diseases of Poultry, H.J. Barnes, J.R. Glisson, A.M. Fadly, L.R. McDougald, and D.E. Swayne, eds. (Ames, Iowa: Iowa State Press), pp. 300–307.

Guy, J.S., Barnes, H.J., Smith, L.G., and Breslin, J. (1997). Antigenic characterization of a turkey coronavirus identified in poult enteritis- and mortality syndrome-affected turkeys. Avian Dis. *41*, 583–590.

Hodgson, T., Britton, P., and Cavanagh, D. (2006). Neither the RNA nor the Proteins of Open Reading Frames 3a and 3b of the Coronavirus Infectious Bronchitis Virus Are Essential for Replication. J.Virol. *80*, 296–305.

Hodgson, T., Casais, R., Dove, B., Britton, P., and Cavanagh, D. (2004). Recombinant infectious bronchitis coronavirus Beaudette with the spike protein gene of the pathogenic M41 strain remains attenuated but induces protective immunity. J. Virol. *78*, 13804–13811.

Hofstad, M.S. (1981). Cross-immunity in chickens using seven isolates of avian infectious bronchitis virus. Avian Dis. *25*, 650–654.

Hofstad, M.S., and Yoder Jr, H.W. (1966). Avian infectious bronchitis—virus distribution in tissues of chicks. Avian Dis. *10*, 230–239.

Ignjatovic, J., and Galli, L. (1994). The S1 glycoprotein but not the N or M proteins of avian infectious bronchitis virus induces protection in vaccinated chickens. Arch. Virol. *138*, 117–134.

Ignjatovic, J., and McWaters, P.G. (1991). Monoclonal-antibodies to 3 structural proteins of avian infectious-bronchitis virus—characterization of epitopes and antigenic differentiation of Australian strains. J. Gen. Virol. *72*, 2915–2922.

Ismail, M.M., Cho, K.O., Ward, L.A., Saif, L.J., and Saif, Y.M. (2001). Experimental bovine coronavirus in turkey poults and young chickens. Avian Dis. *45*, 157–163.

Ismail, M.M., Tang, A.Y., and Saif, Y.M. (2003). Pathogenicity of turkey coronavirus in turkeys and chickens. Avian Dis. *47*, 515–522.

Ito, N.M.K., Miyaji, C.I., and Capellaro, C.E.M.P.D.M. (1991). Studies on broiler's IBV and IB-like virus from guinea fowl. In II International Symposium on Infectious Bronchitis,, E.F. Kaleta, and U. Heffels-Redmann, eds. (Giessen: Justus Leibig University), pp. 302–307.

Johnson, M.A., Pooley, C., Ignjatovic, J., and Tyack, S.G. (2003). A recombinant fowl adenovirus expressing the S1 gene of infectious bronchitis virus protects against challenge with infectious bronchitis virus. Vaccine *21*, 2730–2736.

Jonassen, C.M., Jonassen, T.O., and Grinde, B. (1998). A common RNA motif in the 3′ end of the genomes of astroviruses, avian infectious bronchitis virus and an equine rhinovirus. J. Gen. Virol. *79*, 715–718.

Jonassen, C.M., Kofstad, T., Larsen, I.L., Lovland, A., Handeland, K., Follestad, A., and Lillehaug, A. (2005). Molecular identification and characterization of novel coronaviruses infecting graylag geese (Anser anser), feral pigeons (Columbia livia) and mallards (Anas platyrhynchos). J. Gen. Virol. *86*, 1597–1607.

Kant, A., Koch, G., van Roozelaar, D.J., Kusters, J.G., Poelwijk, F.A.J., and van der Zeijst, B.A.M. (1992). Location of antigenic sites defined by neutralising monoclonal antibodies on the S1 avian infectious bronchitis virus glycopolypeptide. J. Gen. Virol. *73*, 591–596.

Kingham, B.F., Keeler Jr, C.L., Nix, W.A., Ladman, B.S., and Gelb Jr, J. (2000). Identification of avian infectious bronchitis virus by direct automated cycle sequencing of the S-1 gene. Avian Dis. *44*, 325–335.

Koch, G., Hartog, L., Kant, A., and van Roozelaar, D.J. (1990). Antigenic domains of the peplomer protein of avian infectious bronchitis virus: correlation with biological function. J. Gen. Virol. *71*, 1929–1935.

Ladman, B.S., Loupos, A.B., and Gelb Jr, J. (2006). Infectious bronchitis virus S1 gene sequence comparison is a better predictor of challenge of immunity in chickens than serotyping by virus neutralization. Avian Pathol. *35*, 127–133.

Lai, M.M., and Cavanagh, D. (1997). The molecular biology of coronaviruses. Adv. Virus Res. *48*, 1–100.

Lambrechts, C., Pensaert, M., and Ducatelle, R. (1993). Challenge experiments to evaluate cross-protection induced at the trachea and kidney level by vaccine strains and Belgian nephropathogenic isolates of avian infectious bronchitis virus. Avian Pathol. *22*, 577–590.

Li, H., and Yang, H.C. (2001). Sequence analysis of nephropathogenic infectious bronchitis virus strains of the Massachusetts genotype in Beijing. Avian Pathol. *30*, 535–541.

Lin, T.L., Loa, C.C., and Wu, C.C. (2004). Complete sequences of 3′ end coding region for structural protein genes of turkey coronavirus. Virus Res. *106*, 61–70.

Lister, S.A., Beer, J.V., Gough, R.E., Holmes, R.G., Jones, J.M., and Orton, R.G. (1985). Outbreaks of nephritis in pheasants (phasianus colchicus) with a possible coronavirus aetiology. Vet. Rec. *117*, 612–613.

Liu, S., Chen, J., Kong, X., Shao, Y., Han, Z., Feng, L., Cai, X., Gu, S., and Liu, M. (2005). Isolation of avian infectious bronchitis coronavirus from domestic peafowl (Pavo cristatus) and teal (Anas). J. Gen. Virol. *86*, 719–725.

Liu, S., and Kong, X. (2004). A new genotype of nephropathogenic infectious bronchitis virus circulating in vaccinated and non-vaccinated flocks in China. Avian Pathol. *33*, 321–327.

Lucio, B., and Fabricant, J. (1990). Tissue tropism of three cloacal isolates and Massachusetts strain of infectious bronchitis virus. Avian Dis. *34*, 865–870.

Marquardt, W.W., Kadavil, S.K., and Snyder, D.B. (1982). Comparison of ciliary activity and virus recovery from tracheas of chickens and humoral immunity after inoculation with serotypes of avian infectious bronchitis virus. Avian Dis. *26*, 828–834.

Martins, N.R., Mockett, A.P., Barrett, A.D., and Cook, J.K. (1991). IgM responses in chicken serum to live and inactivated infectious bronchitis virus vaccines. Avian Dis. *35*, 470–475.

Matthijs, M.G., van Eck, J.H., Landman, W.J., and Stegeman, J.A. (2003). Ability of Massachusetts-type infectious bronchitis virus to increase colibacillosis susceptibility in commercial broilers: a comparison between vaccine and virulent field virus. Avian Pathol. *32*, 473–481.

McDougall, J.S. (1969). Avian infectious bronchitis virus: the protection afforded by an inactivated virus vaccine. Vet. Rec. *85*, 378–381.

Meulemans, G., Boschmans, M., Decaesstecker, M., van den Berg, T.P., Denis, P., and Cavanagh, D. (2001). Epidemiology of infectious bronchitis virus in Belgian broilers: a retrospective study, 1986 to 1995. Avian Pathol. *30*, 411–421.

Muneer, M.A., Newman, J.A., Halvorson, D.A., Sivanandan, V., and Coon, C.N. (1987). Effects of avian infectious bronchitis virus (Arkansas strain) on vaccinated laying chickens. Avian Dis. *31*, 820–828.

Nix, W.A., Troeber, D.S., Kingham, B.F., Keeler Jr, C.L., and Gelb Jr, J. (2000). Emergence of subtype strains of the Arkansas serotype of infectious bronchitis virus in Delmarva broiler chickens. Avian Dis. *44*, 568–581.

Nuttall, P.A., and Harrap, K.A. (1982). Isolation of a coronavirus during studies on puffinosis, a disease of the Manx shearwater (Puffinus puffinus). Arch. Virol. *73*, 1–13.

Otsuki, K., Huggins, M.B., and Cook, J.K.A. (1990). Comparison of the susceptibility to infectious bronchitis virus infection of two inbred lines of White Leghorn chickens. Avian Pathol. *19*, 467–475.

Otsuki, K., Nakamura, T., Kubota, N., Kawaoka, Y., and Tsubokura, M. (1987). Comparison of two strains of avian infectious bronchitis virus for their interferon induction, viral growth and development of virus-neutralising antibody in experimentally-infected chickens. Vet. Microbiol. *15*, 31–40.

Parsons, D., Ellis, M.M., Cavanagh, D., and Cook, J.K.A. (1992). Characterisation of an avian infectious bronchitis virus isolated from IB-vaccinated broiler breeder flocks. Vet. Rec. *131*, 408–411.

Pei, J., Briles, W.E., and Collisson, E.W. (2003). Memory T cells protect chicks from acute infectious bronchitis virus infection. Virology *306*, 376–384.

Pennycott, T.W. (2000). Causes of mortality and culling in adult pheasants. Vet. Rec. *146*, 273–278.

Pensaert, M., and Lambrechts, C. (1994). Vaccination of chickens against a Belgian nephropathogenic strain of infectious bronchitis virus B1648 using attenuated homologous and heterologous strains. Avian Pathol. *23*, 631–641.

Picault, J.P., Drouin, P., Guittet, M., Bennejean, G., Protais, J., L'Hosptialier, R., Gillet, J.P., Lamande, J., and Le Bachelier, A. (1986). Isolation, characterisation and preliminary cross-protection studies with a new pathogenic avian infectious bronchitis virus (strain PL-84084). Avian Pathol. *15*, 367–383.

Raggi, L.G., and Lee, G.G. (1965). Lack of correlation between infectivity, serological response and challenge results in immunization with an avian infectious bronchitis vaccine. J. Immunol. *94*, 538–543.

Rosenberger, J.K., Alphin, R.L., and Krauss, W.C. (1976). Cross-protection studies with a Holland strain (Nobilis H-52) of infectious bronchitis virus. Avian Dis. *20*, 199–201.

Schultze, B., Cavanagh, D., and Herrler, G. (1992). Neuraminidase treatment of avian infectious bronchitis coronavirus reveals a hemagglutinating activity that is dependent on sialic acid-containing receptors on erythrocytes. Virology *189*, 792–794.

Seo, S.H., and Collisson, E.W. (1997). Specific cytotoxic T lymphocytes are involved in in vivo clearance of infectious bronchitis virus. J. Virol. *71*, 5173–5177.

Seo, S.H., Pei, J., Briles, W.E., Dzielawa, J., and Collisson, E.W. (2000). Adoptive transfer of infectious bronchitis virus primed alphabeta T cells bearing CD8 antigen protects chicks from acute infection. Virology *269*, 183–189.

Smith, H.W., Cook, J.K., and Parsell, Z.E. (1985). The experimental infection of chickens with mixtures of infectious bronchitis virus and Escherichia coli. J. Gen. Virol. *66*, 777–786.

Song, C.S., Lee, Y.J., Lee, C.W., Sung, H.W., Kim, J.H., Mo, I.P., Izumiya, Y., Jang, H.K., and Mikami, T. (1998). Induction of protective immunity in chickens vaccinated with infectious bronchitis virus S1 glycoprotein expressed by a recombinant baculovirus. J. Gen. Virol. *79*, 719–723.

Spackman, D., and Cameron, I.R. (1983). Isolation of infectious bronchitis virus from pheasants. Vet. Rec. *113*, 354–355.

Tarpey, I., Orbell, S.J., Britton, P., Casais, R., Hodgson, T., Lin, F., Hogan, E., and Cavanagh, D. (2006). Safety and efficacy of an infectious bronchitis virus used for chicken embryo vaccination. Vaccine *24*, 6830–6838.

Teixeira, M.C.B., Luvizotto, M.C.R., Ferrari, H.F., Mendes, A.R., da Silva, S.E.L., and Cardoso, T.C. (2007). Detection of turkey coronavirus in commercial turkey poults in Brazil. Avian Pathol. *36*, 29–33.

Thompson, J.D., Gibson, T.J., Plewniak, F., Jeanmougin, F., and Higgins, D.G. (1997). The Clustal_X windows interface: flexible strategies for multiple sequence alignment aided by quality analysis tools. Nucl. Acids Res. *25*, 4876–4882.

Tomley, F.M., Mockett, A.P., Boursnell, M.E., Binns, M.M., Cook, J.K.A., Brown, T.D.K., and Smith, G.L. (1987). Expression of the infectious bronchitis virus spike protein by recombinant vaccinia virus and induction of neutralizing antibodies in vaccinated mice. J. Gen. Virol. *68*, 2291–2298.

Toro, H., Godoy, V., Larenas, J., Reyes, E., and Kaleta, E.F. (1996). Avian infectious bronchitis: viral persistence in the harderian gland and histological changes after eyedrop vaccination. Avian Dis. *40*, 114–120.

Traavik, T., Mehl, R., and Kjeldsberg, E. (1977). "Runde" viurs, a coronavirus-like agent associated with seabirds and ticks. Arch. Virol. *55*, 25–38.

Wang, C.H., Hong, C.C., and Seak, J.C. (2002a). An ELISA for antibodies against infectious bronchitis virus using an S1 spike polypeptide. Vet. Microbiol. *85*, 333–342.

Wang, X.Q., Schnitzlein, W.M., Tripathy, D.N., Girshick, T., and Khan, M.I. (2002b). Construction and immunogenicity studies of recombinant fowl poxvirus containing the S1 gene of Massachusetts 41 strain of infectious bronchitis virus. Avian Dis. *46*, 831–838.

Welchman Dde, B., Bradbury, J.M., Cavanagh, D., and Aebischer, N.J. (2002). Infectious agents associated with respiratory disease in pheasants. Vet. Rec. *150*, 658–664.

Winterfield, R.W., Fadly, A.M., and Hoerr, F.J. (1976). Immunity to infectious bronchitis virus from spray vaccination with derivatives of a Holland strain. Avian Dis. *20*, 42–48.

Yu, L., Jiang, Y., Low, S., Wang, Z., Nam, S.J., Liu, W., and Kwangac, J. (2001). Characterization of three infectious bronchitis virus isolates from China associated with proventriculus in vaccinated chickens. Avian Dis. *45*, 416–424.

Feline Coronaviruses: A Tale of Two-Faced Types

9

Bert Jan Haijema, Peter J.M. Rottier, and Raoul J. de Groot

Abstract

Feline coronaviruses (FCoVs) are common pathogens of cats. Two serotypes (I and II) have been recognized and, to add further spice, each serotype occurs in two markedly different pathotypes. The most common pathotype, enteric FCoV (*e*FCoV) causes a mild, often unapparent enteritis, the other one, feline infectious peritonitis virus (FIPV), a devastating, highly lethal systemic infection. During the last ten years, our knowledge of the pathogenesis of feline infectious peritonitis (FIP) and of the relationships between the different FCoV sero- and pathotypes has increased considerably. In this chapter, we will focus on (i) the pathogenesis and natural history of FCoV infections, (ii) the molecular biology and molecular genetics of FCoVs, and (iii) the development of new and effective vaccines.

Patterns of disease

Enteric FCoV infections: mild but persistent

Coronavirus-infections occur in many exotic members of the *Felidae* family, amongst many others in cheetahs (O'Brien *et al.*, 1985; Heeney *et al.*, 1990; Munson *et al.*, 2004) mountain lions (Paul-Murphy *et al.*, 1994), European wildcats (Leutenegger *et al.*, 1999) and lions (Hofmann-Lehmann *et al.*, 1996). Here, we will restrict ourselves to the coronaviruses of domestic cats.

FCoV serotypes I and II belong to coronavirus subgroup 1. Genetically and antigenically, they are closely related to canine coronavirus and to transmissible gastroenteritis virus of swine (84 and 80% overall nucleotide sequence identity, respectively) and more distantly to human coronaviruses (HCoVs) 229E and NL63 and to porcine epidemic diarrhea virus (PEDV) (overall nucleotide sequence similarities of 35, 35 and 36%, respectively).

FCoVs have a world-wide distribution and are highly prevalent: 20 to 60% of domestic cats test seropositive by virus-neutralization, immunofluorescence assay and/or ELISA (Horzinek and Osterhaus, 1979); in multicat households and breeding facilities, seropositivity may approach 90%. FCoV serotype I prevails in the field, causing 80% to 95% of the infections (Hohdatsu *et al.*, 1992; Kennedy *et al.*, 2002; Addie *et al.*, 2003; Kummrow *et al.*, 2005). This is true both for *e*FCoV and for FIPV infections (70 to 85% type I) (Hohdatsu *et al.*, 1992; Benetka *et al.*, 2004). Moreover, the vast majority of natural infections is caused by eFCoV. FIP is a sporadic disease and as will be discussed in further detail below, FIPVs are in fact rare virulence mutants of enteric FCoVs.

Often, cats become eFCoV-infected already at a tender age as maternal immunity wanes (Pedersen *et al.*, 1981; Addie and Jarrett, 1990; Addie and Jarrett, 1992). Transmission is via the oral–fecal route (Pedersen *et al.*, 1981). eFCoVs are highly contagious: naïve sentinel animals contract the infection within days after exposure to shedders (Pedersen *et al.*, 2004). There are indications that eFCoV may occasionally cause severe enteritis in juvenile and adult cats (Kipar *et al.*, 1998b), but in most cases the course of the infection is benign (Pedersen *et al.*, 1981).

Not only is FCoV seroprevalence high (Horzinek and Osterhaus, 1979; Pedersen *et al.*, 1981; Hohdatsu *et al.*, 1992; Kennedy *et al.*, 2002), at any time a remarkably large proportion of the cat population (30 to 90%) has detectable virus in the feces (Herrewegh *et al.*, 1995a; Herrewegh *et al.*, 1997; Foley *et al.*, 1997a; Harpold *et al.*, 1999; Addie and Jarrett, 2001; Kennedy *et al.*, 2002; Benetka *et al.*, 2004; Pedersen *et al.*, 2004) and/or in the blood (Herrewegh *et al.*, 1997; Gunn-Moore *et al.*, 1998; Meli *et al.*, 2004; Simons *et al.*, 2005). Apparently, eFCoV endemicity is maintained through chronically infected, asymptomatic carriers (Foley *et al.*, 1997a; Herrewegh *et al.*, 1997; Addie and Jarrett, 2001; Addie *et al.*, 2003). Definitive proof for viral persistence came from a study in which cats, naturally infected with eFCoV, were placed in strict isolation. Virus shedding -albeit in decreasing amounts and intermittently continued for up to 300 days (i.e. until termination of the experiment) (Herrewegh *et al.*, 1997).

In specific pathogen free (SPF) kittens inoculated oronasally with eFCoVs the most consistent clinical sign is a moderate reduction in growth rate during the first 8 days of infection (Figure 9.1; L. Vogel *et al.*, unpublished), which may be accompanied by a transient fever (Meli *et al.*, 2004). Seroconversion occurs relatively late (around 3 weeks post infection [p.i.]) and systemic antibody titers remain low. Virus shedding can be detected by RT-PCR from day 1 to 3 onwards. During the acute phase of the infection, large quantities of virus are shed, peaking at day 5 to 7 p.i. with up to 10^6–10^8 genome equivalents per microgram of feces. Virus shedding subsides by day 16 to 23 p.i. to continue at a low but constant level (10^3–10^5 genome equivalents per microgram feces) for at least 9 weeks (Figure 9.1). In chronically infected animals, FCoV RNA can be detected throughout the gastrointestinal tract and in various other organs and tissues (Herrewegh *et al.*, 1997; Meli *et al.*, 2004). More importantly however, viral mRNA, indicative of replication, was found only in the lower part of the intestinal tract (ileum, colon and rectum) (Herrewegh *et al.*, 1997); infected cells were detected in these tissues by immunohistochemistry in the glands of Lieberkühn and in gut-associated lymphoid tissue (GALT) (Pedersen *et al.*, 1981; Herrewegh *et al.*, 1997). How eFCoVs avoid clearance is not known. However, sequence analysis of FCoVs shed during persistent infection, revealed an increasing number of mutations over time in the N-terminal hypervariable region of the viral spike (S) protein, indicative of viral escape in the face of potent humoral and/or cellular immune responses (Herrewegh *et al.*, 1997). An even more important question is why eFCoV replication remains largely confined to the gut (despite the fact that even eFCoVs may gain access to other tissues via blood-borne monocytes), whereas FIPVs are able to produce a systemic-infection with multiorgan involvement.

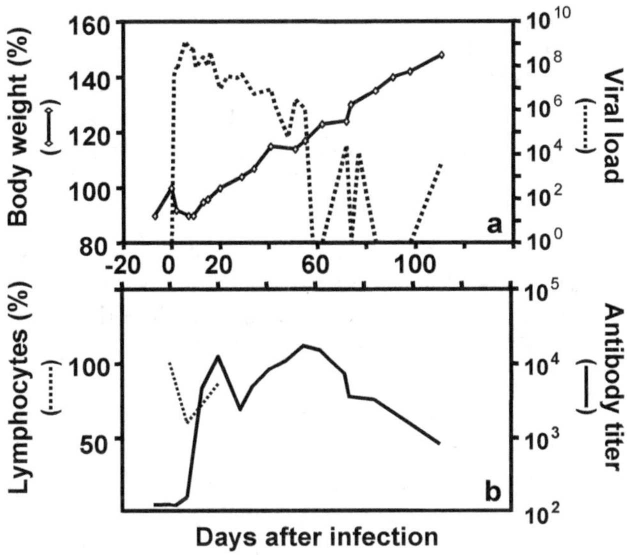

Figure 9.1 Typical example of an experimental infection with an enteric FCoV. A cat, inoculated oronasally with eFCoV strain UCD, was monitored for up to 110 days for body weight (solid line, open circles) and for viral RNA in the feces (dotted line; panel a). In addition, total lymphocyte counts (dotted line) and serum antibody titers (solid line) were determined (panel b). Body weights are expressed as the percentage of weight at day 0. Viral RNA titers determined by real-time RT-PCR are expressed as genome equivalents. Total lymphocyte counts are presented as percentages of the cell counts determined at day 0. Serum antibody titers were determined by ELISA (Vogel, Rottier *et al.*, unpublished results).

Feline infectious peritonitis: rare but deadly

A minority of eFCoV-infected animals (up to 10%) (Addie and Jarrett, 1992; Foley *et al.*, 1997b) proceeds to develop FIP, a progressive debilitating condition characterized by high morbidity and high mortality. The cats affected present with fibrinous-granulomatous serositis often with protein-rich pleural and/or peritoneal effusions, granulomatous-necrotising phlebitis and periphlebitis, and multifocal (pyo)granulomatous inflammatory lesions in various organs (lung, liver, spleen, omentum, brain, kidney and enteric tract) (Hayashi *et al.*, 1977; Weiss and Scott, 1981a; Weiss and Scott, 1981c; Pedersen, 1987; Kipar *et al.*, 1998a; Kipar *et al.*, 2005b). The pathogenetic mechanism is highly complex. Disease progression is accompanied by and presumably the consequence of a dramatic disregulation of the adaptive immune system as evidenced by the common occurrence of hypergammaglobulinemia (Ward *et al.*, 1974; Paltrinieri *et al.*, 1998) and of a profound depletion of T cells from the periphery (de Groot-Mijnes *et al.*, 2005) as well as from lymphoid tissues (Haagmans *et al.*, 1996). In experimentally inoculated cats with end-stage FIP, there is a massive atrophy of the T-cell zones in spleen, mesenteric lymph nodes and thymus (Haagmans *et al.*, 1996; Kipar *et al.*, 2001). Depletion also of B-cells from lymphatic tissues has been reported in cats with naturally acquired FIP (Kipar *et al.*, 2001). Numbers of macrophages, however, are consistently increased. In the course of

FIP, these cells accumulate in the splenic red pulp, apparently as a result of increased local proliferation (Kipar *et al.*, 2001).

It is believed that at the bottom of all these phenomena is the infection of monocytes and macrophages. These are the main target cells of FIPV (Ward, 1970). As there are no indications that lymphocytes are susceptible to infection, their depletion most likely occurs through indirect effects (Haagmans *et al.*, 1996). In HIV-infected patients, up-regulation of death ligands like FasL and TRAIL on the surface of infected macrophages apparently contributes to T-cell apoptosis (for a review see Miura and Koyanagi, 2005). In other examples of virus-induced T-cell depletion, the infection of antigen-presenting cells, and of dendritic cells in particular, is thought to play a central role in this process (Okada *et al.*, 2000; Palucka and Banchereau, 2002; Schneider-Schaulies *et al.*, 2002; Servet-Delprat *et al.*, 2003).

The morphological aspects of FIP lesions have been studied in considerable detail and the mechanisms that lead to their development are relatively well understood. Within the lesions, activated macrophages are the predominant cell-type; many of these macrophages are infected and infection is restricted to these cells (Kipar *et al.*, 1998a; Paltrinieri *et al.*, 2001; Berg *et al.*, 2005) (Figure 9.2a–b). Puzzlingly, the macrophages that during FIP amass in the splenic red pulp remain largely uninfected (Kipar *et al.*, 2001).

Circulating FIPV-infected monocytes are thought to disseminate the virus throughout the body. Infected monocytes adhere to the endothelium of small and medium-sized veins and may give rise to formation of aggregates of infected and non-infected monocytes that occlude the vascular lumen (Figure 9.2c–d). Upon emigration from the vessels, FCoV-infected monocytes initiate vascular and perivascular lesions. These are characterized by macrophage-dominated circular and focal infiltrates and often associated with destruction of the lamina propria. Occasionally, the walls of affected veins are completely replaced by inflammatory cells (Hayashi *et al.*, 1977; Weiss and Scott, 1981b; Weiss and Scott, 1981c; Kipar *et al.*, 2005b).

Most insight into the natural history of FIP comes from longitudinal infection experiments. Best studied in this respect is the highly virulent type II strain WSU FIPV 79–1146 (McKeirnan A.J., 1981; Pedersen *et al.*, 1984b), but disease progression in animals infected with serotype I FIPV variants (UCD1 and UCD 8; Pedersen, 1976; Dean *et al.*, 2003; Haijema *et al.*, unpublished) is very similar. The course of FIP is characteristically multiphasic (Ward *et al.*, 1974; Weiss and Scott, 1981a; Stoddart *et al.*, 1988; Vennema *et al.*, 1990a; Kiss *et al.*, 2004; de Groot-Mijnes *et al.*, 2005). Oronasal inoculation with FIPV WSU 79–1146 produces a recurrent infection with 100% morbidity and 80% mortality (de Groot-Mijnes *et al.*, 2005; Figure 9.3). The animals initially present with the characteristic signs of acute infection. A transient fever develops (day 2 to 4) accompanied by loss of body weight (approximately 10% by day 8), jaundice, acute lymphopenia and detectable viral RNA in blood cells and plasma. Then, by day 7 to 8 p.i., disease seemingly resolves and animals recover: fever subsides, body weight stabilizes or increases, and the amount of viral RNA in the blood decreases below detection levels. This recovery, however, is only temporary as most animals suffer a relapse around day 10. Viral recrudescence, as evidenced by rising amounts of viral RNA in blood and plasma, is accompanied by a second bout of fever and renewed weight loss while peripheral $CD4^+$ and $CD8^+$ T-cell counts are even further reduced. Of the cats that develop lethal disease the average survival time

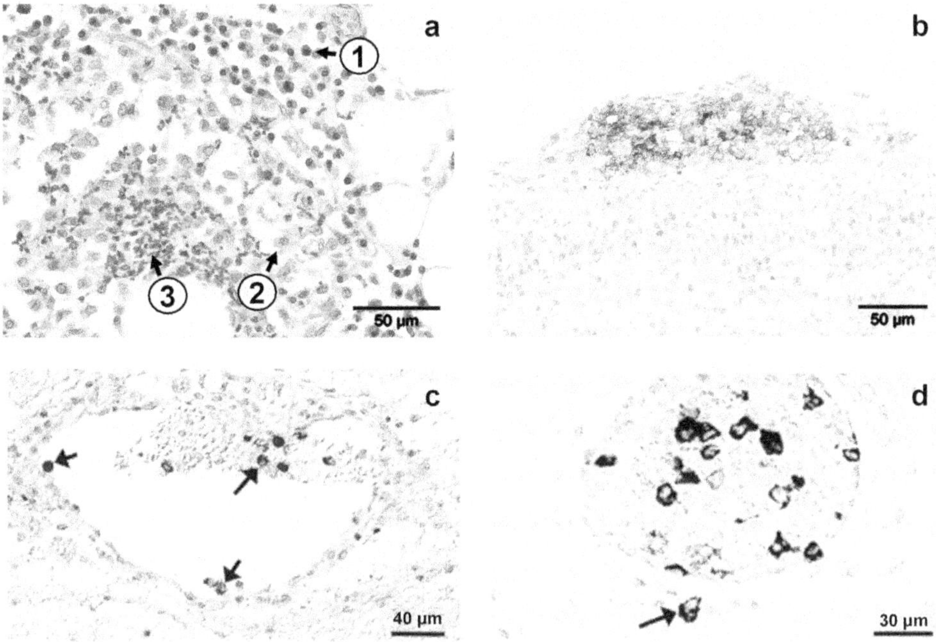

Figure 9.2 FIP histopathology. a, Tissue section of the mesenterium showing a granulomatous lesion, typical for FIP. A zone of plasma cells (1) is surrounding a core of monocytes/macrophages (2) and neutrophils (3) (Haijema and Grinwis, unpublished); b, Granulomatous lesion on the surface of the spleen. Immunohistochemical staining reveals monocytes/macrophages positive for FCoV antigen (Haijema and Grinwis, unpublished); c, FCoV-infected monocytes in the lumen of a vein, focally attaching to endothelial cells (arrows). Tissue section of the brain parenchyma of a cat with FIP, subjected to RNA-*in situ* hybridization with a virus-specific probe and counterstained with Papanicolaou's hematoxylin; reprinted from Kipar *et al.* (2005) with permission; d, Occlusion of a vein by aggregated monocytes. Tissue section of the brain parenchyma of a cat with FIP, subjected to immunohistochemistry. Several monocytes within the vascular lumen contain FCoV-antigens. The arrow points to an emigrated, FCoV-infected monocyte (arrow); reprinted from Kipar *et al.* (2005) with permission.

is 28 days with many animals (~40%) succumbing to FIP within 21–24 days p.i. ("rapid progressors"); some, however, may suffer multiple relapses and yet survive for months (de Groot-Mijnes *et al.*, 2005).

The minority of cats that do make full recovery and survive beyond day 100 p.i. demonstrate that FIPV infection can be controlled at least temporarily. How this is achieved is not known. Humoral immunity does not seem to be protective. In cats with end-stage FIP, the virus replicates seemingly unperturbed, towering titers of neutralizing antibody notwithstanding. In fact, virus-specific antibodies, in particular those directed against the spike protein, can be detrimental. FCoV-seropositive cats, whether they acquired their antibodies through active or passive immunization, are prone to developing "early death syndrome" (EDS), a more fulminating and drastically shortened course of the disease as compared to that seen in naive animals (Pedersen, 1980; Weiss and Scott, 1981a). In one of the most dramatic examples of EDS documented, cats immunized with a recombinant

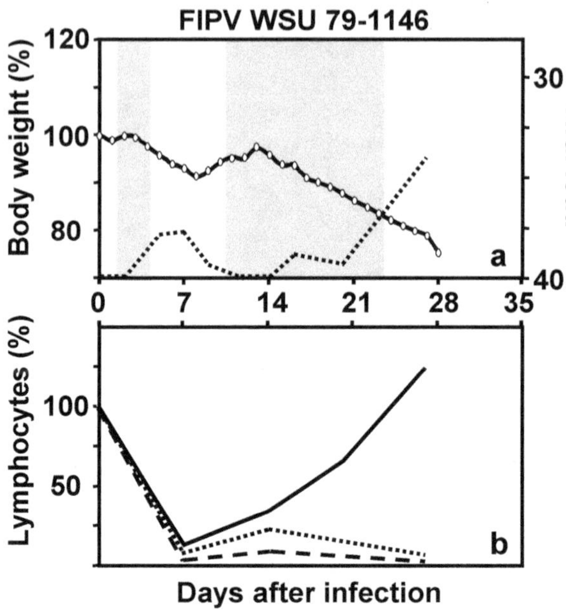

Figure 9.3 Disease progression in cats experimentally infected with FIPV WSU 79–1146. The animals were inoculated oronasally and subsequently monitored for fever, body weight, viral RNA in white blood cells (panel a) and lymphocyte counts (panel b). Data acquired for animals with highly similar disease progression ("rapid progressors;" n = 7) were averaged to obtain trend curves. Periods of fever (body temperature > 39.7°C) are indicated by grey shading. Body weight (black line, open circles) is expressed as the percentage of the weight at day 0. Viral RNA titers as determined by real-time RT-PCR are expressed as C_T values on an inverted *y* axis (dotted line). Total lymphocyte (solid line), $CD4^+$ T-cell (dotted line) and $CD8^+$ T-cell counts (dashed line) are presented as percentages of the cell counts determined at day 0 (adapted from de Groot-Mijnes *et al.*, 2005).

vaccinia virus expressing the FIPV S protein were moribund already 7 days after FIPV challenge versus an average survival time of 28 days in the non-immunized control group (Vennema *et al.*, 1990b). This phenomenon is attributed to antibody-dependent enhancement. Antibodies directed against the spike protein, when present at subneutralizing titers, apparently opsonize the virus and enhance the infection of macrophages via Fc receptor-mediated attachment (Vennema *et al.*, 1990a; Hohdatsu *et al.*, 1991; Olsen *et al.*, 1992; Corapi *et al.*, 1992;Olsen and Scott, 1993; Olsen *et al.*, 1993; Corapi *et al.*, 1995; for a review, see de Groot and Horzinek, 1995). Although EDS can be observed only in experimentally infected animals (for the obvious reason that to identify a case of EDS one has to know the exact time of FIPV-infection), there is little doubt that ADE contributes to disease progression also during natural FIP. It is at the least intriguing that in immunologically naïve cats with experimentally induced FIP the onset of renewed disease temporally correlates with the emergence of virus-specific humoral immunity.

It is commonly believed—though not proven—that the control of infection and FIPV clearance are primarily achieved through cell-mediated immunity (CMI) (Hayashi *et al.*,

1983; Pedersen and Floyd, 1985; Weiss and Cox, 1989; Poland *et al.*, 1996; de Groot-Mijnes *et al.*, 2005). Consistently, in FIP-survivors virus-specific CD4+ and CD8+ T cells, mainly directed against the S protein, can be detected by flow cytometry (de Groot-Mijnes *et al.*, 2004; de Groot-Mijnes *et al.*, 2005). The efficacy of the primary T-cell responses most likely determines to what extent the initial wave(s) of infection can be contained and hence may be the decisive factor in disease progression. If this line of defense fails, however, FIPV-induced T-cell depletion may rapidly lead to a melt-down of the adaptive immune system allowing the virus to replicate unchecked. The combined data indicate that disease progression in FIP is driven by virus replication. One important consequence is that FIP should thus be amenable to therapeutic intervention. Moreover, experimentally induced FIP should provide a well-defined and relevant model system for *in vivo* testing of broad-spectrum anticoronaviral drugs (de Groot-Mijnes *et al.*, 2005).

Molecular biology of FCoVs

The best characterized FCoV to date is FIPV WSU 79–1146, a serotype II laboratory strain isolated by McKeirnan *et al.* (1981). The origin of this strain is somewhat obscure. Though reportedly isolated from a case of neonatal death involving a 4-day-old kitten, FIPV WSU 79–1146 is genetically virtually identical to FIPV NOR15 (Herrewegh *et al.*, 1995b), a virus from the spleen of a cat with FIP, isolated by the same group around the same time (Evermann *et al.*, 1981).

FIPV WSU 79–1146 was the first FCoV for which molecular data and nucleotide sequences became available (de Groot *et al.*, 1987a; de Groot *et al.*, 1987b; de Groot *et al.*, 1988; Vennema *et al.*, 1991). In the mid-1980s, during the advent of coronavirus genomics, we selected this strain for analysis not only because of its high virulence (Pedersen *et al.*, 1984b), but, above all, because of its excellent *in vitro* growth properties. Inadvertently, this choice born out of pragmatism, has biased the FCoV field: FIPV WSU 79–1146, which may in fact be atypical because of serotype, growth properties and extreme pathogenicity has become the "gold standard" both for molecular biological analysis and *in vivo* studies.

FCoV genome organization and gene repertoire

The nucleotide sequence of the FIPV WSU 79–1146 genome was recently completed independently by us (Haijema *et al.*, 2005) and by others (Dye and Siddell, 2005). The viral RNA, as determined in Utrecht, measures 29355 nucleotides (nts) in length. It carries a 5′-cap structure, as demonstrated by RNA-immuno affinity purification with a cap-specific antibody (van Vliet *et al.*, 2002; A. van Vliet *et al.*, unpublished) and a poly(A) tail. Non-translated regions (NTRs) at the 5′- and 3′-ends, measuring 310 and 275 nts, respectively, presumably contain essential signals for (+) and (–) strand RNA synthesis. More than two-thirds of the FIPV WSU 79–1146 genome is occupied by the replicase gene, comprised of the overlapping open reading frames (ORFs) 1a and 1b. The encoded polyproteins, pp1a and pp1ab are predicted to be proteolytically processed by virus-encoded proteinases to yield 16 mature proteins and an unknown number of intermediate products (Dye and Siddell, 2005; Haijema *et al.*, 2005; see also the chapter by Ziebuhr) (Figure 9.4, upper panel). In addition to the replicase gene, there are 9 additional ORFs. These encode the inevitable set of structural proteins S, E, M and N as well as a number of (sub)group specific accessory proteins. The ORFs for the latter are organized in two clusters, 3a-c and 7a-b (Figure 9.4, lower panel).

The genes downstream of ORF1b are expressed from a nested set of subgenomic mRNAs each provided with a 90-nt leader. mRNA synthesis is controlled by a conserved transcription-regulating sequence (TRS) with 5′-AACUAAAC-3′ as core consensus. Copies of the TRS are found upstream of the genes for S, 3a, E, M, N and 7a (Figure 9.4) as well as at the extreme 5′-end of the genome (nts 91–98). Although for FIPV 79–1146 only six subgenomic mRNAs are produced (de Groot *et al.*, 1987b; de Groot and Horzinek, 1995; Dye and Siddell, 2005), it is likely that this laboratory strain arose from a parental virus that produced one additional RNA species to express ORF3c. In FIPV WSU 79–1146, ORF3c is a pseudogene, interrupted by a nonsense mutation and with its dedicated TRS mutated (TTCUAAAC → TTCUAAA*U*) to silence transcription. Yet, in *e*FCoV strains RM, NTU and UCD as well as in the *e*FCoV strain RM-derivatives FIPV UCD8 and UCD10, ORF3c is intact and (in most cases) preceded by a canonical TRS (Vennema *et al.*, 1998; Hsieh *et al.*, 2005; H. Vennema, personal communication). ORF3c must therefore be considered part of the standard FCoV gene repertoire.

Our FCoV WSU 79–1146 sequence differs in minor yet intriguing details from the 29125-nt sequence recently published by Dye and Siddell (2005). Whereas we analyzed

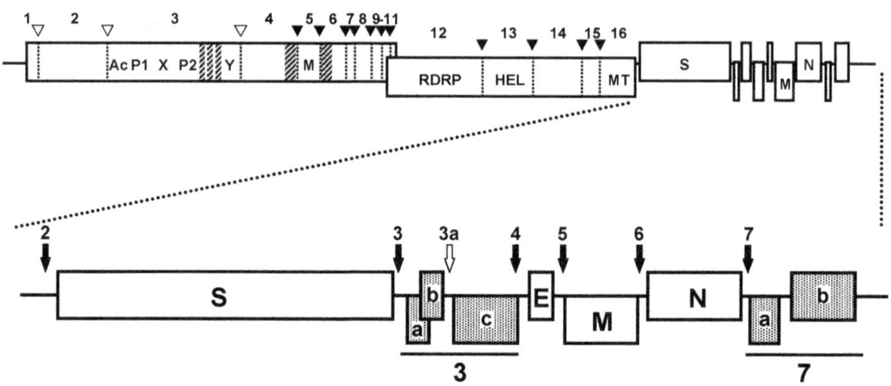

Figure 9.4 Schematic representation of the FCoV genome organization. The upper panel shows the complete genome drawn to scale with the various genes represented by boxes. Also shown is the predicted processing scheme of replicase proteins pp1a and pp1ab with arrowheads indicating cleavage-sites of papain-like proteinases (open arrowheads) or the 3C-like cysteine proteinase (black arrowheads). The predicted cleavage products are provisionally numbered nsp1–nsp16 (Dye and Siddell, 2005) with the various domains indicated as follows: Ac, acidic domain; P1 and P2, papain-like accessory proteinases; X, predicted adenosine diphosphate-ribose 1″-phosphatase; Y, putative metal binding domain; M, main proteinase; RDRP, RNA-dependent RNA polymerase; HEL, superfamily 1 helicase; MT, 2′-O-methyltransferase; cross-hatched regions, predicted transmembrane domains (Gorbalenya *et al.*, 1991; Snijder *et al.*, 2003; Ziebuhr *et al.*, 2001). The lower panel depicts a blow-up of the 3′-terminal part of the FCoV genome drawn to scale with structural protein genes indicated by white boxes and accessory genes indicated by shaded boxes. Locations of TRSs that direct subgenomic RNA synthesis are indicated by arrows numbered 2 through 7 (de Groot *et al.*, 1987b; de Groot and Horzinek, 1995; Dye and Siddell, 2005). The open arrow indicates a putative TRS, which is defect in some FCoV strains and presumably active in others.

a highly virulent low passage derivative obtained from Niels Pedersen (Pedersen *et al.*, 1984b), Dye and Siddell obtained their virus through the ATCC, possibly from a high passage stock. There are 18 single nucleotide differences, most of which are synonymous. More interestingly, in the "Bristol sequence" ORF7b is disrupted by a nonsense mutation truncating the encoded protein at residue 49. Moreover, there is a 231-nt in-frame deletion (corresponding to nts 3355–3586 of the Utrecht sequence) in the ORF1a nsP3-coding sequence (Figure 9.4, upper panel). As a result, the "Bristol variant" would produce an nsP3 protein from which residues 136–212, located within the N-terminal "acidic (Ac) domain" (Ziebuhr *et al.*, 2001) are missing; note that part of the deleted sequence is actually conserved between the Utrecht variant of FIPV-WSU 79–1146 and TGEV. Ziebuhr *et al.* (2001) previously observed sequence heterogeneity in this region among different coronavirus species; as it now appears, heterogeneity and even relatively large deletions may occur in the Ac domain (and are perhaps even selected for) during adaptation to and passage in tissue culture cells. It will be interesting and important to determine to what extent deletions in nsP3 affect virulence.

Relationships between FCoV serotypes

Seminal work by Pedersen and coworkers in the early 80s revealed that FCoVs occur in two distinct serotypes that can readily be distinguished by virus-neutralization assay. These studies also showed a close relationship—as indicated by cross-neutralization—between serotype II FCoVs and CCoV. Based upon these observations, it was assumed that cats could be infected with two antigenically and genetically distinct coronavirus species, one of which CCoV-like. Consistent with this notion and with the cross-neutralization patterns, the spike proteins of serotype II strains were found to be far more closely related to those of CCoV and TGEV than to those of serotype I strains with sequence identities of 91, 81 versus 46%, respectively (de Groot *et al.*, 1989; Motokawa *et al.*, 1995; Vennema *et al.*, 1995; Motokawa *et al.*, 1996).

However, comparative sequence analysis of other parts of the genome brought an unexpected twist: serotype II FCoVs are actually recombinant viruses that have arisen from double recombination events with FCoV serotype I and CCoV variants as parental viruses (Herrewegh *et al.*, 1998; Vennema, 1999). As a result of the genetic exchanges, the serotype I S gene and its flanking regions were replaced by the corresponding CCoV sequences. Even more surprisingly, analysis of different type II strains (WSU 79–1146, Wellcome, UCD6 and WSU 79–1683) showed the 5′- and 3′-recombination sites (i.e. the locations at which the template switches occurred) to be different for each virus (Figure 9.5); multiple sequence alignments allowed the 3′-recombination sites for each type II virus to be mapped within 10–30 nucleotides. The important implication is that these serotype II variants must have originated from independent recombination events (Herrewegh *et al.*, 1998; Vennema, 1999). These observations imply frequent cross-species transmission of FCoV and/or CCoV as well as frequent co-infection, most likely in the cat. Clearly, in this process FCoV persistence in the gut might be a critical factor as it would significantly increase the odds of double infections and hence recombination events to occur.

The findings do raise an intriguing question. What selective advantage do FCoV-CCoV recombinants have over their parental viruses? CCoV can infect cats (Stoddart *et al.*, 1988) but—as far as we are aware—has not established itself in the feline population.

The genome structure of serotype II strains WSU 79–1146 and 79–1683 might offer a clue. Remarkably, these viruses have retained most of the FCoV-specific coding sequences. Perhaps, host species preference is determined by coding and/or non-coding sequences at either end of the FCoV genome. The most striking corollary of FCoV-CCoV recombination is the exchange of the S protein, both a critical factor in virus attachment and cell tropism as well as the main inducer of neutralizing antibodies and prime target for the humoral immune response. Hence, the uptake of a CCoV S gene into the FCoV genome would result not only in an antigenic shift allowing the virus to evade existing serotype I-specific immunity, but—potentially—also in a shift in receptor specificity (*vide infra*). Either aspect might have been decisive for selection of FCoV-CCoV recombinant progeny in individual hosts.

FCoV pathotypes: the strange case of Dr. Jekyll and Mr. Hyde revisited

Early epidemiological studies showed that FIP occurs only in a small proportion of FCoV-infected cats, suggesting that this condition is an uncommon, secondary manifestation of a common, unapparent primary infection (Pedersen *et al.*, 1981). Yet, FCoVs isolated from cats with FIP invariably proved to be highly lethal; depending on dose and inoculation route, mortality rates of up to 100% can be achieved in experimentally infected animals. To explain these paradoxical findings, it was proposed that a non-pathogenic virus serologically related to FIPV might circulate in the cat population. The latter view, supported by the isolation of non-pathogenic FCoVs from apparently healthy cats, was welcomed to such an extent that "feline enteric coronaviruses" and "feline infectious peritonitis viruses" initially were treated as distinct virus species. Comparative sequence analysis, however, indicated that *e*FCoV and FIPV are in fact virulence variants of the same virus (de Groot and Horzinek, 1995; Herrewegh *et al.*, 1995b; Vennema *et al.*, 1995; Vennema *et al.*, 1998). The high genetic similarity between *e*FCoVs and FIPVs isolated from the same geographic area pointed to a common ancestry with deletions in the FIPV genomes indicating the order of descent: from *e*FCoV to FIPV (Vennema *et al.*, 1998).

Direct and conclusive evidence came from a series of elegant animal experiments by Amy Poland and Harry Vennema in the Pedersen lab (Poland *et al.*, 1996; Vennema *et al.*, 1998). Building on observations that concomitant infections with immunosuppressive pathogens like feline leukemia virus promote development of FIP (Cotter *et al.*, 1975; Pedersen and Floyd, 1985), cats in the mid to late stages of feline immunodeficiency virus (FIV) infection were super-infected with the *e*FCoV strain RM. In healthy animals, this isolate consistently causes only a mild enteric infection with low grade persistence (Vennema *et al.*, 1998) but no FIP. In contrast, of twenty *e*FCoV RM-inoculated FIV-infected cats, two developed FIP 8 to 10 weeks p.i. The FCoVs isolated from these cats (FIPV UCD9 and UCD10) were genetically almost identical to *e*FCoV strain RM (> 99% identity), but, in striking contrast to their parental virus, readily induced FIP in SPF cats.

The data paint a picture in which non-pathogenic enteric viruses (*e*FCoVs) through spontaneous mutation give rise to highly pathogenic variants (FIPVs). In immunocompetent cats persistently infected with *e*FCoV, the virus is kept at bay and although the virus may not be cleared, the viral load is kept to a minimum. However, immuno-suppression ensuing from super-infection with other viruses (FeLV, FIV) or from stress might unleash FCoV replication and elicit an expansion of the viral quasi-species cloud, thus raising the odds of generating a "killer virus." This phenomenon occurs in individual hosts time and time again and, ironically, from the perspectives of both host and virus leads to a dead-end. Although there are reports of clustered FIP cases (Foley *et al.*, 1997a), there is little evidence for direct transmission of FIPV.

The FCoV case attests to the pathogenic potential of coronaviruses in general. Also from this perspective, it will be important to determine the molecular basis for the shift in FCoV virulence. Viruses like *e*FCoV that aim at establishing a long term relationship with their host are as a rule non-pathogenic and might have evolved mechanisms to keep themselves in check. It is a tantalizing question whether the transition from *e*FCoV to FIPV results from a gain in function (enhanced replication and/or altered cell tropism allowing systemic infection) or from a loss of function, an accident due to a faulty brake.

Host cell tropism

As for most other coronaviruses, the attachment of FCoVs to host cells and their subsequent entry through membrane fusion are mediated by the S protein. Serotype II FCoVs employ feline aminopeptidase N (APN, CD13) as a receptor in cultured cells (Tresnan *et al.*, 1996; Tresnan and Holmes, 1998). APN is a class II membrane glycoprotein belonging to the family of membrane-bound metalloproteases (Olsen *et al.*, 1988). In intact hosts, it is expressed in a variety of tissues, amongst others in enterocytes and in cells of the monocyte lineage, explaining—at least in part—FCoV host cell tropism. It has been proposed that serotype I FCoVs also may enter cells via APN (Tresnan *et al.*, 1996). However, as already explained the spike proteins of the FCoV serotypes are very different, especially in their N-terminal receptor-binding domains, where sequence identity is only 31%. Moreover, whereas host cell attachment of type II FCoVs can be blocked by the fAPN-specific monoclonal antibody R-G-4, infection of tissue culture cells by the type I FCoV KU-2, Black and UCD I was not affected at all (Hohdatsu *et al.*, 1998). Additional research is required to firmly establish whether the serotype I strains indeed use APN or an alternative receptor yet. In the latter case, the peculiar situation would arise of two viruses with different receptor-specificity causing the same disease.

Both for serotype I and II FCoVs, the ability to *efficiently* infect and replicate in macrophages and monocytes appears to be a distinguishing factor between virulent and avirulent variants. As paraphrased from Kipar *et al.* (2005a), it seems that "quantitative differences regarding the degree of monocyte infection occur at some stage in the transition from *e*FCoV infection to FIP." Support for this view comes from a series of experiments in which different FCoV strains were compared for their ability to infect and replicate in cultured feline peritoneal macrophages. Avirulent strains consistently infected fewer macrophages, produced lower virus titers, were less able to sustain viral replication, and spread less efficiently to other susceptible macrophages (Stoddart and Scott, 1988; Stoddart and Scott, 1989). Similar observations were made when FIPV WSU 79–1146 and of FCoV WSU 79–1683 (a closely related serotype II enteric variant) were compared with regard to their growth properties in cultured bone marrow macrophages and in isolated peripheral blood monocytes (Rottier *et al.*, 2005; Dewerchin *et al.*, 2005). By using recently established protocols for the genetic modification of the FIPV genome (Haijema *et al.*, 2003), the spike protein was identified as a determinant for efficient macrophage infection. The exchange in FIPV WSU 79–1146 of the autologous S protein for that of FCoV WSU 79–1683 results in a loss of infectivity in cultured bone marrow macrophages. Unexpectedly, the ability to efficiently infect macrophages is determined not by the N-terminal domain of the WSU 79–1146 S protein, which is involved in receptor-binding, but rather by the C-terminal domain, implicated in membrane fusion. Thus, enhanced macrophage/ monocyte infectivity of FIPV WSU 79–1146 as compared to FCoV WSU 79–1683 may well be determined at a post-attachment level and be related to differences in fusogenicity (Rottier *et al.*, 2005). So far, only two FCoV laboratory strains have been studied and it is evidently too early to draw any general conclusions.

Accessory proteins: dispensable in cell culture, essential during natural infection

The structural proteins have been studied in detail and their roles in virion assembly, attachment and entry are well defined and discussed elsewhere in this volume (see chapters Chapters 1 and 6). In contrast, no function has been established for any of the FCoV group-specific gene products.

ORFs 3a and b are predicted to encode proteins of 72 residues each. Although there is as yet no direct evidence that these proteins are actually produced in infected cells, there is good reason to assume they are: (i) ORFs 3a and b form a separate transcription unit; ORF3a is preceded by a TRS directing the synthesis of RNA3; (ii) both ORFs are conserved not only among FCoVs but also among TGEV and CCoV strains. As neither of the predicted proteins contains hydrophobic segments that might serve as signal peptide or transmembrane domain, they are supposed to be cytosolic and thus to exert their functions intracellularly.

ORF3c is the only FCoV accessory gene for which (intact) homologues have been found in all other group 1 coronaviruses; in the cell-adapted 229E isolate the gene is present but disrupted by nonsense mutations and deletions (indicated as 3c in Figure 9.5). From its amino acid sequence, 3c appears to be a class III triple spanning membrane protein, 238 residues in length, with a predicted topology similar to that of the M protein (Oostra *et al.*, 2006).

ORFs 7a and b are located downstream of the N gene (de Groot *et al.*, 1987b; de Groot *et al.*, 1988). ORF7a is also found in the genomes of TGEV (Kapke and Brian, 1986) and CCoV (Horsburgh *et al.*, 1992; Vennema *et al.*, 1992) but no representatives are present in the HCoVs. It encodes a small, ~10 kDa membrane protein with an N-terminal cleavable signal sequence and a C-terminal transmembrane domain, the expression of which has actually been shown in TGEV-infected cells (Britton *et al.*, 1988; Garwes *et al.*, 1989; Tung *et al.*, 1992). ORF7b is unique to FCoV, CCoV and the recently identified ferret coronavirus (Wise *et al.*, 2006). Its product, gp7b, is a soluble glycoprotein of ~24 kDa, which is secreted from tissue culture cells albeit slowly because of the presence of a KDEL-like ER retention signal (Vennema *et al.*, 1993). Antibodies against gp7b detected in cats with FIP indicate that the protein is also expressed during natural infection (Vennema *et al.*, 1992; Vennema *et al.*, 1993; Herrewegh *et al.*, 1995b). Proteins 7a and gp7b are dispensable in cell-culture. Inactivation or deletion of their genes, either through spontaneous mutation or by reverse genetics has no effect on replication *in vitro* (Herrewegh *et al.*, 1995a; Haijema *et al.*, 2004). In fact, during propagation of FCoVs in tissue culture cells, viral mutants with deletions in ORF7b arise spontaneously and

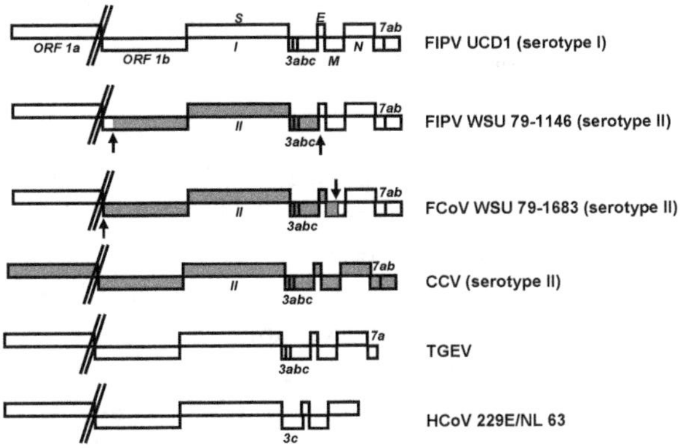

Figure 9.5 Recombination patterns of the FCoV serotype II strains WSU 79–1146 and 79–1683. The genomic organization of several group 1 coronaviruses is schematically presented (not to scale) with boxes representing the various genes. In the genomes of FCoV strains WSU 79–1146 and 79–1683, the regions depicted in grey represent sequences apparently obtained from CCoV. Recombination sites are indicated by arrows.

are selected for (Herrewegh *et al.*, 1995b), suggesting that loss of 7b expression actually increases viral fitness *in vitro*. Conversely, however, in naturally occurring *e*FCoVs, ORF7b is maintained and in the case of FIPV laboratory strains, loss of 7b expression apparently correlates with loss of virulence (Herrewegh *et al.*, 1995b). As gp7b presumably exerts its function in the extracellular milieu, it might be a virus-encoded modulator of the immune system or a chemo-attractant (Herrewegh *et al.*, 1995b).

Though the FCoV accessory proteins may be dispensable *in vitro*, they do most likely play a crucial role during the natural infection. Support for this notion comes from recent reverse genetics studies (Haijema *et al.*, 2004). Deletion of gene clusters 3 or 7 from the FIPV 79–1146 genome which did not affect replication in cell culture had a dramatic effect on virulence as demonstrated by experimental infection of cats; either deletion resulted in complete attenuation (Haijema *et al.*, 2004). As genes were removed in concert, the importance of individual genes/gene products cannot be assessed. Indeed, not all accessory genes are required to establish FIP. ORF7a-deletion mutants were detected by RT-PCR in cats with FIP suggesting that 7a expression is dispensable for systemic infection (Kennedy *et al.*, 2001). For protein 3c, the evidence is even more compelling. As mentioned, FIPV 79–1146 does not express 3c (Yamanaka *et al.*, 1998) and yet is highly pathogenic (Pedersen *et al.*, 1984b); RT-PCR analysis of virus recovered from experimentally infected cats with end-stage FIP did not yield any indication for restoration of ORF3c expression during the course of the disease (Herrewegh and de Groot, unpublished). Actually, ORF3c is most often—though not invariably—inactivated in FIPV variants (Vennema *et al.*, 1998; H. Vennema, personal communication). In contrast, the *e*FCoV isolates studied so far (UCD, RM, NTU2) all carry an intact gene (Vennema *et al.*, 1998; Hsieh *et al.*, 2005).

The disease patterns of *e*FCoV and FIPV are very different. It is quite plausible that certain accessory genes that are required for replication and persistence in the gut and/or for efficient oral–fecal transmission become dispensable if not a burden during rampant dead-end systemic infection. The availability of reverse genetics protocols for FCoV will be an asset in future molecular pathogenesis studies aimed to elucidate the importance of each of the accessory proteins for either pathotype.

Vaccines

With no treatment available, prevention presently seems the only remedy against FIP. Though not a cause of much concern themselves, the avirulent *e*FCoVs appear the logical targets for vaccination. If their infection and spread could be thwarted, the emergence of virulent pathotypes would be effectively prohibited. However, this approach has not yet been explored. Instead, the development of vaccines against FCoVs has so far focused entirely on protection against FIPVs.

A wide range of vaccination strategies against FIPV has been described, conventional as well as modern. As these were mostly unsuccessful and have been reviewed earlier (de Groot and Horzinek, 1995), they are only briefly listed here. Practically all classical approaches have been tested, including inactivated FCoV preparations (Scott, 1987) and various life viruses: *e*FCoV (Pedersen *et al.*, 1984b), low-virulent FIPV (Pedersen and Floyd, 1985), sublethal doses of virulent FIPV (Pedersen and Black, 1983), FIPVs attenuated by repeated propagation through cell culture (Pedersen and Black, 1983), and closely related coronaviruses (Woods, 1979; Barlough *et al.*, 1984; Barlough *et al.*, 1985; Stoddart *et al.*, 1988). Different poxviruses were used as vectors to endogenously express in cats the FCoV structural proteins S (Vennema *et al.*, 1990a; Klepfer *et al.*, 1995), M (Vennema *et al.*, 1991; Hebben *et al.*, 2004) or N (Wasmoen *et al.*, 1995) or the S proteins from the related group 1 coronaviruses CCoV and TGEV (Chalmers *et al.*, 1993). Such recombinant viruses were also used in prime-boost protocols in combination with DNA vaccines; plasmids expressing the FCoV M or N genes (Glansbeek *et al.*, 2002) or (parts of) the polymerase gene (de Groot-Mijnes *et al.*, 2005) were administered first, followed by vaccinia viruses expressing the same genes. These approaches either did not protect at all against virulent FIPV challenge, or protection was not significant or consistent, or the vaccination had adverse effects eliciting EDS in those cases where antibodies reactive to the FIPV S protein had been induced.

The only FIPV vaccine currently on the market consists of a temperature-sensitive strain of a serotype II FIPV. Although its ability to protect cats against FIPV was reported (Gerber *et al.*, 1990a; Gerber *et al.*, 1990b; Gerber, 1995; Fehr *et al.*, 1997), its efficacy has been questioned (McArdle *et al.*, 1995; Scott *et al.*, 1995) and, hence, its applicability remains a matter of dispute.

Recently, two new vaccination approaches were described that seem more encouraging. One study by Hohdatsu *et al.* (2003) uses the N protein, produced in a baculovirus expression system. The protein, included as a crude insect cell extract in an adjuvanted preparation, was administered subcutaneously to cats thrice before the animals were challenged with FIPV WSU 79–1146. A strong protective effect was observed: 75% of the vaccinated cats survived against 12.5% of the control animals. The vaccination had elicited a delayed type hypersensitivity skin response to the N protein, indicative of a CMI response against the FIPV antigen and consistent with the view that cell-mediated rather than humoral immunity is critical for protection against FIPV. While the observations support the potency of the N protein as a candidate vaccine antigen, it remains puzzling why earlier plasmid (Glansbeek *et al.*, 2002) and vaccinia virus (Vennema *et al.*, 1991) formulations of this protein were unprotective.

The other study involved the use of a rationally designed live attenuated vaccine. As mentioned earlier, the deletion of the accessory gene clusters 3abc or 7ab from FIPV WSU 79–1146 by reverse genetics (Haijema *et al.*, 2003) rendered this lethal virus a fully innocuous derivative not even causing fever in infected kittens (Haijema *et al.*, 2004). When used as vaccines against challenge infection with the parental virus both deletion mutants proved to be highly efficacious (Haijema *et al.*, 2004). The most extensively tested mutant, i.e. the virus lacking genes 3abc, provided 100% protection against both mortality and morbidity in four independent experiments. Remarkably, the vaccinated animals did not develop EDS upon challenge despite efficient induction of neutralizing antibodies. The reasons for this are unclear but apparently somehow relate to beneficial effects on the quality of the immune responses provided by the lack of the accessory genes.

The FCoV serotypes differ primarily in their S proteins and to such an extent that there is virtually no cross-neutralization (Pedersen *et al.*, 1984a). Hence, it will be interesting to establish whether the serotype II deletion mutant vaccine also protects against a serotype I challenge infection as this may shed light on the contributions of the cell-mediated immunity versus the virus-neutralizing activity to protection. Such studies are important not only for the further development of vaccines against the predominant and thus most relevant FCoV serotype in the field, the results are likely to be relevant as well for the development of vaccines against other pathogenic coronaviruses including the one causing the severe acute respiratory syndrome (SARS) in humans.

Concluding remarks

Research on feline coronaviruses has obviously made considerable progress. Much has been learned about the biology of the viruses and about the diseases they cause and important new genetic and immunological tools have been developed. Yet, what this review most particularly reveals is that the really essential questions still have to be answered. How do the avirulent viruses escape from immune clearance and establish persistent infection? How and by which mutations do avirulent viruses convert to the virulent pathotype? Are these processes identical for both serotypes? How do these viruses affect the cat immune system? What are the functions of the accessory proteins? What is the serotype I virus cell receptor? With these and many more issues still to be tackled the future of feline coronavirus research will continue to excite us.

References

Addie, D.D., and Jarrett, O. (1990). Control of feline coronavirus infection in kittens. Vet. Rec. *126*, 164.

Addie, D.D., and Jarrett, O. (1992). A study of naturally occurring feline coronavirus infections in kittens. Vet. Rec. *130*, 133–137.

Addie, D.D., and Jarrett, O. (2001). Use of a reverse-transcriptase polymerase chain reaction for monitoring the shedding of feline coronavirus by healthy cats. Vet. Rec. *148*, 649–653.

Addie, D.D., Schaap, I A., Nicolson, L., and Jarrett, O. (2003). Persistence and transmission of natural type I feline coronavirus infection. J. Gen. Virol. *84*, 2735–2744.

Barlough, J.E., Johnson, Lussenburg, C.M., Stoddart, C.A., Jacobson, R.H., and Scott, F.W. (1985). Experimental inoculation of cats with human coronavirus 229E and subsequent challenge with feline infectious peritonitis virus. Avian. Dis. *29*, 613–616.

Barlough, J.E., Stoddart, C.A., Sorresso, G.P., Jacobson, R.H., and Scott, F.W. (1984). Experimental inoculation of cats with canine coronavirus and subsequent challenge with feline infectious peritonitis virus. Lab. Anim. Sci. *34*, 592–597.

Benetka, V., Kubber-Heiss, A., Kolodziejek, J., Nowotny, N., Hofmann-Parisot, M., and Mostl, K. (2004). Prevalence of feline coronavirus types I and II in cats with histopathologically verified feline infectious peritonitis. Vet. Microbiol. *99*, 31–42.

Berg, A.L., Ekman, K., Belak, S., and Berg, M. (2005). Cellular composition and interferon-gamma expression of the local inflammatory response in feline infectious peritonitis (FIP). Vet. Microbiol. *111*, 15–23.

Britton, P., Carmenes, R.S., Page, K.W., Garwes, D.J., and Parra, F. (1988). Sequence of the nucleoprotein gene from a virulent British field isolate of transmissible gastroenteritis virus and its expression in Saccharomyces cerevisiae. Mol. Microbiol. *2*, 89–99.

Chalmers, W.S., Horsburgh, B.C., Baxendale, W., and Brown, T.D. (1993). Enhancement of FIP in cats immunised with vaccinia virus recombinants expressing CCV and TGEV spike glycoproteins. Adv. Exp. Med. Biol. *342*, 455–462.

Corapi, W.V., Darteil, R.J., Audonnet, J.C., and Chappuis, G.E. (1995). Localization of antigenic sites of the S glycoprotein of feline infectious peritonitis virus involved in neutralization and antibody-dependent enhancement. J. Virol. *69*, 2858–2862.

Corapi, W.V., Olsen, C.W., and Scott, F.W. (1992). Monoclonal antibody analysis of neutralization and antibody-dependent enhancement of feline infectious peritonitis virus. J. Virol. *66*, 6695–6705.

Cotter, S.M., Hardy, W.D., Jr., and Essex, M. (1975). Association of feline leukemia virus with lymphosarcoma and other disorders in the cat. J. Am. Vet. Med. Assoc. *166*, 449–454.

de Groot, R.J., Andeweg, A.C., Horzinek, M.C., and Spaan, W.J. (1988). Sequence analysis of the 3′-end of the feline coronavirus FIPV 79–1146 genome: comparison with the genome of porcine coronavirus TGEV reveals large insertions. Virology *167*, 370–376.

de Groot, R.J., Maduro, J., Lenstra, J.A., Horzinek, M.C., van der Zeijst, B.A., and Spaan, W.J. (1987a). cDNA cloning and sequence analysis of the gene encoding the peplomer protein of feline infectious peritonitis virus. J. Gen. Virol. *68*, 2639–2646.

de Groot, R.J., ter Haar, R.J., Horzinek, M.C., and van der Zeijst, B.A. (1987b). Intracellular RNAs of the feline infectious peritonitis coronavirus strain 79–1146. J. Gen. Virol. *68*, 995–1002.

de Groot, R.J., Van Leen, R.W., Dalderup, M.J., Vennema, H., Horzinek, M.C., and Spaan, W.J. (1989). Stably expressed FIPV peplomer protein induces cell fusion and elicits neutralizing antibodies in mice. Virology *171*, 493–502.

de Groot, R.J., Horzinek, M.C. (1995). Feline Infectious Peritonitis. In The Coronaviridae, S.G. Siddell, ed. (New York, Plenum Press), pp. 293–309.

de Groot-Mijnes, J.D., van der Most, R.G., van Dun, J.M., te Lintelo, E.G., Schuurman, N.M., Egberink, H.F., and de Groot, R.J. (2004). Three-color flow cytometry detection of virus-specific CD4+ and CD8+ T cells in the cat. J. Immunol. Methods *285*, 41–54.

de Groot-Mijnes, J.D., van Dun, J.M., van der Most, R.G., and de Groot, R.J. (2005). Natural history of a recurrent feline coronavirus infection and the role of cellular immunity in survival and disease. J. Virol. *79*, 1036–1044.

Dean, G.A., Olivry, T., Stanton, C., and Pedersen, N.C. (2003). In vivo cytokine response to experimental feline infectious peritonitis virus infection. Vet. Microbiol. *97*, 1–12.

Dewerchin, H.L., Cornelissen, E., and Nauwynck, H.J. (2005). Replication of feline coronaviruses in peripheral blood monocytes. Arch. Virol. *150*, 2483–2500.

Dye, C., and Siddell, S.G. (2005). Genomic RNA sequence of Feline coronavirus strain FIPV WSU-79/1146. J. Gen. Virol. *86*, 2249–2253.

Evermann, J.F., Baumgartener, L., Ott, R.L., Davis, E.V., and McKeirnan, A.J. (1981). Characterization of a feline infectious peritonitis virus isolate. Vet. Pathol. *18*, 256–265.

Fehr, D., Holznagel, E., Bolla, S., Hauser, B., Herrewegh, A.A., Horzinek, M.C., and Lutz, H. (1997). Placebo-controlled evaluation of a modified life virus vaccine against feline infectious peritonitis: safety and efficacy under field conditions. Vaccine *15*, 1101–1109.

Foley, J.E., Poland, A., Carlson, J., and Pedersen, N.C. (1997a). Patterns of feline coronavirus infection and fecal shedding from cats in multiple-cat environments. J. Am. Vet. Med. Assoc. *210*, 1307–1312.

Foley, J.E., Poland, A., Carlson, J., and Pedersen, N.C. (1997b). Risk factors for feline infectious peritonitis among cats in multiple-cat environments with endemic feline enteric coronavirus. J. Am. Vet. Med. Assoc. *210*, 1313–1318.

Garwes, D.J., Stewart, F., and Britton, P. (1989). The polypeptide of Mr 14000 of porcine transmissible gastroenteritis virus: gene assignment and intracellular location. J. Gen. Virol. *70*, 2495–2499.

Gerber, J.D. (1995). Overview of the development of a modified live temperature-sensitive FIP virus vaccine. Feline Pract. *23*, 62–66.

Gerber, J.D., Ingersoll, J.D., Gast, A.M., Christianson, K.K., Selzer, N.L., Landon, R.M., Pfeiffer, N.E., Sharpee, R.L., and Beckenhauer, W.H. (1990a). Protection against feline infectious peritonitis by intranasal inoculation of a temperature-sensitive FIPV vaccine. Vaccine *8*, 536–542.

Gerber, J.D., Pfeiffer, N.E., Ingersoll, J.D., Christianson, K.K., Landon, R.M., Selzer, N.L., and Beckenhauer, W.H. (1990b). Characterization of an attenuated temperature sensitive feline infectious peritonitis vaccine virus. Adv. Exp. Med. Biol. *276*, 481–489.

Glansbeek, H.L., Haagmans, B.L., te Lintelo, E.G., Egberink, H.F., Duquesne, V., Aubert, A., Horzinek, M.C., and Rottier, P.J. (2002). Adverse effects of feline IL-12 during DNA vaccination against feline infectious peritonitis virus. J. Gen. Virol. *83*, 1–10.

Gorbalenya, A.E., Koonin, E.V., and Lai, M.M. (1991). Putative papain-related thiol proteases of positive-strand RNA viruses. Identification of rubi- and aphthovirus proteases and delineation of a novel conserved domain associated with proteases of rubi-, alpha- and coronaviruses. FEBS Lett. *288*, 201–205.

Gunn-Moore, D.A., Gruffydd-Jones, T.J., and Harbour, D.A. (1998). Detection of feline coronaviruses by culture and reverse transcriptase-polymerase chain reaction of blood samples from healthy cats and cats with clinical feline infectious peritonitis. Vet. Microbiol. *62*, 193–205.

Haagmans, B.L., Egberink, H.F., and Horzinek, M.C. (1996). Apoptosis and T-cell depletion during feline infectious peritonitis. J. Virol. *70*, 8977–8983.

Haijema, B.J., de Groot-Mijnes, J.D., Vennema, H., Raamsman, M.J., Rottier, P.J., and de Groot, R. J. (2005). GenBank accession number AY994055.

Haijema, B.J., Volders, H., and Rottier, P.J. (2003). Switching species tropism: an effective way to manipulate the feline coronavirus genome. J. Virol. *77*, 4528–4538.

Haijema, B.J., Volders, H., and Rottier, P.J. (2004). Live, attenuated coronavirus vaccines through the directed deletion of group-specific genes provide protection against feline infectious peritonitis. J. Virol. *78*, 3863–3871.

Harpold, L.M., Legendre, A.M., Kennedy, M.A., Plummer, P.J., Millsaps, K., and Rohrbach, B. (1999). Fecal shedding of feline coronavirus in adult cats and kittens in an Abyssinian cattery. J. Am. Vet. Med. Assoc. *215*, 948–951.

Hayashi, T., Goto, N., Takahashi, R., and Fujiwara, K. (1977). Systemic vascular lesions in feline infectious peritonitis. Nippon Juigaku Zasshi *39*, 365–377.

Hayashi, T., Sasaki, N., Ami, Y., and Fujiwara, K. (1983). Role of thymus-dependent lymphocytes and antibodies in feline infectious peritonitis after oral infection. Nippon Juigaku Zasshi *45*, 759–766.

Hebben, M., Duquesne, V., Cronier, J., Rossi, B., and Aubert, A. (2004). Modified vaccinia virus Ankara as a vaccine against feline coronavirus: immunogenicity and efficacy. J. Feline. Med. Surg. *6*, 111–118.

Heeney, J.L., Evermann, J.F., McKeirnan, A.J., Marker-Kraus, L., Roelke, M.E., Bush, M., Wildt, D.E., Meltzer, D.G., Colly, L., Lukas, J., and *et al.* (1990). Prevalence and implications of feline coronavirus infections of captive and free-ranging cheetahs (Acinonyx jubatus). J. Virol. *64*, 1964–1972.

Herrewegh, A.A., de Groot, R.J., Cepica, A., Egberink, H.F., Horzinek, M.C., and Rottier, P.J. (1995a). Detection of feline coronavirus RNA in feces, tissues, and body fluids of naturally infected cats by reverse transcriptase PCR. J. Clin. Microbiol. *33*, 684–689.

Herrewegh, A.A., Mahler, M., Hedrich, H.J., Haagmans, B.L., Egberink, H.F., Horzinek, M.C., Rottier, P.J., and de Groot, R.J. (1997). Persistence and evolution of feline coronavirus in a closed cat-breeding colony. Virology *234*, 349–363.

Herrewegh, A.A., Smeenk, I., Horzinek, M.C., Rottier, P.J., and de Groot, R.J. (1998). Feline coronavirus type II strains 79–1683 and 79–1146 originate from a double recombination between feline coronavirus type I and canine coronavirus. J. Virol. *72*, 4508–4514.

Herrewegh, A.A., Vennema, H., Horzinek, M.C., Rottier, P.J., and de Groot, R.J. (1995b). The molecular genetics of feline coronaviruses: comparative sequence analysis of the ORF7a/7b transcription unit of different biotypes. Virology *212*, 622–631.

Hofmann-Lehmann, R., Fehr, D., Grob, M., Elgizoli, M., Packer, C., Martenson, J.S., O'Brien, S.J., and Lutz, H. (1996). Prevalence of antibodies to feline parvovirus, calicivirus, herpesvirus, coronavirus, and immunodeficiency virus and of feline leukemia virus antigen and the interrelationship of these viral infections in free-ranging lions in east Africa. Clin. Diagn. Lab. Immunol. *3*, 554–562.

Hohdatsu, T., Izumiya, Y., Yokoyama, Y., Kida, K., and Koyama, H. (1998). Differences in virus receptor for type I and type II feline infectious peritonitis virus. Arch. Virol. *143*, 839–850.

Hohdatsu, T., Nakamura, M., Ishizuka, Y., Yamada, H., and Koyama, H. (1991). A study on the mechanism of antibody-dependent enhancement of feline infectious peritonitis virus infection in feline macrophages by monoclonal antibodies. Arch. Virol. *120*, 207–217.

Hohdatsu, T., Okada, S., Ishizuka, Y., Yamada, H., and Koyama, H. (1992). The prevalence of types I and II feline coronavirus infections in cats. J. Vet. Med. Sci. *54*, 557–562.

Hohdatsu, T., Yamato, H., Ohkawa, T., Kaneko, M., Motokawa, K., Kusuhara, H., Kaneshima, T., Arai, S., and Koyama, H. (2003). Vaccine efficacy of a cell lysate with recombinant baculovirus-expressed feline infectious peritonitis (FIP) virus nucleocapsid protein against progression of FIP. Vet. Microbiol. *97*, 31–44.

Horsburgh, B.C., Brierley, I., and Brown, T.D. (1992). Analysis of a 9.6 kb sequence from the 3′ end of canine coronavirus genomic RNA. J. Gen. Virol. *73*, 2849–2862.

Horzinek, M.C., and Osterhaus, A.D. (1979). Feline infectious peritonitis: a worldwide serosurvey. Am. J. Vet. Res. *40*, 1487–1492.

Hsieh, M.W., Chueh, L.L., and Su, B. (2005). Genbank accession number DQ160294.

Kapke, P.A., and Brian, D.A. (1986). Sequence analysis of the porcine transmissible gastroenteritis coronavirus nucleocapsid protein gene. Diagn. Microbiol. Infect. Dis. *5*, 39–54.

Kennedy, M., Boedeker, N., Gibbs, P., and Kania, S. (2001). Deletions in the 7a ORF of feline coronavirus associated with an epidemic of feline infectious peritonitis. Vet. Microbiol. *81*, 227–234.

Kennedy, M., Citino, S., McNabb, A.H., Moffatt, A.S., Gertz, K., and Kania, S. (2002). Detection of feline coronavirus in captive Felidae in the USA. J. Vet. Diagn. Invest. *14*, 520–522.

Kipar, A., Baptiste, K., Barth, A., and Reinacher, M. (2005a). Natural FCoV infection: cats with FIP exhibit significantly higher viral loads than healthy infected cats. J. Feline. Med. Surg. *8*, 69–72.

Kipar, A., Bellmann, S., Kremendahl, J., Kohler, K., and Reinacher, M. (1998a). Cellular composition, coronavirus antigen expression and production of specific antibodies in lesions in feline infectious peritonitis. Vet. Immunol. Immunopathol. *65*, 243–257.

Kipar, A., Kohler, K., Leukert, W., and Reinacher, M. (2001). A comparison of lymphatic tissues from cats with spontaneous feline infectious peritonitis (FIP), cats with FIP virus infection but no FIP, and cats with no infection. J. Comp. Pathol. *125*, 182–191.

Kipar, A., Kremendahl, J., Addie, D.D., Leukert, W., Grant, C.K., and Reinacher, M. (1998b). Fatal enteritis associated with coronavirus infection in cats. J. Comp. Pathol. *119*, 1–14.

Kipar, A., May, H., Menger, S., Weber, M., Leukert, W., and Reinacher, M. (2005b). Morphologic features and development of granulomatous vasculitis in feline infectious peritonitis. Vet. Pathol. *42*, 321–330.

Kiss, I., Poland, A.M., and Pedersen, N.C. (2004). Disease outcome and cytokine responses in cats immunized with an avirulent feline infectious peritonitis virus (FIPV)-UCD1 and challenge-exposed with virulent FIPV-UCD8. J. Feline. Med. Surg. *6*, 89–97.

Klepfer, S., Reed, A.P., Martinez, M., Bhogal, B., Jones, E., and Miller, T.J. (1995). Cloning and expression of FECV spike gene in vaccinia virus. Immunization with FECV S causes early death after FIPV challenge. Adv. Exp. Med. Biol. *380*, 235–241.

Kummrow, M., Meli, M.L., Haessig, M., Goenczi, E., Poland, A., Pedersen, N.C., Hofmann-Lehmann, R., and Lutz, H. (2005). Feline coronavirus serotypes 1 and 2: seroprevalence and association with disease in Switzerland. Clin. Diagn. Lab. Immunol. *12*, 1209–1215.

Leutenegger, C.M., Hofmann-Lehmann, R., Riols, C., Liberek, M., Worel, G., Lups, P., Fehr, D., Hartmann, M., Weilenmann, P., and Lutz, H. (1999). Viral infections in free-living populations of the European wildcat. J. Wildl. Dis. *35*, 678–686.

McArdle, F., Tennant, B., Bennett, M., Kelly, D.F., Gaskell, C.J., and Gaskell, R.M. (1995). Independent evaluation of a modified live FIPV vaccine under experimental conditions (University of Liverpool experience). Feline Pract. Santa Barbara, CA: Veterinary Practice Pub. Co. *23*, 67–71.

McKeirnan A.J., Evermann, J.F., Hargis, A., Miller L.M. Ott R.L. (1981). Isolation of feline coronaviruses from two cats with diverse disease manifestations. Feline Pract. *11*, 16–20.

Meli, M., Kipar, A., Muller, C., Jenal, K., Gonczi, E., Borel, N., Gunn-Moore, D., Chalmers, S., Lin, F., Reinacher, M., and Lutz, H. (2004). High viral loads despite absence of clinical and pathological findings in cats experimentally infected with feline coronavirus (FCoV) type I and in naturally FCoV-infected cats. J. Feline Med. Surg. *6*, 69–81.

Miura, Y., and Koyanagi, Y. (2005). Death ligand-mediated apoptosis in HIV infection. Rev. Med. Virol. *15*, 169–178.

Motokawa, K., Hohdatsu, T., Aizawa, C., Koyama, H., and Hashimoto, H. (1995). Molecular cloning and sequence determination of the peplomer protein gene of feline infectious peritonitis virus type I. Arch. Virol. *140*, 469–480.

Motokawa, K., Hohdatsu, T., Hashimoto, H., and Koyama, H. (1996). Comparison of the amino acid sequence and phylogenetic analysis of the peplomer, integral membrane and nucleocapsid proteins of feline, canine and porcine coronaviruses. Microbiol. Immunol. *40*, 425–433.

Munson, L., Marker, L., Dubovi, E., Spencer, J.A., Evermann, J.F., and O'Brien, S.J. (2004). Serosurvey of viral infections in free-ranging Namibian cheetahs (Acinonyx jubatus). J. Wildl. Dis. *40*, 23–31.

O'Brien, S.J., Roelke, M.E., Marker, L., Newman, A., Winkler, C.A., Meltzer, D., Colly, L., Evermann, J.F., Bush, M., and Wildt, D.E. (1985). Genetic basis for species vulnerability in the cheetah. J. Gen. Virol. *66*, 719–726.

Okada, H., Kobune, F., Sato, T.A., Kohama, T., Takeuchi, Y., Abe, T., Takayama, N., Tsuchiya, T., and Tashiro, M. (2000). Extensive lymphopenia due to apoptosis of uninfected lymphocytes in acute measles patients. Arch. Virol. *145*, 905–920.

Olsen, C., and Scott, F. (1993). Evaluation of antibody-dependent enhancement of feline infectious peritonitis virus infectivity using in situ hybridization. Microb. Pathog. *14*, 275–285.

Olsen, C.W., Corapi, W.V., Jacobson, R.H., Simkins, R.A., Saif, L.J., and Scott, F.W. (1993). Identification of antigenic sites mediating antibody-dependent enhancement of feline infectious peritonitis virus infectivity. J. Gen. Virol. *74*, 745–749.

Olsen, C.W., Corapi, W.V., Ngichabe, C.K., Baines, J.D., and Scott, F.W. (1992). Monoclonal antibodies to the spike protein of feline infectious peritonitis virus mediate antibody-dependent enhancement of infection of feline macrophages. J. Virol. *66*, 956–965.

Olsen, J., Cowell, G.M., Konigshofer, E., Danielsen, E.M., Moller, J., Laustsen, L., Hansen, O.C., Welinder, K.G., Engberg, J., Hunziker, W., and *et al.* (1988). Complete amino acid sequence of human intestinal aminopeptidase N as deduced from cloned cDNA. FEBS Lett. *238*, 307–314.

Oostra, M., de Haan, C.A., de Groot, R.J., and Rottier, P.J. (2006). Glycosylation of the severe acute respiratory syndrome coronavirus triple-spanning membrane proteins 3a and M. J. Virol. *80*, 2326–2336.

Paltrinieri, S., Cammarata, M.P., Cammarata, G., and Comazzi, S. (1998). Some aspects of humoral and cellular immunity in naturally occuring feline infectious peritonitis. Vet. Immunol. Immunopathol. *65*, 205–220.

Paltrinieri, S., Grieco, V., Comazzi, S., and Cammarata Parodi, M. (2001). Laboratory profiles in cats with different pathological and immunohistochemical findings due to feline infectious peritonitis (FIP). J. Feline Med. Surg. *3*, 149–159.

Palucka, K., and Banchereau, J. (2002). How dendritic cells and microbes interact to elicit or subvert protective immune responses. Curr. Opin. Immunol. *14*, 420–431.

Paul-Murphy, J., Work, T., Hunter, D., McFie, E., and Fjelline, D. (1994). Serologic survey and serum biochemical reference ranges of the free-ranging mountain lion (Felis concolor) in California. J. Wildl. Dis. *30*, 205–215.

Pedersen, N.C. (1976). Morphologic and physical characteristics of feline infectious peritonitis virus and its growth in autochthonous peritoneal cell cultures. Am. J. Vet. Res. *37*, 567–572.

Pedersen, N.C. (1987). Virologic and immunologic aspects of feline infectious peritonitis virus infection. Adv. Exp. Med. Biol. *218*, 529–550.

Pedersen, N.C., and Black, J.W. (1983). Attempted immunization of cats against feline infectious peritonitis, using avirulent live virus or sublethal amounts of virulent virus. Am. J. Vet. Res. *44*, 229–234.

Pedersen, N.C., Black, J.W., Boyle, J.F., Evermann, J.F., McKeirnan, A.J., and Ott, R.L. (1984a). Pathogenic differences between various feline coronavirus isolates. Adv. Exp. Med. Biol. *173*, 365–380.

Pedersen, N.C., Boyle, J.F., Floyd, K., Fudge, A., and Barker, J. (1981). An enteric coronavirus infection of cats and its relationship to feline infectious peritonitis. Am. J. Vet. Res. *42*, 368–377.

Pedersen, N.C., Evermann, J.F., McKeirnan, A.J., and Ott, R.L. (1984b). Pathogenicity studies of feline coronavirus isolates 79–1146 and 79–1683. Am. J. Vet. Res. *45*, 2580–2585.

Pedersen, N.C., and Floyd, K. (1985). Experimental studies with three new strains of feline infectious peritonitis virus: FIPV-UCD2, FIPV-UCD3, and FIPV-UCD4. Compend. Contin. Educ. Pract. Vet. *7*, 1001–1011.

Pedersen, N.C., Sato, R., Foley, J.E., and Poland, A.M. (2004). Common virus infections in cats, before and after being placed in shelters, with emphasis on feline enteric coronavirus. J. Feline. Med. Surg. *6*, 83–88.

Pedersen, N.C.B., J.F. (1980). Immunologic phenomena in the effusive form of feline infectious peritonitis. Am. J. Vet. Res. *41*, 868–876.

Poland, A.M., Vennema, H., Foley, J.E., and Pedersen, N.C. (1996). Two related strains of feline infectious peritonitis virus isolated from immunocompromised cats infected with a feline enteric coronavirus. J. Clin. Microbiol. *34*, 3180–3184.

Rottier, P.J., Nakamura, K., Schellen, P., Volders, H., and Haijema, B.J. (2005). Acquisition of macrophage tropism during the pathogenesis of feline infectious peritonitis is determined by mutations in the feline coronavirus spike protein. J. Virol. *79*, 14122–14130.

Schneider-Schaulies, S., Bieback, K., Avota, E., Klagge, I., and ter Meulen, V. (2002). Regulation of gene expression in lymphocytes and antigen-presenting cells by measles virus: consequences for immunomodulation. J. Mol. Med. *80*, 73–85.

Scott, F.W. (1987). Immunization against feline coronaviruses. Adv. Exp. Med. Biol. *218*, 569–576.

Scott, F.W., Corapi, W.V., and Olsen, C.W. (1995). Independent evaluation of a modified live FIPV vaccine under experimental conditions (Cornell experience). Feline Pract. Santa Barbara, CA: Veterinary Practice Pub. Co. *23*, 74–76.

Servet-Delprat, C., Vidalain, P.O., Valentin, H., and Rabourdin-Combe, C. (2003). Measles virus and dendritic cell functions: how specific response cohabits with immunosuppression. Curr. Top. Microbiol. Immunol. *276*, 103–123.

Simons, F.A., Vennema, H., Rofina, J.E., Pol, J.M., Horzinek, M.C., Rottier, P.J., and Egberink, H.F. (2005). A mRNA PCR for the diagnosis of feline infectious peritonitis. J. Virol. Methods *124*, 111–116.

Snijder, E.J., Bredenbeek, P.J., Dobbe, J.C., Thiel, V., Ziebuhr, J., Poon, L.L., Guan, Y., Rozanov, M., Spaan, W.J., and Gorbalenya, A.E. (2003). Unique and conserved features of genome and proteome of SARS-coronavirus, an early split-off from the coronavirus group 2 lineage. J. Mol. Biol. *331*, 991–1004.

Stoddart, C.A., Barlough, J.E., Baldwin, C.A., and Scott, F.W. (1988). Attempted immunisation of cats against feline infectious peritonitis using canine coronavirus. Res. Vet. Sci. *45*, 383–388.

Stoddart, C.A., and Scott, F.W. (1988). Isolation and identification of feline peritoneal macrophages for in vitro studies of coronavirus-macrophage interactions. J. Leukoc. Biol. *44*, 319–328.

Stoddart, C.A., and Scott, F.W. (1989). Intrinsic resistance of feline peritoneal macrophages to coronavirus infection correlates with in vivo virulence. J. Virol. *63*, 436–440.

Stoddart, M.E., Whicher, J.T., and Harbour D.A. (1988). Cats inoculated with feline infectious peritonitis virus exhibit a biphasic acute phase plasma protein response. Vet. Rec. *123*, 622–624.

Tresnan, D.B., and Holmes, K.V. (1998). Feline aminopeptidase N is a receptor for all group I coronaviruses. Adv. Exp. Med. Biol. *440*, 69–75.

Tresnan, D.B., Levis, R., and Holmes, K.V. (1996). Feline aminopeptidase N serves as a receptor for feline, canine, porcine, and human coronaviruses in serogroup I. J. Virol. *70*, 8669–8674.

Tung, F.Y., Abraham, S., Sethna, M., Hung, S. L., Sethna, P., Hogue, B.G., and Brian, D.A. (1992). The 9-kDa hydrophobic protein encoded at the 3′ end of the porcine transmissible gastroenteritis coronavirus genome is membrane-associated. Virology *186*, 676–683.

van Vliet, A.L., Smits, S.L., Rottier, P.J., and de Groot, R.J. (2002). Discontinuous and non-discontinuous subgenomic RNA transcription in a nidovirus. EMBO J. *21*, 6571–6580.

Vennema, H. (1999). Genetic drift and genetic shift during feline coronavirus evolution. Vet. Microbiol. *69*, 139–141.

Vennema, H., de Groot, R.J., Harbour, D.A., Dalderup, M., Gruffydd-Jones, T., Horzinek, M.C., and Spaan, W.J. (1990a). Early death after feline infectious peritonitis virus challenge due to recombinant vaccinia virus immunization. J. Virol. *64*, 1407–1409.

Vennema, H., de Groot, R.J., Harbour, D.A., Dalderup, M., Gruffydd-Jones, T., Horzinek, M.C., and Spaan, W.J. (1990b). Immunogenicity of recombinant feline infectious peritonitis virus spike protein in mice and kittens. Adv. Exp. Med. Biol. *276*, 217–222.

Vennema, H., de Groot, R.J., Harbour, D.A., Horzinek, M.C., and Spaan, W.J. (1991). Primary structure of the membrane and nucleocapsid protein genes of feline infectious peritonitis virus and immunogenicity of recombinant vaccinia viruses in kittens. Virology *181*, 327–335.

Vennema, H., Heijnen, L., Rottier, P.J., Horzinek, M.C., and Spaan, W.J. (1993). A novel glycoprotein of feline infectious peritonitis coronavirus contains a KDEL-like endoplasmic reticulum retention signal. Adv. Exp. Med. Biol. *342*, 209–214.

Vennema, H., Poland, A., Floyd Hawkins, K., and Pedersen, N.C. (1995). A comparison of the genomes of FECVs and FIPVs and what they tell us about the relationships between feline coronaviruses and their evolution. Feline Pract. *23*, 40–44.

Vennema, H., Poland, A., Foley, J., and Pedersen, N.C. (1998). Feline infectious peritonitis viruses arise by mutation from endemic feline enteric coronaviruses. Virology *243*, 150–157.

Vennema, H., Rossen, J.W., Wesseling, J., Horzinek, M.C., and Rottier, P.J. (1992). Genomic organization and expression of the 3′ end of the canine and feline enteric coronaviruses. Virology *191*, 134–140.

Ward, J.M. (1970). Morphogenesis of a virus in cats with experimental feline infectious peritonitis. Virology *41*, 191–194.

Ward, J.M., Gribble, D.H., and Dungworth, D.L. (1974). Feline infectious peritonitis: experimental evidence for its multiphasic nature. Am. J. Vet. Res. *35*, 1271–1275.

Wasmoen, T.L., Kadakia, N.P., Unfer, R.C., Fickbohm, B.L., Cook, C.P., Chu, H.J., and Acree, W.M. (1995). Protection of cats from infectious peritonitis by vaccination with a recombinant raccoon poxvirus expressing the nucleocapsid gene of feline infectious peritonitis virus. Adv. Exp. Med. Biol. *380*, 221–228.

Weiss, R.C., and Cox, N.R. (1989). Evaluation of immunity to feline infectious peritonitis in cats with cutaneous viral-induced delayed hypersensitivity. Vet. Immunol. Immunopathol. *21*, 293–309.

Weiss, R.C., and Scott, F.W. (1981a). Antibody-mediated enhancement of disease in feline infectious peritonitis: comparisons with dengue hemorrhagic fever. Comp. Immunol. Microbiol. Infect. Dis. *4*, 175–189.

Weiss, R.C., and Scott, F.W. (1981b). Pathogenesis of feline infectious peritonitis: nature and development of viremia. Am. J. Vet. Res. 42, 382–390.

Weiss, R.C., and Scott, F.W. (1981c). Pathogenesis of feline infetious peritonitis: pathologic changes and immunofluorescence. Am. J. Vet. Res. *42*, 2036–2048.

Wise, A.G., Kiupel, M., and Maes, R.K. (2006). Molecular characterization of a novel coronavirus associated with epizootic catarrhal enteritis (ECE) in ferrets. Virology *349*164–174.

Woods, R.D., Pedersen, N.C. (1979). Cross-protection studies between feline infectious peritonitis virus and porcine transmissible gastroenteritis viruses. Vet. Microbiol. *4*, 11–16.

Yamanaka, M., Crisp, T., Brown, R., and Dale, B. (1998). Nucleotide sequence of the inter-structural gene region of feline infectious peritonitis virus. Virus Genes *16*, 317–318.

Ziebuhr, J., Thiel, V., and Gorbalenya, A.E. (2001). The autocatalytic release of a putative RNA virus transcription factor from its polyprotein precursor involves two paralogous papain-like proteases that cleave the same peptide bond. J. Biol. Chem. *276*, 33220–33232.

Control of Neurotropic MHV by Multifactorial Mechanisms

10

Cornelia C. Bergmann and Stephen A. Stohlman

Abstract

Infection of the central nervous system (CNS) by mouse hepatitis virus (MHV) can result in fatal encephalitis. Rapid viral growth especially in neurons leads to death prior to effective immunity. Alternatively, MHV can induce an acute encephalomyelitis that resolves into a persistent infection under immune pressure. During both infections there is a coordinated innate immune response which, although unable to control virus, facilitates adaptive immunity during non-fatal infections. During non-lethal infection, virus is controlled by $CD8^+$ T cells which use both perforin mediated cytolysis and IFN-γ, depending upon the specific type of cell infected. In contrast to innate and antiviral cellular mechanisms evident during acute infection, persistence is maintained via virus neutralizing antibody. $CD8^+$ T cells loose antiviral function as virus is controlled prior to complete virus elimination. Even memory T cells are insufficient to control persisting virus in the absence of antibody. Retention of virus specific antibody secreting within the CNS, despite their delayed appearance relative to T cells, implies local antibody production in providing protection throughout persistence.

Introduction

Mouse hepatitis viruses (MHV) are causative agents of respiratory infections, hepatitis, enteritis and encephalitis in susceptible rodents. Respiratory, enteric or central nervous system (CNS) infections depend on the strain of virus and route of inoculation (Compton *et al.*, 1993; Houtman and Fleming, 1996; Perlman, 1998; Buchmeier and Lane, 1999; Stohlman *et al.*, 1999). The prototype virus isolate was derived from the brain of a Swiss white mouse with spontaneous paralysis. Serial passage in mouse brain enhanced virulence, with early passages producing paralytic disease, and later passages causing a lethal encephalitis. The extremely neurovirulent virus was named JHM virus (JHMV) after Professor John Howard Mueller (1949) and is also known as MHV-4. Subsequent isolation and derivation of numerous related MHV strains and variants, associated with different tropism and pathogenesis, has led to a classification based on serological relatedness, tropism and genetic relatedness. Hepatotropic strains are represented by MHV-1, -2 and -3. JHMV or MHV-4 strains are prominently neurotropic. The MHV-A59 strain is both hepato- and neurotropic, causing milder CNS pathology than the JHMV variants. However, tissue tropism is not absolute, as all prototype strains, i.e. MHV-1, -3, -A59 and

-JHM infect upper respiratory tract mucosa and can disseminate to other organs (Compton *et al.*, 1993; Perlman, 1998). Enterotropic MHV infections remain a major concern in many breeding and research mouse colonies due to their efficient transmission and immune modulation. Severe hepatitis caused by MHV-3 provides a model for acute human hepatitis (Marsden *et al.*, 2003). The neurotropic variants are the most studied viruses, as they provide excellent models to study several disease phenotypes at the viral genomic, immunological and pathogenic levels. These range from transient asymptomatic infection to acute lethal encephalitis, and persistent CNS infections associated with chronic myelin loss (Houtman and Fleming, 1996; Perlman, 1998; Buchmeier and Lane, 1999; Stohlman *et al.*, 1999). Similar to Theiler's murine encephalomyelitis virus, a non-enveloped Picornavirus, MHV induced demyelination constitutes a versatile rodent model for the human demyelinating disease multiple sclerosis (MS) (Ercolini and Miller, 2006). Although the human coronavirus (CoV) OC-43 has been implicated as a causative microbial agent for MS, the evidence is circumstantial. Nevertheless, recent data revealing detection of SARS-CoV in CNS autopsy tissues are compelling, as they clearly establish the potential for neuroinvasive properties of human CoV (Gu *et al.*, 2005).

Viral and host factors influencing pathogenesis

Genetic variability, virulence, tropism and course of infection

The infidelity of CoV polymerases and a high recombination rate during replication results in considerable variability and adaptability of MHV viral genomes both *in vitro* and *in vivo* (Lai and Cavanagh, 1997). Although most neurotropic strains used in research laboratories worldwide are derived from the original isolate, propagation at geographically distinct sites has led to multitudinous MHV strains and variants with vastly different pathologies. Despite the same terminology for individual viruses in the literature, the reader is thus cautioned that JHMV and its derivatives in different laboratories are not identical and may constitute a different composition of prominent quasispecies (Adami *et al.*, 1995; Houtman and Fleming, 1996; Stohlman *et al.*, 1999). Irrespective of differences in JHMV induced disease, a common trait of mice surviving acute infection is establishment of persistence associated with ongoing, primary demyelination (Houtman and Fleming, 1996; Perlman, 1998; Buchmeier and Lane, 1999; Stohlman *et al.*, 1999). In immunocompetent mice virus titers in the brain generally peak at 5 days post inoculation (p.i.) and survivors have no detectable infectious virus by approx. 2 weeks p.i. (Fleming *et al.*, 1986). However, viral antigen and more prominently RNA remain detectable within the CNS, especially the spinal cord, up to 2 years p.i. (Adami *et al.*, 1995; Rowe *et al.*, 1997a; Marten *et al.*, 2000). Encephalitis is pronounced during the acute phase, whereas paralysis and demyelination is most severe between days 10–18 p.i., when virus replication is controlled. Clinical recovery is generally evident by day 21 p.i., although histological examination shows ongoing primary CNS demyelination as a hallmark of the persisting infection (Perlman, 1998; Buchmeier and Lane, 1999; Stohlman *et al.*, 1999). Persisting virus cannot be recovered from CNS explants or from immunosuppressed mice perfused to remove neutralizing antibody (Ab) prior to explant. Although the mechanisms of virus persistence remain to be resolved, persistence in a replication competent form is indicated by several observations. Virus recrudesces in mice genetically impaired in antiviral Ab production, despite a period of transient clearance of infectious virus (Lin *et al.*, 1999; Ramakrishna *et al.*, 2002). Ab transfer prior to recrudescence delays but does not prevent recrudescence (Ramakrishna *et al.*, 2002). Lastly, persistence established by the JHMV-DS plaque size variant is associated with progressive accumulation of mutations and deletions in the hypervariable region of the viral spike (S) protein (Adami *et al.*, 1995; Rowe *et al.*, 1997a). Collectively these data suggest that both immune pressure as well as viral replication strategies contribute to a mutual state of host-virus co-existence.

Derivation of attenuated mutants that cause less encephalitis enabled the study of various factors associated with increased survival and the demyelinating process (Houtman and Fleming, 1996, Perlman, 1998; Stohlman *et al.*, 1999). While fatal encephalitis was generally correlated with prominent infection of neurons, reduced pathogenicity coincided with preferential tropism for glial cells and establishment of persistent CNS infection. Pathogenic outcome was also reflected by distinct plaque size variants (Stohlman *et al.*, 1982). Commonly used mutant isolates, including temperature sensitive mutants and monoclonal antibody derived variants are summarized in several chapters (Perlman, 1998; Stohlman *et al.*, 1999). Genetic analyses implicate the viral S protein as the main determinant of cell tropism and pathogenicity (Fleming *et al.*, 1986; Parker *et al.*, 1989; Houtman and Fleming, 1996; Perlman, 1998). Both S protein deletions and point mutations show dramatically different tropisms and spread (Fleming *et al.*, 1986; Parker *et al.*, 1989; Phillips *et al.*, 1999; Ontiveros *et al.*, 2003; Navas-Martin *et al.*, 2005). Recent studies using recombinant viruses expressing distinct S proteins on either MHV-JHM or -A59 viral backgrounds confirm that the JHM S protein confers neurovirulence (Phillips *et al.*, 1999). Interestingly, JHM background genes override A59 S protein in determining hepatotropism (Navas *et al.*, 2003). Identification of genetic regions contributing to altered phenotypes *in vivo* is complex (Navas *et al.*, 2003; Chua *et al.*, 2004; MacNamara *et al.*, 2005), particularly as S proteins vary in their immunogenic regions; e.g. the A59 S protein lacks a dominant $CD8^+$ T cell epitope present in the JHM variant. It thus remains challenging to evaluate whether genetic alterations in proteins with pleiotropic functions exert their affects primarily via receptor usage, interaction with other viral or host genes, or immune responses. Inflammatory mediators may also manipulate the reservoir of susceptible cells during the course of infection, as indicated by decreased CEACAM-1 receptor expression on microglia during JHMV induced inflammation (Ramakrishna *et al.*, 2004a).

Immune responses and the CNS

Infections of the CNS commonly result in persistence (Lipton and Gilden, 1997; Rall and Oldstone; 1997; Griffin *et al.*, 2003; Fazakerley and Walker, 2003). This is attributed to the unique composition and communication network not only between highly specialized resident cells but also interaction with the immune system (Fabry *et al.*, 1994; Hickey, 2001; Aloisi *et al.*, 2000). In the quiescent state immune surveillance and immunological activity in the CNS is restrained by the absence of a dedicated lymphatic drainage system, paucity of major histocompatibility expression on resident glia and neurons, and constitutive secretion of anti-inflammatory neurotrophins and TGF-β (Griffin *et al.*, 2003; Ransohoff at el., 2003). The blood brain barrier (BBB) formed by tight junctions between endothelial cells and their limited expression of adhesion molecules further limits large molecules, e.g. antibodies, as well as T cells from entering the CNS. The small number of activated/memory T cells randomly patrolling the normal CNS, are not retained in the absence of antigen recognition (Hickey *et al.*, 2001). However, many CNS infections dramatically alter the factors maintaining the steady state. MHV infection, similar to other neurotropic viruses induce strong cytokine and chemokine responses in resident cells, which promote anti viral inflammatory responses, but also immune pathology (Houtman and Fleming, 1996; Fazakerley *et al.*, 2003; Griffin *et al.*, 2003; Ercolini and Miller, 2006).

The distinct genetic makeup of MHV variants obscures direct comparison of immune components involved in viral control, persistence and pathogenesis. To highlight the complex interplay of innate and adaptive host responses to acute infection leading to persistence, data discussed in this chapter pertain largely to infections favoring survival, persistence, and demyelination, as exemplified by a monoclonal Ab neutralization JHMV escape mutant, designated 2.2v-1 (Fleming *et al.*, 1986). Primary targets of infection are glia, composed of microglia, astrocytes and oligodendroglia. Neurons are only rarely infected, sparing mice from death due to neuronal dysfunction. Furthermore, hepatitis is extremely rare following intracerebral infection, restricting immune responses largely to the CNS. Variance and similarities of immune responses to lethal neurotropic infection as well as dualtropic MHV-A59 infection are discussed to emphasize the plasticity of the host response despite limited variability in viral genome.

Dynamics of proinflammatory responses and composition of infiltrates

Chemokines and proteases as mediators of inflammation

Early viral replication within the CNS induces glial cell activation and alterations in BBB integrity, even in the absence of overt clinical disease (Bergmann *et al.*, 2006). This is manifested by coordinated upregulation of chemokines, pro-inflammatory cytokines, matrix metalloproteinases (MMPs) and tissue inhibitors of MMPs (TIMPs). Their overall signal strength and patterns have mostly been characterized by mRNA analysis of infected brains from untreated or irradiated mice to distinguish expression by resident and infiltrating cells. Furthermore, a limited number of studies clearly indicate that both virus infected and uninfected resident glial cells, most prominently astrocytes, are early responders (Lane *et al.*, 1998; Lane *et al.*, 2000; Zhou *et al.*, 2005).

Of the multitudinous MMP proteins associated with cytokine secretion, influx of inflammatory cells, and CNS damage (Yong *et al.*, 2001), JHMV infection only induces expression of MMP-3, -9 and -12 (Zhou *et al.*, 2002; Zhou *et al.*, 2005). Expression of MMP-3 and -12 is independent of the inflammatory response and prominent in astrocytes and oligodendroglia, respectively. The onset of inflammation is further associated with detection of MMP9 activity (Zhou *et al.*, 2002; Zhou *et al.*, 2003). Flow cytometric analysis of cell preparations from infected brains revealed that neutrophils, and macrophages are the initial inflammatory cells recruited into the infected CNS (Bergmann *et al.*, 2003; Zhou *et al.*, 2003) with natural killer (NK) cells following closely behind (Figure 10.1). Neutrophils appear to play a key role in propagating CNS infiltration by other cell types via secretion of pre-packaged MMP-9 (Zhou *et al.*, 2003). Increased replication and virulence by lethal compared to sublethal JHMV infection correlates with enhanced

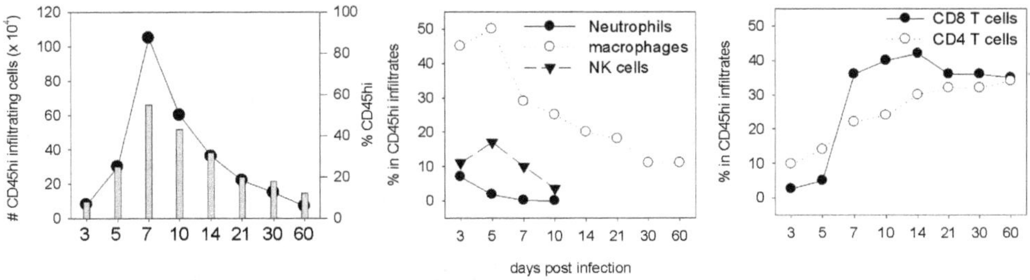

Figure 10.1 Kinetics of innate and adaptive cellular infiltrates following sublethal JHMV CNS infection. CNS infiltrating cells derived from mechanic disruption of brain tissue and concentration of mononuclear cells by gradient centrifugation following sublethal neurotropic JHMV variant V2.2v-1 CNS infection are identified by their borne marrow derived $CD45^{hi}$ phenotype using flow cytometric analysis. Resident cells express either a $CD45^{lo}$ phenotype, characteristic of resident macrophages, termed microglia, or a $CD45^{-}$ phenotype. Right panel: Symbols depict numbers of $CD45^{hi}$ CNS infiltrating cells as determined by flow cytometric analysis of live cells at various times p.i. (x-axis). The bars (y-axis) depict relative percentages of $CD45^{hi}$ infiltrates within total brain cells. The composition and dynamics of individual innate and T cell populations within $CD45^{hi}$ infiltrates are depicted in the middle and left panels as indicated. Neutrophils were characterized by staining for Ly-6G, (A18 mAb) but not class II or F4/80. Macrophages were identified by staining with F4/80, and NK cells by reactivity with mAb DX5. T cells were identified by staining with anti CD4 or CD8 mAb. Macrophages make up the vast majority of infiltrates early during infection, while T cells are most abundant during peak inflammation and thereafter.

MMP-9 activity, and increased MMP-3, -12, and TIMP-1 mRNA levels, but not cell infiltration (Zhou *et al.*, 2002). MMP-3 and -12 mRNAs all decrease either at the peak of JHMV-induced inflammation or rapidly thereafter (Zhou *et al.*, 2002; Zhou *et al.*, 2005), supporting an early role in shaping the CNS environment. However, unlike MMP-9, their role(s) in innate inflammatory cell trafficking and CNS pathology are unclear. A potential mechanism for counteracting MMP-mediated CNS pathology may reside in sustained elevation of mRNA encoding the MMP inhibitor, TIMP-1 (Zhou *et al.*, 2002; Zhou *et al.*, 2005).

Analysis of brain mRNA revealed CXCL10 and CCL3 as the earliest and most abundant chemokines induced following infection (Lane *et al.*, 1998; Trifilo *et al.*, 2003). CXCL10 is expressed by both infected and uninfected glial cells as early as 1 day post inoculation and generally acts on activated lymphocytes expressing the CXCR3 receptor. NK cells are the prominent lymphocyte population recruited within 5 days p.i. (Trifilo *et al.*, 2004), although T cells and plasmablasts, members of the B cell linage, are also responders to CXCL10 (Liu *et al.*, 2000; Hauser *et al.*, 2002; Christensen *et al.*, 2004; De Lemos et. al, 2005). CCL3 upregulation appears to link the innate and adaptive immune responses by stimulating recruitment, activation and maturation of dendritic cells and T cells (Trifilo *et al.*, 2003; Trifilo *et al.*, 2004). Infection also induces CCL2, CCL4, and CCL5 (Zhou *et al.*, 2002). CCL2 and CCL5 upregulation enhance accumulation of macrophages (Lane *et al.*, 2000), which comprise the largest fraction of innate infiltrates (Figure 10.1). Lethal JHMV infection is associated with slightly higher levels of these chemokines compared to sublethal infection (Zhou *et al.*, 2002). However, significantly increased CXCL1 mRNA levels constitute a prominent difference in brains of mice infected with the more virulent JHMV (Zhou *et al.*, 2002). CXCL1 is a potent neutrophils chemotractant, supporting increased inflammation by neutrophils and MMP-9 activity.

Chemokines and T cells during acute inflammation

The CXCR3 ligands CXCL9 and CXCL11 are induced delayed compared to CXCL10 (Lane at al., 1998; Stiles *et al.*, 2006) and coincide with increased CNS infiltration by T cells. Virus specific T cells acquire CXCR3 expression and increased responsiveness to CXCR3 ligands during activation and expansion in secondary lymphoid organs (Christensen *et al.*, 2004). The majority of early T cell infiltrates are unbiased memory T cells of unknown specificity, supporting chemokine-mediated, antigen-independent T cell recruitment (Chen *et al.*, 2005). However, as virus specific T cells expand and exit lymphoid organs, they gradually overwhelm the irrelevant T cell population. Increasing accumulation of T cells also coincides with a decline in neutrophils and NK cells whereas macrophages persist (Figure 10.1). T cells constitute 60–70% of infiltrates during peak CNS inflammation between days 6–8, with the majority of $CD8^+$ and $CD4^+$ T cells being virus-specific (Bergmann *et al.*, 1999; Marten *et al.*, 2000; Haring *et al.*, 2001). CXCL9, CXCL10, CXCL11, CCL2, CCL3 and CCL5 chemokine expression is sustained in the CNS during the adaptive immune response, concomitant with expression of the chemokine receptors: CCR1, CCR5, and CXCR3 (Lane *et al.*, 1998). Prominent CXCL9 and CXCL10 expression throughout the acute infection is consistent with their promotion of neuroinflammation (Liu *et al.*, 2000; Liu *et al.*, 2001; Trifilo *et al.*, 2004). Inhibition of CXCL9 and CXCL10 increases MHV-induced mortality and supports their anti viral role by effector cell recruitment (Liu *et al.*, 2000; Liu *et al.*, 2001). This CNS chemokine pattern is not unique to MHV infection or dependent on inflammation, as microglia and astrocytes synthesize chemokines following infection with both RNA and DNA viruses, even in the absence of inflammatory cells (Asenio and Campbell, 2001; Lane and Buchmeier, 2002).

Interestingly, recruitment of $CD4^+$ and $CD8^+$ T cells during JHMV infection appears to be regulated differently at least at two levels. First, although both $CD4^+$ T and $CD8^+$ T cells cross the BBB, $CD8^+$ T cells enter the parenchyma, while $CD4^+$ T cells accumulate around blood vessels (Stohlman *et al.*, 1998). Differential trafficking through the infected tissue is associated with expression of the MMP inhibitor TIMP-1 by $CD4^+$ T cells but not $CD8^+$ T cells (Zhou *et al.*, 2005). This distinction is most evident early during inflammation, as $CD4^+$ T cells do eventually localize to the parenchyma. Trafficking into the CNS parenchyma is thus not only controlled by proteases that promote migration, but also by selective expression of protease inhibitors stalling migration. Second, although both $CD4^+$ and $CD8^+$ T cells within the CNS express similar CXCR3 levels, migration of $CD4^+$ T cells appears more stringently dependent on CXCR3-ligand interactions compared to $CD8^+$ T cells. This is indicated by a significant reduction of $CD4^+$, but not $CD8^+$ T cells in the CNS following anti CXCR3 Ab blocking (Stiles *et al.*, 2006).

The identity and origin of antigen presenting cells responsible for T cell activation following CNS infections is unknown. Virus-specific T cells are detected in the cervical lymph nodes (CLN) prior to the CNS or spleen (Marten *et al.*, 2003), suggesting initial T cell activation occurs in the CLN. Antigen may drain into the CLN via the cerebrospinal fluid (CSF), which connects the CNS to the lymphatic system (Dorries, 2001; Ransohoff *et al.*, 2003). It remains to be elucidated whether infectious virus or only viral antigen is present in CLN. Recent evidence suggests that both MHV-A59 and -JHM strains can productively infect dendritic cells (DC) *in vitro* (Cervantes-Barragan *et al.*, 2006; Turner *et al.*, 2004; Zhou *et al.*, 2006). MHV-A59 replicates less efficiently in plasmacytoid DCs compared to conventional DCs, coincident with high IFN-α production (Cervantes-Barragan *et al.*, 2006). JHMV preferentially targets mature DC, resulting in suboptimal $CD8^+$ T cell stimulation *in vitro* (Zhou *et al.*, 2006). Although it cannot be excluded that dysregulation of DC may thus modulate T cell activation *in vivo*, JHMV replication in lymphoid organs *in vivo* could not been demonstrated (Zhou *et al.*, 2006). Nevertheless, cells with a DC-like phenotype in the CNS parenchyma and CLN are detected as early as day 2 p.i. (Trifilo *et al.*, 2004). This implies the possibility that antigen presenting cells acquire viral antigens within the CNS and subsequently enter the CLN. CNS presentation may also occur by bone-marrow-derived circulating monocytes recruited into the CNS and subsequently differentiating into macrophages or DC (Greter *et al.*, 2005).

Proinflammatory cytokines

The most rapidly induced proinflammatory cytokines within the MHV infected CNS are IL-1α, IL-1β, IL-6 and IL-12 (Pearce *et al.*, 1994; Parra *et al.*, 1997; Rempel *et al.*, 2004b; Rempel *et al.*, 2005). Similar to chemokines, production of these cytokines is predominant in astrocytes and microglia and is a signature of innate responses to many other viral CNS infections (Biron, 1998; Lane and Buchmeier, 2002; Asenio and Campbell, 2001). IL-6, in conjunction with upregulation of adhesion molecules on CNS endothelium and MMP-9 activity, might enhance inflammatory cell passage across the BBB (Ishihara and Hirano, 2002). TNF-α is also prominently produced in resident CNS cells, albeit slightly delayed (Parra *et al.*, 1997). Although virus-specific T cells can secrete TNF-α, TNF-α mRNA decreases prior to IFN-γ (Parra *et al.*, 1997) supporting microglia and macrophages as major early sources of TNF-α production. Early induction of these cytokines is consistent with their potential to mediate MMP induction (Zhou *et al.*, 2002). IFN-γ production is generally delayed and coincides with T cell inflammation (Parra *et al.*, 1997; Bergmann *et al.*, 2003; Zuo *et al.*, 2006).

Distinct MHV variants are associated with overall similar cytokine patterns, albeit varying expression levels (Li *et al.*, 2004; Rempel *et al.*, 2004a). Efforts to correlate enhanced neurovirulence of lethal JHMV compared to MHV-A59 to specific immune factors revealed no correlation with TNF-α, CCL5 or CCL2 mRNA expression (Rempel *et al.*, 2004a; Rempel *et al.*, 2004b). This is consistent with unaltered JHMV replication *in vivo* or CNS pathology in the absence of TNF-α (Stohlman *et al.*, 1995; Pewe *et al.*, 2002). However, lethal JHMV induces higher CCL3, CCL4 and MIP-2, as well as IL-6, IL-1α, IL-1β, and IFN-β levels compared to MHV-A59, which in turn is characterized by prominent IFN-γ mRNA induction (Rempel *et al.*, 2004a; Rempel *et al.*, 2004b). MHV-JHM/A59 recombinants using a highly virulent JHMV parent are beginning to decipher the role of the S protein and other viral proteins in pathogenesis (Phillips *et al.*, 1999; Rempel *et al.*, 2004a; MacNamara *et al.*, 2005). Surprisingly, IL-6 induction correlates with JHMV background genes, rather than the S protein (Rempel *et al.*, 2004a).

In contrast to the unique chemokine and cytokine induction associated with MHV-JHM and -A59 CNS infection, non-lethal JHMV variants, differing only in S protein, induce no significant differences, except for magnitude (Parra *et al.*, 1997).

Changes associated with viral control

Decreased viral burden coincident with T cell accumulation during sublethal JHMV infection is reflected by altered MMP, chemokine and cytokine patterns. Transcripts for both pro-inflammatory cytokines (IL-1α, IL-1β, IL-6, IL-12, IFN-β) and chemokines CCL2, CCL3, CCL7, CXCL9 and CXCL11, are notably reduced (Parra *et al.*, 1997; Lane *et al.*, 1998). By contrast, the T-cell chemoattractant chemokines CXCL10 and CCL5 remain elevated (Lane *et al.*, 1998) correlating with increased T-cell numbers and IFN-γ expression (Parra *et al.*, 1997; Liu *et al.*, 2000; Glass *et al.*, 2004). Similar to acute infection, blocking of CXCR3-ligand interactions after virus control selectively diminishes CD4 T cells, but ameliorates clinical disease, implicating ongoing CXCL10 expression in T cell mediated pathology (Stiles *et al.*, 2006).

Innate and adaptive anti viral effector mechanisms

Contributions of antiviral type 1 IFN

The failure of SCID, $RAG^{-/-}$, and nude mice to clear infectious virus, despite CNS recruitment of neutrophils, macrophages and NK cells, implies that innate responses are insufficient to control infection (Houtman and Fleming, 1996; Wu and Perlman, 1999; Bergmann *et al.*, 2004). However, as discussed above, induction of innate factors are absolutely necessary as a first line of defense by facilitating inflammation of innate responder cells, which in turn amplify local immunity. In addition, analysis of innate cytokine mRNA levels both *in vitro* using primary neuronal cultures (Li *et al.*, 2004; Rempel *et al.*, 2005), as well as *in vivo* using CNS samples demonstrate induction of IFN-β mRNA (Rempel *et al.*, 2004a). Functional type I IFN is supported by MHC class I upregulation in microglia following sublethal JHMV infection of IFN-γ deficient mice (Bergmann *et al.*, 2003). Gene profiling of JHMV infected neuronal cultures revealed upregulation of transcription factors IRF7, STAT-1 and ISGF3, induction of class I heavy chains, multiple cofactors involved in class I antigen processing and non-classical MHC class I molecules (Rempel *et al.*, 2005). Concomitant IFN-β and TNF-α induction implied virus induced upregulation of select antigen presentation molecules in the absence of inflammation, potentially enhancing both innate and adaptive lymphocyte function within the CNS. Distinct responsiveness of different cell types to MHV infection *in vitro* and *in vivo* is emphasized by potent IFN-α/β upregulation in MHV-A59 infected plasmacytoid DCs, but only low levels in conventional DCs (Cervantes-Barragan *et al.*, 2006). MHV-A59 infection of L cells elicited no detectable induction of IFN-α/β (Versteeg *et al.*, 2006). These results all support a role of IFN-α/β in host defense. Although IFN-β treatment only modestly affects JHMV infection of bone marrow-derived DC *in vitro* (Zhou and Perlman, 2006), the extent to which type I IFNs influence spread of MHV in the CNS has not been studied extensively. Intranasal administration of IFN-α/β prevented spread of intranasally administered JHMV to the CNS, but not dissemination to other organs (Smith *et al.*, 1987). Furthermore, administration of anti-type I IFN Ab enhanced MHV-A59 virulence (Lavi and Wang, 1995) and mice genetically impaired in IFNα/β responsiveness are highly susceptible to viral spread and succumb to MHV-A59 infection (Cervantes-Barragan *et al.*, 2006); protection in the latter model is largely mediated by plasmacytoid DC derived IFN-α/β. Unlike IFN antagonists uncovered in SARS-CoV (Spiegel *et al.*, 2005), MHV viral gene products antagonizing IFN-α/β function have as yet not been identified. The mechanisms of direct antiviral effects of IFN-α/β within the CNS also remain to be established.

Contributions of TNF-α, iNOS and IFN-γ

TNF-α and nitric oxide synthase-2 (iNOS, the inducible NOS isoform) influence immunity to other CNS viral infections (Biron, 1998; Asenio and Campbell, 2001; Lane and Buchmeier, 2002). Although both anti-viral molecules are induced, they appear to have no *in vivo* role in the anti-MHV host response (Lane *et al.*, 1997; Lane *et al.*, 1999, Wu *et al.*, 2000). Neither MHV replication *in vivo* nor CNS pathology are altered in the absence of TNF-α (Stohlman *et al.*, 1995; Pewe and Perlman, 2002). Although TNF-α translation is inhibited in MHV-infected cells (Stohlman *et al.*, 1995), the small fraction of MHV-infected microglia *in vivo* presumably carries little weight relative to TNF-α production by uninfected microglia.

JHMV pathogenesis in mice deficient in individual antiviral effector functions revealed that the absence of IFN-γ results in one of the most pronounced disease phenotypes (Lin *et al.*, 1997; Parra *et al.*, 2000; Parra *et al.*, 2000). IFN-γ has direct anti-viral activity and induces MHC expression on CNS resident cells, facilitating T cell effector function. In the absence of IFN-γ, MHC class I expression is reduced and MHC class II remains undetectable on microglia and most CNS macrophages (Bergman *et al.*, 2003; Bergman *et al.*, 2004). In JHMV infected wild type mice, IFN-γ protein is functionally evident in the inflamed CNS by maximal expression of both MHC class I and II on microglia coincident with peak T cell infiltration (Bergman *et al.*, 2003). Despite peak accumulation of NK cells prior to T cells in immune competent mice, there is little evidence for direct or indirect NK cell dependent IFN-γ or perforin anti-viral function (Bergman *et al.*, 2004). IL-15 deficient mice, which lack NK cells and are impaired in homeostasis of memory CD8 T cells (Kennedy *et al.*, 2000), control sublethal JHMV infection with similar kinetics as wild type mice and exhibit no delays in either class I nor class II expression on microglia (Zuo *et al.*, 2006). Nevertheless, infection with a recombinant MHV expressing high CXCL10 levels leads to extensive accumulation of NK cells in the CNS of immunodeficient $RAG1^{-/-}$ mice (Trifilo *et al.*, 2004). Antiviral function of NK cells can thus be forced under targeted conditions.

T cell specificities and effector functions

Depletion of T cell subsets, adoptive T cell transfers, as well as infection of Ab deficient mice all demonstrate that T cells are critical anti-viral effectors in controlling acute JHMV replication (Lin *et al.*, 1999; Bergmann *et al.*, 2003; Bergmann *et al.*, 2004). While $CD8^+$ T cells exert direct antiviral activity, $CD4^+$ T cells provide crucial accessory function by enhancing virus specific $CD8^+$ T cell expansion and maintaining $CD8^+$ T cell viability within the CNS (Stohlman *et al.*, 1998; Zhou *et al.*, 2005). Parenchymal $CD8^+$ T-cells are dramatically decreased in the absence of $CD4^+$ T cells, potentially due to increased apoptosis (Stohlman *et al.*, 1998). How CD4$^+$ T cells provide help, despite their perivascular location in the CNS remains unclear, although IL-2 as an obvious candidate has been excluded (Zhou *et al.*, 2005).

Several MHV CD8$^+$ and CD4$^+$ T cell epitopes have been mapped and previously reviewed (Perlman, 1998). To date they are all localized within the structural N, M and S proteins. BALB/c mice mount an L^d restricted immunodominant CD8 response to the JHMV N protein (N318), which is crossreactive with many MHV variants (Bergmann *et al.*, 1993). Mutations in this epitope throughout infection are rare (Bergmann *et al.*, 1998). By contrast, C57BL/6 mice respond to two epitopes within the polymorphic JHMV S protein (Perlman, 1998). The immunodominant epitope is D^b restricted (S510) and resides in the hypervariable domain, whereas the K^b restricted subdominant epitope (S598) is more conserved. Class I tetramer technology designed to detect T cell receptors specific for the N318 or S510 epitope revealed that at least 40–50% of CNS CD8 T cells are virus specific during and following resolution of acute infection (Bergmann *et al.*, 1998; Marten *et al.*, 2000; Ramakrishna *et al.*, 2004b).

It is critical to note that the S510 epitope is deleted in several commonly used MHV strains, including the avirulent JHM 2.2/7.2v-2 variant and the dualtropic MHV-A59 strain. Furthermore, infection of C57BL/6 mice with sublethal JHMV viruses containing this epitope can result in progressive accumulation of quasispecies with both deletions as well as point mutations in this epitope (Rowe *et al.*, 1997a). Secondary structure analysis revealed that the deleted domains reside in a stem loop structure, which constitutes a hot spot for RNA recombination (Rowe *et al.*, 1997b). However, the relative contribution of immune pressure or replication strategies to the emergence of S mutants is unclear. Mutations in this S epitope also arise following infection of neonatal mice protected by maternal antibody (Pewe *et al.*, 1996); in this model, mutations are clearly associated with CD8$^+$ T cell pressure and increased infectious virus in the CNS. Nevertheless, such mutations are not as evident in adult mice (Parra *et al.*, 1999; Ramakrishna *et al.*, 2006). A potential for immune escape is also demonstrated in T cell pre-immune mice challenged with an A59 recombinant virus expressing a foreign epitope tagged to enhanced green fluorescent protein from the accessory gene 4 locus (Chua *et al.*, 2004). Taken together, these data suggest that T cell escape variants do not play a prominent role in primary infection of naïve, adult mice, but readily emerge in genome regions not affecting viral fitness, especially under conditions of preexisting antibody or T cell memory.

Virus-specific CD8$^+$ T cells accumulate to 10-fold higher frequencies in the CNS compared with the periphery and express the CD44^{hi}, CD62L$^{-/lo}$, CD11a^{hi}, and CD49d$^+$ (VLA-4) activation/memory phenotypic markers (Bergmann *et al.*, 1999). CD43^{hi}, CD127$^{-/lo}$ expression discriminates virus-specific CD8$^+$ T cells within the CNS from "bystander" T cells, which retain a CD43^{int}, CD127$^+$ phenotype (Chen *et al.*, 2005). CD8$^+$ T cells recruited into the CNS during JHMV infection retain expression of the early activation marker CD69 (Bergmann *et al.*, 1999), similar to other CNS inflammation models (Hawke *et al.*, 1998). Consistent with their activated effector phenotype, virus-specific CD8$^+$ T cells from the acutely inflamed CNS express granzyme B, exert cytolytic effector function, produce IFN-γ and to a lesser degree TNF-α (Bergmann *et al.*, 1999; Ramakrishna *et al.*, 2004; Ramakrishna *et al.*, 2006). Fas/FasL interactions do not appear to play a role in virus clearance or pathology (Parra *et al.*, 2000). The more critical role of IFN-γ, compared to perforin, is evident by diminished virus control and enhanced mortality in the absence of either function (Lin *et al.*, 1997; Parra *et al.*, 1999). The focused study of antiviral mechanisms uncovered a previously unappreciated aspect of viral clearance from the CNS, namely that susceptibility to distinct T cell effector function is target cell type specific. In the absence of IFN-γ, replication in astrocytes and microglia, but not oligodendrocytes is controlled via perforin mediated cytolysis (Parra *et al.*, 1999). By contrast, in the absence of perforin, IFN-γ is capable of affecting replication in oligodendrocytes, but is insufficient for virus elimination in astrocytes and microglia (Lin *et al.*, 1997). Glial cell type dependent susceptibility to T cell mediated antiviral functions was confirmed by adoptive transfer of virus specific memory $CD8^+$ T cells deficient in either cytolytic activity or IFN-γ secretion into infected immunodeficient hosts (Bergmann *et al.*, 2003; Bergmann *et al.*, 2004). Finally, infection of mice with an IFN-γ signaling defect selectively in oligodendroglia supported the importance of IFN-γ signaling in this cell type for controlling oligodendroglial viral clearance (Gonzales *et al.*, 2005; Gonzales *et al.*, 2006). These data implied either differences in antigen presentation by glial cell subsets and/or inherent resistance of oligodendrocytes to cytolysis. Although previously controversial, class I upregulation on oligodendrocytes has recently clearly been shown in two independent models of neurotropic MHV infection (Redwine *et al.*, 2001; Ramakrishna *et al.*, 2006). Even though detection of class I expression on astrocytes has been elusive (Redwine *et al.*, 2001), perforin-mediated control of virus replication in these cells suggests sufficient MHC class I engagement to trigger effector function. More detailed studies are required to assess the possibility of impaired antigen processing or distinct thresholds of oligodendrocytes to initiate and sustain antigen presentation.

T cell effector functions during persistence

After infectious virus is eliminated, viral mRNA and inflammatory cells persist (Bergmann *et al.*, 1999; Marten *et al.*, 2000). The percentages of virus specific T cells within the $CD8^+$ compartment remain remarkably stable throughout infection (Bergmann *et al.*, 1999; Marten *et al.*, 2000; Ramakrishna *et al.*, 2004b) suggesting indiscriminate homeostatic survival. T cell retention within the CNS may be supported by decreased presence of apoptosis inducing factors. This notion is supported by overall higher apoptosis of T cells in lymphoid compared to non-lymphoid tissues (Wang *et al.*, 2003), as well as low potential for TNF-α secretion by $CD8^+$ T cells throughout JHMV infection (Ramakrishna *et al.*, 2004). There is no evidence that T cells exert antiviral function during persistence. Cytolytic activity by virus-specific $CD8^+$ T cells is lost by 2 weeks p.i., concomitant with virus clearance (Bergmann *et al.*, 1999; Marten *et al.*, 2000; Ramakrishna *et al.*, 2004) and IFN-γ mRNA levels decline (Parra *et al.*, 1997). Loss of cytolytic function is independent of demyelination associated factors (Marten *et al.*, 2000) and cannot be recovered during virus recrudescence in B cell deficient mice (Ramakrishna *et al.*, 2002). However, there is no evident impairment of IFN-γ secretion upon Ag re-exposure *in vitro*, suggesting $CD8^+$ T cells are not energized (Bergmann *et al.*, 1999; Ramakrishna *et al.*, 2002). A comparative study of MHV-specific primary versus memory $CD8^+$ T cells in response to infection revealed that CNS derived $CD8^+$ T cells from immunized mice exhibit increased IFN-γ and granzyme B production, as well as enhanced cytolysis at a single cell level compared to naïve mice after challenge (Ramakrishna *et al.*, 2004b). Importantly, reactivated memory $CD8^+$ T cells retained cytolytic function coincident with increased granzyme B levels for prolonged periods compared to primary CD8$^+$ T cells. Loss of virus-specific cytolytic function thus reflects distinct differentiation states of primary compared to memory CD8$^+$ T cells rather than an intrinsic property of the inflamed CNS environment. Cytolytic function was also retained in reactivated memory cells following an otherwise lethal neurotropic influenza virus challenge (Hawke *et al.*, 1998) with no evidence for pathology. Although enhanced effector function by reactivated memory CD8$^+$ T cells directly translated to more effective virus control upon challenge, their persistence at higher numbers and a pre-armed state was insufficient to prevent virus reactivation in the absence of antiviral Ab (Ramakrishna *et al.*, 2006). MHC expression on glial cells at the time of recrudescence was severely reduced compared to the acute response. Memory T cells are thus not responsive *in vivo* in an environment presenting few if any target MHC molecules, despite being fully armed to exert anti-viral activity upon Ag exposure *in vitro*. Vaccine induced T cell memory alone may thus be unable to control persisting virus in a tissue with strict IFN-dependent MHC regulation, as evident in immune privileged sites.

Although persisting T cells in the CNS following control of viral replication may merely constitute bystanders, several observations indicate active maintenance. Following infection with a neurotropic MHV not associated with persistence or myelin loss T cells ultimately drop to background levels found in the naïve CNS (Marten *et al.*, 2000). Persistence is associated with selection of CD8$^+$ T cell populations with limited T cell receptor specificities compared to the acute infection (Marten *et al.*, 1999). The contribution of local homeostatic proliferation or ongoing recruitment to T cell maintenance in the CNS remains unclear. Memory cells traffic poorly into the quiescent CNS (Masopust *et al.*, 2004) and activated T cells are only retained within the CNS upon cognate antigen recognition (Hickey, 2001; Chen *et al.*, 2005). Preliminary evidence suggests that IL-15, which regulates homeostasis of CD8 memory cells in lymphoid organs (Masopust and Ahmed, 2004), is not required (Zuo and Bergmann, unpublished). Furthermore, blockade of CXCR3 signaling early during persistence selectively reduces CD4$^+$ T cell but not CD8$^+$ T cells (Stiles *et al.*, 2006). These observations suggest that neither peripheral chemokine-mediated recruitment nor local division contribute significantly to CNS CD8$^+$ T cell maintenance.

Humoral immunity

Ab mediated protection against acute MHV CNS infection has been reviewed previously (Houtman and Fleming; 1996; Stohlman *et al.*, 1999). The following discussion focuses on the role of Ab during MHV persistence. Mice devoid of antiviral Ab secretion mount a normal inflammatory response that controls sublethal infectious JHMV with kinetics similar to immune competent mice. However, infectious virus re-emergence within the CNS is associated with increased mortality (Lin *et al.*, 1997; Ramakrishna *et al.*, 2002). This dependence on Ab may be unique to control CNS infection, as infection of Ab impaired mice with the dualtropic MHV A59 strain, fails to reactivate in the liver (Matthews *et al.*, 2001). Transfer of polyclonal Ab at the time infectious virus is initially cleared in sublethal JHMV infection can prevent recrudescence in the CNS of Ab deficient mice, albeit only transiently (Ramakrishna *et al.*, 2003). Dissection of the specificities and mechanisms of Ab mediated protection revealed that only S protein-specific neutralizing

IgG prevented recrudescence. Non-neutralizing Ab specific for S, matrix or nucleocapsid proteins had no anti-viral effect (Ramakrishna *et al.*, 2003), distinct from the apparent protective role for non-neutralizing Ab prior to infection (Stohlman *et al.*, 1999). Loss of antiviral control as transferred Ab decayed was the first evidence suggesting that virus is maintained in a replication competent form within the CNS and that a renewable source of Ab is required to maintain control. Detection of virus-specific Ab secreting cells (ASC) in the CNS following viral clearance confirmed the implication of sustained intrathecal Ab (Tschen *et al.*, 2003; Tschen *et al.*, 2006).

Although virus-specific ASC are barely detectable in the CNS during the virus clearance phase, they accumulate prominently in the CNS approximately one week after clearance of infectious virus (Tschen *et al.*, 2003). The delayed peak in virus-specific IgG ASC in the CNS compared to cervical lymph nodes and spleen suggests maturation in secondary lymphoid organs precedes migration into the CNS. Consistent with the meager presence of virus-specific ASC in either the CNS or lymphoid organs during acute infection, virus specific serum Ab, including neutralizing Ab, is barely detectable and limited to IgM prior to complete elimination of infectious virus (Tschen *et al.*, 2003). Neutralizing serum Ab reach a plateau delayed compared to total virus specific Ig, but levels are sustained following clearance of infectious virus, supporting differentiation to long lived plasma cells to maintain serum levels. Following many peripheral infections, plasmablasts preferentially traffic to bone marrow, where they differentiate into long lived, sessile plasma cells (Slifka and Ahmed, 1998). However, following JHMV infection, virus-specific ASC are retained in the CNS at high frequencies for at least 3 months p.i. with a paucity of bone marrow accumulation (Tschen *et al.*, 2006). Although virus-specific ASC progressively decline, they are maintained at higher numbers than virus-specific T cells (Bergmann *et al.*, 2006). The recruitment signals as well as survival factors for virus specific ASC in the CNS are just beginning to be explored, although intrathecal Ab synthesis is common during persisting infections (Griffin *et al.*, 2003). The CNS as a survival niche for ASC has received renewed attention by the discovery of ectopic follicle formation and de novo ASC differentiation during chronic inflammatory conditions (Uccelli *et al.*, 2005). Such structures have not been reported following MHV infections. However, ongoing local B cell differentiation within the brains and spinal cords is supported by a progressive decline of immature class II^+ $CD138^+$ $CD19^+$ cells in favor of differentiated class II^- $CD138^+$ plasma cells (Tschen *et al.*, 2006). Maintenance of virus induced CXCL9 and CXCL10 mRNA within the CNS following viral clearance implicates CXCR3 ligands as possible mediators of ASC recruitment. Lastly, upregulation of mRNA encoding the plasma cell survival factor BAFF precedes ASC accumulation and remains detectable throughout ASC persistence (Tschen *et al.*, 2006). These data suggest that virus induced CNS inflammation maintains an environment that sustains local protective ASC, in addition to sustained serum Ab.

Bystander recruitment and pathology

A unifying mechanism defining the roles of individual effectors in the demyelinating process has not emerged (Buchmeier and Lane, 1999; Stohlman and Hinton, 2001; Ercolini and Miller, 2006). T cells are prominent mediators of demyelination despite their protective role in controlling acute MHV replication (Stohlman *et al.*, 1999; Buchmeier and Lane, 1999, Lin *et al.*, 2001; Wu *et al.*, 2000). Immune compromised SCID and

RAG1$^{-/-}$ mice develop little if any demyelination, despite uncontrolled virus replication (Houtman and Fleming, 1996; Wu *et al.*, 2000b). Mice deficient for IFN-γ, perforin, inducible nitric oxide synthase, CD4$^+$, CD8$^+$ or B cells all develop demyelination (Lin *et al.*, 1997; Parra *et al.*, 1999; Stohlman *et al.*, 1999; Wu *et al.*, 2000a; Wu *et al.*, 2000b). Similarly, immunodeficient recipients of CD4$^+$ T cells, CD8$^+$ T cells, or γδ T cells are all susceptible to immunopathology (Wu *et al.*, 2000b; Bergmann *et al.*, 2003; Pewe *et al.*, 2002; Pewe and Perlman 2002, Dandekar *et al.*, 2005). Nevertheless, the distinct abilities of transferred populations to control virus preempt conclusive interpretation of the role of viral Ag load versus immune function. Similarly, effects on macrophages recruitment and microglial activation, both of which appear to be at least one requirement for myelin loss, are just beginning to be characterized (Kim and Perlman, 2005). The common theme emerging is that recruitment of most immune cell populations as a single entity, or over expression of specific chemokines can lead to demyelination. Oligodendroglia, a prominent target of JHMV infection, are the CNS cells that synthesize and maintain the myelin sheath surrounding neuronal axons. Infection of oligodendroglia appears to be one requirement for demyelination (Houtman and Fleming, 1996). Although reducing the ability of oligodendroglia to respond to IFN-γ dramatically increases the number of infected oligodendroglia, no associated increase in myelin loss was detected (Gonzales *et al.*, 2006). Therefore, although infection of oligodendroglia is required for demyelination, it appears that additional factors must contribute to or regulate the extent of myelin loss. A separate issue still unresolved is the role of bystander cells. Bystander cells of irrelevant specificity are commonly recruited by chemokines. Their activation within the CNS environment following microbial infection may influence subsequent CNS inflammation and/or enhance pathogenesis. Enhanced CNS pathology via activation of bystander CD8$^+$ T cells has recently been demonstrated following JHMV infection of LCMV specific TCR/Rag2$^{-/-}$ mice (Haring *et al.*, 2002). By contrast, analysis of JHMV induced heterologous memory T cells recruited in mice with a wild type TCR repertoire, demonstrated they are neither activated nor acquire expression of *ex vivo* cytolytic activity in the process (Chen *et al.*, 2005). These results are consistent with the requirement for a cognate or cross reactive antigen driven component in contributing to CD8$^+$ T cell mediated CNS damage in other models (McGavern, 2005).

Conclusions

The study of MHV-induced CNS infection has revealed several novel concepts in neuro inflammation and local antiviral immunity. Resident CNS cells send potent alarm signals to recruit innate and adaptive immune mediators in an interdependent fashion. Innate immune cells pave the way for effector T cells, but regress as T cells accumulate. Migration of CD4$^+$ and CD8$^+$ T cell subsets is differentially regulated at both the chemokine and extracellular matrix level. Whereas CD8$^+$ T cells exert antiviral function at the site of infection, CD4$^+$ T cells provide accessory function. Although T cell effector functions are readily expressed *in vitro*, the *in vivo* execution is dependent on MHC expression, which in turn appears regulated tightly by IFNγ. Sustained antibody secretion within the CNS is thus absolutely critical to prevent virus recrudescence from persistently infected cells. This function is aided by sustained expression of chemokines and B cell survival factors, which preferentially maintain virus-specific ASC in the CNS.

Acknowledgments

This research was supported by the National Institutes of Health Grants NS18146, NS 40667 and AI47249.

References

Adami, C., Pooley, J., Glomb, J., Stecker, E., Fazal, F., Fleming, J.O., and Baker, S.C. (1995). Evolution of mouse hepatitis virus (MHV) during chronic infection: quasispecies nature of the persisting MHV RNA. Virology *209*, 337–346.

Aloisi, F., Ria, F., and Adorini, L. (2000). Regulation of T-cell responses by CNS antigen-presenting cells: different roles for microglia and astrocytes. Immunol. Today *21*, 141–147.

Asensio, V.C., and Campbell, I.L. (2001). Chemokines and viral diseases of the central nervous system. Adv. Virus Res. *56*, 127–173.

Bergmann, C., McMillan, M., and Stohlman, S. (1993). Characterization of the Ld restricted CTL epitope in the mouse hepatitis virus nucleocapsid protein. J. Virol. *67*, 7041–7049.

Bergmann, C.C., Dimacali, E., Stohl, S., Wei, W., Lai, M.C.C., Tahara, S., and Marten, N. (1998). Variability of persisting MHV RNA sequences constituting immune and replication relevant domains. Virology *244*, 563–572.

Bergmann, C.C., Altman, J., Hinton, D., and Stohlman, S.A. (1999). Inverted immunodominance and Impaired cytolytic function of CD8+ T cells during viral persistence in the CNS. J. Immunol. *163*, 3379–3387.

Bergmann, C., Parra, B., Hinton, D., Ramakrishna. C., Morrison, M., and Stohlman, S. (2003). Perforin mediated effector function within the CNS requires IFN-γ mediated MHC upregulation. J. Immunol. *170*, 3204–3213.

Bergmann, C., Parra, B., Hinton, D.R., Ramakrishna, C., Dowdell, K.C., and Stohlman, S. (2004). Perforin and Interferon Gamma Mediated Control of Coronavirus Central Nervous System Infection by CD8 T Cells in the Absence of CD4 T Cells. J. Virol. *78*, 1739–1750.

Bergmann, C.C., Lane, T.E., and Stohlman, S.A. (2006). Coronavirus infection of the CNS: Host-virus standoff. Nature Reviews Microbiol. *4*:121–132.

Biron, C.A. (1998). Role of early cytokines, including alpha and beta interferons (IFN-alpha/beta), in innate and adaptive immune responses to viral infections. Semin. Immunol. *10*, 383–390.

Buchmeier, M.J., and Lane, T. (1999). Viral-induced neurodegenerative disease. Current Opin. Microbiol. 2, 398–402.

Cabarrocas, J., Bauer, J., Piaggio, E., Liblau, R., and Lassman, H. (2003). Effective and selective immune surveillance of the brain by MHC class I-restricted cytotoxic T lymphocytes. Eur. J. Immunol. *33*, 1174–1182.

Cervantes-Barragan, L., Zust, R., Weber F., Spiegel, M., Lang, K.S., Akira, S., Thiel, V., and Ludewig, B. (2006). Control of coronavirus infection through plasmacytoid dendritic cell-derived type I interferon. Blood. Sep 19; [Epub ahead of print]

Chen, A.M., Khanna, N., Stohlman, S.A., and Bergmann, C.C. (2005). Virus-specific and bystander CD8 T cells recruited during virus-induced encephalomyelitis. J. Virol. *79*, 4700–4708.

Christensen, J.E., Nansen, A., Moos, T., Lu, B., Gerard, C., Christensen J.P., and Thomsen, A.R. (2004). Efficient T-cell surveillance of the CNS requires expression of the CXC chemokine receptor 3. J. Neurosci. *24*, 4849–4858.

Chua, M., MacNamara, K., Mateo, L., Shen, H., and Weiss, S. (2004). Effects of an epitope specific CD8+ T cell response on murine central nervous system disease: Protection from virus replication and antigen spread and selection of epitope escape mutants. J. Virol. *78*, 1150–1159.

Dandekar, A.A., O'Malley, K., and Perlman, S. (2005). Important roles for gamma interferon and NKG2D in gammadelta T-cell-induced demyelination in T-cell receptor beta-deficient mice infected with a coronavirus. J. Virol. *79*, 9388–9396.

De Lemos, C., Christensen, J.E., Nansen, A., Moos, T., Lu, B., Gerard, C., Christensen, J.P., and Thomsen, A,R. (2005). Opposing effects of CXCR3 and CCR5 deficiency on CD8+ T cell-mediated inflammation in the central nervous system of virus-infected mice. J. Immunol. *175*, 1767–1775.

Dorries, R. (2001). The role of T-cell-mediated mechanisms in virus infections of the nervous system. Curr. Top. Microbiol. Immunol. *253*, 219–245.

Ercolini, A.M., and Miller, S.D. (2006). Mechanisms of immunopathology in murine models of central nervous system demyelinating disease. J. Immunol. *176*, 3293–3298.

Fabry, Z., Raine, C.S., and Hart, M.N. (1994). Nervous tissue as an immune compartment: the dialect of the immune response in the CNS. Immunol. Today *15*, 218–224.

Fazakerley, J., and Walker, R. (2003). Virus demyelination. J. Neurovirol. *9*, 148–164.

Fleming, J.O., Trousdale, M., El-Zaatari, F., Stohlman, S.A., and Weiner, L.P. (1986). Pathogenicity of antigenic variants of murine coronavirus JHM selected with monoclonal antibodies. J. Virol. *58*, 869–875.

Fu, L., Gonzales, D.M., Sarma, J.D., and Lavi, E. (2004). A combination of mutations in the S1 part of the spike glycoprotein gene of coronavirus MHV-A59 abolishes demyelination. J. Neurovirol. *10*, 41–51.

Glass, W.G., Hickey, M.J., Hardison, J.L., Liu, M.T., Manning, J.E., and Lane, T.E. (2004). Antibody targeting of the CC chemokine ligand 5 (CCL5) results in diminished leukocyte infiltration into the central nervous system and reduced neurologic disease in a viral model of multiple sclerosis. J. Immunol. *172*, 4018–4025.

Gonzalez, J.M., Bergmann, C.C., Fuss, B., Hinton, D.R., Macklin, W.B., and Stohlman, S.A. (2005). Expression of a dominant negative IFN-γ receptor on mouse oligodendrocytes. Glia *51*, 22–34.

Gonzalez, J. M., Bergmann, C.C., Ramakrishna, C., Hoskin, J., Macklin, W.B., Stohlman, S.A. (2006). Inhibition of IFN-γ signaling in oligodendroglia delays coronavirus clearance without altering demyelination. J. Am. Pathol. *168*, 796–804.

Greter, M., Heppner, F.L., Lemos, M.P., Odermatt, B.M., Goebels, N., Laufer, T., Noelle, R.J., and Becher, B. (2005). Dendritic cells permit immune invasion of the CNS in an animal model of multiple sclerosis. Nat. Med. *11*, 328–334.

Griffin, D.E. (2003). Immune responses to RNA-virus infections of the CNS. Nat. Rev. Immunol. *3*, 493–502.

Gu, J., Gong, E., Zhang, B., Zheng, J., Gao, Z., Zhong, Y., Zou, W., Zhan, J., Wang, S., Xe, Z., Zhuang, H., Wu, B., Zhong, H., Shao, H., Fang, W. Gao, D., Pei, F., Li, X., Xu, D., Shi, X., Anderson, V., Leong, A. (2005). Multiple organ infection and the pathogenesis of SARS. J. Exp. Med. *202*, 415–424.

Haring, J.S., Pewe, L.L., and Perlman, S. (2001). High-magnitude, virus-specific CD4 T-cell response in the central nervous system of coronavirus-infected mice. J. Virol. *75*, 3043–3047.

Haring, J.S., Pewe, L.L., and Perlman, S. (2002). Bystander CD8+ T cell-mediated demyelination after viral infection of the central nervous system. J. Immunol. *169*, 1550–1555.

Hauser, A.E., Debes, G.F., Arce, S., Cassese, G., Hamann, A., Radbruch, A., and Manz, R.A. (2002). Chemotactic responsiveness toward ligands for CXCR3 and CXCR4 is regulated on plasma blasts during the time course of a memory immune response. J. Immunol. *169*, 1277–1282.

Hawke, S., Stevenson, P.G., Freeman, S., and Bangham, C.R.M. (1998). Long term persistence of activated cytotoxic T lymphocytes after viral infection of the central nervous system. J. Exp. Med. *187*, 1575–1582.

Hickey, W.F. (2001). Basic principles of immunological surveillance of the normal central nervous system. Glia *36*, 118–124.

Houtman, J.J., and Fleming, J.O. (1996a). Dissociation of demyelination and viral clearance in congenitally immunodeficient mice infected with murine coronavirus JHM. J. Neurovirol. *2*, 101–110.

Houtman, J.J., and Fleming, J.O. (1996b). Pathogenesis of mouse hepatitis virus-induced demyelination. J. Neurovirol. *2*, 361–376.

Ishihara, K., and Hirano, T. (2002). IL-6 in autoimmune disease and chronic inflammatory proliferative disease. Cytokine Growth Factor Rev. *13*, 357–368.

Kennedy, M.K., Glaccum, M., Brown, S.N., Butz, E.A., Viney, J.L., Embers, M., Matsuki, N., Charrier, K., Sedger, L., Willis, C.R., Brasel, K., Morrissey, P.J., Stocking, K., Schuh, J.C., Joyce, S., and Peschon, J.J. 2000. Reversible defects in natural killer and memory CD8 T cell lineages in interleukin 15-deficient mice. J. Exp. Med. *191*, 771–780.

Kim, T.S., and Perlman, S. (2005). Viral expression of CCL2 is sufficient to induce demyelination in RAG1−/− mice infected with a neurotropic coronavirus. J. Virol. *79*, 7113–7120.

Lai, M., and Cavanagh, D. (1997). The molecular biology of coronaviruses. Adv. Virus. Res. *48*, 1–100.

Lane, T.E., Paoletti, A.D., and Buchmeier, M.J. (1997). Disassociation between the in vitro and in vivo effects of nitric oxide on a neurotropic murine coronavirus. J. Virol. *71*, 2202–2210.

Lane, T.E., Asensio, V.C., Yu, N., Paoletti, A.D., Campbell, I.L., and Buchmeier, M.J. (1998). Dynamic regulation of alpha- and beta-chemokine expression in the central nervous system during mouse hepatitis virus-induced demyelinating disease. J. Immunol. *160*, 970–978.

Lane, T.E., Fox, H.S., and Buchmeier, M.J. (1999). Inhibition of nitric oxide synthase-2 reduces the severity of mouse hepatitis virus-induced demyelination: implications for NOS2/NO regulation of chemokine expression and inflammation. J. Neurovirol. *5*, 48–54.

Lane, T.E., Liu, M.T., Chen, B.P., Asensio, V.C., Samawi, R.M., Paoletti, A.D., Campbell, I.L., Kunkel, S.L., Fox, H.S., and Buchmeier, M.J. (2000). A central role for CD4(+) T cells and RANTES in virus-induced central nervous system inflammation and demyelination. J. Virol. *74*, 1415–1424.

Lane, T.E., and Buchmeier M.J. (2002). Chemokine responses in virus-induced neurologic disease: Balancing host defense and neuropathology. In Universe in delicate balance: Chemokines and the nervous system. Ed. R. Ransohoff, Elsevier Press. pp. 191–202.

Lavi, E., and Wang, Q. (1995). The protective role of cytotoxic T cells and interferon against coronavirus invasion of the brain. Adv. Exp. Med. Biol. *380*, 145–149.

Li, Y., Fu L, Gonzales, D.M., and Lavi, E. (2004). Coronavirus neurovirulence correlates with the ability of the virus to induce proinflammatory cytokine signals from astrocytes and microglia. J Virol. *78*, 3398–3406.

Lin, M., Stohlman, S., and Hinton, D. (1997). Mouse hepatitis virus is cleared from the central nervous system of mice lacking perforin-mediated cytolysis. J. Virol. *71*, 383–391.

Lin, M.T., Hinton, D.R., Marten, N.W., Bergmann, C.C., and Stohlman, S.A. (1999). Antibody prevents virus reactivation within the central nervous system. J. Immunol. *162*, 7358–7368.

Lipton, H., and Gilden, D. (1997). Viral Diseases of the Central Nervous System: Persistent Infections. In Viral Pathogenesis, N. Nathanson, ed. (Philadelphia, PA: Lippincott-Rave), pp. 855–870.

Liu, M.T., Chen, B.P., Oertel, P, Buchmeier, M.J., Armstrong, D., Hamilton, T.A and Lane, T.E. (2000). Cutting edge: The T cell chemoattractant IFN-inducible protein 10 (IP-10) is essential in host defense against viral-induced neurologic disease. J. Immunol. *165*, 2327–2330.

Liu, M.T., Armstrong, D., Hamilton, D.A., and Lane, T.E. (2001). Expressing of Mig (monokine induced by interferon gamma) is important in T lymphocyte recruitment and host defense following viral infection of the central nervous system. J. Immunol. *166*, 1790–1795.

MacNamara, K., Chua, M., Phillips, J., and Weiss, S. (2005). Contributions of viral genetic background and a single amino acid substitution in an immunodominant CD8+ T cell epitope to murine coronavirus neurovirulence. J. Virol. *79*, 9108–9118.

Marsden, P.A., Ning, Q., Fung, L.S., Luo, X., Chen, Y., Mendicino, M., Ghanekar, A., Scott, J.A., Miller, T., Chan, C.W., Chan, M.W., He, W., Gorczynski, R.M., Grant, D.R., Clark, D.A., Phillips, M.J., and Levy. G.A. (2003). The Fgl2/fibroleukin prothrombinase contributes to immunologically mediated thrombosis in experimental and human viral hepatitis. J. Clin. Invest. *112*, 58–66.

Marten, N., Stohlman, S., Smith-Begolka, W., Miller, S., Dimicali, M., Yao, Q., Stohl, S., Goverman, J., and Bergmann, C. (1999). Selection of CD8+ T cells with highly focused specificity during viral persistence in the central nervous system. J. Immunol. *162*, 3905–3914.

Marten, N.W., Stohlman, S.A., Atkinson, R., Hinton, D.A., and Bergmann, C.C. (2000a). Contributions of CD8+ T cells and viral spread to demyelinating disease. J. Immunol. *164*, 4080–4088.

Marten, N., Stohlman, S., and Bergmann, C. (2000b). Role of Viral Persistence in Retaining CD8+ T cells within the central nervous system. J. Virol. *74*, 7903–7910.

Marten, N., Stohlman, S., Zhou, Z., and Bergmann, C. (2003). Kinetics of virus specific CD8+ T cell expansion and trafficking following central nervous system infection. J. Virol. *77*, 2775–2778.

Masopust, D., Vezys, V,. Usherwood, E.J., Cauley, L.S., Olson, S., Marzo, A.L., Ward, R.L., Woodland, D.L., and Lefrancois, L. (2004). Activated primary and memory CD8 T cells migrate to nonlymphoid tissues regardless of site of activation or tissue of origin. J. Immunol. *172*, 4875–82.

Masopust, D., and Ahmed R. (2004). Reflections on CD8 T-cell activation and memory. Immunol. Res. *29*, 151–160.

Matthews, A.E., Weiss, S.R., Shlomchik, M.J., Hannum, L.G., Gombold, J.L., and Paterson, Y. (2001). Antibody is required for clearance of infectious murine hepatitis virus A59 from the central nervous system, but not the liver. J. Immunol. *167*, 5254–5263.

McGavern, D.B. (2005). The role of bystander T cells in CNS pathology and pathogen clearance. Crit. Rev. Immunol. *25*, 289–303.

Navas, S., and Weiss, S.R. (2003). Murine coronavirus-induced hepatitis:JHM genetic background eliminates A59 spike-determined hepatotropism. J. Virol. *77*, 4972–4978.

Ontiveros, E., Kim, T.S., Gallagher, T.M., and Perlman, S. (2003). Enhanced virulence mediated by the murine coronavirus, mouse hepatitis strain JHM, is associated with a glycine at residue 310 of the spike glycoprotein. J. Virol. *77*, 10260–10269.

Parker, S.E., Gallagher, T.M., and Buchmeier, M.J. (1989). Sequence analysis reveals extensive polymorphism and evidence of deletions within the E2 glycoprotein gene of several strains of murine hepatitis virus. Virol. *173*, 664–673.

Parra, B., Hinton, D.R., Lin, M.T., Cua, D.J., and Stohlman, S.A. (1997). Kinetics of cytokine mRNA expression in the CNS following lethal and sublethal coronavirus-induced encephalomyelitis. Virology *233*, 260–270.

Parra, B., Hinton, D.R. Marten, N., Bergmann, C.C., Lin, M. Yang, C., and Stohlman, S.A. (1999). Gamma interferon is required for viral clearance from central nervous system oligodendroglia. J. Immunol. *162*, 1641–1647.

Parra, B., Lin, M., Stohlman, S., Bergmann, C., Atkinson, R., and Hinton, D. (2000). Contributions of Fas-Fas ligand interactions to the pathogenesis of mouse hepatitis virus in the central nervous system. J. Virol. *74*, 2447–2450.

Pearce, B.D., Hobbs, M.V., McGraw, T.S., and M.J. Buchmeier. (1994). Cytokine induction during T-cell-mediated clearance of mouse hepatitis virus from neurons in vivo. J. Virol. *68*, 5483–5495.

Perlman, S. (1998). Pathogenesis of Coronavirus-induced infections: Review of Pathological and Immunological Aspects. In Coronaviruses and Arteriviruses. L. Enjuanes, S. Siddell, and W. Spaan, eds. (N.Y., and London: Plenum Press), Adv. Exp. Med., and Biology 440, pp. 503–513.

Pewe, L., Wu, G.F., Barnett, E.M., Castro, R.F., and Perlman, S. (1996). Cytotoxic T cell-resistant variants are selected in a virus-induced demyelinating disease. Immunity *5*, 253–262.

Pewe, L., and Perlman, S. (2002). Cutting edge: CD8 T cell-mediated demyelination is IFN-gamma dependent in mice infected with a neurotropic coronavirus. J. Immunol. *168*, 1547–1551.

Pewe, L., Haring, J., and Perlman, S. (2002). CD4 T-cell-mediated demyelination is increased in the absence of gamma interferon in mice infected with mouse hepatitis virus. J. Virol. *76*, 7329–7333.

Phillips, J.J., Chua, M.M., Lavi, E., and Weiss, S.R. (1999). Pathogenesis of chimeric MHV4/MHV-A59 recombinant viruses: the murine coronavirus spike protein is a major determinant of neurovirulence. J. Virol. *73*, 7752–7760.

Rall, G., and Oldstone, M. (1997). Viral Persistence in the Central Nervous System. In Defense of the Brain: Current concepts in the immunopathogenesis and clinical aspects of CNS infection. (Malden, MA: Blackwell Science), pp. 273–289.

Ramakrishna, C., Stohlman, S.A., Atkinson, R., Schlomchik, M., and Bergmann, C.C. (2002). Mechanisms of central nervous system viral persistence: Critical role of antibody and B cells. J. Immunol. *168*, 1204–1211.

Ramakrishna, C., Bergmann, C., Holmes, K.V., and Stohlman S. (2004a). Expression of the mouse hepatitis virus receptor by central nervous system microglia. J. Virol. *78*, 7828–7832.

Ramakrishna, C., Stohlman, S., Atkinson, R., Hinton, D.H., and Bergmann, C.C. (2004b). Differential regulation of primary and secondary CD8+ T cells in the CNS. J. Immunol. *173*, 6265–6273.

Ramakrishna, C., R.A. Atkinson, S.A. Stohlman, and Bergmann C.C. (2006). Vaccine Induced Memory CD8+ T Cells Cannot Prevent CNS Viral Persistence and Reactivation. J. Immunol. *176*, 3062–3069.

Ransohoff, R.M., Kivisakk, P., and Kidd, G. (2003). Three or more routes for leukocyte migration into the central nervous system. Nat. Rev. Immunol. *3*, 569–581.

Redwine, J.M., Buchmeier, M.J., and Evans, C.F. (2001). In vivo expression of major histocompatibility complex molecules on oligodendrocytes and neurons during viral infection. Amer. J. Pathol. *159*, 1219–1224.

Rempel, J.D., Murray, S.J., Meisner, J., and Buchmeier M.J. (2004a). Mouse hepatitis virus neurovirulence: evidence of a linkage between S glycoprotein expression and immunopathology. Virology *318*, 45–54.

Rempel, J.D., Murray, S.J., Meisner, J., and Buchmeier, M.J. (2004b). Differential regulation of innate and adaptive immune responses in viral encephalitis. Virology *318*, 381–392.

Rempel, J.D., Quina, L.A., Blakelu-Gonzales, P.K., Buchmeier, M.J., and Gruol, D.L. (2005). Viral induction of central nervous system innate immune responses. J. Virol. *79*, 4369–4381.

Rowe, C.L., Baker, S.C., Nathan, M.J., and Fleming, J.O. (1997a). Evolution of mouse hepatitis virus: detection and characterization of spike deletion variants during persistent infection. J Virol. *71*, 2959–2969.

Rowe, C.L., Fleming, J.O., Nathan, M.J., Sgro, J.-Y., Palmenberg, A.C., and Baker, S.C. (1997b). Generation of Coronavirus spike deletion variants by high frequency recombination at regions of predicted RNA secondary structure. J. Virol. *71*, 6183–6190.

Slifka, M.K., and Ahmed, R. (1998). Long-lived plasma cells: a mechanism for maintaining persistent antibody production. Curr. Opin. Immunol. *10*, 252–258.

Smith, A.L,, Barthold, S.W., and Beck D.S. (1987). Intranasally administered alpha/beta interferon prevents extension of mouse hepatitis virus, strain JHM, into the brains of BALB/cByJ mice. Antiviral Res. *8*, 239–245.

Spiegel, M., Pichlmair, A., Martinez-Sobrido, L., Cros, J., Garcia-Sastre, A., Haller, O., and Weber, F. (2005). Inhibition of Beta interferon induction by severe acute respiratory syndrome coronavirus suggests a two-step model for activation of interferon regulatory factor 3. J. Virol. *79*, 2079–2086.

Stiles, L.N., Hosking, M.P., Edwards, R.A. Strieter, R.M., Lane, T.E. (2006). Differential roles for CXCR3 in CD4(+) and CD8(+) T cell trafficking following viral infection of the CNS. Eur. J. Immunol. *36*, 613–622.

Stohlman, S.A., Fleming, J.O., Brayton, P.R., Weiner, L.P., and Lai, M.M.C. (1982). Murine coronaviruses: Isolation and characterization of two plaque morphology variants of the JHM neurotropic strain. J. Gen. Virol. *63*, 265–275.

Stohlman, S.A., Hinton, D.R., Cua, D., Dimacali, E., Sensintaffar, J., Tahara, S., Hofman, F., and Yao, Q. (1995). Tumor necrosis factor expression during mouse hepatitis virus induced demyelination. J. Virol. *69*, 5898–5903.

Stohlman, S.A., Bergmann, C.C., Lin, M.T., Cua, D.J., and Hinton, D.R. (1998). CTL effector function within the CNS requires CD4+ T cells. J. Immunol. *160*, 2896–2904.

Stohlman, S., Bergmann, C., and Perlman, S. (1999). Mouse Hepatitis Virus. In Persistent Viral Infection R. Ahmed and I. Chen, eds. (N.Y.: John Wiley), pp. 537–558.

Stohlman, S.A., and Hinton, D.R. (2001). Viral induced demyelination. Brain Pathol. *11*, 92–106.

Trifilo, M.J., Bergmann, C.C., Kuziel, W.A., and Lane, T.E. (2003). CC chemokine ligand 3 (CCL3) regulates CD8+-T-cell effector function and migration following viral infection. J. Virol. *77*, 4004–4014.

Trifilo, M.J., Montalto-Morrison, C., Stiles, L.N., Hurst, K.R., Hardison, J.L., Manning, J.E. Masters, P.S., and Lane, T.E. (2004). CXC chemokine ligand 10 controls viral infection in the central nervous system: Evidence for a role in innate immune response through recruitment and activation of natural killer cells. J. Virol. *78*, 585–594.

Trifilo, J.J., and Lane, T.E. (2004). The CC chemokine ligand 3 regulates CD11c+CD11b+CD8alpha-dendritic cell maturation and activation following viral infection of the central nervous system: implications or a role in T cell activation. Virology *327*, 8–15.

Tschen, S.I., Bergmann, C., Ramakrishna, C., Atkinson, R., and Stohlman, S. (2002). Recruitment kinetics of antibody secreting cells within the CNS following viral encephalomyelitis. J. Immunol. *168*, 2922–2929.

Tschen, S.-I, Ramakrishna, C. Atkinson, R. Hinton, D.R. Stohlman, S.A. Bergmann, C.C. (2006). Virus CNS infection diverts homing of antibody secreting cells from lymphoid organs to the CNS. Eur. J. Immunol. *36*, 603–612.

Turner, B.C., Hemmila, E.M., Beauchemin, N., and Holmes, K.V. (2004). Receptor-dependent coronavirus infection of dendritic cells. J. Virol. *78*, 5486–5490.

Uccelli, A., Aloisi, F., and Pistoia, V. (2005). Unveiling the enigma of the CNS as a B-cell fostering environment. Trends Immunol. *26*, 254–259.

Versteeg, G.A., Slobodskaya, O., and Spaan, W.J. (2006). Transcriptional profiling of acute cytopathic murine hepatitis virus infection in fibroblast-like cells.

J Gen Virol. *87*, 1961–75.

Wang, X.Z., Stepp, S.E., Brehm, M.A., Chen, H.D., Selin, L.K., and Welsh, R.M. (2003). Virus-specific CD8 T cells in peripheral tissues are more resistant to apoptosis than those in lymphoid organs. Immunity *18*, 631–42.

Wu, G.F., Perlman S. (1999). Macrophage infiltration, but not apoptosis, is correlated with immune-mediated demyelination following murine infection with a neurotropic coronavirus. J. Virol. *73*, 8771–8780.

Wu, G.F., Pewe, L., and Perlman, S. (2000a). Coronavirus-induced demyelination occurs in the absence of inducible nitric oxide synthase. J. Virol. *74*, 7683–7686.

Wu, G.F., Dandekar, A.A., Pewe, L., and Perlman, S. (2000b). CD4 and CD8 T cells have redundant but not identical roles in virus-induced demyelination. J. Immunol. *165*, 2278–2286.

Yong, V.W., Power, C., Forsyth, P., and Edwards, D.R. (2001). Metalloproteinases in biology and pathology of the nervous system. Nat. Rev. Neurosci. *2*, 502–511.

Zhou, J., Stohlman, S., Atkinson, R., Hinton, D., and Marten, N. (2002). Matrix metalloproteinase expression correlates with virulence following neurotropic mouse hepatitis virus infection. J. Virol. *76*, 7373–7384.

Zhou, J., Stohlman, S.A., Hinton, D.R., and Marten, N.W. (2003). Neutrophils promote mononuclear cell infiltration during viral-induced encephalitis. J. Immunol. *170*, 3331–3336.

Zhou, J., Marten, N.W., Bergmann, C.C., Macklin, W.B., Hinton, D.R., and Stohlman, S.A. (2005a). Expression of matrix metalloproteinases and their tissue inhibitor during viral encephalitis. J. Virol. *79*, 4764–4773.

Zhou, J., Hinton, D.R., Stohlman, S.A., and Marten, N. (2005b). Maintenance of CD8+ T cells during acute viral infection of the central nervous system requires CD4+ T cells but not interleukin-2. Viral Immunol. *18*, 162–169.

Zhou, H., and Perlman, S. (2006). Preferential infection of mature dendritic cells by mouse hepatitis virus strain JHM. J. Virol. *80*, 2506–14.

Zuo, J., Stohlman, S.A, Hoskin, J.B., Hinton, D.R., Atkinson, R., and Bergmann, C.C. (2006). Mouse hepatitis virus pathogenesis in the central nervous system is independent of IL-15 and natural killer cells. Virology *350*, 206–215.

SARS Coronavirus: Pathogenesis and Correlation with Clinical Disease 11

John M. Nicholls and J.S. Malik Peiris

Abstract

Human infection by SARS coronavirus appears to be limited to the respiratory tract where infection of susceptible cells leads to damage to the pneumocytes resulting in a histological picture of diffuse alveolar damage and a clinical picture of adult respiratory distress syndrome. Diarrhea is also present but there is limited evidence of damage to the intestinal epithelium. The damage to the respiratory tree appears limited to the lower respiratory tract and there is evidence that the immune response plays a part in the outcome of patients with SARS.

The search for a new agent

During the early stages of the outbreak of SARS in 2002–3 many institutions around the world collaborated with the World Health Organization in a search for the cause of this rapid outbreak, and a number of possible agents were detected in the course of investigating patients with suspected SARS, including human metapneumovirus and chlamydia (WHO, 2003). The problem was that these were only detected in a few patient cohorts. In March, 2003, a novel coronavirus was isolated in culture in Vero-E6 cells and FRhK-4 cells and its presence was consistently demonstrated by molecular detection and serology in patients with SARS in many parts of the world (Drosten, *et al.*, 2003; Ksiazek, *et al.*, 2003; Kuiken *et al.*, 2003; Peiris *et al.*, 2003b). To fulfill Koch's postulates for disease etiology, experimental infection of cynomolgus macaques provided evidence of virus replication and significant lung pathology by this novel coronavirus (Fouchier *et al.*, 2003). The sharing of information and viruses within the WHO laboratory network allowed a consensus to be reached that this coronavirus was the etiological agent of SARS.

The two previously recognized human coronaviruses 229E and OC43 belong to groups 1 and 2 respectively (Holmes, 2001). These viruses in the past have been identified as causes of upper respiratory disease and the common cold, but were rarely identified as a cause of severe human disease. Questions were thus raised as to why this novel SARS coronavirus (SARS-CoV) was causing so much disease severity in humans. What cell was SARS-CoV infecting, and was the severity due to a direct viral effect, or the body's reaction to this novel virus?

The genetic sequence of the SARS coronavirus was fully determined within weeks of the virus being isolated (Marra *et al.*, 2003; Zeng *et al.*, 2003) and this confirmed that

SARS-CoV was a novel coronavirus, distinct from all other human and animal coronaviruses but distantly related to the group 2 coronaviruses (Snider *et al.*, 2003). It is now designated as a group 2b coronavirus. When we tested the lung tissue of patients with SARS with antibodies to existing coronaviruses there was consistently no reaction. Patients with SARS invariably seroconverted to this novel virus in immunofluorescent assays, ELISA tests and neutralization tests (Peiris *et al.*, 2003b; Ksiazek *et al.*, 2003; Rota *et al.*, 2003). However, when the general population was examined there was little evidence of pre-existing antibody to SARS-CoV suggesting that this was indeed a novel virus that recently emerged in the human population, very likely from an animal reservoir. Furthermore, subsequent studies after the 2003 outbreak showed some of the patients with SARS-like disease in the early stages of the outbreak had occupational exposure to live game animals held within markets in Guangdong (Xu *et al.*, 2004). Testing of animals in these live markets in Guangdong led to the detection of a SARS-like coronavirus in a number of small mammal species including Himalayan palm civets (*Paguma larvata*) and racoon dogs (*Nyctereutes procyonides*) (Guan *et al.*, 2003). In addition, those handlers directly involved in the live animal trade within these markets had a high seroprevalence to SARS-like viruses, although they gave no history of SARS-like disease (Guan *et al.*, 2003; Tu *et al.*, 2004). These findings suggested that the animal markets could be the interface at which inter-species transmission to humans occurred. Civets are susceptible to experimental infection with the human SARS coronavirus (Wu *et al.*, 2005). However, SARS-CoV antibody has not been consistently detected in the farms that supply civets to the live animal markets. There is little data from wild caught civets, but from the limited studies it does not appear that these animals are endemically infected in the wild and other animal species may harbor SARS- like viruses (Poon *et al.*, 2005). It has been proposed that the civets and racoon dogs amplify and perpetuate the circulation of the virus within the live animal markets setting but are not the natural reservoir of the virus, which still remains to be identified. In such markets, animals are continually removed for consumption while new animals are added providing a continued supply of susceptible animals.

Studies on the molecular evolution of SARS-CoV revealed that viruses in the early phase of the human SARS outbreak were more closely related to viruses detected in palm civets and other small mammals in the live animal markets. The virus was under strong positive selection in the early phase of outbreak suggesting a process of adaptation to the new human host (Song *et al.*, 2005; Guan *et al.*, 2004; Yeh *et al.*, 2004). Interestingly, the virus was also evolving rapidly in palm civets, an observation that is compatible with the idea that civets may not be the natural reservoir of the virus. In the second episode of SARS in December 2003 and January 2004, the virus detected in the four patients with SARS-CoV infection was phylogenetically more closely related to SARS-like viruses identified in civets rather than to the viruses associated with the global outbreak in 2003, suggesting that the patients with these infections were new introductions from the animal reservoir.

SARS-CoV entry and replication

The molecular biology of the coronaviruses is discussed elsewhere in this monograph but of the various proteins produced by the virus, the spike protein mediates virus attachment and entry intro susceptible cells (Simmons *et al.*, 2004) and attention has focused on the interaction of this spike protein with human cells. A receptor for SARS-CoV has been identified as angiotensin converting enzyme 2 (ACE2) (Li *et al.*, 2003; Wang *et al.*, 2004). Evidence for this is that SARS-CoV replication in the lungs of ACE2 knock-out mice is markedly suppressed, proving the importance of this receptor *in vivo*, at least in mice (Kuba *et al.*, 2005). Furthermore, ACE2 expression is detectable on pneumocytes, enterocytes, vascular endothelial cells and smooth muscle cells (Hamming *et al.*, 2004; To *et al.*, 2004) and this finding of ACE2 on respiratory and intestinal epithelium is compatible with the known tropism of the virus. However, not all cells that express ACE2 are susceptible to virus infection (e.g. endothelial cells) and a number of other molecules such as L-SIGN and DC-SIGN have been demonstrated to bind SARS-CoV (Jeffers *et al.*, 2004; Chan *et al.*, 2006; Yang *et al.*, 2004). DC-SIGN serves as a binding receptor and may be relevant in transferring infectious virus bound to the surface of dendritic cells from one site to another. L-SIGN mediates both binding and entry of SARS-CoV, and cells expressing L-SIGN (especially those which are homozygous) have a higher capacity for SARS-CoV degradation allowing a protective role (Chan *et al.*, 2006).

The spike protein of SARS-CoV also displays a degree of species specificity. For instance, the spike protein of SARS-CoV (Tor 2), a virus from the later phase of the outbreak in 2003, interacts efficiently with ACE2 of human and civet origin (Li *et al.*, 2005) allowing infection of both species. In contrast, the SARS CoV spike protein of civet cat origin binds well to ACE2 of civet cats but poorly to human ACE2 and this suggests that, without prior adaptation, the civet virus is not well adapted to replicate in human cells. This also explains why the SARS -like coronavirus found in civet cats cannot be maintained in culture *in vitro* in primate cell lines that effectively support the replication of the human SARS-CoV. It may also explain the mild disease and low transmissibility observed when SARS-CoV re-emerged from its animal reservoir to infect humans in December 2003 and January 2004. Furthermore, asymptomatic infections of SARS-CoV were uncommon during the outbreak of SARS in 2003 (Lee *et al.*, 2003a). Most persons at high risk through family or health care exposure either developed overt disease or did not seroconvert at all. This is in marked contrast to exposure to the precursor animal virus where asymptomatic infections were common (Leung *et al.*, 2004b).

Quantitative studies on viral load in the upper respiratory tract and feces of patients infected with SARS-CoV reveals a progressive increase in viral load, peaking around day 10 after onset of disease symptoms (Lu *et al.*, 2006; Cheng *et al.*, 2004; Peiris *et al.*, 2003a). This explains the epidemiological observation that transmission mainly occurs after the fifth day of illness and is less common in the early stage of the illness (Lipsitch *et al.*, 2005; Riley *et al.*, 2003). This unusual feature of SARS allowed public health measures to be so dramatically successful in interrupting disease transmission. Although virus RNA was detectable by RT-PCR for many weeks into convalescence, virus isolation after the third week of illness was rare and this positive RNA may therefore represent residual viral genome rather than active viral replication.

Clinical features

The clinical features of SARS have been reviewed in other reports (Christian *et al.*, 2004; Peiris *et al.*, 2003c) and are mentioned briefly. As befits its name, SARS typically presents as an illness with acute onset, fever, myalgia, malaise and chills or rigor, with the rhinorrhea and sore throat that are more a feature of influenza being less common features (Peiris *et al.*,2003b; Booth *et al.*, 2003; Tsang *et al.*, 2003). A dry cough is common but shortness of breath, tachypnea or pleurisy is only prominent later in the course of the illness. About one third of patients improve with resolution of radiographic changes. Others progress to have increasing shortness of breath, tachypnoea, oxygen desaturation, worsening of chest signs on physical examination and onset of diarrhea. About 20 to 30 percent of all patients need observation in intensive care and most of these will require mechanical ventilation. Overall case fatality was 9.6% and the terminal events were severe respiratory failure, multiple organ failure, sepsis and intercurrent medical illness such as an acute myocardial infarction. Age and the presence of co-morbid conditions appeared to have prognostic significance regarding disease severity and clinical outcome.

A watery diarrhea occurs in some patients, typically associated with clinical deterioration in the second week of the illness. There are few reports of the tropism of SARS-CoV for the gastrointestinal tract and one case report showed evidence of viral replication in columnar epithelial cells of the gastro-intestinal tract (Leung *et al.*, 2003). However, there is minimal cellular infiltrate or disruption of the intestinal architecture. The reasons for this and the mechanism underlying the pathogenesis of diarrhea remains unclear but it is worthy of note that the symptoms and findings are similar to the coronavirus infections of cows. It may be relevant that some human intestinal epithelial cell lines induce anti-apoptotic cellular responses and support the persistent replication of SARS-CoV *in vitro* (Cinatl *et al.*, 2004).

Other extrapulmonary findings include hepatic dysfunction and central nervous system manifestations (Lau *et al.*, 2004) but these may be related to therapies instituted for the treatment of SARS.

The initial chest radiograph is abnormal in 60 to 100 percent of the cases at initial presentation, depending on the duration of illness and typically shows ground glass opacities and focal consolidation over the periphery and in the subpleural regions of the lower zones of the lung. Residual impairment of lung function persists in convalescence in over 20% of patients (Ong *et al.*, 2004).

Retrospective studies have now identified a number of clinical and laboratory findings that predict outcome from SARS. Age, presence of co-morbidities, more extensive lung involvement and abnormal lymphocyte subsets are poor prognostic indicators (Chau *et al.*, 2004). High viral loads in nasopharynx and serum early in the illness are an independent risk factor for mortality (Chu *et al.*, 2004). Similarly, viral load in the nasopharynx and serum between day 10–15 of clinical illness also correlated with clinical outcome including the need for mechanical ventilation and death (Hung *et al.*, 2004).

In contrast to adults, children have a much milder clinical course than adults (Leung *et al.*, 2004a). During the SARS outbreak in 2003, few children with SARS required intensive care, mortality was exceedingly rare, and it was rare for children to transmit infection to others. Atypical presentations of SARS can occur in the elderly and in the immunocompromised. Patients may be afebrile and may present with non-specific clinical symptoms such as decreased appetite or a general deterioration in their clinical condition. Such cases have led to nosocomial disease outbreaks with devastating consequences.

Lymphopenia involving both $CD4^+$ and $CD8^+$ T-cells is a common finding during the acute phase of SARS and is associated with an adverse outcome (Li *et al.*, 2004). This lymphopenia is rapidly reversed during convalescence. Some patients also have other hematological abnormalities such as a low platelet count and activated plasma thromboplastin time. At autopsy some patients had a reduction in the white pulp of the spleen but as many of these patients were on high dose steroids it is difficult to determine if this is a direct viral effect.

Pathological findings

The signs and symptoms of SARS in the early phase indicated that the lower respiratory tract, and in particular the lung, rather than the upper respiratory tract is the target for the symptomatology. Within the acinus of the lung there is a respiratory bronchiole, its distal alveolar duct and alveolar sacs. Two types of epithelial cells line the alveoli; the flattened type 1 pneumocyte and the more cuboidal type 2. The type 1 pneumocytes are less numerous than the type 2 pneumocytes but have a large surface area. The type 2 pneumocytes are more abundant but have a smaller surface area than the type 1 pneumocytes. The epithelial cells are connected by tight junctions but the underlying endothelial cells have leaky junctions that allow the extravasation of fluid. The type 1 cells are regarded as non-replicating whilst the type 2 are the progenitor cells of type 1 and replicate after injury to replace the type 1 cells. This type 2 hyperplasia is interpreted as a non-specific marker of alveolar injury and repair. In addition to the epithelial cells, alveolar macrophages are identified attached to the wall of the alveolus. These are non-specific scavenger cells that are derived from monocytes that have emigrated from the blood and differ in function from the interstitial macrophages.

Autopsy findings in patients dying within the first 10 days of illness revealed diffuse alveolar damage (DAD), desquamation of pneumocytes, an inflammatory infiltrate, edema and hyaline membrane formation (Figure 11.1a). From a pathogenic perspective this can be explained by damage to type 1 pneumocytes allowing the leakage of plasma proteins and inflammatory cells to produce the hyaline membranes. Patients dying later in the illness had an organizing DAD (Figure 11.1b) together with squamous metaplasia and multinucleated giant cells of either macrophage or epithelial cell origin (Franks *et al.*, 2003; Ding *et al.*, 2003; Lee *et al.*, 2003b; Nicholls *et al.*, 2003). Focal thrombosis of pulmonary vessels was seen (Figure 11.1c). Viral RNA was detectable by RT-PCR at high viral load in lung, bowel and lymph node, but was also detectable in spleen, liver and kidney, suggesting multiorgan involvement of the disease (Farcas *et al.*, 2005). When immunohistochemistry and in-situ hybridization for SARS-CoV were used, virus-infected cells in the lung were identified to be primarily alveolar epithelial cells and macrophages. Scanty positive bronchial epithelial cells have also been detected (Nicholls *et al.*, 2006) (Figure 11.1d) and this has also been seen in a bronchial culture model (Sims *et al.*, 2005). We were, however, unable to detect virus antigen in lymph nodes draining the lung. The immunohistochemistry and in-situ hybridization findings were present in lung biopsy or autopsy tissue of patients within the first 10 day of illness but rarely after then (Nicholls *et al.*, 2006). In the lungs, the infected cells demonstrated by IHC are similar to those seen with *in situ hybridization*, that is, they are mainly flattened pneumocytes, i.e. those of type 1 morphology, and less often macrophages. Unlike non-avian influenza infection there is little histological evidence of bronchial epithelial cells necrosis. Though nasopharyngeal aspirates were able to detect SARS-CoV by PCR and culture, there have been no reports

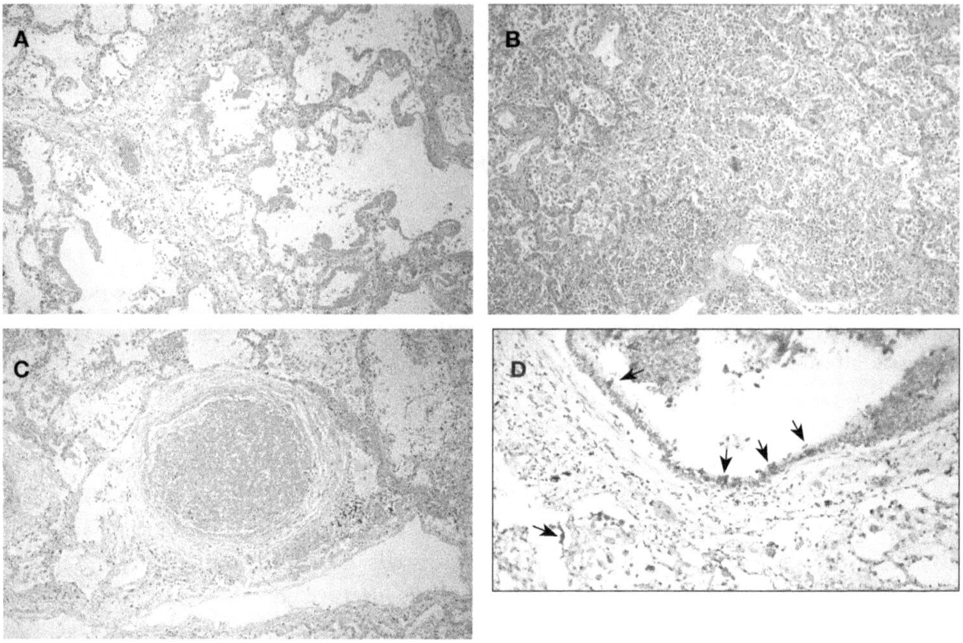

Figure 11.1 A, Early phase of SARS pneumonia showing pink hyaline membranes within the alveoli and increased inflammatory cells in the interstitium (H and E ×200). B, Organizing phase of SARS. There is obliteration of the alveoli by fibrous tissue and reactive pneumocytes. A giant cell is identified in the centre of the field (H and E ×20. C, Organizing phase of SARS showing thrombosis of a pulmonary vessel (H and E ×200). D, Early SARS pneumonia. In situ hybridization with pooled SARS-CoV genes showing positive staining of the bronchial epithelium, flattened pneumocytes and macrophages (AEC with hematoxylin counterstain ×200).

of the extent of cellular infection of the upper respiratory tract by SARS-CoV. The closest histological correlate for cellular tropism and lung damage was in the cynomolgus macaques but upper respiratory tract involvement was not reported. Laboratory infection of golden Syrian hamsters also shows a similar cellular pattern of infection, though hyaline membranes are not a conspicuous finding (Roberts *et al.*, 2005), however it should be noted that occasional squamous cells and columnar cells of the upper respiratory tract in these hamsters did show the presence of SARS-CoV by immunohistochemistry (Figure 11.2 a-d).

It has been previously mentioned that ACE2 is one of the receptors for SARS-CoV. This enzyme is involved in the renin–angiotensin system and it may have a protective effect in acute lung injury through its counterbalance of the effects of the stimulatory ACE1 and angiotensin (Kuba *et al.*, 2005). It may play a role in the pathogenesis of SARS and other infectious agents associated with acute lung injury leading to acute respiratory distress syndrome (ARDS). In mouse models, treatment with recombinant human ACE2, or with inhibitors of the renin–angiotensin system appear to mitigate the effects of acute lung injury. These may provide promising novel approaches for the management of ARDS (Kuba *et al.*, 2005).

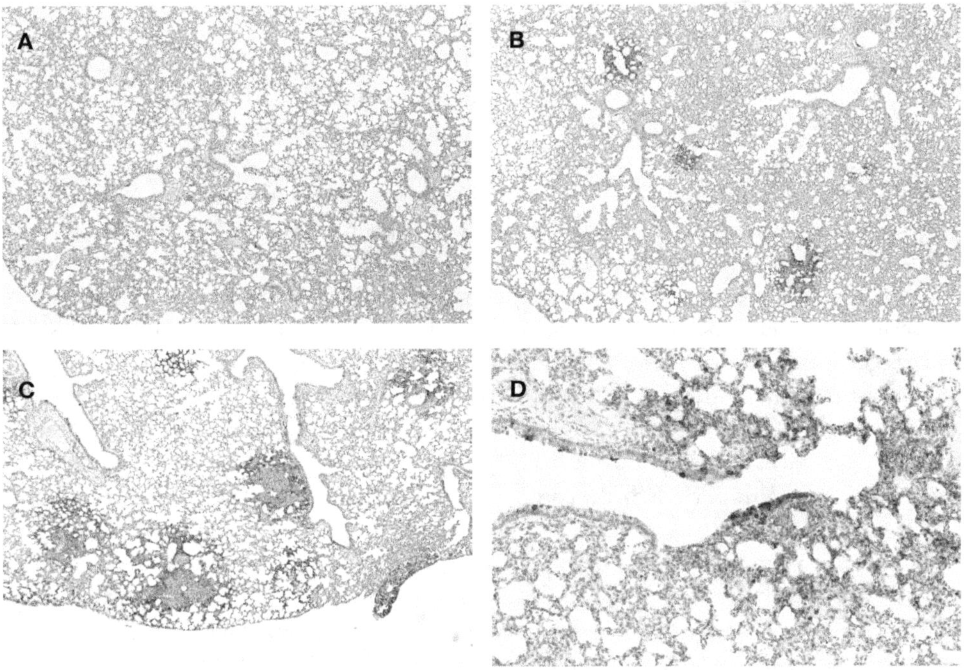

Figure 11.2 SARS-CoV infection of Syrian Hamster. A, Day 2, B, Day 3, C, Day 4. Infection is maximal at the terminal bronchiole and extends outwards to involve the alveoli and pleura. D, Only scanty bronchial epithelial cells are positive and show no evidence of necrosis. Monoclonal antibody to N protein. AEC with hematoxylin counterstain ×40 and ×200. Original tissue courtesy of Dr M Huerre, Insitut Pasteur, Paris, France.

Although critical clinical symptoms relate to the respiratory tract, there have been contrasting reports on the extent of extra-pulmonary dissemination. Evidence of SARS-CoV replication in the lungs and intestine is provided by detection of virus particles by electron microscopy, virus isolation and by the detection of viral antigens and nucleic acid by immuno-histology and in-situ hybridization, respectively (Ksiazek *et al.*, 2003; Peiris *et al.*, 2003b). Viral RNA has also been detected using RT-PCR in lymph nodes, spleen, liver, heart, kidney and skeletal muscle suggesting multiorgan dissemination of the virus, but in-situ hybridization studies and immunhistochemistry do not fully support these findings (Nicholls *et al.*, 2006). Whilst one report identified virus particles by electron microscopy in lymphocytes (Gu *et al.*, 2005) this has not been found by other investigators. As mentioned previously, viral load in the upper respiratory tract or in the serum in the first few days of illness is an independent predictor of clinical outcome (Chu *et al.*, 2004). This suggests that the final outcome of the disease is determined early in the course of the illness. Thus, the dose of the infecting virus inoculum, the early innate defenses of the host, or both, are critical in determining the disease prognosis. The fact that children had a much better outcome than adults also needs to be taken into consideration (Leung *et al.*, 2004a).

Immune response

Genetic polymorhisms associated with susceptibility and disease severity to SARS have been identified. HLA-B*4601 was reported to associate with both disease susceptibility and severity in a Chinese population in Taiwan but this association was not found in Hong Kong Chinese patients (Ng *et al.*, 2004). In contrast, among Hong Kong Chinese, HLA-B*0703 rather than HLA-B*4601 was found to be associated with SARS. A number of polymorphisms of the innate immune response have been studied (Chan *et al.* 2006).

Virus neutralization studies suggest that SARS-CoV may bind to receptors other than ACE2. This is because many of the prototype SARS-CoV isolates commonly studied derive from the global outbreak arising from the super-spreading event in Hong Kong and are genetically and antigenically very similar (Yang *et al.*, 2005). As virus isolates from the earliest patients in the course of the outbreak in Guangdong or isolates of the animal precursor virus detected in civet cats are not available, virus neutralization of these animal precursor-like viruses has been explored by generating lentivirus pseudotypes incorporating the spike proteins of SARS-like coronaviruses SZ3 and SZ16 (detected in civet cats), GD03 (a civet cat-like virus isolated from humans following the re-emergence of SARS in December 2003) and the human SARS-CoV strains (e.g. Urbani) isolated during the major SARS outbreak during early 2003. The prototype Urbani virus isolated from a patient from the major human outbreak in 2003 was neutralized efficiently with homologous antibody as well as antibody to the animal-like SARS-CoV GD03. Surprisingly, the animal precursor-like virus pseudotypes were refractory to neutralization with both homologous antibody as well as with antibody to the prototype Urbani virus. Indeed, the antibody to the Urbani virus enhanced virus entry of the civet SARS-CoV SZ3 and SZ16. The evasion of neutralization exhibited by the animal precursor-like SARS-CoV correlated with poor binding to the ACE2 receptor. Thus it appears likely that the animal precursor SARS-CoV uses an alternative receptor for viral entry and is possibly adapted to evade neutralization by host antibody.

Apart from the pro-inflammatory aspects of the renin–angiotensin system, the body's immune response to the virus needs to be taken into consideration. HLA-A2 restricted T-cell epitopes have been identified that are capable of eliciting a cytotoxic $CD8^+$ T cell response in patients who have recovered from SARS and both chemokines (IL-8, MCP-1, IP-10) and pro-inflammatory cytokines (IL-1, IL-6, IL-12) have been found to be elevated in patients with SARS (Jiang *et al.*, 2004 Cheung *et al.*, 2005). Prolonged dysregulation of cytokine production has also been demonstrated by ELISPOT assays *in vitro* (Law *et al.*, 2005; Jones *et al.*, 2004).

The immunohistochemical and in situ hybridization data indicates that the primary mechanism of pathogenesis in the lung is likely to be the direct virus damage to primarily type 1 and to a lesser extent type 2 pneumocytes (Nicholls *et al.*, 2006). The infiltration of macrophages into the lung is a prominent manifestation of the disease. Viral infection of monocyte derived macrophages *in vitro* leads to the release of monocyte attractant chemokines such as MCP-1 and IP-10 (Wong *et al.*, 2004) and patients with SARS also have high levels of these chemokines present in the serum. In contrast, SARS-CoV infection of primary human macrophages, dendritic cells and the transformed kidney epithelial cell line 293, have failed to reveal evidence of induction of a type 1 interferon response

(Cheung *et al.*, 2005). Furthermore, in contrast to patients with influenza infection, gene expression profiling of peripheral blood mononuclear cells from patients with SARS did not reveal induction of type 1 interferon responses. Thus it remains possible that a failure of the innate antiviral immune defenses of the host in the early stages of the disease may contribute to the pathogenesis and may also explain the progressive increase in viral load to peak at day 10, the stage of illness when the adoptive immune responses become activated, presumably controlling further virus replication.

Necrosis and atrophy of the lymphoid tissue of lymph nodes and the white pulp of the spleen is observed at autopsy. B and T lymphocytes including both $CD4^+$ and $CD8^+$ T lymphocyte subsets, natural killer cells and dendritic cells appear decreased during the acute stage of SARS. The mechanism underlying the lymphopenia seen in SARS remains obscure. Lymphocytes do not carry ACE2 on their surface and appear to be refractory to virus infection and therefore apoptosis of bystander uninfected lymphocytes as happens with other vial infections remains a possibility. However, one report suggests that virus particles can be detected by electron microscopy in a high proportion of lymphocytes from patients with SARS (Gu *et al.*, 2005). If confirmed, this would provide an alternative explanation for the lymphopenia seen with SARS.

Conclusion

Clinical and pathological studies of SARS cases indicate that the lower respiratory tract is the prime target of SARS-CoV, and in particular the type 1 pneumocytes with macrophages subsequently being infected. Though ACE2 has been shown to be a receptor for SARS-CoV, the existence of other receptors such as L-SIGN and DC-SIGN suggest that different pathways, possibly with a genetic basis, may be invoked by the virus as a protective response. The initial phase of lung damage exists for 10–14 days, after which many of the changes present can be attributed to the sequelae of diffuse lung damage. The epithelial lung damage together with macrophage activation results in increased levels of pro-inflammatory cytokines. As with many other new emerging infections, SARS had zoonotic origins highlighting the need for a better understanding of the virus ecology of animals, both domestic animals and wildlife. As many of these zoonoses do not cause disease in their natural host, surveillance needs to encompass viruses causing subclinical infections as well as those causing overt disease and will require the co-operation of multiple agencies including those dealing with public health, veterinary medicine and wildlife or the environment (Kuiken *et al.*, 2005).

Acknowledgments

Drs Y. Guan, L.L.M. Poon, V.C.C. Cheng,, and Professor K.Y. Yuen for providing valuable references into the still rapidly evolving field of SARS pathogenesis. Mr Kevin Fung for technical assistance in the SARS in situ hybridization and immunohistochemistry.

References

Booth, C.M., Matukas, L.M., Tomlinson, G.A., Rachlis, A.R., Rose, D.B., Dwosh, H.A., Walmsley, S.L., Mazzulli, T., Avendano, M., Derkach, P., Ephtimios, I.E., Kitai, I., Mederski, B.D., Shadowitz, S.B., Gold, W.L., Hawryluck, L.A., Rea, E., Chenkin, J.S., Cescon, D.W., Poutanen, S.M., and Detsky, A.S. (2003). Clinical features and short-term outcomes of 144 patients with SARS in the greater Toronto area. JAMA. *289*, 2801–2809.

Chan, V.S., Chan, K.Y., Chen, Y., Poon, L.L., Cheung, A.N., Zheng, B., Chan, K.H., Mak, W., Ngan, H.Y., Xu, X., Screaton, G., Tam, P.K., Austyn, J.M., Chan, L.C., Yip, S.P., Peiris, M., Khoo, U.S., and Lin, C.L. (2006). Homozygous L-SIGN (CLEC4M) plays a protective role in SARS coronavirus infection. Nat. Genet. *38*, 38–46.

Chau, T.N., Lee, P.O., Choi, K.W., Lee, C.M., Ma, K.F., Tsang, T.Y., Tso, Y.K., Chiu, M.C., Tong, W.L., Yu, W.C., and Lai, S.T. (2004). Value of initial chest radiographs for predicting clinical outcomes in patients with severe acute respiratory syndrome. Am.J. Med. *117*, 249–54.

Cheng, P.K., Wong, D.A., Tong, L.K., Ip, S.M., Lo, A.C., Lau, C.S., Yeung, E.Y., and Lim, W.W. (2004). Viral shedding patterns of coronavirus in patients with probable severe acute respiratory syndrome. Lancet *363*, 1699–1700.

Cheung, C.Y., Poon, L.L.M., Ng, I.H.Y., Luk, W., Sia, S.F., Wu, M.H.S., Chan, K.H., Yuen, K.Y., Gordon, S., Guan, Y., and Peiris, J.S.M. (2005). SARS-Coronavirus is a Poor Inducer of Interferon-β in Macrophages: Its Possible Relevance to Pathogenesis. J. Virol. *79*, 7819–26.

Cinatl, J.Jr., Hoever, G., Morgenstern, B., Preiser, W., Vogel, J.U., Hofmann, W.K., Bauer, G., Michaelis, M., Rabenau, H.F., and Doerr. H.W. (2004). Infection of cultured intestinal epithelial cells with severe acute respiratory syndrome coronavirus. Cell. Mol. Life Sci. *61*, 2100–2112.

Christian, M.D., Poutanen, S.M., Loufty, M.R., Mueller, M.P., and Low, D.E. (2004). Severe acute respiratory syndrome. Clin. Infect. Dis. *38*, 1420–1427.

Chu, C.M., Poon, L.L., Cheng, V.C., Chan, K.S., Hung, I.F., Wong, M.M., Chan, K.H., Leung, W.S., Tang, B.S., Chan, V.L., Ng, W.L., Sim, T.C., Ng, P.W., Law, K.I., Tse, D.M., Peiris, J.S., and Yuen, K.Y. (2004). Initial viral load and the outcomes of SARS. CMAJ *171*, 1349–1352.

Ding, Y., Wang, H., Shen, H., Li, Z., Geng, J., Han, H., Cai, J., Li, X., Kang, W., Weng, D., Lu, Y., Wu, D., He, L., and Yao, K. (2003). The clinical pathology of severe acute respiratory syndrome (SARS): a report from China. J. Pathol. *200*, 282–289.

Drosten, C., Gunther S., Preiser, W., van der Werf, S., Brodt, H.R., Becker, S., Rabenau, H., Panning, M., Kolesnikova, L., Fouchier, R.A., Berger, A., Burguiere, A.M., Cinatl, J., Eickmann, M., Escriou, N., Grywna, K., Kramme, S., Manuguerra, J.C., Muller, S., Rickerts, V., Sturmer, M., Vieth, S., Klenk, H.D., Osterhaus, A.D., Schmitz, H., and Doerr, H.W. (2003). Identification of a novel coronavirus in patients with severe acute respiratory syndrome. N. Engl. J. Med. *348*, 1967–1976.

Farcas, G.A., Poutanen, S.M., Mazzulli, T., Willey, B.M., Butany, J., Asa, S.L., Faure, P., Akhavan, P., Low, D.E., and Kain, K.C. (2005). Fatal severe acute respiratory syndrome is associated with multiorgan involvement by coronavirus. J. Infec. Dis. *191*, 193–197.

Fouchier, R.A., Kuiken, T., Schutten, M., van Amerongen, G., van Doornum, G.J., van den Hoogen, B.G., Peiris, M., Lim, W., Stohr, K., Osterhaus. A.D. (2003). Aetiology: Koch's postulates fulfilled for SARS virus. Nature *423*, 240.

Franks, T.J., Chong, P.Y., Chui, P., Galvin, J.R., Lourens, R.M., Reid, A.H., Selbs, E., McEvoy, C.P., Hayden, C.D., Fukuoka, J., Taubenberger, J.K., and Travis, W.D. (2003). Lung pathology of severe acute respiratory syndrome (SARS): a study of 8 autopsy cases from Singapore. Hum. Pathol. *34*, 743–748.

Gu, J., Gong, E., Zhang, B., Zheng, J., Gao, Z., Zhong, Y., Zou, W., Zhan, J., Wang, S., Xie, Z., Zhuang, H., Wu, B., Zhong, H., Shao, H., Fang, W., Gao, D., Pei, F., Li, X., He, Z., Xu, D., Shi, X., Anderson, V.M., and Leong, A.S. (2005). Multiple organ infection and the pathogenesis of SARS. J. Exp. Med. *202*, 415–424.

Guan, Y., Zheng, B.J., He, Y.Q., Liu, X.L., Zhuang, Z.X., Cheung, C.L., Luo, S.W., Li, P.H., Zhang, L.J., Guan, Y.J., Butt, K.M., Wong, K.L., Chan, K.W., Lim, W., Shortridge, K.F., Yuen, K.Y., Peiris, J.S., and Poon, L.L. (2003). Isolation and characterization of viruses related to the SARS coronavirus from animals in southern China. Science *302*, 276–278.

Guan, Y., Peiris, J.S., Zheng, B., Poon, L.L., Chan, K.H., Zeng, F.Y., Chan, C.W., Chan, M.N., Chen, J.D., Chow, K.Y., Hon, C.C., Hui, K.H., Li, J., Li, V.Y., Wang, Y., Leung, S.W., Yuen, K.Y., and Leung, F.C. (2004). Molecular epidemiology of the novel coronavirus that causes severe acute respiratory syndrome. Lancet *363*, 99–104.

Hamming, I., Timens, W., Bulthuis, M.L., Lely, A.T., Navis, G.J., and van Goor, H. (2004). Tissue distribution of ACE2 protein, the functional receptor for SARS coronavirus. A first step in understanding SARS pathogenesis. J. Pathol. *203*, 631–637.

Holmes, K.V. (2001). Coronaviruses. *In* Knipe DM, Howley PM (eds) *Fields Virology*. Lippincott Williams & Wilkins, Philadelphia PA, ch 36, 1187–1203.

Hung, I.F.N., Cheng, V.C.C., Wu, A.K.L., Tang, B.S.F., Chan, K.H., Chu, C.M., Wong, M.M.L., Hui, W.T., Poon, L.L.M., Tse, D.M.W., Chan, K.S., Woo, P.C.Y., Lau, S.K.P., Peiris, J.S.M., and Yuen, K.Y. (2004). Viral loads in clinical specimens and SARS manifestations. Emerg. Infect. Dis. *10*, 1550–1557.

Jeffers, S.A., Tusell, S.M., Gillim-Ross, L., Hemmila, E.M., Achenbach, J.E., Babcock, G.J., Thomas Jr, W.D., Thackray, L.B., Young, M.D., Mason, R.J., Ambrosino, D.M., Wentworth, D.E., Demartini, J.C., and Holmes, K.V. (2004). CD209L (L-SIGN) is a receptor for severe acute respiratory syndrome coronavirus. Proc. Natl. Acad. Sci. USA. *101*, 15748–15753.

Jiang, Y., Xu, J., Zhou, C., Wu, Z., Zhong, S., Liu, J., Luo, W., Chen, T., Qin, Q., and Deng, P. (2005). Characterization of cytokine and chemokine profiles of severe acute respiratory syndrome. Am. J. Respir. Crit. Care. Med. *171*, 850–857.

Jones, B.M., Ma, E.S., Peiris, J.S., Wong, P.C., Ho, J.C., Lam, B., Lai, K.N., and Tsang, K.W. (2004). Prolonged disturbances of in vitro cytokine production in patients with severe acute respiratory syndrome (SARS) treated with ribavirin and steroids. Clin. Exp. Immunol. *135*, 467–473.

Ksiazek, T.G., Erdman, D., Goldsmith, C.S., Zaki, S.R. Peret, T., Emery, S., Tong, S., Urbani, C., Comer, J.A., Lim, W., Rollin, P.E., Dowell, S.F., Ling, A.E., Humphrey, C.D., Shieh, W.J., Guarner, J., Paddock, C.D., Rota, P., Fields, B., DeRisi, J., Yang, J.Y., Cox, N., Hughes, J.M., LeDuc, J.W., Bellini, W.J., Anderson, L.J., and SARS Working Group. (2003). A novel coronavirus associated with severe acute respiratory syndrome. N. Engl. J. Med. *348*, 1953–1966.

Kuba, K., Imai, Y., Rao, S., Gao, H., Guo, F., Guan, B., Huan, Y., Yang, P., Zhang, Y., Deng, W., Bao, L., Zhang, B., Liu, G., Wang, Z., Chappell, M., Liu, Y., Zheng, D., Leibbrandt, A., Wada, T., Slutsky, A.S., Liu, D., Qin, C., Jiang, C., and Penninger, J.M. (2005). A crucial role of angiotensin converting enzyme 2 (ACE2) in SARS coronavirus-induced lung injury. Nat. Med. *11*, 875–879.

Kuiken, T., Fouchier, R.A., Schutten, M., Rimmelzwaan, G.F., van Amerongen, G., van Riel, D., Laman, J.D., de Jong, T., van Doornum, G., Lim, W., Ling, A.E., Chan, P.K., Tam, J.S., Zambon, M.C., Gopal, R., Drosten, C., van der Werf, S., Escriou, N., Manuguerra, J.C., Stohr, K., Peiris, J.S., and Osterhaus, A.D. (2003). Newly discovered coronavirus as the primary cause of severe acute respiratory syndrome. Lancet *362*, 263–270.

Kuiken, T., Leighton, F.A., Fouchier, R.A.M., LeDUc, J.W., Peiris, J.S.M., Schudel, A., Stohr, K., and Osterhaus, A.D.M.E. (2005). Pathogen surveillance in animals. Science *310*, 974.

Lau, K.K., Yu, W.C., Chu, C.M., Lau, S.T., Sheng, B., and Yuen, K.Y. (2004). Possible central nervous system infection by SARS coronavirus. Emerg. Infect. Dis. *10*, 342–344.

Law, H.K., Cheung, C.Y., Ng, H.Y., Sia, S.F., Chan, Y.O., Luk, W., Nicholls, J.M., Peiris, J.S., and Lau, Y.L. (2005). Chemokine upregulation in SARS coronavirus infected human monocyte derived dendritic cells. Blood. *106*, 2366–2374.

Lee, H.K., Tso, E.Y., Chau, T.N., Tsang, O.T., Choi, K.W., and Lai, T.S. (2003a). Asymptomatic severe acute respiratory syndrome-associated coronavirus infection. Emerg. Infect. Dis. *9*, 1491–1492.

Lee, N., Hui, D., Wu, A., Chan, P., Cameron, P., Joynt, G.M., Ahuja, A., Yung, M.Y., Leung, C.B., To, K.F., Lui, S.F., Szeto, C.C., Chung, S., and Sung, J.J. (2003b). A major outbreak of severe acute respiratory syndrome in Hong Kong. N. Engl. J. Med. *348*, 1986–1994.

Leung, C.W., Kwan, Y.W., Ko, P.W., Chiu, S.S., Loung, P.Y., Fong, N.C., Lee, L.P., Hui, Y.W., Law, H., Wong, W.H., Chan, K.H., Peiris, J.S., Lim, W.W., Lau, Y.L., and Chiu, M.C. (2004a). Severe acute respiratory syndrome among children. Pediatrics *113*, e535–543.

Leung, G.M., Chung, P.H., Tsang, T., Lim, W., Chan, S.K., Chau, P., Donnelly, C.A., Ghani, A.C., Fraser, C., Riley, S., Ferguson, N.M., Anderson, R.M., Law, Y.L., Mok, T., Ng, T., Fu, A., Leung, P.Y., Peiris, J.S., Lam, T.H., and Hedley, A.J. (2004b). SARS-CoV antibody prevalence in all Hong Kong patient contacts. Emerg. Infect. Dis. *10*, 1653–1656.

Leung, W.K., To, K.F., Chan, P.K., Chan, H.L., Wu, A.K., Lee, N., Yuen, K.Y., and Sung, J.J. (2003). Enteric involvement of severe acute respiratory syndrome-associated coronavirus infection. Gastroenterology *125*, 1011–1017.

Li, T., Qiu, Z., Zhang, L., Han, Y., He, W., Liu, Z., Ma, X., Fan, H., Lu, W., Xie, J., Wang, H., Deng, G., and Wang, A. (2004). Significant changes of peripheral T lymphocyte subsets in patients with severe acute respiratory syndrome. J. Infect. Dis. *189*, 648–651.

Li, W., Moore, M.J., Vasilieva, N., Sui, J., Wong, S.K., Berne, M.A., Somasundaran, M., Sullivan, J.L., Luzuriaga, K., Greenough, T.C., Choe, H., and Farzan, M. (2003). Angiotensin-converting enzyme 2 is a functional receptor for the SARS coronavirus. Nature *426*, 450–454.

Li, W., Zhang, C., Sui, J., Kuhn, J.H., Moore, M.J., Luo, S., Wong, S.K., Huang, I.C., Xu, K., Vasilieva, N., Murakami, A., He, Y., Marasco, W.A., Guan, Y., Choe, H., and Farzan, M. (2005). Receptor and viral determinants of SARS-coronavirus adaptation to human ACE2. EMBO J. *24*, 1634–1643.

Lipsitch, M., Cohen, T., Cooper, B., Robins, J.M., Ma, S., James, L., Gopalakrishna, G., Chew, S.K., Tan, C.C., Samore, M.H., Fisman, D., and Murray, M. (2003). Transmission dynamics and control of severe acute respiratory syndrome. Science *300*, 1966–1970

Lu, Y.T., Chen, P.J., Sheu, C.Y., Liu, C.L. (2006). Viral load and outcome in SARS infection: The role of personal protective equipment in the emergency department. J. Emerg. Med. *30*, 7–15

Marra, M.A., Jones, S.J., Astell, C.R., Holt, R.A., Brooks-Wilson, A., Butterfield, Y.S., Khattra, J., Asano, J.K., Barber, S.A., Chan, S.Y., Cloutier, A., Coughlin, S.M., Freeman, D., Girn, N., Griffith, O.L., Leach, S.R., Mayo, M., McDonald, H., Montgomery, S.B., Pandoh, P.K., Petrescu, A.S., Robertson, A.G., Schein, J.E., Siddiqui, A., Smailus, D.E., Stott, J.M., Yang, G.S., Plummer, F., Andonov, A., Artsob, H., Bastien, N., Bernard, K., Booth, T.F., Bowness, D., Czub, M., Drebot, M., Fernando, L., Flick, R., Garbutt, M., Gray, M., Grolla, A., Jones, S., Feldmann, H., Meyers, A., Kabani, A., Li, Y., Normand, S., Stroher, U., Tipples, G.A., Tyler, S., Vogrig, R., Ward, D., Watson, B., Brunham, R.C., Krajden, M., Petric, M., Skowronski, D.M., Upton, C., and Roper, R.L. (2003). The Genome sequence of the SARS-associated coronavirus. Science *300*, 1399–1404.

Ng, M.H., Lau, K.M., Li, L., Cheng, S.H., Chan, W.Y., Hui, P.K., Zee, B., Leung, C.B., and Sung, J.J. (2004). Association of human-leukocyte-antigen class I (B*0703) and class II (DRB1*0301) genotypes with susceptibility and resistance to the development of severe acute respiratory syndrome. J. Infect. Dis. *190*, 515–518.

Nicholls, J.M., Poon, L.L., Lee, K.C., Ng, W.F., Lai, S.T., Leung, C.Y., Chu, C.M., Hui, P.K., Mak, K.L., Lim, W., Yan, K.W., Chan, K.H., Tsang, N.C., Guan, Y., Yuen, K.Y., and Peiris, J.S. (2003). Lung pathology of fatal severe acute respiratory syndrome. Lancet *361*, 1773–1778.

Nicholls, J.M., Butany, J., Poon, L.L., Chan, K.H., Beh, S.L., Poutanen, S., Peiris, J.S., Wong, M. (2006). Time Course and Cellular Localization of SARS-CoV Nucleoprotein and RNA in Lungs from Fatal Cases of SARS. PLoS Med. 3, e27

Ong, K.C., Ng, A.W., Lee, L.S., Kaw, G., Kwek, S.K., Leow, M.K., and Earnest, A. (2004). Pulmonary function and exercise capacity in survivors of severe acute respiratory syndrome. Eur. Respir. J. *24*, 436–442.

Peiris, J.S., Chu, C.M., Cheng, V.C., Chan, K.S., Hung, I.F., Poon, L.L., Law, K.I., Tang, B.S., Hon, T.Y., Chan, C.S., Chan, K.H., Ng, J.S., Zheng, B.J., Ng, W.L., Lai, R.W., Guan, Y., Yuen, K.Y., and HKU/UCH SARS Study Group. (2003a). Clinical progression and viral load in a community outbreak of coronavirus-associated SARS pneumonia: a prospective study. Lancet *361*, 1767–1772.

Peiris, J.S., Lai, S.T., Poon, L.L., Guan, Y., Yam, L.Y., Lim, W., Nicholls, J., Yee, W.K., Yan, W.W., Cheung, M.T., Cheng, V.C., Chan, K.H., Tsang, D.N., Yung, R.W., Ng, T.K., Yuen, K.Y., and SARS study group. (2003b). Coronavirus as a possible cause of severe acute respiratory syndrome. Lancet *361*, 1319–1325.

Peiris, J.S.M., Yuen, K.Y., Osterhaus, A.D.M.E., Stohr, K. (2003c). The Severe Acute Respiratory Syndrome. N. Engl. J. Med. *349*, 2431–2441.

Poon, L.L., Chu, D.K., Chan, K.H., Wong, O.K., Ellis, T.M., Leung, Y.H., Lau, S.K., Woo, P.C., Suen, K.Y., Yuen, K.Y., Guan, Y., and Peiris, J.S. (2005). Identification of a novel coronavirus in bats. J. Virol. *79*, 2001–2009.

Poutanen, S.M., Low, D.E., Henry, B., Finkelstein, S., Rose, D., Green, K., Tellier, R., Draker, R., Adachi, D., Ayers, M., Chan, A.K., Skowronski, D.M., Salit, I., Simor, A.E., Slutsky, A.S., Doyle, P.W., Krajden, M., Petric, M., Brunham, R.C., McGeer, A.J., National Microbiology Laboratory, Canada, and Canadian Severe Acute Respiratory Syndrome Study Team. (2003). Identification of severe acute respiratory syndrome in Canada. N. Engl. J. Med. *348*, 1995–2005.

Riley, S., Fraser, C., Donnelly, C.A., Ghani, A.C., Abu-Raddad, L.J., Hedley, A.J., Leung, G.M., Ho, L.M., Lam, T.H., Thach, T.Q., Chau, P., Chan, K.P., Lo, S.V., Leung, P.Y., Tsang, T., Ho, W., Lee, K.H., Lau, E.M., Ferguson, N.M, and Anderson, R.M. (2003). Transmission dynamics of the etiological agent of SARS in Hong Kong: impact of public health interventions. Science *300*, 1961–1966.

Roberts, A., Vogel, L., Guarner, J., Hayes, N., Murphy, B., Zaki, S., and Subbarao, K. (2005). Severe acute respiratory syndrome coronavirus infection of golden Syrian hamsters. J. Virol. *79*, 503–511.

Rota, P.A., Oberste, M.S., Monroe, S.S., Nix, W.A., Campagnoli, R., Icenogle, J.P., Penaranda, S., Bankamp, B., Maher, K., Chen, M.H., Tong, S., Tamin, A., Lowe, L., Frace, M., DeRisi, J.L., Chen, Q., Wang, D., Erdman, D.D., Peret, T.C., Burns, C., Ksiazek, T.G., Rollin, P.E., Sanchez, A., Liffick, S., Holloway, B., Limor, J., McCaustland, K., Olsen-Rasmussen, M., Fouchier, R., Gunther, S., Osterhaus, A.D., Drosten, C., Pallansch, M.A., Anderson, L.J., and Bellini, W.J. (2003). Characterization of a novel coronavirus associated with severe acute respiratory syndrome. Science *300*, 1394–1399.

Simmons, G., Reeves, J.D., Rennekamp, A.J., Amberg, S.M., Piefer, A.J., and Bates, P. (2004). Characterization of severe acute respiratory syndrome-associated coronavirus (SARS-CoV) spike glycoprotein-mediated viral entry. Proc. Natl. Acad. Sci. USA. *101*, 4240–4245.

Sims, A.C., Baric, R.S., Yount, B., Burkett, S.E., Collins, P.L., Pickles, R.J. (2005). Severe acute respiratory syndrome coronavirus infection of human ciliated airway epithelia: role of ciliated cells in viral spread in the conducting airways of the lungs. J. Virol. 79, 15511–15524.

Snijder, E.J., Bredenbeek, P.J., Dobbe, J.C., Thiel, V., Ziebuhr, J., Poon, L.L., Guan, Y., Rozanov, M., Spaan, W.J., and Gorbalenya, A.E. (2003). Unique and conserved features of genome and proteome of SARS-coronavirus, an early split-off from the coronavirus group 2 lineage. J. Mol. Biol. *331*, 991–1004.

Song, H.D., Tu, C.C., Zhang, G.W., Wang, S.Y., Zheng, K., Lei, L.C., Chen, Q.X., Gao, Y.W., Zhou, H.Q., Xiang, H., Zheng, H.J., Chern, S.W., Cheng, F., Pan, C.M., Xuan, H., Chen, S.J., Luo, H.M., Zhou, D.H., Liu, Y.F., He, J.F., Qin, P.Z., Li, L.H., Ren, Y.Q., Liang, W.J., Yu, Y.D., Anderson, L., Wang, M., Xu, R.H., Wu, X.W., Zheng, H.Y., Chen, J.D., Liang, G., Gao, Y., Liao, M., Fang, L., Jiang, L.Y., Li, H., Chen, F., Di, B., He, L.J., Lin, J.Y., Tong, S., Kong, X., Du, L., Hao, P., Tang, H., Bernini, A., Yu, X.J., Spiga, O., Guo, Z.M., Pan, H.Y., He, W.Z., Manuguerra, J.C., Fontanet, A., Danchin, A., Niccolai, N., Li, Y.X., Wu, C.I., and Zhao, G.P. (2005). Cross-host evolution of severe acute respiratory syndrome coronavirus in palm civet and human. Proc. Natl. Acad. Sci. USA. *102*, 2430–2435.

To, K.F., and Lo, A.W. (2004). Exploring the pathogenesis of severe acute respiratory syndrome (SARS): the tissue distribution of the coronavirus (SARS-CoV) and its putative receptor, angiotensin-converting enzyme 2 (ACE2). J. Pathol. *203*, 740–743.

Tsang, K.W., Ho, P.L., Ooi, G.C., Yee, W.K., Wang, T., Chan-Yeung, M., Lam, W.K., Seto, W.H., Yam, L.Y., Cheung, T.M., Wong, P.C., Lam, B., Ip, M.S., Chan, J., Yuen, K.Y., and Lai, K.N. (2003). A cluster of cases of severe acute respiratory syndrome in Hong Kong. N. Engl. J. Med. *348*, 1977–1985.

Tu, C., Crameri, G., Kong, X., Chen, J., Sun, Y., Yu, M., Xiang, H., Xia, X., Liu, S., Ren, T., Yu, Y., Eaton, B.T., Xuan, H., and Wang, L.F. (2004). Antibodies to SARS coronavirus in civets. Emerg. Infect. Dis. *10*, 2244–2248.

Wang, P., Chen, J., Zheng, A., Nie, Y., Shi, X., Wang, W., Wang, G., Luo, M., Liu, H., Tan, L., Song, X., Wang, Z., Yin, X., Qu, X., Wang, X., Qing, T., Ding, M., and Deng, H. (2004). Expression cloning of functional receptor used by SARS coronavirus. Biochem. Biophys. Res. Commun. *315*, 439–444.

Wong, C.K., Lam, C.W., Wu, A.K., Ip, W.K., Lee, N.L., Chan, I.H., Lit, L.C., Hui, D.S., Chan, M.H., Chung, S.S., and Sung, J.J. (2004). Plasma inflammatory cytokines and chemokines in severe acute respiratory syndrome. Clin. Exp. Immunol. *136*, 95–103.

World Health Organization. (2003). A multicentre collaboration to investigate the cause of severe acute respiratory syndrome. Lancet *361*, 1730–1733.

Wu, D., Tu, C., Xin, C., Xuan, H., Meng, Q., Liu, Y., Yu, Y., Guan, Y., Jiang, Y., Yin, X., Crameri, G., Wang, M., Li, C., Liu, S., Liao, M., Feng, L., Xiang, H., Sun, J., Chen, J., Sun, Y., Gu, S., Liu, N., Fu, D., Eaton, B.T., Wang, L.F., and Kong, X. (2005). Civets are equally susceptible to experimental infection by two different severe acute respiratory syndrome coronavirus isolates. J. Virol. *79*, 2620–2625.

Xu, R.H., He, J.F., Evans, M.R., Peng, G.W., Field, H.E., Yu, D.W., Lee, C.K., Luo, H.M., Lin, W.S., Lin, P., Li, L.H., Liang, W.J., Lin, J.Y., and Schnur, A. (2004). Epidemiologic clues to SARS origin in China. Emerg. Infect. Dis. *10*, 1030–1037.

Yang, Z.Y., Huang, Y., Ganesh, L., Leung, K., Kong, W.P., Schwartz, O., Subbarao, K., and Nabel, G.J. (2004). pH-dependent entry of severe acute respiratory syndrome coronavirus is mediated by the spike glycoprotein and enhanced by dendritic cell transfer through DC-SIGN. J. Virol. *78*, 5642–5650.

Yang, Z.Y., Werner, H.C., Kong, W.P., Leung, K., Traggiai, E., Lanzavecchia, A., and Nabel, G.J. (2005). Evasion of antibody neutralization in emerging severe acute respiratory syndrome coronaviruses. Proc. Natl. Acad. Sci. USA. *102*, 797–801.

Yeh, S.H., Wang, H.Y., Tsai, C.Y., Kao, C.L., Yang, J.Y., Liu, H.W., Su, I.J., Tsai, S.F., Chen, D.S., Chen, P.J., and National Taiwan University SARS Research Team. (2004). Characterization of severe acute respiratory syndrome coronavirus genomes in Taiwan: molecular epidemiology and genome evolution. Proc. Natl. Acad. Sci. USA. *101*, 2542–2547.

Zeng, F.Y., Chan, C.W., Chan, M.N., Chen, J.D., Chow, K.Y., Hon, C.C., Hui, K.H., Li, J., Li, V.Y., Wang, C.Y., Wang, P.Y., Guan, Y., Zheng, B., Poon, L.L., Chan, K.H., Yuen, K.Y., Peiris, J.S., Leung, F.C. (2003). The complete genome sequence of severe acute respiratory syndrome coronavirus strain HKU-39849 (HK-39). Exp. Biol. Med. (Maywood). *228*, 866–873.

SARS Coronavirus and the Antiviral Cytokine Response

12

Martin Spiegel and Friedemann Weber

Abstract

Intracellular pathogens such as SARS Coronavirus (SARS-CoV) have to cope intensively with the innate immune system. Type I interferons (IFN-α/β) represent an important part of this host defense. IFNs are potent, antivirally active cytokines which can be produced by all nucleated cells in response to infection. Moreover, professional IFN producer cells such as dendritic cells are able to secrete immense amounts of IFNs within short time. IFNs trigger the synthesis of antivirally active proteins and shape adaptive immunity. Thus, in order to establish infection, viruses were forced to evolve efficient anti-IFN strategies. There is growing evidence that SARS-CoV is no exception to that rule. Here, we attempt to summarize the current state of knowledge regarding the interaction SARS-CoV with components of the IFN system, its connection to the induction of other cytokines, and the consequences for the pathogenesis of SARS.

Interferons as anti-SARS therapy

SARS-CoV is known to be sensitive to the antiviral action of type I interferons (IFN-α/β). In cell culture, virus growth is efficiently inhibited by exogenously added IFN-β, and, somewhat less efficiently, also by IFN-α (Cinatl *et al.*, 2004b; Cinatl *et al.*, 2003; Haagmans *et al.*, 2004; Hensley *et al.*, 2004; Spiegel *et al.*, 2004; Stroher *et al.*, 2004; Zheng *et al.*, 2004). Consequently, mice lacking a functional type I IFN system are more prone to virus-induced lung damages (Hogan *et al.*, 2004), and treatment with IFN-α can protect experimentally infected macaques (Haagmans *et al.*, 2004). IFNs may also alleviate symptoms in SARS patients (Loutfy *et al.*, 2003). In the macaque model, IFNs were most efficient if given before infection, but still have a certain antiviral effect if given after exposure to virus (Haagmans *et al.*, 2004). Thus, IFNs, which are an approved medication against several viral diseases (Samuel, 2001), offer the possibility both of prevention and treatment of SARS with a licensed drug (Cinatl *et al.*, 2004b).

The interferon system

IFNs provide a powerful and universal intracellular defense system against viruses. Accordingly, knockout mice which are unresponsive to IFN-α/β due to targeted deletions in the type I IFN receptor (Muller *et al.*, 1994) quickly succumb to viral infections although they have a regular adaptive immune system (reviewed in Weber *et al.*, 2004). Likewise, humans with genetic defects in the IFN system die of viral disease at an early age (Dupuis *et al.*, 2003).

Interferon induction

As far as it is known, all nucleated cells of the mammalian body are able to synthesize and secrete type I IFNs in response to virus infection. Secreted IFNs are then recognized by neighboring cells and cause them to express potent antiviral proteins (Samuel, 2001; Weber *et al.*, 2004). As a result, virus multiplication is slowed down or even stopped, buying the organism time for the establishment of an adaptive immune response.

Type I IFNs are classified according to their amino acid sequence and comprise a large number (at least 13) of IFN-α subtypes and a single IFN-β (Stark *et al.*, 1998), as well as some additional family members (Roberts *et al.*, 2003; van Pesch *et al.*, 2004). Expression patterns, i.e. which IFNs will be synthesized at which time point, mostly depend on the particular cell type.

Fibroblasts secrete mainly IFN-β as an initial response to infection but switch to IFN-α during the subsequent amplification phase of the IFN response (Marie *et al.*, 1998). By contrast, dendritic cells, which play an important role in immunosurveillance, directly secrete high levels of IFN-α subtypes (Colonna *et al.*, 2002; Diebold *et al.*, 2003).

On the molecular level, induction of type I IFN gene expression by the so-called "classical pathway" is best understood for IFN-β in fibroblasts (Figure 12.1). In infected cells, a signaling chain is activated by double-stranded RNA (dsRNA) molecules which are generated by most viruses as intermediates of viral transcription (Weber *et al.*, 2006). Two intracellular RNA helicases, RIG-I (Yoneyama *et al.*, 2004) and MDA-5 (Andrejeva *et al.*, 2004), act as sentinels for viral dsRNA (Kato *et al.*, 2005; Yoneyama *et al.*, 2005). RIG-I and MDA-5 both contain N-terminal caspase-recruiting domain (CARD)-like regions and C-terminal DExD/H box RNA helicase domains. dsRNA binding to the helicase domain most likely induces a conformational change which enables the CARD domain to interact with CARD-like domains of downstream signaling partners. A recently discovered adaptor protein which contains a RIG-I/MDA-5-binding CARD domain mediates the signal to downstream factors. This adaptor protein is called either IPS-1 for "interferon-β-promoter stimulator 1" (Kawai *et al.*, 2005), MAVS for "mitochondrial antiviral signaling" molecule (Seth *et al.*, 2005), VISA for "virus-induced signaling adaptor" (Xu *et al.*, 2005), or Cardif for "CARD adaptor inducing IFN-β" (Meylan *et al.*, 2005). IPS-1 activates the two IκB kinase (IKK)-related kinases, IKKε and TANK-binding kinase-1 (TBK-1), which phosphorylate the transcription factor IRF-3 (Fitzgerald *et al.*, 2003; Sharma *et al.*, 2003). IRF-3 is a member of the IFN regulatory factor (IRF) family and plays a central role in the activation of the IFN-β promoter (Lin *et al.*, 1998; Schafer *et al.*, 1998; Wathelet *et al.*, 1998; Weaver *et al.*, 1998; Yoneyama *et al.*, 1998). Phosphorylated IRF-3 homo-dimerizes and moves into the nucleus where it recruits the transcriptional coactivators p300 and CREB-binding protein (CBP) to initiate IFN-β mRNA synthesis (Hiscott *et al.*, 1999; Suhara *et al.*, 2002). This first-wave IFN triggers expression of a related factor, IRF-7, which in fibroblasts is only present in low amounts (Sato *et al.*, 2000). IRF-7 can a be activated the same way as IRF-3 (Iwamura *et al.*, 2001; Smith *et al.*, 2001; tenOever *et al.*, 2004), leading to a positive-feedback loop that initiates the synthesis of several IFN-α subtypes as the second-wave IFNs (Marie *et al.*, 1998; Sato *et al.*, 1998). In addition, NF-κB and AP-1 are activated in a dsRNA-dependent way (Chu *et al.*, 1999; Yang *et al.*, 1995). Together these transcription factors strongly upregulate IFN-β gene expression.

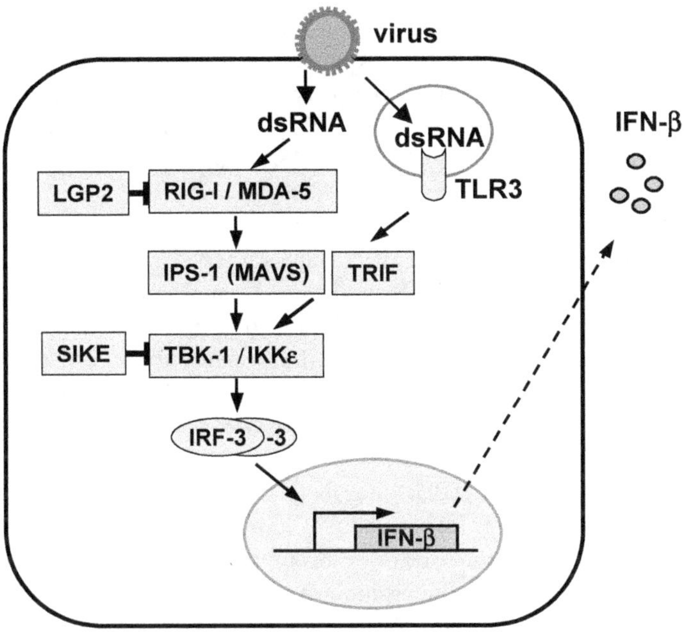

Figure 12.1 Classical pathway of type I IFN induction. dsRNA, a characteristic by-product of virus replication, leads to activation of the transcription factor IRF-3. Cooperative action with two other transcription factors, NF-kB and AP-1 (not shown), is required for full activation of the IFN-β promoter, but IRF-3 is the main player. IRF-3 is phosphorylated by the kinases IKKε and TBK-1 which in turn are activated by the RNA-sensing complex of RIG-I, MDA-5 and IPS-1/MAVS. A second signaling pathway involves endosomal TLR3 and TRIF.

Among the cells of the lymphatic system, myeloid dendritic cells (mDCs) (Diebold *et al.*, 2003) and, most prominently, plasmacytoid dendritic cells (pDCs) (Colonna *et al.*, 2002) are the main IFN producers. In addition to the classical, intracellular pathway of IFN induction described above, pDCs sense the presence of viruses by the extracellular or endosomal toll-like receptors (TLRs) (Beutler, 2004; Bowie and Haga, 2005). It is thought that TLRs serve as sensors for viral infection of phagocytosed cells (Schulz *et al.*, 2005). Human pDCs mostly express TLR7 and TLR9 which recognize viral single-stranded (ss) RNA and dsDNA, respectively (Iwasaki and Medzhitov, 2004), whereas myeloid DCs (mDCs) express TLR3 which responds to dsRNA (Alexopoulou *et al.*, 2001). Upon activation, TLRs signal through different intracellular adaptor molecules such as MyD88 (TLR7 and 9) or TRIF (TLR3) to induce IFN transcription (Iwasaki and Medzhitov, 2004). Interestingly, pDCs already contain high levels of IRF-7 (Kerkmann *et al.*, 2003; Prakash *et al.*, 2005), and TLRs are retained in endosomes to prolong signaling (Honda *et al.*, 2005). These features may explain the extraordinary ability of pDCs to rapidly produce high amounts of IFN-α.

It is not surprising that such a rapidly and heavily responding signaling cascade needs a counterbalance. Two molecules, named LGP2 and SIKE, were recently identified as negative regulators of the IFN induction pathway (see Figure 12.1). LGP2 shares the DExD/H box RNA helicase motif with the dsRNA detectors RIG-I and MDA-5. Since LGP2 however lacks the CARD motif necessary to relay the signal to IPS-1, it acts as a dominant-negative inhibitor interrupting the dsRNA-IPS-1 signal transduction (Rothenfusser *et al.*, 2005; Yoneyama *et al.*, 2005). SIKE inhibits a subsequent step by associating and inactivating the IRF-3 kinases TBK1 and IKKε (Huang *et al.*, 2005a).

Interferon signaling

All IFN-α/β subtypes bind to and activate a common type I IFN receptor. It consists of two subunits (IFNAR-1 and IFNAR-2) and is present on virtually all host cells (Samuel, 2001; Stark *et al.*, 1998). Binding of IFN-α/β leads to heterodimerization of the IFNAR subunits and to conformational changes in the intracellular parts of the receptor which activate the so-called JAK-STAT signaling pathway (Figure 12.2). The signal transducer and activator of transcription (STAT) proteins are latent cytoplasmic transcription factors which become phosphorylated by the Janus kinase (JAK) family members JAK-1 and TYK-2 (Levy and Darnell, 2002). Phosphorylated STAT-1 and STAT-2 recruit a third factor, IRF-9 (also called p48), to form a complex known as IFN stimulated gene factor 3 (ISGF-3). The ISGF-3 heterotrimer translocates to the nucleus and binds to IFN-stimulated response elements (ISRE) in the promoter regions of IFN-stimulated genes (ISGs), thereby inducing their transcription.

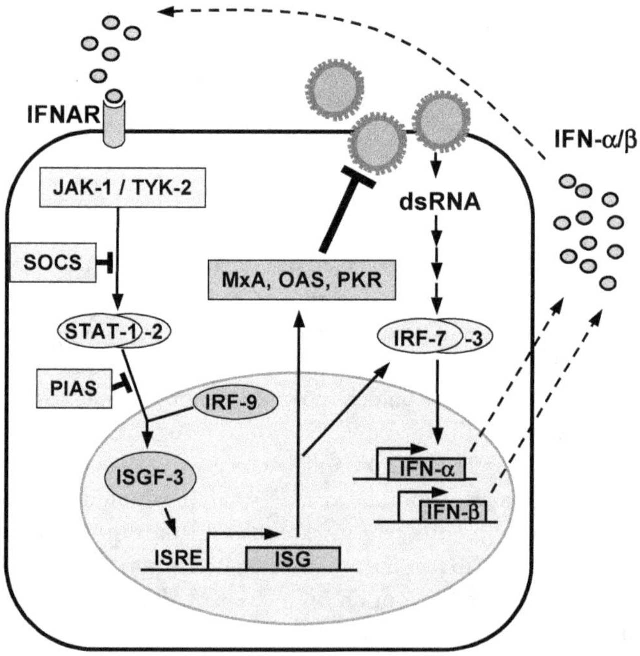

Figure 12.2 Type I IFN signaling and action. Newly synthesized IFN-β binds to the type I IFN receptor (IFNAR) and activates the expression of numerous ISGs via the JAK/STAT pathway. IRF-7 amplifies the IFN response by inducing the expression of several IFN-α subtypes. SOCS and PIAS are negative regulators of the JAK-STAT pathway. Mx, OAS and PKR are examples of proteins with antiviral activity. For details see text.

Several specialized proteins serve as negative regulators and inhibitors of the JAK-STAT pathway. For example, the suppressor of cytokine signaling (SOCS) proteins specifically prevent STAT activation by binding to activated cytokine receptors, inhibiting the activity of JAKs, and targeting bound signaling proteins for proteasomal degradation (Kubo *et al.*, 2003). Also, the protein inhibitor of activated STAT (PIAS) family members function as small ubiquitin-like modifier (SUMO) E3 ligases and inhibit the transcriptional activity of STATs (Shuai and Liu, 2005).

Interferon effector proteins

IFN-α/β activates the expression of more than 300 IFN-stimulated genes (ISGs) which have antiviral, antiproliferative, and immunomodulatory functions (de Veer *et al.*, 2001; Der *et al.*, 1998). IFN-induced proteins include enzymes, transcription factors, cell surface glycoproteins, cytokines, chemokines and a large number of factors with unknown function. Up to now, only a few antiviral proteins have been characterized in detail. The best studied IFN-induced antiviral proteins are the Mx GTPases (Haller and Kochs, 2002), the 2′–5′ oligoadenylate synthetases (2–5 OAS)/RNaseL (Silverman, 1994), and the protein kinase R (PKR) (Williams, 1999). PKR and 2–5 OAS are constitutively expressed in normal cells in a latent, inactive form. Basal mRNA levels are upregulated by IFN-α/β and these enzymes need to be activated by viral dsRNA. PKR is a serine-threonine kinase that phosphorylates the alpha subunit of the eukaryotic translation initiation factor eIF2 (Williams, 1999). As a consequence, translation of cellular and viral mRNAs is blocked. The 2′–5′ OAS catalyses the synthesis of short 2′–5′ oligoadenylates that activate the latent endoribonuclease RNaseL (Silverman, 1994). RNaseL, in turn, then degrades both viral and cellular RNAs, leading to viral inhibition (Zhou *et al.*, 1997).

The relevance of the Mx, OAS/RNaseL and PKR systems in the IFN response to viral infection is well documented both in tissue culture and animal experiments. Mice lacking one of these components show increased virus susceptibilities (Haller and Kochs, 2002; Zhou *et al.*, 1997; Zhou *et al.*, 1999). Interestingly, cells from so-called triple knockout mice lacking PKR, RNaseL and Mx still mount a limited IFN-induced antiviral state, indicating that additional antiviral pathways exist (Zhou *et al.*, 1999). Additional proteins with known antiviral activities are P56 (Guo *et al.*, 2000; Hui *et al.*, 2003), ISG20 (Espert *et al.*, 2003), promyelocytic leukemia protein (PML) (Regad and Chelbi-Alix, 2001) and guanylate-binding protein 1 (GBP-1) (Anderson *et al.*, 1999).

How does SARS-CoV cope with the IFN system?

SARS-CoV is highly pathogenic despite its pronounced sensitivity against IFN-α/β. This implies that the virus avoids activation of the IFN system to a large extent. Indeed, in fibroblast cells which are infected with SARS-CoV, no detectable induction of IFN occurs (Figure 12.3). pDCs, by contrast, but not mDCs, secrete substantial amounts of IFN in response to SARS-CoV (Cervantes-Barragan *et al.*, 2007). A lack of transcriptional induction in fibroblasts was also observed for IFN-α, IFN-λ, RANTES and IL-6 (Spiegel and Weber, 2006), suggesting that SARS-CoV-infected tissue cells are severely impaired in the production of a wide range of cytokines. In agreement with this, we have recently shown that SARS-CoV inhibits the full activation of IRF-3 (Spiegel *et al.*, 2005). Early after infection, IRF-3 is transiently transported to the nucleus in a similar way as

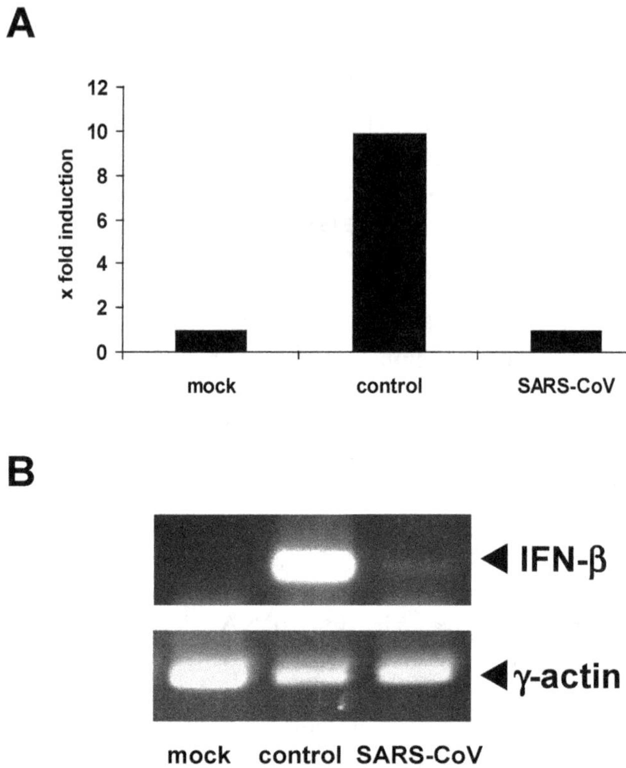

Figure 12.3 Absence of IFN-β induction in SARS-CoV-infected cells. A, Reporter assays. Human 293 cells were transfected with an IFN-β promoter reporter plasmid expressing firefly luciferase (Yoneyama *et al.*, 1998), together with an internal control construct encoding Renilla luciferase. At 6 hours post-transfection, cells were infected with 5 plaque forming units per cell of SARS-CoV or the IFN-inducing control virus BUNdelNSs (Weber *et al.*, 2002), or were left uninfected (mock). After overnight incubation, cells were lysed and luciferase activities were measured. Firefly luciferase counts were normalized to Renilla luciferase. B, RT-PCR analysis. Total RNA from infected 293 cells were tested by RT-PCR for the presence of IFN-β mRNA and γ-actin mRNA as described previously (Spiegel *et al.*, 2005).

for IFN-inducing viruses. At later time points, however, IRF-3 unexpectedly returns to the cytoplasm (see Figure 12.4). Furthermore, all other features of IRF-3 activation such as hyperphosphorylation, dimerization and interaction with the cofactor CBP are absent. Thus, SARS-CoV counteracts IRF-3 activation to avoid induction of antiviral cytokines.

These data pose several important questions about the underlying mechanisms both for the viral and the cellular side. First, is it really an active inhibition of IRF-3 in fibroblasts or is it just a passive hiding mechanism by SARS-CoV? Second, if there is an inhibition, which viral factor(s) may confer the inhibition of IRF-3? Third, which of the signal molecules involved in IRF-3 activation is actually affected? Fourth, what triggers the transient transport of IRF-3 to the nucleus early after infection? And fifth, what enables pDCs to circumvent the virally induced block of IFN induction.

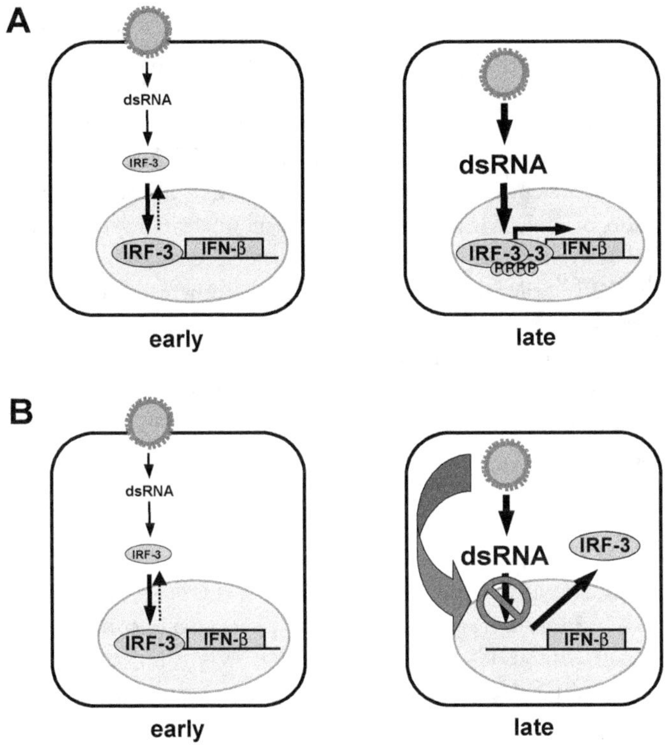

Figure 12.4 Model for the incomplete activation of IRF-3 in SARS-CoV-infected cells (Spiegel *et al.*, 2005). A, Full IRF-3 response to an IFN-inducing virus. The incoming virus produces dsRNA which immediately triggers transport of IRF-3 into the nucleus. IRF-3 however is a shuttling protein and needs to be hyperphosphorylated by TBK1 or IKKε in order to stay in the nucleus. Hyperphosphorylated IRF-3 also dimerizes and strongly activates the IFN-β promoter. B, Abortive IRF-3 response in SARS-CoV-infected cells. Initial nuclear transport is triggered by the virus, but no subsequent hyperphosphorylation or dimerization occurs. IRF-3 therefore shuttles out of the nucleus again without transactivating the IFN-β promoter.

The first question concerning an active inhibition vs. a passive avoidance mechanism by the virus can be approached by circumstantial evidence. SARS-CoV has a huge RNA genome which contains several secondary structures with dsRNA motifs (Plant *et al.*, 2005; Su *et al.*, 2005). Also, the dsRNA-responsive enzyme PKR is activated by SARS-CoV (Verena Krähling and Elke Mühlberger, personal communication). Furthermore, dsRNA can be directly detected in SARS-CoV-infected cells using a specific antibody (Weber *et al.*, 2006). The strength of the dsRNA signal is comparable to the signals observed in cells infected with other positive-strand RNA viruses. Thus, it can be concluded that SARS-CoV produces dsRNA sufficient for IFN induction, suggesting that it actively inhibits the IRF-3 pathway. This leads to the question of which viral protein mediates the IRF-3 inhibition, and which molecular step in IRF-3 activation is affected. It is meanwhile well known that many viruses express proteins inhibiting the induction of IFN by IRF-3, and that these IFN antagonists confer an important part of viral virulence (Garcia-Sastre, 2004; Haller *et al.*, 2006; Weber *et al.*, 2004). IFN antagonists of viruses can interfere either with the presentation of dsRNA, with the subsequent signal transducers MDA-5, IPS-1, or TBK-1, or with IRF-3 itself. For SARS-CoV, the nsp1 protein was recently identified as a general host cell shutoff factor (Kamitani *et al.*, 2006), but neither a viral IRF antagonist nor its cellular target are known so far. It is however feasible to propose that SARS-CoV expresses at least one protein which specifically affects IRF-3 activation. Apparently, the IRF-3 inhibition is sufficient to suppress IFN induction in fibroblasts, but not to prevent an initial activation of IRF-3 leading to a transient nuclear import. The nature of this incomplete IRF-3 activation, which is independent of the well-characterized hyperphosphorylation by TBK-1 (Spiegel *et al.*, 2005), remains to be established. Interestingly, the TLR3-pathway requires not only TBK1 but also PI3 kinase for the activation of IRF-3 (Sarkar *et al.*, 2004). Therefore, we have tested the PI3 kinase inhibitor LY294002 as well as inhibitors for MAPKK/Erk1,2 (U0126) and p38 MAPK (SB203580), but did not find any influence on the initial IRF-3 nuclear import by SARS-CoV (M. Spiegel and F. Weber, unpublished results).

Lastly, the fully functional IFN production by pDCs in response to SARS-CoV (Cervantes-Barragan *et al.*, 2007) again argues for a highly specific IRF-3 escape mechanism by SARS-CoV. IFN induction in pDCs mostly depends on TLR7 and TLR9 which activate the IFN transcription factor IRF-7 via the adaptor MyD88 and the kinases IRAK4 and IRAK1 (Honda *et al.*, 2005). Thus, apparently, in contrast to the ubiquitous IRF-3 signaling chain, the pDC-restricted IRF-7 pathway is not affected by the virus.

Taken together, SARS-CoV infection of fibroblasts results in the production of biologically active dsRNA and perhaps other danger signal molecules which then lead to an initial activation of IRF-3. However, all further steps of IRF-3 activity are inhibited by a yet to be identified viral factor, resulting in the absence of an IFN response by the infected cells. This efficient suppression of IFN induction almost certainly increases virulence as it was previously described for influenza virus and for Ebola virus (Geiss *et al.*, 2002; Kash *et al.*, 2006). However, at least in some cells such as pDCs the IFN escape strategy of SARS-CoV is somewhat leaky. Accordingly, mice deficient in IFN signaling are still more prone to SARS-CoV infection than IFN-competent animals (Hogan *et al.*, 2004). This well-known phenomenon of imperfect IFN inhibition was also shown for other viruses, suggesting that pathogens modulate rather than abrogate the IFN response (reviewed in Weber *et al.*, 2004).

Cytokines and chemokines directly induced by SARS-CoV

IRF-3 is not only crucial for IFN induction, but also participates in transactivation of the genes for RANTES (Lin *et al.*, 1999) and IP-10 (Sakaguchi *et al.*, 2003; Yamamoto *et al.*, 2003). Thus, it could be expected that SARS-CoV also suppresses production of these chemokines (also termed CCL5 and CXCL10, respectively), and recent data demonstrate that RANTES transcripts are indeed absent in productively infected tissue cells (Cinatl *et al.*, 2004a; Spiegel and Weber, 2006; Tang *et al.*, 2005a), as well as in SARS patients (Wong *et al.*, 2004). IP-10, however, is clearly upregulated in SARS-CoV-infected fibroblasts (Cinatl *et al.*, 2004a; Spiegel and Weber, 2006; Tang *et al.*, 2005a), indicating that this chemokine is less dependent on IRF-3 than RANTES. Indeed, transactivation of the IP-10 gene is not only triggered by IRF-3, but also dependent on the important transcription factor NF-κB (Leung *et al.*, 2004). Since NF-κB is strongly activated by SARS-CoV

(Cinatl *et al.*, 2004a; Cinatl *et al.*, 2005), this might provide an explanation for the upregulation of the IP-10 gene in infected cells despite the abortive activation of IRF-3.

SARS-CoV-infected cells also produce significant amounts of the chemokine IL-8 (CXCL8) (Cinatl *et al.*, 2004a; Tang *et al.*, 2005a; Wang *et al.*, 2004), an activity which was traced back to the viral spike and nucleocapsid proteins (Chang *et al.*, 2004; He *et al.*, 2003). Expression of IL-8 is dependent on the transcription factor AP-1, and molecular analyses revealed that AP-1 is strongly activated by SARS-CoV (Cinatl *et al.*, 2005). Importantly, IL-8 is able to inhibit the antiviral action of IFN (Polyak *et al.*, 2001). Therefore, besides the direct inhibition of IFN induction by SARS-CoV (see above), the upregulation of IL-8 might further add to the lack of an IFN response in cells and patients.

Taken together, the *in vitro* cytokine profile of SARS-CoV infection appears to be mainly based on the activation of the transcription factors NF-κB and AP-1 which in turn activate the expression of IP-10 and IL-8. IRF-3-depending genes for antiviral cytokines such as type I IFNs, by contrast, remain silent during the course of infection.

Interestingly, IP-10 is an excellent prognostic marker for SARS disease progression (Huang *et al.*, 2005b; Jiang *et al.*, 2005; Tang *et al.*, 2005b), implicating that the findings in cell culture reflect the *in vivo* situation to a considerable extent.

In vivo cytokine induction and SARS pathology

In agreement with the findings in cell culture (Cheung *et al.*, 2005; Law *et al.*, 2005; Spiegel *et al.*, 2005; Yilla *et al.*, 2005), no significant upregulation of α/β-IFNs or IFN-induced genes was reported in peripheral blood mononuclear cells or serum from SARS patients (Huang *et al.*, 2005b; Jones *et al.*, 2004; Reghunathan *et al.*, 2005; SARS, 2003; Wong *et al.*, 2004; Yu *et al.*, 2005). By contrast, high levels of the chemokines IL-8 and IP-10 along with the proinflammatory cytokine IL-6 (Hsueh *et al.*, 2004; Huang *et al.*, 2005b; SARS, 2003; Wang *et al.*, 2004; Wong *et al.*, 2004; Xu *et al.*, 2004; Yu *et al.*, 2005; Zhang *et al.*, 2004) were detected in the patients. The origin of this "cytokine storm" is largely unknown, but at least for IL-8 and IP-10 a direct production by virus-infected cells is conceivable (see above). IL-6, however, is not detected in productively infected tissue cells (Cinatl *et al.*, 2004a; Ng *et al.*, 2004; Tang *et al.*, 2005a) but moderately upregulated in abortively infected macrophages and DCs (Law *et al.*, 2005). Interestingly, IL-6 induction by SARS-CoV could be strongly boosted by priming of macrophages with bacterial LPS (Tang *et al.*, 2005a). Thus, it can be speculated that IL-8 and IP-10 in SARS patients are directly produced by virus-infected cells, whereas upregulation of the proinflammatory cytokine IL-6 is more likely a secondary response due to the activation of the immune system.

High viral load, systemic infection, influx of monocytes and macrophages in the infected tissues, and rapid depletion of T cells are typical for full-blown SARS (Ding *et al.*, 2004; Drosten *et al.*, 2003; Farcas *et al.*, 2005; Gu *et al.*, 2005; Lau and Peiris, 2005; Li *et al.*, 2004; Peiris *et al.*, 2003; Perlman and Dandekar, 2005; Tsui *et al.*, 2003; Weiss and Navas-Martin, 2005; Wong *et al.*, 2003). It is still not clear whether the disease is caused by the virus or is the result of a dysregulated immune response. Immediately after infection, SARS-CoV grows at a fast rate and spreads to different organs, including the lungs (Chan *et al.*, 2004; Gu *et al.*, 2005; Peiris *et al.*, 2003). Autopsies from deceased patients revealed severe damage of the lungs and lymphatic tissues, accompanied by infiltrations of monocytic cells (Ding *et al.*, 2003; Lang *et al.*, 2003; Nicholls *et al.*, 2003). This may indicate that immunopathogenesis is involved in the severe outcome of the disease, providing the rationale for SARS therapy with immunosuppressant corticosteroids (Bermejo and Munoz-Fernandez, 2004). On the other hand, there is evidence that cell damages could be directly caused by the virus, since SARS-CoV is cytolytic (Ng *et al.*, 2003), and high titers of virus have been found in several organs of deceased patients (Farcas *et al.*, 2005; Gu *et al.*, 2005; Mazzulli *et al.*, 2004). In addition, signs of necrosis were found besides virus particles in affected tissues (Nicholls *et al.*, 2003), and high viral loads are predictive of adverse clinical outcome (Hung *et al.*, 2004).

Integrating all the data listed above might help to reveal the role of the virus-cytokine interaction in the pathology of lethal SARS. In the initial phase of infection, the dampening of the antiviral IFN response allows the virus to grow exponentially and spread to different organs, including the lungs (Gu *et al.*, 2005; Peiris *et al.*, 2004). Thus, the virus buys time during the first, critical phase of infection in order to establish itself in the host. At the same time, however, the virus-induced chemokines IP-10 and IL-8 attract immune cells. These invading cells can themselves be infected (Gu *et al.*, 2005) and might produce even more chemokines and cytokines such as the proinflammatory cytokine IL-6 (Law *et al.*, 2005; Tang *et al.*, 2005a) and possibly also the anti-inflammatory cytokine TGF-β (Huang *et al.*, 2005b; Lee *et al.*, 2004; SARS, 2003; Yu *et al.*, 2005). Possibly, this mixture of high-level virus replication followed by the invasion of activated immune cells and production of both pro- and anti-inflammatory cytokines results in a strong cytokine storm, leading to an exhaustion and destruction of the immune system which culminates in the severe and often fatal respiratory distress, the hallmark of full-blown SARS.

The fact that a considerable fraction of patients survive infection with SARS-CoV may be due, at least in part, to a rapid production of sufficient amounts of IFN by pDCs (Cervantes-Barragan *et al.*, 2007). Although IFN has not been detected in SARS patients, local production of this short-lived cytokine by pDCs may be sufficient to slow down infection.

Open questions

Much has been learned about SARS-CoV and its interaction with the IFN and cytokine responses, but a lot of questions remain to be answered. As mentioned above, it is still unknown which viral gene product targets the IRF-3 pathway. For many viruses, the identity or even the molecular mechanism of the IFN antagonist are well characterized. Not so for SARS-CoV. Furthermore, the identity of the cellular target is unsolved. Is it IRF-3 itself? Or is it one of the proteins transferring the signal from viral dsRNA to IRF-3? Also, it would be interesting to know how IFN production by pDCs contributes to pathology or protection from SARS *in vivo*. Importantly, we also do not know whether SARS-CoV has also an impact on IFN signaling. Pathogenic viruses often disturb not only the induction of IFN, but also the signaling chain elicited by IFN. It is therefore conceivable the SARS-CoV has a similar activity. Lastly, an urgent issue is the identification of the ISG responsible for the clear effect of IFN on virus growth. The moderate inhibiting effect of IFN-γ is most likely caused by inducible NO synthetase (Akerstrom *et al.*, 2005), but which ISG confers the much stronger effect of IFN-α/β is unknown.

Future experiments approaching these issues will certainly improve our knowledge of SARS pathology and open up new ways for the prevention and treatment of SARS and possibly other coronaviral diseases.

Acknowledgments

Our own work described in the text was supported by the grants GZ Nr. 239 (202/12) from the Sino-German Center for Research promotion and We 2616/4 from the Deutsche Forschungsgemeinschaft (F.W.).

References

Akerstrom, S., Mousavi-Jazi, M., Klingstrom, J., Leijon, M., Lundkvist, A., and Mirazimi, A. (2005). Nitric oxide inhibits the replication cycle of severe acute respiratory syndrome coronavirus. J. Virol. *79*, 1966–1969.

Alexopoulou, L., Holt, A.C., Medzhitov, R., and Flavell, R.A. (2001). Recognition of double-stranded RNA and activation of NF-kappaB by Toll-like receptor 3. Nature *413*, 732–738.

Anderson, S.L., Carton, J.M., Lou, J., Xing, L., and Rubin, B.Y. (1999). Interferon-induced guanylate binding protein-1 (GBP-1) mediates an antiviral effect against vesicular stomatitis virus and encephalomyocarditis virus. Virology *256*, 8–14.

Andrejeva, J., Childs, K.S., Young, D.F., Carlos, T.S., Stock, N., Goodbourn, S., and Randall, R.E. (2004). The V proteins of paramyxoviruses bind the IFN-inducible RNA helicase, mda-5, and inhibit its activation of the IFN-beta promoter. Proc. Natl. Acad. Sci. USA *101*, 17264–17269.

Bermejo, J.F., and Munoz-Fernandez, M.A. (2004). Severe acute respiratory syndrome, a pathological immune response to the new coronavirus--implications for understanding of pathogenesis, therapy, design of vaccines, and epidemiology. Viral Immunol. *17*, 535–544.

Beutler, B. (2004). Inferences, questions and possibilities in Toll-like receptor signalling. Nature *430*, 257–263.

Bowie, A.G., and Haga, I.R. (2005). The role of Toll-like receptors in the host response to viruses. Mol. Immunol. *42*, 859–867.

Chan, K.H., Poon, L.L., Cheng, V.C., Guan, Y., Hung, I.F., Kong, J., Yam, L.Y., Seto, W.H., Yuen, K.Y., and Peiris, J.S. (2004). Detection of SARS coronavirus in patients with suspected SARS. Emerg. Infect. Dis. *10*, 294–299.

Chang, Y.J., Liu, C.Y., Chiang, B.L., Chao, Y.C., and Chen, C.C. (2004). Induction of IL-8 release in lung cells via activator protein-1 by recombinant baculovirus displaying severe acute respiratory syndrome-coronavirus spike proteins: identification of two functional regions. J. Immunol. *173*, 7602–7614.

Cheung, C.Y., Poon, L.L., Ng, I.H., Luk, W., Sia, S.F., Wu, M.H., Chan, K.H., Yuen, K.Y., Gordon, S., Guan, Y., and Peiris, J.S. (2005). Cytokine responses in severe acute respiratory syndrome coronavirus-infected macrophages in vitro: possible relevance to pathogenesis. J. Virol. *79*, 7819–7826.

Chu, W.M., Ostertag, D., Li, Z.W., Chang, L., Chen, Y., Hu, Y., Williams, B., Perrault, J., and Karin, M. (1999). JNK2 and IKKbeta are required for activating the innate response to viral infection. Immunity *11*, 721–731.

Cervantes-Barragan, L., Zust, R., Weber, F., Spiegel, M., Lang, K.S., Akira, S., Thiel, V., Ludewig, B. (2007). Control of coronavirus infection through plasmacytoid dendritic cell-derived type I interferon. Blood *109*, 1131–1137.

Cinatl, J., Jr., Hoever, G., Morgenstern, B., Preiser, W., Vogel, J.U., Hofmann, W.K., Bauer, G., Michaelis, M., Rabenau, H.F., and Doerr, H.W. (2004a). Infection of cultured intestinal epithelial cells with severe acute respiratory syndrome coronavirus. Cell. Mol. Life Sci. *61*, 2100–2112.

Cinatl, J., Jr., Michaelis, M., Morgenstern, B., and Doerr, H.W. (2005). High-dose hydrocortisone reduces expression of the pro-inflammatory chemokines CXCL8 and CXCL10 in SARS coronavirus-infected intestinal cells. Int. J. Mol. Med. *15*, 323–327.

Cinatl, J., Jr., Michaelis, M., Scholz, M., and Doerr, H.W. (2004b). Role of interferons in the treatment of severe acute respiratory syndrome. Expert Opin. Biol. Ther. *4*, 827–836.

Cinatl, J., Morgenstern, B., Bauer, G., Chandra, P., Rabenau, H., and Doerr, H.W. (2003). Treatment of SARS with human interferons. Lancet *362*, 293–294.

Colonna, M., Krug, A., and Cella, M. (2002). Interferon-producing cells: on the front line in immune responses against pathogens. Curr. Opin. Immunol. *14*, 373–379.

de Veer, M.J., Holko, M., Frevel, M., Walker, E., Der, S., Paranjape, J.M., Silverman, R.H., and Williams, B.R. (2001). Functional classification of interferon-stimulated genes identified using microarrays. J. Leukoc. Biol. *69*, 912–920.

Der, S.D., Zhou, A., Williams, B.R., and Silverman, R.H. (1998). Identification of genes differentially regulated by interferon alpha, beta, or gamma using oligonucleotide arrays. Proc. Natl. Acad. Sci. USA *95*, 15623–15628.

Diebold, S.S., Montoya, M., Unger, H., Alexopoulou, L., Roy, P., Haswell, L.E., Al-Shamkhani, A., Flavell, R., Borrow, P., and Reis e Sousa, C. (2003). Viral infection switches non-plasmacytoid dendritic cells into high interferon producers. Nature *424*, 324–328.

Ding, Y., He, L., Zhang, Q., Huang, Z., Che, X., Hou, J., Wang, H., Shen, H., Qiu, L., Li, Z., *et al.* (2004). Organ distribution of severe acute respiratory syndrome (SARS) associated coronavirus (SARS-CoV) in SARS patients: implications for pathogenesis and virus transmission pathways. J. Pathol. *203*, 622–630.

Ding, Y., Wang, H., Shen, H., Li, Z., Geng, J., Han, H., Cai, J., Li, X., Kang, W., Weng, D., *et al.* (2003). The clinical pathology of severe acute respiratory syndrome (SARS): a report from China. J. Pathol. *200*, 282–289.

Drosten, C., Gunther, S., Preiser, W., van der Werf, S., Brodt, H.R., Becker, S., Rabenau, H., Panning, M., Kolesnikova, L., Fouchier, R.A., *et al.* (2003). Identification of a novel coronavirus in patients with severe acute respiratory syndrome. N. Engl. J. Med. *348*, 1967–1976.

Dupuis, S., Jouanguy, E., Al-Hajjar, S., Fieschi, C., Al-Mohsen, I. Z., Al-Jumaah, S., Yang, K., Chapgier, A., Eidenschenk, C., Eid, P., *et al.* (2003). Impaired response to interferon-alpha/beta and lethal viral disease in human STAT1 deficiency. Nat. Genet. *33*, 388–391.

Espert, L., Degols, G., Gongora, C., Blondel, D., Williams, B.R., Silverman, R.H., and Mechti, N. (2003). ISG20, a new interferon-induced RNase specific for single-stranded RNA, defines an alternative antiviral pathway against RNA genomic viruses. J. Biol. Chem. *278*, 16151–16158.

Farcas, G.A., Poutanen, S.M., Mazzulli, T., Willey, B.M., Butany, J., Asa, S.L., Faure, P., Akhavan, P., Low, D.E., and Kain, K.C. (2005). Fatal severe acute respiratory syndrome is associated with multiorgan involvement by coronavirus. J. Infect. Dis. *191*, 193–197.

Fitzgerald, K.A., McWhirter, S.M., Faia, K.L., Rowe, D.C., Latz, E., Golenbock, D.T., Coyle, A.J., Liao, S.M., and Maniatis, T. (2003). IKKepsilon and TBK1 are essential components of the IRF3 signaling pathway. Nat. Immunol. *4*, 491–496.

Garcia-Sastre, A. (2004). Identification and characterization of viral antagonists of type I interferon in negative-strand RNA viruses. Curr. Top. Microbiol. Immunol. *283*, 249–280.

Geiss, G.K., Salvatore, M., Tumpey, T.M., Carter, V.S., Wang, X., Basler, C.F., Taubenberger, J.K., Bumgarner, R.E., Palese, P., Katze, M.G., and Garcia-Sastre, A. (2002). Cellular transcriptional profiling in influenza A virus-infected lung epithelial cells: the role of the nonstructural NS1 protein in the evasion of the host innate defense and its potential contribution to pandemic influenza. Proc. Natl. Acad. Sci. USA *99*, 10736–10741.

Gu, J., Gong, E., Zhang, B., Zheng, J., Gao, Z., Zhong, Y., Zou, W., Zhan, J., Wang, S., Xie, Z., *et al.* (2005). Multiple organ infection and the pathogenesis of SARS. J. Exp. Med. *202*, 415–424.

Guo, J., Hui, D.J., Merrick, W.C., and Sen, G.C. (2000). A new pathway of translational regulation mediated by eukaryotic initiation factor 3. EMBO J. *19*, 6891–6899.

Haagmans, B.L., Kuiken, T., Martina, B.E., Fouchier, R.A., Rimmelzwaan, G.F., Van Amerongen, G., Van Riel, D., De Jong, T., Itamura, S., Chan, K.H., *et al.* (2004). Pegylated interferon-alpha protects type 1 pneumocytes against SARS coronavirus infection in macaques. Nat. Med. *10*, 290–293.

Haller, O., and Kochs, G. (2002). Interferon-induced Mx proteins: dynamin-like GTPases with antiviral activity. Traffic *3*, 710–717.

Haller, O., Kochs, G., and Weber, F. (2006). The interferon response circuit: Induction and suppression by pathogenic viruses. Virology *344*, 119–130.

He, R., Leeson, A., Andonov, A., Li, Y., Bastien, N., Cao, J., Osiowy, C., Dobie, F., Cutts, T., Ballantine, M., and Li, X. (2003). Activation of AP-1 signal transduction pathway by SARS coronavirus nucleocapsid protein. Biochem. Biophys. Res. Commun. *311*, 870–876.

Hensley, L.E., Fritz, L.E., Jahrling, P.B., Karp, C.L., Huggins, J.W., and Geisbert, T.W. (2004). Interferon-beta 1a and SARS coronavirus replication. Emerg. Infect. Dis. *10*, 317–319.

Hiscott, J., Pitha, P., Genin, P., Nguyen, H., Heylbroeck, C., Mamane, Y., Algarte, M., and Lin, R. (1999). Triggering the interferon response: the role of IRF-3 transcription factor. J. Interferon Cytokine Res. *19*, 1–13.

Hogan, R.J., Gao, G., Rowe, T., Bell, P., Flieder, D., Paragas, J., Kobinger, G.P., Wivel, N.A., Crystal, R.G., Boyer, J., *et al.* (2004). Resolution of primary severe acute respiratory syndrome-associated coronavirus infection requires Stat1. J. Virol. *78*, 11416–11421.

Honda, K., Ohba, Y., Yanai, H., Negishi, H., Mizutani, T., Takaoka, A., Taya, C., Taniguchi, T. (2005). Spatiotemporal regulation of MyD88-IRF-7 signalling for robust type-I interferon induction. Nature *434*1035–1040.

Hsueh, P.R., Chen, P.J., Hsiao, C.H., Yeh, S.H., Cheng, W.C., Wang, J.L., Chiang, B.L., Chang, S.C., Chang, F.Y., Wong, W.W., *et al.* (2004). Patient data, early SARS epidemic, Taiwan. Emerg. Infect. Dis. *10*, 489–493.

Huang, J., Liu, T., Xu, L. G., Chen, D., Zhai, Z., and Shu, H. B. (2005a). SIKE is an IKK epsilon/TBK1-associated suppressor of TLR3- and virus-triggered IRF-3 activation pathways. EMBO J. *24*, 4018–4028.

Huang, K.J., Su, I.J., Theron, M., Wu, Y.C., Lai, S.K., Liu, C.C., and Lei, H.Y. (2005b). An interferon-gamma-related cytokine storm in SARS patients. J. Med. Virol. *75*, 185–194.

Hui, D.J., Bhasker, C.R., Merrick, W.C., and Sen, G.C. (2003). Viral stress-inducible protein p56 inhibits translation by blocking the interaction of eIF3 with the ternary complex eIF2.GTP.Met-tRNAi. J. Biol. Chem. *278*, 39477–39482.

Hung, I.F., Cheng, V.C., Wu, A.K., Tang, B.S., Chan, K.H., Chu, C.M., Wong, M.M., Hui, W.T., Poon, L.L., Tse, D.M., *et al.* (2004). Viral loads in clinical specimens and SARS manifestations. Emerg. Infect. Dis. *10*, 1550–1557.

Iwamura, T., Yoneyama, M., Yamaguchi, K., Suhara, W., Mori, W., Shiota, K., Okabe, Y., Namiki, H., and Fujita, T. (2001). Induction of IRF-3/-7 kinase and NF-kappaB in response to double- stranded RNA and virus infection: common and unique pathways. Genes Cells *6*, 375–388.

Iwasaki, A., and Medzhitov, R. (2004). Toll-like receptor control of the adaptive immune responses. Nat. Immunol. *5*, 987–995.

Jiang, Y., Xu, J., Zhou, C., Wu, Z., Zhong, S., Liu, J., Luo, W., Chen, T., Qin, Q., and Deng, P. (2005). Characterization of cytokine/chemokine profiles of severe acute respiratory syndrome. Am. J. Respir. Crit. Care Med. *171*, 850–857.

Jones, B.M., Ma, E.S., Peiris, J.S., Wong, P.C., Ho, J.C., Lam, B., Lai, K.N., and Tsang, K.W. (2004). Prolonged disturbances of in vitro cytokine production in patients with severe acute respiratory syndrome (SARS) treated with ribavirin and steroids. Clin. Exp. Immunol. *135*, 467–473.

Kamitani, W., Narayanan, K., Huang, C., Lokugamage, K., Ikegami, T., Ito, N., Kubo, H., Makino, S. (2006). Severe acute respiratory syndrome coronavirus nsp1 protein suppresses host gene expression by promoting host mRNA degradation. Proc. Natl. Acad. Sci. USA *103*, 12885–90.

Kash, J.C., Mühlberger, E., Carter, V., Grosch, M., Perwitasari, O., Proll, S.C., Thomas, M.J., Weber, F., Klenk, H.-D., and Katze, M.G. (2006). Global Suppression of the Host Anti-Viral Response by Ebola and Marburg Viruses: Increased Antagonism of the Type I IFN response is Associated with Enhanced Virulence. J. Virol. *80*, 3009–3020.

Kato, H., Sato, S., Yoneyama, M., Yamamoto, M., Uematsu, S., Matsui, K., Tsujimura, T., Takeda, K., Fujita, T., Takeuchi, O., and Akira, S. (2005). Cell Type-Specific Involvement of RIG-I in Antiviral Response. Immunity *23*, 19–28.

Kawai, T., Takahashi, K., Sato, S., Coban, C., Kumar, H., Kato, H., Ishii, K.J., Takeuchi, O., and Akira, S. (2005). IPS-1, an adaptor triggering RIG-I- and Mda5-mediated type I interferon induction. Nat. Immunol. *6*, 981–988.

Kerkmann, M., Rothenfusser, S., Hornung, V., Towarowski, A., Wagner, M., Sarris, A., Giese, T., Endres, S., and Hartmann, G. (2003). Activation with CpG-A and CpG-B oligonucleotides reveals two distinct regulatory pathways of type I IFN synthesis in human plasmacytoid dendritic cells. J. Immunol. *170*, 4465–4474.

Kubo, M., Hanada, T., and Yoshimura, A. (2003). Suppressors of cytokine signaling and immunity. Nat. Immunol. *4*, 1169–1176.

Lang, Z.W., Zhang, L.J., Zhang, S.J., Meng, X., Li, J.Q., Song, C.Z., Sun, L., Zhou, Y.S., and Dwyer, D.E. (2003). A clinicopathological study of three cases of severe acute respiratory syndrome (SARS). Pathology *35*, 526–531.

Lau, Y.L., and Peiris, J.S. (2005). Pathogenesis of severe acute respiratory syndrome. Curr. Opin. Immunol. *17*, 404–410.

Law, H.K., Cheung, C.Y., Ng, H.Y., Sia, S.F., Chan, Y.O., Luk, W., Nicholls, J.M., Peiris, J.S., and Lau, Y.L. (2005). Chemokine upregulation in SARS coronavirus infected human monocyte derived dendritic cells. Blood *106*, 2366–2374.

Lee, C.H., Chen, R.F., Liu, J.W., Yeh, W.T., Chang, J.C., Liu, P.M., Eng, H.L., Lin, M.C., and Yang, K.D. (2004). Altered p38 mitogen-activated protein kinase expression in different leukocytes with increment of immunosuppressive mediators in patients with severe acute respiratory syndrome. J. Immunol. *172*, 7841–7847.

Leung, T.H., Hoffmann, A., and Baltimore, D. (2004). One nucleotide in a kappaB site can determine cofactor specificity for NF-kappaB dimers. Cell *118*, 453–464.

Levy, D.E., and Darnell, J.E., Jr. (2002). Stats: transcriptional control and biological impact. Nat. Rev. Mol. Cell Biol. *3*, 651–662.

Li, T., Qiu, Z., Zhang, L., Han, Y., He, W., Liu, Z., Ma, X., Fan, H., Lu, W., Xie, J., *et al.* (2004). Significant changes of peripheral T lymphocyte subsets in patients with severe acute respiratory syndrome. J. Infect. Dis. *189*, 648–651.

Lin, R., Heylbroeck, C., Genin, P., Pitha, P.M., and Hiscott, J. (1999). Essential role of interferon regulatory factor 3 in direct activation of RANTES chemokine transcription. Mol. Cell Biol. *19*, 959–966.

Lin, R., Heylbroeck, C., Pitha, P.M., and Hiscott, J. (1998). Virus-dependent phosphorylation of the IRF-3 transcription factor regulates nuclear translocation, transactivation potential, and proteasome-mediated degradation. Mol. Cell Biol. *18*, 2986–2996.

Loutfy, M.R., Blatt, L.M., Siminovitch, K.A., Ward, S., Wolff, B., Lho, H., Pham, D.H., Deif, H., LaMere, E.A., Chang, M., *et al.* (2003). Interferon alfacon-1 plus corticosteroids in severe acute respiratory syndrome: a preliminary study. JAMA *290*, 3222–3228.

Marie, I., Durbin, J.E., and Levy, D.E. (1998). Differential viral induction of distinct interferon-alpha genes by positive feedback through interferon regulatory factor-7. EMBO J. *17*, 6660–6669.

Mazzulli, T., Farcas, G.A., Poutanen, S.M., Willey, B.M., Low, D.E., Butany, J., Asa, S.L., and Kain, K.C. (2004). Severe acute respiratory syndrome-associated coronavirus in lung tissue. Emerg. Infect. Dis. *10*, 20–24.

Meylan, E., Curran, J., Hofmann, K., Moradpour, D., Binder, M., Bartenschlager, R., and Tschopp, J. (2005). Cardif is an adaptor protein in the RIG-I antiviral pathway and is targeted by hepatitis C virus. Nature *437*, 1167–1172.

Muller, U., Steinhoff, U., Reis, L.F., Hemmi, S., Pavlovic, J., Zinkernagel, R.M., and Aguet, M. (1994). Functional role of type I and type II interferons in antiviral defense. Science *264*, 1918–1921.

Ng, L.F., Hibberd, M.L., Ooi, E.E., Tang, K.F., Neo, S.Y., Tan, J., Murthy, K.R., Vega, V.B., Chia, J.M., Liu, E.T., and Ren, E.C. (2004). A human in vitro model system for investigating genome-wide host responses to SARS coronavirus infection. BMC Infect. Dis. *4*, 34.

Ng, M. L., Tan, S. H., See, E. E., Ooi, E. E., and Ling, A. E. (2003). Proliferative growth of SARS coronavirus in Vero E6 cells. J Gen Virol *84*, 3291–3303.

Nicholls, J.M., Poon, L.L., Lee, K.C., Ng, W.F., Lai, S.T., Leung, C.Y., Chu, C.M., Hui, P.K., Mak, K.L., Lim, W., *et al.* (2003). Lung pathology of fatal severe acute respiratory syndrome. Lancet *361*, 1773–1778.

Peiris, J.S., Chu, C.M., Cheng, V.C., Chan, K.S., Hung, I.F., Poon, L.L., Law, K.I., Tang, B.S., Hon, T.Y., Chan, C.S., *et al.* (2003). Clinical progression and viral load in a community outbreak of coronavirus-associated SARS pneumonia: a prospective study. Lancet *361*, 1767–1772.

Peiris, J.S., Guan, Y., and Yuen, K.Y. (2004). Severe acute respiratory syndrome. Nat. Med. *10*, S88–97.

Perlman, S., and Dandekar, A.A. (2005). Immunopathogenesis of coronavirus infections: implications for SARS. Nat. Rev. Immunol. *5*, 917–927.

Plant, E.P., Perez-Alvarado, G.C., Jacobs, J. L., Mukhopadhyay, B., Hennig, M., and Dinman, J.D. (2005). A three-stemmed mRNA pseudoknot in the SARS coronavirus frameshift signal. PLoS Biol. *3*, e172.

Polyak, S.J., Khabar, K.S., Paschal, D.M., Ezelle, H.J., Duverlie, G., Barber, G.N., Levy, D.E., Mukaida, N., and Gretch, D.R. (2001). Hepatitis C virus nonstructural 5A protein induces interleukin-8, leading to partial inhibition of the interferon-induced antiviral response. J. Virol. *75*, 6095–6106.

Prakash, A., Smith, E., Lee, C.K., and Levy, D.E. (2005). Tissue-specific positive feedback requirements for production of type I interferon following virus infection. J. Biol. Chem. *280*, 18651–18657.

Regad, T., and Chelbi-Alix, M.K. (2001). Role and fate of PML nuclear bodies in response to interferon and viral infections. Oncogene *20*, 7274–7286.

Reghunathan, R., Jayapal, M., Hsu, L.Y., Chng, H.H., Tai, D., Leung, B.P., and Melendez, A.J. (2005). Expression profile of immune response genes in patients with Severe Acute Respiratory Syndrome. BMC Immunol. *6*, 2.

Roberts, R.M., Ezashi, T., Rosenfeld, C.S., Ealy, A.D., and Kubisch, H.M. (2003). Evolution of the interferon tau genes and their promoters, and maternal-trophoblast interactions in control of their expression. Reprod. Suppl. *61*, 239–251.

Rothenfusser, S., Goutagny, N., DiPerna, G., Gong, M., Monks, B.G., Schoenemeyer, A., Yamamoto, M., Akira, S., and Fitzgerald, K.A. (2005). The RNA helicase Lgp2 inhibits TLR-independent sensing of viral replication by retinoic acid-inducible gene-I. J. Immunol. *175*, 5260–5268.

Sakaguchi, S., Negishi, H., Asagiri, M., Nakajima, C., Mizutani, T., Takaoka, A., Honda, K., and Taniguchi, T. (2003). Essential role of IRF-3 in lipopolysaccharide-induced interferon-beta gene expression and endotoxin shock. Biochem. Biophys. Res. Commun. *306*, 860–866.

Samuel, C.E. (2001). Antiviral actions of interferons. Clin. Microbiol. Rev. *14*, 778–809.

Sarkar, S.N., Peters, K.L., Elco, C.P., Sakamoto, S., Pal, S., and Sen, G.C. (2004). Novel roles of TLR3 tyrosine phosphorylation and PI3 kinase in double-stranded RNA signaling. Nat. Struct. Mol. Biol. *11*, 1060–1067.

SARS, B. G. o. N. R. P. f. (2003). Dynamic changes in blood cytokine levels as clinical indicators in severe acute respiratory syndrome. Chin. Med. J. (Engl.) *116*, 1283–1287.

Sato, M., Hata, N., Asagiri, M., Nakaya, T., Taniguchi, T., and Tanaka, N. (1998). Positive feedback regulation of type I IFN genes by the IFN-inducible transcription factor IRF-7. FEBS Lett. *441*, 106–110.

Sato, M., Suemori, H., Hata, N., Asagiri, M., Ogasawara, K., Nakao, K., Nakaya, T., Katsuki, M., Noguchi, S., Tanaka, N., and Taniguchi, T. (2000). Distinct and essential roles of transcription factors IRF-3 and IRF-7 in response to viruses for IFN-alpha/beta gene induction. Immunity *13*, 539–548.

Schafer, S.L., Lin, R., Moore, P.A., Hiscott, J., and Pitha, P.M. (1998). Regulation of type I interferon gene expression by interferon regulatory factor-3. J. Biol. Chem. *273*, 2714–2720.

Schulz, O., Diebold, S.S., Chen, M., Naslund, T.I., Nolte, M.A., Alexopoulou, L., Azuma, Y.T., Flavell, R.A., Liljestrom, P., and Reis e Sousa, C. (2005). Toll-like receptor 3 promotes cross-priming to virus-infected cells. Nature *433*, 887–892.

Seth, R.B., Sun, L., Ea, C-K., and Chen, Z.J. (2005). Identification and Characterization of MAVS, a Mitochondrial Antiviral Signaling Protein that Activates NF-κB and IRF3. Cell *122*, 669–682.

Sharma, S., TenOever, B.R., Grandvaux, N., Zhou, G.P., Lin, R., and Hiscott, J. (2003). Triggering the Interferon Antiviral Response Through an IKK-Related Pathway. Science *300*, 1148–1151.

Shuai, K., and Liu, B. (2005). Regulation of gene-activation pathways by PIAS proteins in the immune system. Nat. Rev. Immunol. *5*, 593–605.

Silverman, R.H. (1994). Fascination with 2–5A-dependent RNase: a unique enzyme that functions in interferon action. J. Interferon Res. *14*, 101–104.

Smith, E.J., Marie, I., Prakash, A., Garcia-Sastre, A., and Levy, D.E. (2001). IRF3 and IRF7 phosphorylation in virus-infected cells does not require double-stranded RNA-dependent protein kinase R or Ikappa B kinase but is blocked by Vaccinia virus E3L protein. J. Biol. Chem. *276*, 8951–8957.

Spiegel, M., Pichlmair, A., Martinez-Sobrido, L., Cros, J., Garcia-Sastre, A., Haller, O., and Weber, F. (2005). Inhibition of Beta interferon induction by severe acute respiratory syndrome coronavirus suggests a two-step model for activation of interferon regulatory factor 3. J. Virol. *79*, 2079–2086.

Spiegel, M., Pichlmair, A., Mühlberger, E., Haller, O., and Weber, F. (2004). The antiviral effect of interferon-beta against SARS-Coronavirus is not mediated by MxA. J. Clinical. Virol. *30*, 211–213.

Spiegel, M., and Weber, F. (2006). Inhibition of cytokine gene expression and induction of chemokine genes in non-lymphatic cells infected with SARS coronavirus. Virol. J. *3*,17.

Stark, G.R., Kerr, I.M., Williams, B.R., Silverman, R.H., and Schreiber, R.D. (1998). How cells respond to interferons. Annu. Rev. Biochem. *67*, 227–264.

Stroher, U., DiCaro, A., Li, Y., Strong, J.E., Aoki, F., Plummer, F., Jones, S.M., and Feldmann, H. (2004). Severe acute respiratory syndrome-related coronavirus is inhibited by interferon- alpha. J. Infect. Dis. *189*, 1164–1167.

Su, M.C., Chang, C.T., Chu, C.H., Tsai, C.H., and Chang, K.Y. (2005). An atypical RNA pseudoknot stimulator and an upstream attenuation signal for -1 ribosomal frameshifting of SARS coronavirus. Nucleic Acids Res. *33*, 4265–4275.

Suhara, W., Yoneyama, M., Kitabayashi, I., and Fujita, T. (2002). Direct involvement of CREB-binding protein/p300 in sequence-specific DNA binding of virus-activated interferon regulatory factor-3 holocomplex. J. Biol. Chem. *277*, 22304–22313.

Tang, B.S., Chan, K.H., Cheng, V.C., Woo, P.C., Lau, S.K., Lam, C.C., Chan, T.L., Wu, A.K., Hung, I.F., Leung, S.Y., and Yuen, K.Y. (2005a). Comparative host gene transcription by microarray analysis early after infection of the Huh7 cell line by severe acute respiratory syndrome coronavirus and human coronavirus 229E. J. Virol. *79*, 6180–6193.

Tang, N.L., Chan, P.K., Wong, C.K., To, K.F., Wu, A.K., Sung, Y.M., Hui, D.S., Sung, J.J., and Lam, C.W. (2005b). Early enhanced expression of interferon-inducible protein-10 (CXCL-10) and other chemokines predicts adverse outcome in severe acute respiratory syndrome. Clin. Chem. *51*, 2333–2340.

tenOever, B.R., Sharma, S., Zou, W., Sun, Q., Grandvaux, N., Julkunen, I., Hemmi, H., Yamamoto, M., Akira, S., Yeh, W.C., *et al.* (2004). Activation of TBK1 and IKK epsilon kinases by vesicular stomatitis virus infection and the role of viral ribonucleoprotein in the development of interferon antiviral immunity. J. Virol. *78*, 10636–10649.

Tsui, P.T., Kwok, M.L., Yuen, H., and Lai, S.T. (2003). Severe acute respiratory syndrome: clinical outcome and prognostic correlates. Emerg. Infect. Dis. *9*, 1064–1069.

van Pesch, V., Lanaya, H., Renauld, J.C., and Michiels, T. (2004). Characterization of the murine alpha interferon gene family. J. Virol. *78*, 8219–8228.

Wang, W.K., Chen, S.Y., Liu, I.J., Kao, C.L., Chen, H.L., Chiang, B.L., Wang, J.T., Sheng, W.H., Hsueh, P.R., Yang, C.F., *et al.* (2004). Temporal relationship of viral load, ribavirin, interleukin (IL)-6, IL-8, and clinical progression in patients with severe acute respiratory syndrome. Clin. Infect. Dis. *39*, 1071–1075.

Wathelet, M.G., Lin, C.H., Parekh, B.S., Ronco, L.V., Howley, P.M., and Maniatis, T. (1998). Virus infection induces the assembly of coordinately activated transcription factors on the IFN-beta enhancer in vivo. Mol. Cell *1*, 507–518.

Weaver, B.K., Kumar, K.P., and Reich, N.C. (1998). Interferon regulatory factor 3 and CREB-binding protein/p300 are subunits of double-stranded RNA-activated transcription factor DRAF1. Mol. Cell Biol. *18*, 1359–1368.

Weber, F., Bridgen, A., Fazakerley, J.K., Streitenfeld, H., Randall, R.E., and Elliott, R.M. (2002). Bunyamwera bunyavirus nonstructural protein NSs counteracts the induction of alpha/beta interferon. J. Virol. *76*, 7949–7955.

Weber, F., Kochs, G., and Haller, O. (2004). Inverse interference: how viruses fight the interferon system. Viral Immunol. *17*, 498–515.

Weber, F., Wagner, V., Rasmussen, S.B., Hartmann, R., and Paludan, S.R. (2006). Double-stranded RNA is produced by positive-strand RNA viruses and DNA viruses, but not in detectable amounts by negative-strand RNA viruses. J. Virol. *80*, 5059–5064

Weiss, S.R., and Navas-Martin, S. (2005). Coronavirus pathogenesis and the emerging pathogen severe acute respiratory syndrome coronavirus. Microbiol. Mol. Biol. Rev. *69*, 635–664.

Williams, B.R. (1999). PKR; a sentinel kinase for cellular stress. Oncogene *18*, 6112–6120.

Wong, C.K., Lam, C.W., Wu, A.K., Ip, W.K., Lee, N.L., Chan, I.H., Lit, L.C., Hui, D.S., Chan, M.H., Chung, S.S., and Sung, J.J. (2004). Plasma inflammatory cytokines and chemokines in severe acute respiratory syndrome. Clin. Exp. Immunol. *136*, 95–103.

Wong, R.S., Wu, A., To, K.F., Lee, N., Lam, C.W., Wong, C.K., Chan, P.K., Ng, M.H., Yu, L.M., Hui, D.S., *et al.* (2003). Haematological manifestations in patients with severe acute respiratory syndrome: retrospective analysis. BMJ *326*, 1358–1362.

Xu, L., Ran, L., Cameron, M.J., Persad, D., Danesh, A., Gold, W., Keshavjee, S., Brunton, J., Loeb, M., and Kelvin, D.J. (2004). Proinflammatory gene expression profiles and severity of disease course in SARS patients. Paper presented at: International Conference on SARS, one year after the (first) outbreak (Luebeck, Germany).

Xu, L.G., Wang, Y.Y., Han, K.J., Li, L.Y., Zhai, Z., and Shu, H.B. (2005). VISA is an adapter protein required for virus-triggered IFN-beta signaling. Mol. Cell *19*, 1–14.

Yamamoto, M., Sato, S., Hemmi, H., Hoshino, K., Kaisho, T., Sanjo, H., Takeuchi, O., Sugiyama, M., Okabe, M., Takeda, K., and Akira, S. (2003). Role of adaptor TRIF in the MyD88-independent toll-like receptor signaling pathway. Science *301*, 640–643.

Yang, Y.L., Reis, L.F., Pavlovic, J., Aguzzi, A., Schafer, R., Kumar, A., Williams, B.R., Aguet, M., and Weissmann, C. (1995). Deficient signaling in mice devoid of double-stranded RNA-dependent protein kinase. EMBO J. *14*, 6095–6106.

Yilla, M., Harcourt, B.H., Hickman, C.J., McGrew, M., Tamin, A., Goldsmith, C.S., Bellini, W.J., and Anderson, L.J. (2005). SARS-coronavirus replication in human peripheral monocytes/macrophages. Virus Res. *107*, 93–101.

Yoneyama, M., Kikuchi, M., Matsumoto, K., Imaizumi, T., Miyagishi, M., Taira, K., Foy, E., Loo, Y.M., Gale, M., Jr., Akira, S., *et al.* (2005). Shared and Unique Functions of the DExD/H-Box Helicases RIG-I, MDA5, and LGP2 in Antiviral Innate Immunity. J. Immunol. *175*, 2851–2858.

Yoneyama, M., Kikuchi, M., Natsukawa, T., Shinobu, N., Imaizumi, T., Miyagishi, M., Taira, K., Akira, S., and Fujita, T. (2004). The RNA helicase RIG-I has an essential function in double-stranded RNA-induced innate antiviral responses. Nat. Immunol. *5*, 730–737.

Yoneyama, M., Suhara, W., Fukuhara, Y., Fukuda, M., Nishida, E., and Fujita, T. (1998). Direct triggering of the type I interferon system by virus infection: activation of a transcription factor complex containing IRF-3 and CBP/p300. EMBO J. *17*, 1087–1095.

Yu, S.Y., Hu, Y.W., Liu, X.Y., Xiong, W., Zhou, Z.T., and Yuan, Z.H. (2005). Gene expression profiles in peripheral blood mononuclear cells of SARS patients. World J. Gastroenterol. *11*, 5037–5043.

Zhang, Y., Li, J., Zhan, Y., Wu, L., Yu, X., Zhang, W., Ye, L., Xu, S., Sun, R., Wang, Y., and Lou, J. (2004). Analysis of serum cytokines in patients with severe acute respiratory syndrome. Infect. Immun. *72*, 4410–4415.

Zheng, B., He, M.L., Wong, K.L., Lum, C.T., Poon, L.L., Peng, Y., Guan, Y., Lin, M.C., and Kung, H.F. (2004). Potent inhibition of SARS-associated coronavirus (SCOV) infection and replication by type I interferons (IFN-alpha/beta) but not by type II interferon (IFN-gamma). J. Interferon Cytokine Res. *24*, 388–390.

Zhou, A., Paranjape, J., Brown, T.L., Nie, H., Naik, S., Dong, B., Chang, A., Trapp, B., Fairchild, R., Colmenares, C., and Silverman, R.H. (1997). Interferon action and apoptosis are defective in mice devoid of 2′,5′- oligoadenylate-dependent RNase L. EMBO J. *16*, 6355–6363.

Zhou, A., Paranjape, J.M., Der, S.D., Williams, B.R., and Silverman, R.H. (1999). Interferon action in triply deficient mice reveals the existence of alternative antiviral pathways. Virology *258*, 435–440.

Grand Challenges in Human Coronavirus Vaccine Development

13

Barry Rockx and Ralph S. Baric

Abstract

The emergence and identification of several common and rare human coronaviruses that cause severe lower respiratory tract infection argues for the judicious development of robust coronavirus vaccines and vector platforms.

Currently, limited information is available on the correlates of protection against SARS-CoV and other severe lower respiratory tract human coronavirus infections, a clear priority for future research. Passive immunization has been successful in establishing protection from SARS-CoV in animal models suggesting an important role for neutralizing antibodies. One important property of future vaccine candidates is the ability to confer protection against multiple variant strains of SARS-CoV, especially in senescent populations that are most at risk for severe disease.

Many vaccine candidates are capable of inducing humoral and cellular responses and protection from subsequent homologous challenge has been reported in animal models. The development of infectious clones for coronaviruses has facilitated the identification of attenuating mutations, deletions and recombinations which could ultimately result in live attenuated vaccine candidates. Stable vaccine platforms should be developed that allow for rapid intervention strategies against any future emerging coronaviruses.

Vaccine correlates that enhance disease after challenge should be thoroughly investigated and mechanisms devised to circumvent vaccine-associated complications.

Introduction: overview of human coronaviruses

Historically, coronavirus infections were associated with mild upper respiratory tract diseases in infants, children and adults, as the prototype strains HCoV-OC43 and HCoV-229E were associated with 10–30% of common colds in winter (Larson *et al.*, 1980). Occasionally, infection was also linked with more severe lower respiratory tract disease, especially in populations with chronic underlying diseases (Gagneur *et al.*, 2002). The 21st century heralded the recognition of the more pathogenic human coronaviruses, like SARS-CoV which is a zoonotic virus that likely originated in bats and circulated between civets and humans (Guan *et al.*, 2003; Lau *et al.*, 2005; Li *et al.*, 2005; Song *et al.*, 2005). Soon after, HCoV-NL63 was identified as an important cause of lower respiratory tract infections in infants, children and to a lesser extent adults (Chiu *et al.*, 2005; van der Hoek *et al.*, 2004). In young children and infants, virus infection oftentimes results in crop and bronchiolitis (van der Hoek *et al.*, 2005). Most recently, the HKU1 isolate was identified from two adult patients with pneumonia, leading to additional epidemiologic studies that suggest an HKU1 infection incidence of about 2–5% of community acquired pneumonia in adults, many of whom present with enteric symptoms (Woo *et al.*, 2005). Human coronavirus research was complicated by the lack of a reverse genetic system or animal models of human disease. These viruses oftentimes propagate slowly or not at all and the number of reagents was limited. Given that coronaviruses are capable of rapid host switching and evolution in changing ecologies and the existence of more pathogenic strains associated with severe lower respiratory tract infection, vaccines and therapeutic developments are warranted. Moreover, the unique genome organization and regulatory mechanisms, tropism to mucosal compartments in the lung and intestine, impart novel characteristics for use in coronavirus vaccine vectors and therapeutics. In this manuscript, we review the grand challenges associated with the development of efficacious human coronavirus vaccines using the SARS-CoV as a model.

SARS-CoV structure and replication

SARS-CoV is an enveloped virus of 80 to 120 nm in diameter containing a positive-strand RNA of approximately 30 kb bound by the nucleocapsid protein (N). The capsid is surrounded by a lipid bilayer that contains at least four structural proteins, spike (S), envelope (E), ORF3a and membrane (M). The replicase gene comprises about two-thirds of the genome (Figure 13.1). Replicase genes are encoded in two large ORFs, designated ORF1a and ORF1b, proteolytically processed into 16 non-structural proteins (nsp) by two proteases, PLP (nsp3) and Mpro (nsp5). The role of the replicase proteins in viral pathogenesis is unknown but replicase likely influences disease outcomes related to its role in RNA synthesis and viral replication and high probability for encoding alleles that influence pathogenesis and virulence, directly. For example, SARS and MHV nsp2 is non-essential for virus replication and mutations in nsp14 have little impact on virus replication *in vitro*, but attenuate *in vivo* pathogenesis (Graham *et al.*, 2005; Sperry *et al.*, 2005). Downstream of the replicase, there are four open-reading frames (ORF) encoding the structural proteins S, E, M and N. In addition there are several group specific ORFs located between the different structural genes (Rota *et al.*, 2003). ORF3a localizes in the rER/golgi compartment, interacts with the S and M glycoproteins, and is incorporated into virions, suggesting that it may function in virion maturation and release (Ito *et al.*, 2005). ORF3b has no described functional activities and ORF6 is largely unstudied. ORF7 is

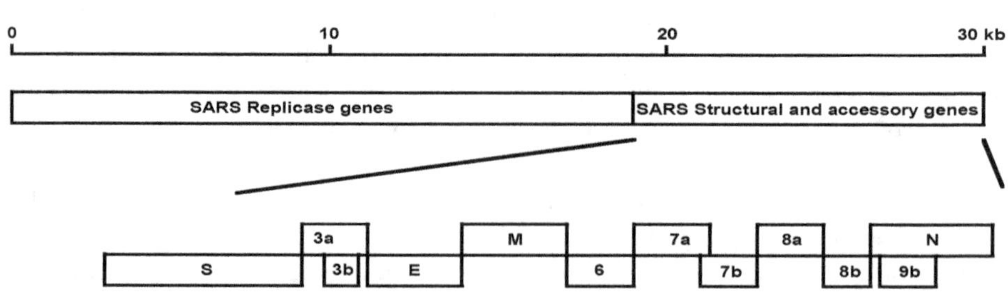

Figure 13.1 Genome organization of SARS-CoV. The SARS-CoV genome comprises of a single positive strand RNA of approximately 30 kb in length. The genome is organized into 14 open reading frames (ORFs).

divided into two ORFs designated ORF7a and 7b. Although no ORF7b product has been described to date, ORF7a encodes a 122-amino-acid type I transmembrane protein and structural studies reveal a compact seven-stranded β sandwich similar in fold and topology to members of the Ig superfamily (Nelson *et al.*, 2005). Sequences encoded in the transmembrane domain and short cytoplasmic tail mediated trafficking into the rER/Golgi. ORF7a protein can be expressed *in vitro* and has been detected *in vivo* (Chen *et al.*, 2005) and in plasmid vectors has been shown to block cell cycle progression at the G0/G1 phase (Yuan *et al.*, 2006) and induce apoptosis (Tan *et al.*, 2004; Tan *et al.*, 2004). Early studies have failed to demonstrate ORF7a in the virion; however, as the E protein was also not detected in virions, additional studies are warranted (Tan *et al.*, 2004). The group specific ORFs, including ORF3a, seem to be non-essential for replication *in vitro* and in mice, but may encode virulence determinants *in vivo* (Yount *et al.*, 2005). For example, chimeric mouse hepatitis virus encoding the SARS-CoV ORF6 was more pathogenic in mice, suggesting a key role for ORF6 in viral pathogenesis (Pewe *et al.*, 2005). The function of ORF6 is unknown.

Replication of the SARS-CoV genome during infection involves the synthesis of a full-length negative-strand RNA that serves as a template for full-length genomic RNA. For the SARS-CoV, nine overlapping 3′-co-terminal subgenomic RNAs serve as mRNAs, as does full-length genomic RNA. Each mRNA has a common leader sequence at the 5′ end derived from the 5′ end of the genomic RNA through a mechanism unique to viruses within the family of *Nidovirales* called discontinuous transcription (Thiel *et al.*, 2003). Discontinuous transcription occurs during the synthesis of the negative-strand RNA by adding an antileader sequence to the 3′ ends of the negative strand RNAs, involving transcription-regulating sequences (TRS), which then serve as a template for the synthesis of mRNAs.

Evolution of the SARS-CoV

The first cases of SARS-CoV infection were detected in late 2002 in Guangdong Province in China. By July 2003 the outbreak was divided into early, middle and late phases based largely on epidemiologic and phylogenetic analyses (Chinese, 2004). The early phase was initially characterized by a series of independent cases, likely due to zoonotic transmission. The middle phase refers to the major outbreak of SARS-CoV in a hospital where a superspreader event was associated with more than 130 primary and secondary infections. The late phase started at Hotel M in Honk Kong from where virus transmission occurred to Vietnam, Singapore, Canada and the USA, eventually spreading to about 30 countries globally (Figure 13.2A).

Both serological, epidemiological and phylogenetical analyses suggested that the virus originated from animals (Yu at al., 2003; Guan *et al.*, 2003; Kan *et al.*, 2005). Multiple studies have shown the presence of phylogenetically related viruses in Himalayan palm civets and raccoon dogs from live animal markets leading to the large scale culling of these animals at the markets that successfully prevented further outbreaks (Guan *et al.*, 2003; Kan *et al.*, 2005). Importantly, subsequent studies did not show a widespread infection in wild civets although a large number of additional zoonotic isolates with unique sequence have been identified during the course of these studies. In addition a recent study showed that civets, experimentally infected with either early or middle phase strains of SARS-

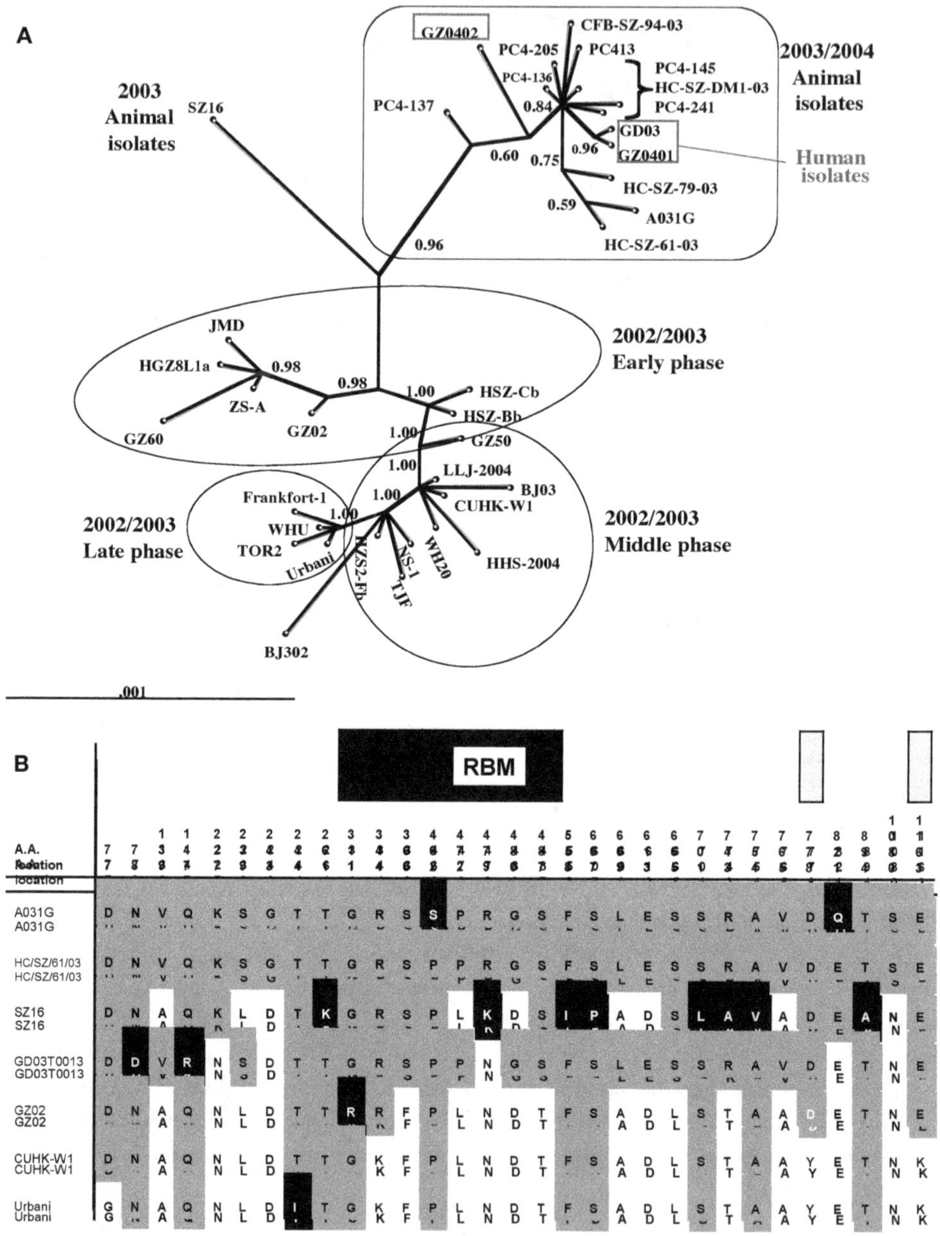

Figure 13.2 Phylogenetic relationships among SARS-CoV. A, An unrooted tree generated by Bayesian Inference using representative SARS spike protein sequences. B, Genetic variation in SARS-CoV Spike glycoprotein. Mutations in Receptor Binding Domain (RBD), Fusion peptide (FP) and Heptad repeat 2 (HR-2) are shown.

CoV, developed clinical symptoms, making it unlikely that these species are the natural reservoir (Wu *et al.*, 2005). All together these data suggest that palm civets may have been the source of the outbreak through zoonotic transmission at live animal markets where both animals and humans are in close contact with each other, facilitating in the establishment of a transmission cycle. Recently two reports have suggested that bats may be the natural reservoirs of SARS-CoV-like viruses (Lau *et al.*, 2005; Li *et al.*, 2005). A high prevalence (23–84%) of SARS-CoV antibodies was found in Chinese horseshoe bats and related species. These findings were confirmed by detection of viral genome in anal swaps and fecal samples. Sequence analysis showed that bat-SARS-CoV was closely related to SARS-CoV from humans and civets with an 88% nucleotide and 93% amino acid similarity to human and civet SARS-CoV isolates. Clearly, the data indicate that zoonotic SARS-CoV strains will continue to circulate, minimally in Southeast Asia, providing a steady source for future human outbreaks. Little is known about SARS-CoV replication and pathogenesis in its natural host, the host range of the bat isolates and their global distribution among new and old world bat species.

Comparative analyses of the SARS-CoV genomes of different isolates from both humans and civets throughout the different phases of the outbreak showed a high mutational rate in S protein that seemed critical for the transition from animal-to-human to human-to-human transmission (Chinese, 2004; Kan *et al.*, 2005; Pavlovic-Lazetic *et al.*, 2005; Song *et al.*, 2005). The spike protein mediates both cell attachment and membrane fusion through binding to its receptor. Angiotensin-converting enzyme 2 (ACE-2) has been identified as the main cellular receptor for SARS-CoV and a receptor-binding domain (RBD) mediating this interaction has been defined (Chakraborti *et al.*, 2005; Li *et al.*, 2003). Within this RBD only seven amino acids differ between the human and civet isolates of SARS-CoV (Figure 13.2B), but a difference in binding affinity to human ACE-2 of up to 1000-fold could be observed (Li *et al.*, 2005). By structure analyses, two of the amino acids (479N and 487T) are located in the binding interface of the S-ACE-2 complex, both evolved during the course of the SARS epidemic, and have been identified to be critical in the binding to ACE-2 (Li *et al.*, 2005). Most of the animal isolates contain an arginine or lysine at position 479 and a serine at position 487, whereas early to late phase human isolates contain an asparagine and threonine respectively (Figure 13.2B). These substitutions are thought to cause steric hindrance thereby dramatically decreasing receptor-binding affinity. In addition to the S glycoprotein, significant evolution has been noted in the replicase polyprotein, ORF3a, M glycoprotein and ORF8, although direct support for a functional role in the expanding epidemic has not been described experimentally (Chinese, 2004; Hu *et al.*, 2003; Hu *et al.*, 2003; Zeng *et al.*, 2004).

Pathogenesis and immune responses

Infections of the human lung are especially common, accounting for about 6% of the total global burden of disease in 2002, consistently more than any other disease identified in humans (Mizgerd, 2006). The overall contribution of human coronaviruses to this global burden is unknown, especially in resource poor nations. SARS-CoV infection resulted in acute interstitial pneumonitis with lung pathology showing diffuse alveolar damage, multinucleate giant-cells and an influx of macrophages in the alveoli and the interstitium of the lung (Ding *et al.*, 2003; Franks *et al.*, 2003; Hsiao *et al.*, 2005; Nicholls *et al.*, 2003).

In addition, SARS-CoV has a widespread multiorgan distribution including intestinal tract, kidney, sweatglands, liver, pancreas and cerebrum (Ding *et al.*, 2004). Although epidemiologic studies suggest that the virus is primarily spread by large droplets in respiratory secretions, virus in feces, urine and sweat provide the opportunity for other transmission modes. Stimulated macrophages in the alveoli are thought to play a role in the pathogenesis of SARS-CoV by releasing proinflammatory cytokines (Wang *et al.*, 2005). Macrophages have been shown to be infected with SARS-CoV *in vitro* leading to initiation of viral transcription and protein synthesis but replication was abortive and no infectious virus was produced (Cheung *et al.*, 2005). SARS-CoV did induce expression of chemokines such as CXCL10/Interferon (IFN)-gamma-inducible protein 10 and CCL2/monocyte chemotactic protein 1 in macrophages but did not induce beta interferon (Cheung *et al.*, 2005; Tseng *et al.*, 2005). SARS pathology is not that dissimilar from findings in H5N1 infected patients and animals and in cases of severe community acquired pneumonias that are unresponsive to antibiotic therapy. In general, acute interstitial pneumonitis is a severe disease, respiratory failure is the rule and fatal outcomes are not uncommon. Pathophysiological mechanisms underlying diffuse alveolar damage and pulmonary fibrosis in clinical and experimental models revealed involvement of the proinflammatory and fibrogenic cytokine pathways (Ware and Matthay, 2000). The relative contributions of direct viral cytopathology versus immunopathology remain unclear (Nicholls *et al.*, 2006). SARS-CoV, as a new human coronavirus, is especially virulent in senescent populations. Age is a major risk factor for severe disease suggesting that senescence of the immune system is a major contributor to disease outcomes (Choi *et al.*, 2003). It is not clear whether previous exposure to other group 2 coronaviruses like HKU1 and OC43 may have contributed to misbalanced immune responses in the elderly.

The mechanism of SARS-CoV spread in the human lung is unknown. One model suggests that SARS-CoV infects lung dendritic cells that transport virus to the alveolar regions allowing for infection of type II pneumocytes. This model is based on studies in which dendritic cells were infected by SARS-CoV via an interaction with high expression levels of DC-SIGN and genetic studies that suggest that polymorphisms in DC-SIGN influence susceptibility to infection (Yang *et al.*, 2004). An alternative model predicts that SARS-CoV infection and spread could occur through some component of the respiratory epithelium. Indeed, consideration of the early pathology of SARS-CoV observed in the bronchiolar regions and the demonstration that SARS-CoV replicates efficiently in ciliated cells isolated from nasal and tracheobronchial airway regions support the latter hypothesis (Sims *et al.*, 2005). Moreover, hACE2 expression occurs throughout the conducting airways of the human lung and provides a suitable substrate for repeated cycles of virus amplification and spread throughout the airways, ultimately spreading to some cell component of the alveolus regions (Nicholls *et al.*, 2006; Sims *et al.*, 2005). Unfortunately, clinical samples are not available to determine virus distribution in the lung and cell targets at early times in infection.

A lymphopenia and increase in viral load for the first 10 days after onset of SARS-CoV infection suggest immune evasion in the early phases of infection (Booth *et al.*, 2003; Hui *et al.*, 2003). One proposed mechanism of immune evasion is the low expression of antiviral cytokines (IFN) through interference of the IFN regulatory factor 3 as well as the upregulation of proinflammatory cytokines (Law *et al.*, 2005; Okabayashi *et al.*, 2006;

Spiegel *et al.*, 2005). The observed lymphopenia is likely the result of pro-apoptotic properties of SARS-CoV rather than the direct infection of lymphocytes since lymphocytes lack expression of the SARS-CoV receptor ACE-2 (Yan *et al.*, 2004). Several SARS-CoV proteins have been implicated in the induction of apoptosis *in vitro*. These include ORF7a, 3a and 3b (Law *et al.*, 2005; Tan *et al.*, 2004; Yuan *et al.*, 2005). In contrast collectins that can bind the glycosylated S protein of SARS-CoV like the mannose-binding lectin (MBL) have been suggested to play a protective role against the development of SARS (Ip *et al.*, 2005). Virus infection of the lung is typically cleared by 14 days after the onset of symptoms (Nicholls *et al.*, 2006) although detection of viral genomes in stool by RT-PCR persists for several more weeks in some patients.

Antibodies against SARS-CoV can not be detected within the first 7 days of illness but both IgG and IgM increase dramatically by day 15 (Mo *et al.*, 2006; Woo *et al.*, 2004). The IgG titer reaches a peak around day 60 and remains high until day 180 after which it slowly declines until day 720. IgM rapidly reaches a peak after day 15 and then declines until it is undetectable at day 180. Antibody reactivity is mainly directed against the S protein and the N protein. The presence of neutralizing antibodies suggests that patients infected with SARS-CoV may be protected from reinfection for up to 2 years (Liu *et al.*, 2006). In contrast, most immunity to other human coronaviruses is short-lived as many volunteers can be experimentally reinfected with HCoV-229E within 1 year of an earlier challenge.

Components of protective immunity

Global humoral and cellular immune responses

After human infection, most patients seroconvert by day 14 after onset of illness, antibody titers peak between 20–30 days and then sustained over 150 days. Most neutralizing activity is attributed to IgG targeting the S glycoprotein. Humoral immune responses have been demonstrated against the SARS N, M, ORF3a, ORF6 and S glycoproteins (Buchholz *et al.*, 2004). Cellular immune responses are targeted primarily against the SARS S (CD8) and N (CD4) structural proteins, although detailed analyses of the replicase and group specific ORFs have not been reported (Chen *et al.*, 2005; Wang *et al.*, 2004; Xu and Gao, 2004). In the mouse, two H-2(b) restricted epitopes (S436–443, S525–532) and one H-2(d) restricted epitope (S366–374) have been mapped in the S glycoprotein (Zhi *et al.*, 2005) and two T cell epitopes have been mapped in the N protein of the SARS-CoV (Gupta *et al.*, 2006; Liu *et al.*, 2006). In humans, the relative contributions of the cellular and humoral immune responses to protective immunity are unknown although most agree the strong neutralizing titers are positively correlated with protection from severe disease or death.

SARS-CoV S glycoprotein antigenicity

Since the SARS-CoV S glycoprotein mediates viral attachment and is the major structural protein in the virion it is not surprising that this protein is highly immunogenic and the most important neutralizing antigen. Although strong immune responses are elicited against both S and N protein (Buchholz *et al.*, 2004; Wang *et al.*, 2005; Zhu, 2004), passive transfer studies have illustrated that only anti-S antibody confers protection from

SARS-CoV replication in the mouse model (Bisht *et al.*, 2005; Sui *et al.*, 2005). Several immunogenic epitopes have been identified within the S protein, however only three domains within the S protein seem to be the main targets for neutralizing antibodies (Figure 13.3). One domain is the RBD, where antibodies are likely to block binding to ACE-2 (He *et al.*, 2005; Ho *et al.*, 2004). Interestingly this is a much conserved domain with only 4 amino acid changes between SARS-CoV isolates of all phases of the epidemic and market animals and therefore a possible target for vaccine development. However a change in the critical amino acid 479N will not only change affinity for binding to ACE-2 but can also significantly reduce binding of antibodies to the RBD, thereby strongly diminishing neutralization (Li *et al.*, 2005). A second domain is localized between amino acid residues 130 and 140 in the S glycoprotein and monoclonal antibodies targeted against this domain strongly neutralize virus infectivity (Greenough *et al.*, 2005). The mechanism by which these antibodies neutralize infectivity is unknown. Finally a third domain targeted by neutralizing antibodies resides in the heptad repeat 2 (HR2), located adjacent to the transmembrane domain. Antibodies directed against this site likely upset the interaction of the HR2 with HR1 thereby abolishing the fusion activity (Lai *et al.*, 2005; Lip *et al.*, 2006).

A panel of human monoclonal antibodies has been developed against late phase SARS-CoV strains (Traggiai *et al.*, 2004). Many of these human monoclonals efficiently neutralize wildtype late phase Urbani and Tor-2 strains, although activity against early, middle and zoonotic strains is less well characterized. Some antibodies appear to have strong activity against pseudotyped lentiviruses bearing various SARS S glycoproteins (Temperton *et al.*, 2005); however, in one instance the monoclonal antibody enhanced binding and infectivity in a human renal adenocarcinoma cell line (Yang *et al.*, 2005). Clearly more detailed studies regarding the antigenic relationships and cross neutralizing capabilities of these monoclonal antibodies is essential for developing effective vaccines and therapeutics. A panel of antigenically unique SARS-CoV that span the epidemic and zoonotic strains is desperately needed to test the efficacy of vaccines and therapeutics against zoonotic strains, the likely source of future outbreak strains.

Components of protective immunity

To date, all studies agree that the induction of neutralizing antibodies against the SARS-CoV S glycoprotein is a critical component of protective immunity in animal models. Passive immunization of mice with neutralizing monoclonal antibodies confers protection against homologous SARS-CoV replication (Subbarao *et al.*, 2004; Sui *et al.*, 2005). In

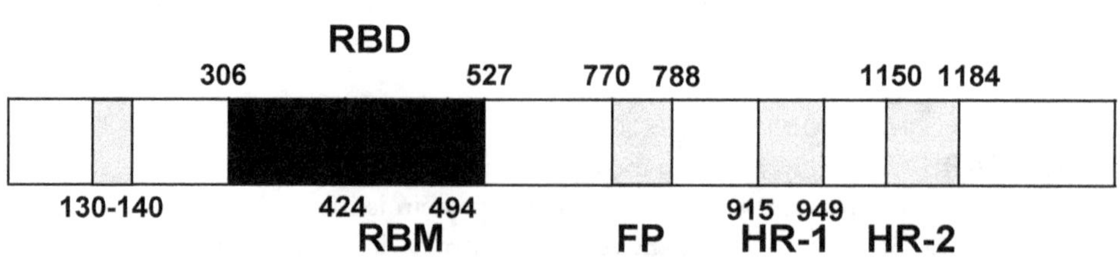

Figure 13.3 Structure of SARS-CoV Spike glycoprotein. Neutralizing epitopes are present within the RBD, HR-1 and HR-2 as well as in an unassigned region (amino acid 130–140).

humans the presence of high levels of neutralizing antibody is correlated with recovery from SARS-CoV infection (Ho *et al.*, 2005). No information on cellular immunological correlates of protection has been described in humans. The relative contribution of other viral structural proteins (M and E) to protection from SARS-CoV infection is thought minimal (Buchholz *et al.*, 2004). The contributions of the group specific ORFs and replicase ORFs to protective immunity are poorly studied.

Human coronavirus vaccines

Classic human coronavirus infection results in mild upper respiratory tract infection which resolves within a few days after onset of symptoms. Human challenge experiments indicate that about 50% of infected volunteers can be reinfected by homologous challenge within 1 year of exposure (Callow, 1985; Callow *et al.*, 1990). Not surprisingly, infection severity limited interest in the development of human coronavirus vaccines. Discovery of the more severe lower respiratory tract pathogens, NL63 and HKU1, and the SARS-CoV has stimulated interest in the development of efficacious human coronavirus vaccines and recombinant vaccine platforms for the treatment of human respiratory disease. As full length molecular clones to HCV-229E and HCoV-OC43 have recently been established (St-Jean *et al.*, 2006; Thiel *et al.*, 2001), platform approaches for developing coronavirus live attenuated vaccines and vaccine vectors for humans is possible. Further development of attenuated coronaviruses is essential for establishing the safety, efficacy and immunogenicity of human coronavirus based vaccines and vaccine platforms.

Given SARS-CoV's high morbidity and mortality in humans, the possibility of laboratory acquired infections, its prevalence in animal reservoirs, its recognized potential as an agent of bioterrorism, and the severe economic effects to communities, understanding SARS-CoV pathogenesis and developing efficacious vaccines is clearly essential for protecting the global public health. Inactivated vaccines suffer from the possibility of incomplete inactivation, short-lived and unbalanced immunity while recombinant vaccines suffer from high expense, limited global applications, and possible unbalanced immune responses (Chanock *et al.*, 1992; Fulginiti *et al.*, 1967; Moss and Polack, 2001; Norrby *et al.*, 1975). For coronaviruses, the MHV model suggests that optimal protective immune responses require both T cell and antibody responses (Lin *et al.*, 1999; Stohlman *et al.*, 1998). In mice infected with MHV-JHM, cytotoxic T cells are critical for initial elimination of virus, but in the absence of antibody, recrudescence and death results (Lin *et al.*, 1999; Stohlman *et al.*, 1998). Given that an immunopathological component contributes to coronavirus pathogenesis in infected cats and mice (Lin *et al.*, 1999; Rottier, 1999; Stohlman *et al.*, 1998), balanced T and antibody responses may be essential for control of SARS-CoV and other severe human coronavirus infections. Given the problems associated with FIPV vaccine development, experts recommend caution when developing SARS-CoV vaccines (Marshall and Enserink, 2004). Because SARS-CoV transmission is highly susceptible to public health intervention strategies, the most likely global scenario is spread of endemic disease in resource poor nations followed by episodic introduction into resource rich nations.

Vaccines face a series of potential concerns including reversion or recombination repair of attenuated vaccine strains, induction of immune-mediated enhancement of pathology, waning immune protection, lack of cross-protection for heterologous strains, and limited vaccine efficacy within senescent populations. Furthermore, immune enhancement has been demonstrated with another coronavirus, feline infectious peritonitis coronavirus (Olsen, 1993; Olsen *et al.*, 1992), and more recently with a modified vaccinia vector expressing SARS-CoV S that exacerbated hepatitis in ferrets while failing to protect from infection (Czub *et al.*, 2005). Notably, antibodies against the epidemic Urbani strain are reported to enhance binding and infectivity of lentiviruses pseudotyped with an animal SARS-CoV S glycoprotein, raising the specter of vaccine mediated immune enhancement following heterotypic challenge (Yang *et al.*, 2005). Another potential problem is that SARS vaccines might fail to induce antibodies which protect from infection with divergent strains of SARS-CoV. The S glycoprotein of SARS-CoV contains about 2–10% amino acid variation between zoonotic and the 2003 epidemic strains (Chinese, 2004), possibly limiting the effectiveness of monotypic SARS-CoV S vaccines. Finally, studies measuring the duration of protective immunity or vaccine efficacy in aged animals have not yet been reported.

Killed and recombinant antigen vaccines

Several inactivated SARS-CoV-based vaccines have been described in the literature (He *et al.*, 2004; Spruth *et al.*, 2006; Tang *et al.*, 2004; Zakhartchouk *et al.*, 2005; Zhang *et al.*, 2004; Zhou *et al.*, 2005) (Table 13.1). One of the advantages of an inactivated vaccine is that all structural proteins of the SARS-CoV are present and may serve as antigens to induce neutralizing antibodies and protective responses. Killed vaccines have been produced by beta-propiolactone, formaldehyde or UV inactivation and were shown effective in mice or non-human primates challenged with homologous SARS-CoV strains (Qin *et al.*, 2006; See *et al.*, 2006; Spruth *et al.*, 2006). They induced protective and potent humoral and mucosal immune responses and to date show no adverse effects in animal models. However the virus inactivation protocols should be well established and tested for SARS-CoV to prevent incomplete inactivation and subsequent inoculation of live virus. This is especially important as the laboratory acquired outbreak of SARS-CoV in China was associated with improperly inactivated SARS-CoV lysates (Li *et al.*, 2004).

A safer method to present antigen would be by recombinant antigen vaccination. Several recombinant antigen vaccines have been described (Bisht *et al.*, 2005; Keng *et al.*, 2005; Zhou *et al.*, 2006). All are based on fragments of the S protein, either peptides of about 30 amino acids or a polypeptide containing amino acids 14 to 762, and induce neutralizing antibodies. In one study the recombinant vaccine also showed complete protection from replication in the lungs of mice (Bisht *et al.*, 2005). However expression of antigen in recombinant protein expression systems may affect the conformation of the protein and may therefore affect the induction of conformation dependent neutralizing antibodies. The formation of virus-like particles (VLP) containing the S, E and M proteins may provide a way of presenting the recombinant proteins in their naïve form, although it is unclear whether sufficient material can be harvested to provide sufficient vaccine for broad human use (Ho *et al.*, 2004).

Long-term efficacy of killed and recombinant antigen vaccines has not been evaluated in any animal models. Such studies should be undertaken as unbalanced Th1/Th2 responses and/or waning of the immune response may occur in elderly populations, as has been described with measles and RSV (Anderson and Heilman, 1995; Glass and Grenfell,

Table 13.1 Current vaccine candidates for SARS-CoV

Vaccine type	Target	Immune response	Protection against SARS-CoV challenge?	Reference
Killed virus vaccines				
Beta-propiolactone inactivated	Whole virus	Neutralizing Ab using pseudo-typed virus	Not tested	(He *et al.*, 2004)
Beta-propiolactone inactivated	Whole virus	Neutralizing Ab	Not tested	(Tang *et al.*, 2004)
Formaldehyde inactivated	Whole virus	Neutralizing Ab	Not tested	(Zhang *et al.*, 2004)
Beta-propiolactone inactivated	Whole virus	Neutralizing Ab T-helper type 2	Prevention of replication in mice	(See *et al.*, 2006)
Formaldehyde and U.V. inactivation	Whole virus	Neutralizing Ab IFN-gamma and IL-4	Prevention of replication in respiratory tract of mice	(Spruth *et al.*, 2006)
Beta-propiolactone inactivated	Whole virus	Neutralizing Ab	Prevention of replication in cynomolgus macaques	(Qin *et al.*, 2006)
Recombinant protein vaccines				
Baculovirus expressed	S protein	Neutralizing Ab	Not tested	(Zhou *et al.*, 2006)
E. coli expressed	S ectodomain (a.a. 1029–1192)	Neutralizing Ab	Not tested	(Keng *et al.*, 2005)
Baculovirus expressed	S Glycoprotein (N-terminal domain a.a. 14–762)	Neutralizing Ab	Complete protection of upper and lower respiratory tract in mice	(Bisht *et al.*, 2005)
DNA vaccines				
Plasmid with CMV promoter and intron A	S protein	T-helper 1	Not tested	(Zakhartchouk *et al.*, 2005)
tPA-optimimize800 DNA	S protein	Neutralizing Ab	Not tested	(Woo *et al.*, 2005)
pcDNA3.1	S, E, M and N	CTL and T-cell proliferation	Not tested	(Okada *et al.*, 2005)

Table 13.1 continued

Vaccine type	Target	Immune response	Protection against SARS-CoV challenge?	Reference
pcD3d	E, M and N	Humoral response, T-cell proliferation, IFN-gamma, DTH responses and cytotoxic T-cells	Not tested	(Jin *et al.*, 2005)
pVAX1	M and N	Presence of M augments T-cell response to N	Protection against lethal SARS-CoV infection in Brandt's vole	(Shi *et al.*, 2006)
pcDN-fn	N	CTL response	Not tested	(Zhu *et al.*, 2004)
pcDNA3.1	N	Humoral response, T-cell response	Reduction in replication of pseudotyped vaccinia virus	(Kim *et al.*, 2004)
Plasmid DNA expression vector	S	Neutralizing Ab T-cell response	Reduction in SARS-CoV replication in mice	(Yang *et al.*, 2004)
pCI-SP$_{pGH}$	S1 and S2	Neutralizing Ab	Not tested	(Zeng *et al.*, 2004)
Vector-based vaccines				
Rhabdovirus-based vector	S and N	Neutralizing Ab	Not tested	(Faber *et al.*, 2005)
Adenovirus vector	S1	Neutralizing Ab T-cell response	Not tested	(Gao *et al.*, 2003)
Modified vaccinia virus Ankara vector	S	Neutralizing Ab	Reduction of SARS-CoV replication in upper and lower respiratory tract of mice	(Bisht *et al.*, 2004)
Vesicular Stomatitis Virus vector	S	Neutralizing Ab	Protection against SARS-CoV challenge in mice at 1 and 4 months post vaccination	(Kapadia *et al.*, 2005)
Parainfluenza virus type 3 vector	S	Neutralizing Ab	Prevention of viral shedding after SARS-CoV challenge in African green monkeys	(Bukreyev *et al.*, 2004)
Venezuelan equine encephalitis replicon vector	S and N	Neutralizing Ab in groups vaccinated against S	Prevention of SARS-CoV replication in mice	Deming (submitted)

2004). Moreover, vaccine efficacy in aged populations should be evaluated as senescent immune responses may be weighted toward Th2 responses including characteristically weak humoral immune responses, especially problematic as neutralizing antibody levels appear the principle component of protective immunity (Hsu *et al.*, 2005; Simons and Reynolds, 1990). As the epidemic strain of SARS-CoV has not been identified in nature for over three years, new human strains will likely evolve from zoonotic reservoirs. Consequently, protection against zoonotic strains especially in more vulnerable elderly populations should receive high priority.

Recombinant DNA vaccines

Several publications have investigated the use of DNA vaccines for prevention of SARS-CoV infection focusing on the S, E, M and N gene (Jin *et al.*, 2005; Kim *et al.*, 2004; Okada *et al.*, 2005; Shi *et al.*, 2006; Woo *et al.*, 2005; Zakhartchouk *et al.*, 2005; Zhu *et al.*, 2004; Yang *et al.*, 2004; Zeng *et al.*, 2004). (Table 13.1). Following vaccination, specific neutralizing antibodies could be detected as well as specific cytotoxic T-cell responses. Only 2 studies using either S or a combination of M and N demonstrated protection from subsequent SARS-CoV challenge (Shi *et al.*, 2006; Yang *et al.*, 2004).

Vectored vaccines

Several different viral vectors are currently in use for the development of vaccine against SARS-CoV including Rhabdovirus (Faber *et al.*, 2005), Adenovirus (Gao *et al.*, 2003; Zakhartchouk *et al.*, 2005), Modified Vaccinia Ankara (MVA) (Bisht *et al.*, 2004), recombinant attenuated Vesicular Stomatitis Virus (VSV) (Kapadia *et al.*, 2005), recombinant attenuated Parainfluenza type 3 (PIV3) (Buchholz *et al.*, 2004; Bukreyev *et al.*, 2004), and Venezuelan Equine Encephalitis (VEE) replicon vectors (Deming *et al.* submitted) (Table 13.1). All studies included constructs for S protein, as well as constructs for M, E or N in 2 studies (Buchholz *et al.*, 2004; Gao *et al.*, 2003). The S protein induced neutralizing antibodies and protection against homologous challenge and the N protein vectored vaccines were potent inducers of cytotoxic and helper T-cell responses, but did not protect against challenge. One study showed extended protection (up to 4 months) in mice after a single dose of recombinant attenuated VSV expressing the SARS-CoV S protein, although true long-term protective experiments (> 1 year) have not been performed (Kapadia *et al.*, 2005). To date, no studies have demonstrated potent cross protection against heterologous strains. Detailed antigenic comparisons are needed of vaccine efficacy against homologous and heterologous challenge are a priority.

Compare and contrast current vaccine candidates and protective effects

A relevant disease model for SARS-CoV pathogenesis has been difficult to identify. In humans, SARS-CoV seems to cause acute infection of type 2 pneumocytes, resulting in diffuse alveolar disease and virus is cleared by 14 days post infection (Nicholls *et al.*, 2006). However, disease and pathology continue after virus clearance suggesting that continuing disease pathology is probably associated with the immune response to infection or pathophysiological complications associated with alveoli pathology and regeneration following virus clearance (Nicholls *et al.*, 2006). A variety of animal models for SARS-CoV have been reported and infection is associated with mild acute disease in the macaque, ferret, cat,

Balb/c and C57BL/6 mice and hamster with little if any mortality (Fouchier *et al.*, 2003; Glass *et al.*, 2004; Martina *et al.*, 2003; Roberts *et al.*, 2005; Subbarao *et al.*, 2004). In these models, virus clearance occurs rapidly following infection. Animal models reminiscent of late stage human disease has not been established, a clear priority for future research. In young Balb/c mice, peak virus replication occurred on day 2 with virus clearance in a week in the absence of any clinical disease or pathology. Viral antigen and RNA were detected in bronchiolar epithelial cells (Subbarao *et al.*, 2004). In senescent mice, virus titers are increased and mild pathologic lesions are noted in the lungs of infected animals at 4 days post infection (Roberts *et al.*, 2005). Some animals may experience mild weight loss, but overall infection remains benign. More robust animal models, especially those that mimic late stage immunopathology and lung disfunction, are needed to evaluate the efficacy of SARS-CoV vaccines and therapeutics, especially in more vulnerable senescent populations. A promising new animal model using Brandt's Vole (*Microtus brandtii*) was recently described (Shi *et al.*, 2006). Upon intranasal SARS-CoV infection in juvenile and adult Brandt's Vole, animals died with pathology in multiple organs. Adult animals seemed more susceptible than juveniles. Although promising, more information is needed to ascertain the robustness as a SARS-CoV animal infection model.

Both DNA and vector based vaccines are effective in induction of the humoral and cellular responses. In the current animal models used for vaccine testing the humoral response appears to be the main immunological response that confers protection against challenge with homologous SARS-CoV strains. The S protein is a potent inducer of neutralizing antibodies which correlate with protection, whereas the N protein primarily induces a specific cytotoxic and helper T-cell responses (Gao *et al.*, 2003).

In a head to head comparison, a killed vaccine elicited a stronger neutralizing response and protective immunity than a adenovirus vectored vaccine, suggesting that some vaccine vectors may not induce robust immune responses (See *et al.*, 2006). No other comparative studies on vaccine efficacy have been described.

Two clinical trials in humans are currently under way, one using a killed vaccine, the other a DNA vaccine (Qin *et al.*, 2006; Yang *et al.*, 2004). No data on either of these vaccine trials have been presented yet.

Grand challenges in vectored vaccine development

One of the main challenges in vector-based vaccine development as well as in all other attempts of vaccine development will be the observation of vaccine induced disease. One vaccination study in ferrets showed inflammation in the liver of all animals vaccinated with S protein expressing MVA and one animal with MVA expressing N protein (Czub *et al.*, 2005; Weingartl *et al.*, 2004). The mechanism of vaccine enhanced pathology was not identified and studies by additional groups are needed to confirm and evaluate the reproducibility of these results. In addition enhanced lung inflammation and pathology including eosinophilia could be observed after challenge with SARS-CoV in senescent mice previously vaccinated with N protein expressed by VEE (Deming *et al.*, submitted). The observations of disease enhancement by vaccination should be addressed in detail using different vaccine platforms, adjuvants and young and senescent animal models of infection.

Live-attenuated vaccines

Live attenuated vaccines have prevented important human infections over the past century because they induce high levels and long lasting immunity. Concerns over risks of reversion to virulence by mutation and recombination, as well as the associated challenges for developing such vaccines are causing live attenuated virus vaccines to lose their appeal as universally protective strategies. This dichotomy between the known efficacy of live attenuated vaccines and the costs and risks for development has been recognized as one of the "Grand challenges in Global Health" by the National Foundation for Infectious Diseases. Currently no studies with live-attenuated human coronavirus vaccines have been reported. In mice, infection of human SARS-CoV is mild and self-limiting. Importantly, infected animals develop strong cellular and humoral immune responses that provide complete protection from subsequent challenge with the homologous virus (Subbarao *et al.*, 2004). Similar findings have been reported in the mildly pathogenic primate model following Urbani strain SARS-CoV infection, suggesting that live virus vaccines will induce robust protective immune responses that protect against homologous challenge (Lawler *et al.*, 2006 In press, Plos Medicine). These data indicate the feasibility of developing live attenuated human coronavirus vaccines and given the high morbidity and mortality, SARS-CoV represents an excellent model system to develop rational strategies to attenuate severe human coronavirus infection of the lung.

The grand challenges of live attenuated virus vaccine development are the prevention of reversion of attenuated viruses to wild-type and recombination with other closely related viruses. Coronaviruses are a family of RNA viruses with a high mutation and recombination rate (Baric *et al.*, 1995; Herrewegh *et al.*, 1998). Ideally, a stable live attenuated coronavirus vaccine would be highly resistant to recombination repair and reversion to virulence following release into environmental settings. Inexpensive SARS vaccines for Southeast Asia would be valuable, although safety and manufacturing concerns would warrant trials with the safer vectored and killed vaccines.

Genetically engineered SARS-CoV vaccines

Rationale strategies are needed to globally attenuate currently circulating human and animal coronaviruses that could be readily applied to any newly emerged coronaviruses in the future. Moreover, these approaches could also function to stabilize platform coronavirus vectored vaccines. Data on correlates of protection against human coronavirus infection is very limited, almost exclusively focused on SARS-CoV in animal models that do not reproduce the classic human disease phenotype. In other coronavirus infections, virus specific and cross-reactive mucosal antibodies provided protection against TGEV challenge (Sestak *et al.*, 1996; VanCott *et al.*, 1994). Data on passive immunization against SARS-CoV infection mice are in agreement with these observations (Roberts *et al.*, 2006; Sui *et al.*, 2005; Traggiai *et al.*, 2004). In addition, several studies on MHV have demonstrated that virus clearance is T-cell mediated (Bergmann *et al.*, 2001; Sussman *et al.*, 1989). Before deciding on strategies of vaccine development, the properties of a good vaccine should be defined. Since information on both humoral and cellular responses is limited, it is difficult to develop and enhance the immunological properties of vaccine candidates.

Coronaviruses have a history of undergoing dramatic host shifts and cross species transmission events and dramatic virulence and tropism shifts including the emergence of porcine epidemic diarrhea virus, porcine respiratory coronavirus, shipping fever and bovine respiratory coronavirus and the SARS-CoV (Bell *et al.*, 2004; Rasschaert *et al.*, 1990). As host shifts by serial passage in cell culture and after *in vivo* passage, are common as well (Baric *et al.*, 1997; Hensley *et al.*, 1998; Schickli *et al.*, 1997), one must conclude that the high mutation and recombination rate, coupled with the plasticity of the coronavirus structural proteins, provides a fertile environment for future cross species transmission events. Coronaviruses have been identified in many mammalian and avian species, including the recent identification of several bat coronaviruses (Guy *et al.*, 2000; Holmes, 2001; Lau *et al.*, 2005; Li *et al.*, 2005; Poon *et al.*, 2005). As host shifts will likely continue in the future, a second grand challenge for coronavirus live virus vaccine development is the development of platform-based live virus or replicon particle vaccines and the identification of universally conserved sites that attenuate coronavirus pathogenesis *in vivo*. To date, universally attenuating sites have not been identified that are conserved and attenuate all the Coronaviridae, a testable hypothesis as more full length molecular clones become available. Interesting, a mutation within a universally conserved site in nsp14 of MHV completely attenuated virus pathogenesis *in vivo*, but had little impact on *in vitro* replication (Sperry *et al.*, 2005). Unfortunately, the mechanism of attenuation is unknown and the pathogenic phenotype of this mutation has not been tested in other group 1–3 coronaviruses. Currently, gene order rearrangements and deletions of non-essential genes represents another broad based candidate approach to rapidly attenuate coronavirus pathogenesis (de Haan *et al.*, 2002; de Haan *et al.*, 2002; Haijema *et al.*, 2004; Sperry *et al.*, 2005; Yount *et al.*, 2005). Deletions of ORFs 2a, HE, 4 and 5a or gene rearrangements of S, E, M and N in MHV resulted in mutants with growth kinetics comparable to wild type, but a LD_{50} greatly reduced *in vivo* (de Haan *et al.*, 2002; de Haan *et al.*, 2002). Similar results were observed when ORF 3abc and/or ORF 5ab were deleted from the FIP genome (Haijema *et al.*, 2004). Surprisingly, deletion of ORF3a, OF3b, ORF6, ORF7a, and ORF7b in SARS-CoV did not result in significant changes in growth either *in vitro* or *in vivo* (Yount *et al.*, 2005). However, since the current mouse model for SARS-CoV is not robust, it is likely that better models will be needed to dissect the genetic determinants of SARS-CoV virulence or pathogenesis in humans.

In order to study the universal effects of gene order rearrangements and deletions of non-essential genes across the Coronaviridae, multiple infectious clones are needed. Infectious cDNA clones have been described for a number of coronaviruses including SARS-CoV, MHV-A59, TGEV, IBV, HCoV-229E and HCoV-OC43 (Curtis *et al.*, 2001; Masters and Rottier, 2005; St-Jean *et al.*, 2006; Thiel *et al.*, 2001; Yount *et al.*, 2003; Yount *et al.*, 2002, Casais *et al.*, 2001). Alternatively, targeted recombination allows for deletion and rearrangement of the 3′ end of the genome as well (Masters and Rottier, 2005). We developed a systematic method to assemble a full-length infectious cDNA of SARS-CoV (icSARS-CoV). Six contiguous cDNA clones that spanned the 30-kb SARS-CoV genome were isolated (Figure 13.4A). The ends of the cDNAs were engineered with unique junctions and assembled with only the adjacent cDNA subclones, resulting in an intact full-length icSARS-CoV cDNA construct. The interconnecting restriction site junctions that are located at the ends of each cDNA are systematically removed during the assembly of the complete full-length cDNA product, allowing reassembly without the introduction of nucleotide changes. RNA transcripts derived from the full-length icSARS-CoV

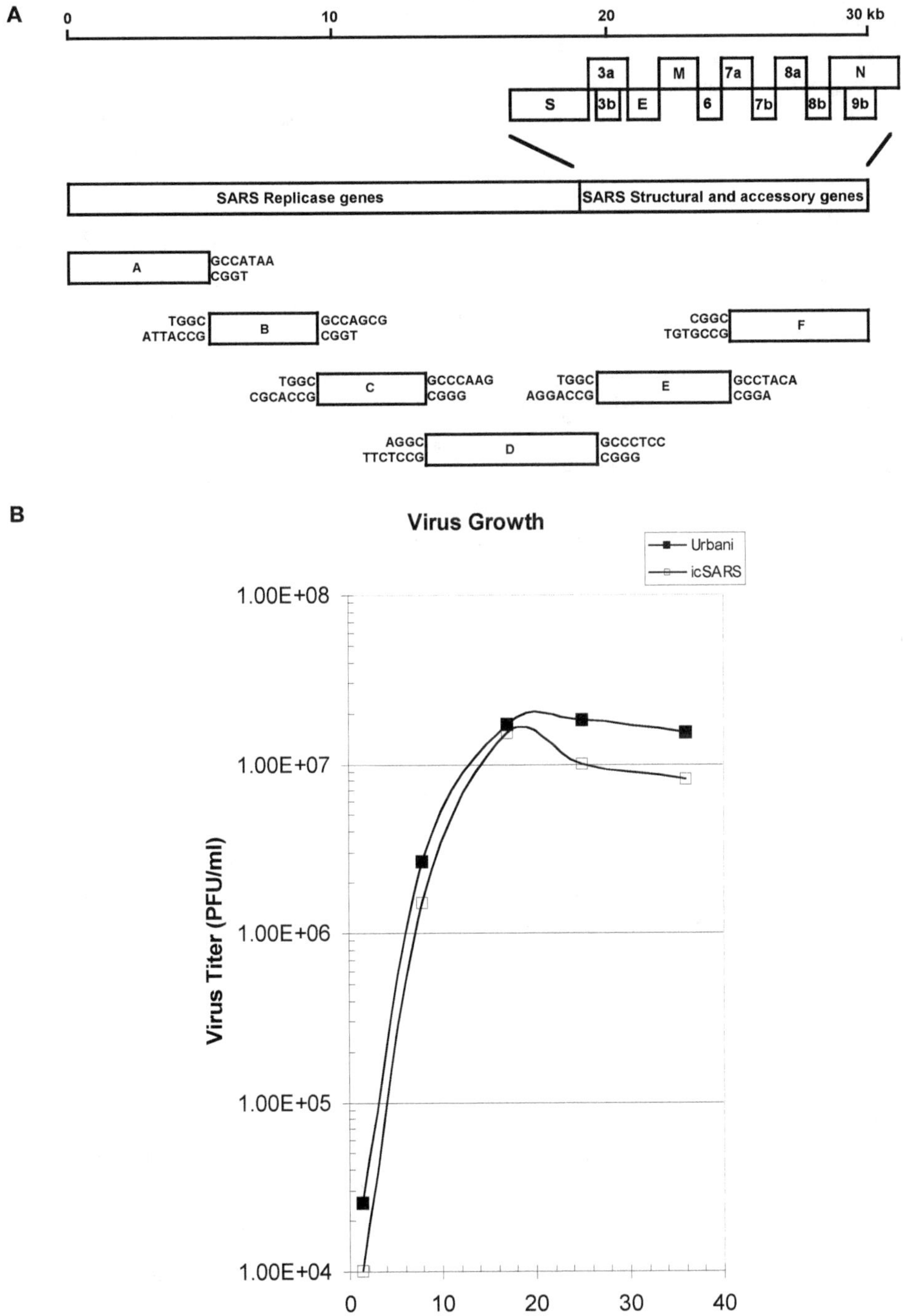

Figure 13.4 A, Assembly of a full-length SARS-CoV cDNA. Six cDNAs spanning the complete SARS-CoV genome were constructed with unique restriction sites for seamless, directional ligation. Full-length cDNA is transcribed and infectious is RNA transfected into Vero E6 cells. B, Comparative growth curves between wild type Urbani and infectious clone Urbani (icSARS).

construct were infectious. The recombinant icSARS-CoV replicated as efficiently as the wild-type SARS-CoV (Figure 13.4B).

Coronavirus group specific ORFs have no described functions, yet passage in cell culture and experiments with recombinant viruses lacking these ORFs have revealed that their loss has little impact on *in vitro* replication. However, MHV recombinant viruses lacking some, but not all, group specific ORFs were attenuated (de Haan *et al.*, 2002). At this time, it is not clear whether the attenuation phenotype was governed by subtle changes in regulation of virus mRNA and negative strand RNA synthesis that were associated with gene deletion, or whether the deletion of the ORF attenuated virus pathogenesis *in vivo*. Recently, our laboratory has shown that many SARS-CoV group specific ORFs (e.g. ORF3a, ORF3b, ORF6, ORF7a, ORF7b) encoded luxury functions that had little impact on *in vitro* and *in vivo* virus growth (Yount *et al.*, 2005). However, the mouse model is a replication model devoid of pathology or clinical disease, making interpretations regarding the role of these alleles in viral pathogenesis less robust. The development of pathogenic mouse adapted strains of SARS-CoV may provide a mechanism for elucidating the role of these group specific ORFs in virus pathogenesis in the future.

Coronaviruses have highly defined gene orders, but studies with Mouse hepatitis virus, transmissible gastroenteritis virus and feline infectious peritonitis virus indicate that gene order is plastic and can be rearranged, attenuating virus pathogenicity *in vivo* (Curtis *et al.*, 2002; de Haan *et al.*, 2002; Haijema *et al.*, 2004). The mechanism of attenuation is unknown but likely relates to disruptions in the carefully orchestrated balance of gene products that are associated with the unique transcriptional strategy of coronaviruses (Pasternak *et al.*, 2001; Schaad and Baric, 1994). Gene order rearrangements provide a mechanism for combating the high frequency of RNA recombination, as a portion of recombinant viruses will be double deletion mutants lacking essential genes for virus replication and particle formation (Figure 13.5A). Another method is reorientation of the S gene to different downstream positions at the 3′ end of the genome the virus should exhibit an attenuated phenotype (Figure 13.5B). In addition, by repositioning the S gene to the 3′ end, the expression of S protein should be increased and theoretically induce a stronger neutralizing response while down-regulating the level of expression of the remaining structural proteins, that do not function as components of protective immunity.

The coronavirus replicase proteins comprise 16 individual proteins which are highly conserved in sequence and gene order (Snijder *et al.*, 2003). Recent studies have demonstrated that the nsp2 replicase protein of MHV and SARS-CoV is not essential for virus replication *in vitro* (Graham *et al.*, 2005). To our knowledge, this is the first example of an RNA virus replicase protein which was completely dispensable for virus replication. This amazing finding suggests that other coronavirus replicase proteins may also encode luxury functions for *in vitro* replication, providing a means of altering the gene order and organization of the coronavirus replicase. Reorganization of the coronavirus replicase proteins provides a novel mechanism for engineering recombination stabilizing mutations into the 5′-most 2/3 of the genome. Coupled with downstream ORF rearrangements, a very stable, highly recombination resistant genome platform can be assembled as a template for the incorporation of other rational attenuating mutations, either in live virus vaccines or replicon particles. Single-hit coronavirus replicon particles can be beneficiary as vaccine candidates as these replicon particles will only be able to infect host cells but not replicate.

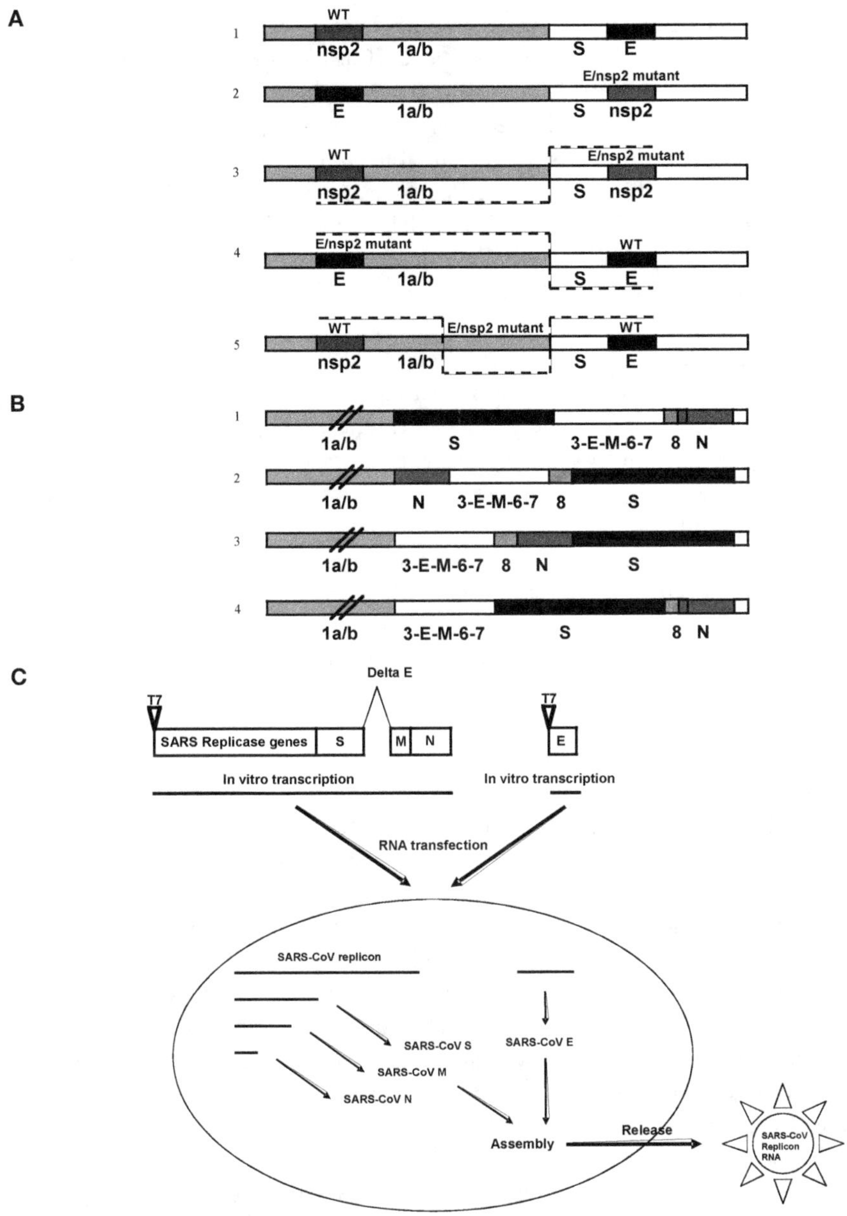

Figure 13.5 Proposed genome reorganization of SARS-CoV for development of live attenuated vaccines. A, Strategy to prevent recombination. By swapping the location of 2 genes nsp2 and E, a mutant is constructed that should be viable but attenuated in growth (2). Any single recombination event with a wild type virus (1) will result in a double mutant (3 or 4). Only a double recombination event (5) will revert the mutant back to wild type. B, Changing Spike location to an upstream position will increase protein expression. In contrast changing the location of N to a downstream location will decrease expression of N and therefore attenuate virus replication. 1: SEMN (Wild Type); 2: NEMS; 3: EMNS; 4: EMSN. C, Strategy for constructing SARS-CoV replicon particles. An E deletion mutant is co-transfected with a packaging vector carrying the E gene. A virus replicon particle, capable of replication but not virus production, is assembled and released.

Coronavirus replicon particles can be constructed by deletion of the E gene as has been described for TGEV (Curtis *et al.*, 2002; Ortego *et al.*, 2002). Co-transfection of the E deletion mutant and the E gene will result in replicon particles that can efficiently replicate in cells without subsequent production of infectious virus particles (Figure 13.5C). Among group 2 coronaviruses, the E protein is non-essential for *in vitro* replication, although severely attenuating the release of infectious particles (Kuo and Masters, 2003). Coronavirus group 2 replicon particles could be constructed by deleting the M gene and providing the M glycoprotein in trans. These replicon particles can serve as vaccine platforms for coronaviruses or can be used to express heterologous proteins.

Grand challenges in SARS-CoV vaccine development

Studies with human and animal coronaviruses strongly argue that protective immune responses against the S glycoprotein, and to a lesser extent the N protein, provide critical components of protective immunity. A major challenge in the development of coronavirus vaccines is the variability of the S glycoprotein. The SARS-CoV S glycoprotein mediates binding to ACE-2 and is a major target for neutralizing antibodies that protect from live virus challenge (Greenough *et al.*, 2005; Sui *et al.*, 2005). Studies with human and animal coronaviruses indicate that rapid evolution (Chinese, 2004), either by point mutation or recombination, provide a ready mechanism for evolving variant viruses that are antigenically distinct. Limited S gene variation was found among the different SARS-CoV isolates throughout the epidemic, although animal strains displayed greater heterogeneity (Chinese, 2004; Song *et al.*, 2005; Tong *et al.*, 2004; Wang *et al.*, 2005). Studies using pseudotyped lentiviruses bearing epidemic and zoonotic SARS-CoV S glycoproteins showed that a single mutation within the RBD can have a drastic effect of antibody binding and subsequent neutralization (Li *et al.*, 2005). In fact antibodies that neutralize most human SARS-CoV S proteins can enhance entry mediated by civet isolate S proteins (Yang *et al.*, 2005). Candidate vaccines should provide protection against all isolates of SARS-CoV throughout the epidemic including the animal isolates to prevent zoonotic transmission. As many animal and early human SARS-CoV isolates were never successfully cultured *in vitro*, an appropriate panel of viruses encoding variant S glycoproteins is desperately needed for testing the efficacy of vaccines and therapeutics against heterotypic challenge. Using the SARS-CoV molecular clone, we have isolated a panel of recombinant viruses bearing early and middle phase epidemic as well as zoonotic S glycoproteins. The availability of these viruses as heterologous challenge inocula will allow vaccine testing against antigenically distinct SARS-CoV strains.

Another challenge is immune enhancement. Antibody-mediated enhancement (ADE) is a mechanism by which the opsonized virus interacts with the Fc receptor on the target cell after binding of antibodies to a virus (Tirado and Yoon, 2003). After binding of the opsonized virus to the Fc receptor, the virus is internalized and released into the cytoplasm. This mechanism has been described for the coronavirus Feline Infectious Peritonitis virus (FIPV) and may explain the observations of enhanced entry of civet SARS-CoV isolates (Rottier *et al.*, 2005). Vaccines in an "aged" population are oftentimes characterized by poor cellular and humoral responses or unbalanced immune responses due to senescence of the immune system (Hsu *et al.*, 2005; Murasko *et al.*, 2002; Simons and Reynolds, 1990). As with many viruses, vaccine efficacy in aged populations is understudied. With newly emerging viruses, this is especially problematic as the lack of a pre-exposure history may contribute to high vulnerability of elderly populations to lethal infection.

Conclusion

The discovery of more severe human coronaviruses that cause severe lower respiratory tract infection provides for new avenues of research to the coronavirus research community. Disease burden associated with human coronavirus infection will likely become clearer over the next few years, as patterns of disease incidence and prevalence become evident with HCoV-NL63, HCoV-OC43, HCoV-229E, and HCoV-HKU1 and perhaps other new human coronavirus infections. The development and testing of vaccine candidates against less pathogenic human coronaviruses will provide significant insight into the duration of protective immune responses and identify determinants of protective immunity. Platform vectors that demonstrate efficacy against these less pathogenic human coronaviruses should be developed and tested, allowing for rapid implementation of vaccines against future emergent highly pathogenic coronaviruses like the SARS-CoV. Implementation of these goals will require the development of animal models against other human coronaviruses, recognizing the recent development of "humanized" mice for HCoV-229E as a major accomplishment and a model for future studies (Lassnig *et al.*, 2005). Finally, due to the high mutation and recombination rates of coronaviruses, vaccine platforms (e.g. live attenuated and replicon particle vaccines) should harness the knowledge of the virus replication and packaging mechanisms and apply reverse genetics approaches to design recombinant genetic backbones that are "recombinant proof" and reversion proof. By identifying conserved alleles within the coronavirus family that attenuate viral pathogenesis, as well as developing universal methods for making coronaviruses highly resistant to recombination repair and reversion to virulent strains, vaccine strategies can be developed for any future emergence of novel coronaviruses.

The mechanisms by which zoonotic viruses like SARS-CoV and HKU1 induce acute interstitial pneumonia and diffuse alveolar disease are unknown and, if robust animal models can be developed, they should serve as excellent model systems for studying the molecular mechanisms of virus induced acute and chronic lung injury. The mechanisms governing human coronavirus pathogenesis are unknown, although several reports indicate that SARS-CoV antagonizes the IFN response (Spiegel *et al.*, 2005; Ziegler *et al.*, 2005). The mechanism and identity of virally encoded gene products that antagonize the IFN system is unknown, providing a new avenue of coronavirus vaccine research. Moreover, a variety of innate immune functions protect the lung from acute lung injury including mucus secretion (Girod *et al.*, 1992), surfactant proteins (Kishore *et al.*, 2005) and collectins (Holmskov, 2000). A dry cough and decreased sputum production associated with SARS-CoV infection is suggestive of an impaired mucus secretory system (Fan *et al.*, 2006). The recent development of an *in vitro* model of human airway epithelium (HAE) enables us to study the pathophysiological consequences of SARS-CoV infections of these cells (Sims *et al.*, 2005). Mucosiliary clearance as well as mucus secretion was significantly reduced after SARS-CoV infection of HAE (R.J. Pickles, unpublished). In addition the composition of the mucus secretions can be determined as well as cellular immune responses.

SARS-CoV evolved numerous changes in the replicase, structural and accessory ORFs during adaptation from animal to human hosts. Tractable models are needed to determine the role of these determinants in emergence, cross species transmission and pathogenesis. SARS-CoV represents the best characterized virus disease emergence model, robust sequence databases, and available strains. An major opportunity is the use of genetic and biochemical approaches, coupled with robust *in vitro* and *in vivo* disease models, to identify the function of discrete gene alterations in virus emergence and pathogenesis, providing a template for understanding future zoonotic reintroductions by the family.

References

Anderson, L.J., and Heilman, C.A. (1995). Protective and disease-enhancing immune responses to respiratory syncytial virus. J. Infect. Dis. *171*, 1–7.

Baric, R.S., Fu, K., Chen, W., and Yount, B. (1995). High recombination and mutation rates in mouse hepatitis virus suggest that coronaviruses may be potentially important emerging viruses. Adv. Exp. Med. Biol. *380*, 571–576.

Baric, R.S., Yount, B., Hensley, L., Peel, S.A., and Chen, W. (1997). Episodic evolution mediates interspecies transfer of a murine coronavirus. J. Virol. *71*, 1946–1955.

Bell, D., Roberton, S., and Hunter, P.R. (2004). Animal origins of SARS coronavirus: possible links with the international trade in small carnivores. Philos. Trans. R. Soc. Lond. B. Biol. Sci. *359*, 1107–1114.

Bergmann, C.C., Marten, N.W., Hinton, D.R., Parra, B., and Stohlman, S.A. (2001). CD8 T cell mediated immunity to neurotropic MHV infection. Adv. Exp. Med. Biol. *494*, 299–308.

Bisht, H., Roberts, A., Vogel, L., Bukreyev, A., Collins, P.L., Murphy, B.R., Subbarao, K., and Moss, B. (2004). Severe acute respiratory syndrome coronavirus spike protein expressed by attenuated vaccinia virus protectively immunizes mice. Proc. Natl. Acad. Sci. USA *101*, 6641–6646.

Bisht, H., Roberts, A., Vogel, L., Subbarao, K., and Moss, B. (2005). Neutralizing antibody and protective immunity to SARS coronavirus infection of mice induced by a soluble recombinant polypeptide containing an N-terminal segment of the spike glycoprotein. Virology *334*, 160–165.

Booth, C.M., Matukas, L.M., Tomlinson, G.A., Rachlis, A.R., Rose, D.B., Dwosh, H.A., Walmsley, S.L., Mazzulli, T., Avendano, M., Derkach, P., *et al.* (2003). Clinical features and short-term outcomes of 144 patients with SARS in the greater Toronto area. JAMA *289*, 2801–2809.

Buchholz, U.J., Bukreyev, A., Yang, L., Lamirande, E.W., Murphy, B.R., Subbarao, K., and Collins, P.L. (2004). Contributions of the structural proteins of severe acute respiratory syndrome coronavirus to protective immunity. Proc. Natl. Acad. Sci. USA *101*, 9804–9809.

Bukreyev, A., Lamirande, E.W., Buchholz, U.J., Vogel, L.N., Elkins, W.R., St Claire, M., Murphy, B.R., Subbarao, K., and Collins, P.L. (2004). Mucosal immunisation of African green monkeys (Cercopithecus aethiops) with an attenuated parainfluenza virus expressing the SARS coronavirus spike protein for the prevention of SARS. Lancet *363*, 2122–2127.

Callow, K.A. (1985). Effect of specific humoral immunity and some non-specific factors on resistance of volunteers to respiratory coronavirus infection. J. Hyg. (Lond) *95*, 173–189.

Callow, K.A., Parry, H.F., Sergeant, M., and Tyrrell, D.A. (1990). The time course of the immune response to experimental coronavirus infection of man. Epidemiol. Infect. *105*, 435–446.

Chakraborti, S., Prabakaran, P., Xiao, X., and Dimitrov, D.S. (2005). The SARS coronavirus S glycoprotein receptor binding domain: fine mapping and functional characterization. Virol. J. *2*, 73.

Chanock, R.M., Parrott, R.H., Connors, M., Collins, P.L., and Murphy, B.R. (1992). Serious respiratory tract disease caused by respiratory syncytial virus: prospects for improved therapy and effective immunization. Pediatrics *90*, 137–143.

Casais, R., Thiel, V., Siddell, S.G., Cavanagh, D., and Britton, P. (2001). Reverse genetics system for the avian coronavirus infectious bronchitis virus. J Virol. *75*12359–12369.

Chen, H., Hou, J., Jiang, X., Ma, S., Meng, M., Wang, B., Zhang, M., Zhang, M., Tang, X., Zhang, F., *et al.* (2005). Response of memory CD8+ T cells to severe acute respiratory syndrome (SARS) coronavirus in recovered SARS patients and healthy individuals. J. Immunol. *175*, 591–598.

Chen, Y.Y., Shuang, B., Tan, Y.X., Meng, M.J., Han, P., Mo, X.N., Song, Q.S., Qiu, X.Y., Luo, X., Gan, Q.N., *et al.* (2005). The protein X4 of severe acute respiratory syndrome-associated coronavirus is expressed on both virus-infected cells and lung tissue of severe acute respiratory syndrome patients and inhibits growth of Balb/c 3T3 cell line. Chin. Med. J. (Engl) *118*, 267–274.

Cheung, C.Y., Poon, L.L., Ng, I.H., Luk, W., Sia, S.F., Wu, M.H., Chan, K.H., Yuen, K.Y., Gordon, S., Guan, Y., and Peiris, J.S. (2005). Cytokine responses in severe acute respiratory syndrome coronavirus-infected macrophages in vitro: possible relevance to pathogenesis. J. Virol. *79*, 7819–7826.

Chinese, S.M.E.C. (2004). Molecular evolution of the SARS coronavirus during the course of the SARS epidemic in China. Science *303*, 1666–1669.

Chiu, S.S., Chan, K.H., Chu, K.W., Kwan, S.W., Guan, Y., Poon, L.L., and Peiris, J.S. (2005). Human coronavirus NL63 infection and other coronavirus infections in children hospitalized with acute respiratory disease in Hong Kong, China. Clin. Infect. Dis. *40*, 1721–1729.

Choi, K.W., Chau, T.N., Tsang, O., Tso, E., Chiu, M.C., Tong, W.L., Lee, P.O., Ng, T.K., Ng, W.F., Lee, K.C., *et al.* (2003). Outcomes and prognostic factors in 267 patients with severe acute respiratory syndrome in Hong Kong. Ann. Intern. Med. *139*, 715–723.

Curtis, K.M., Yount, B., and Baric, R.S. (2001). A simple strategy to assemble infectious RNA and DNA clones. Adv. Exp. Med. Biol. *494*, 475–481.

Curtis, K.M., Yount, B., and Baric, R.S. (2002). Heterologous gene expression from transmissible gastroenteritis virus replicon particles. J. Virol. *76*, 1422–1434.

Czub, M., Weingartl, H., Czub, S., He, R., and Cao, J. (2005). Evaluation of modified vaccinia virus Ankara based recombinant SARS vaccine in ferrets. Vaccine *23*, 2273–2279.

de Haan, C.A., Masters, P.S., Shen, X., Weiss, S., and Rottier, P.J. (2002). The group-specific murine coronavirus genes are not essential, but their deletion, by reverse genetics, is attenuating in the natural host. Virology *296*, 177–189.

de Haan, C.A., Volders, H., Koetzner, C.A., Masters, P.S., and Rottier, P.J. (2002). Coronaviruses maintain viability despite dramatic rearrangements of the strictly conserved genome organization. J. Virol. *76*, 12491–12502.

Ding, Y., He, L., Zhang, Q., Huang, Z., Che, X., Hou, J., Wang, H., Shen, H., Qiu, L., Li, Z., *et al.* (2004). Organ distribution of severe acute respiratory syndrome (SARS) associated coronavirus (SARS-CoV) in SARS patients: implications for pathogenesis and virus transmission pathways. J. Pathol. *203*, 622–630.

Ding, Y., Wang, H., Shen, H., Li, Z., Geng, J., Han, H., Cai, J., Li, X., Kang, W., Weng, D., *et al.* (2003). The clinical pathology of severe acute respiratory syndrome (SARS): a report from China. J. Pathol. *200*, 282–289.

Faber, M., Lamirande, E.W., Roberts, A., Rice, A.B., Koprowski, H., Dietzschold, B., and Schnell, M.J. (2005). A single immunization with a rhabdovirus-based vector expressing severe acute respiratory syndrome coronavirus (SARS-CoV) S protein results in the production of high levels of SARS-CoV-neutralizing antibodies. J. Gen. Virol. *86*, 1435–1440.

Fan, C.K., Yieh, K.M., Peng, M.Y., Lin, J.C., Wang, N.C., and Chang, F.Y. (2006). Clinical and laboratory features in the early stage of severe acute respiratory syndrome. J. Microbiol. Immunol. Infect. *39*, 45–53.

Fouchier, R.A., Kuiken, T., Schutten, M., van Amerongen, G., van Doornum, G.J., van den Hoogen, B.G., Peiris, M., Lim, W., Stohr, K., and Osterhaus, A.D. (2003). Aetiology: Koch's postulates fulfilled for SARS virus. Nature *423*, 240.

Franks, T.J., Chong, P.Y., Chui, P., Galvin, J.R., Lourens, R.M., Reid, A.H., Selbs, E., McEvoy, C.P., Hayden, C.D., Fukuoka, J., *et al.* (2003). Lung pathology of severe acute respiratory syndrome (SARS): a study of 8 autopsy cases from Singapore. Hum. Pathol. *34*, 743–748.

Fulginiti, V.A., Eller, J.J., Downie, A.W., and Kempe, C.H. (1967). Altered reactivity to measles virus. Atypical measles in children previously immunized with inactivated measles virus vaccines. JAMA *202*, 1075–1080.

Gagneur, A., Sizun, J., Vallet, S., Legr, M.C., Picard, B., and Talbot, P.J. (2002). Coronavirus-related nosocomial viral respiratory infections in a neonatal and paediatric intensive care unit: a prospective study. J. Hosp. Infect. *51*, 59–64.

Gao, W., Tamin, A., Soloff, A., D'Aiuto, L., Nwanegbo, E., Robbins, P.D., Bellini, W.J., Barratt-Boyes, S., and Gambotto, A. (2003). Effects of a SARS-associated coronavirus vaccine in monkeys. Lancet *362*, 1895–1896.

Girod, S., Zahm, J.M., Plotkowski, C., Beck, G., and Puchelle, E. (1992). Role of the physiochemical properties of mucus in the protection of the respiratory epithelium. Eur. Respir. J. *5*, 477–487.

Glass, K., and Grenfell, B.T. (2004). Waning immunity and subclinical measles infections in England. Vaccine *22*, 4110–4116.

Glass, W.G., Subbarao, K., Murphy, B., and Murphy, P.M. (2004). Mechanisms of host defense following severe acute respiratory syndrome-coronavirus (SARS-CoV) pulmonary infection of mice. J. Immunol. *173*, 4030–4039.

Graham, R.L., Sims, A.C., Brockway, S.M., Baric, R.S., and Denison, M.R. (2005). The nsp2 replicase proteins of murine hepatitis virus and severe acute respiratory syndrome coronavirus are dispensable for viral replication. J. Virol. *79*, 13399–13411.

Greenough, T.C., Babcock, G.J., Roberts, A., Hernandez, H.J., Thomas, W.D., Jr., Coccia, J.A., Graziano, R.F., Srinivasan, M., Lowy, I., Finberg, R.W., *et al.* (2005). Development and characterization of a severe acute respiratory syndrome-associated coronavirus-neutralizing human monoclonal antibody that provides effective immunoprophylaxis in mice. J. Infect. Dis. *191*, 507–514.

Guan, Y., Zheng, B.J., He, Y.Q., Liu, X.L., Zhuang, Z.X., Cheung, C.L., Luo, S.W., Li, P.H., Zhang, L.J., Guan, Y.J., *et al.* (2003). Isolation and characterization of viruses related to the SARS coronavirus from animals in southern China. Science *302*, 276–278.

Gupta, V., Tabiin, T.M., Sun, K., Chandrasekaran, A., Anwar, A., Yang, K., Chikhlikar, P., Salmon, J., Brusic, V., Marques, E.T., *et al.* (2006). SARS coronavirus nucleocapsid immunodominant T-cell epitope cluster is common to both exogenous recombinant and endogenous DNA-encoded immunogens. Virology *347*, 127–139.

Guy, J.S., Breslin, J.J., Breuhaus, B., Vivrette, S., and Smith, L.G. (2000). Characterization of a coronavirus isolated from a diarrheic foal. J. Clin. Microbiol. *38*, 4523–4526.

Haijema, B.J., Volders, H., and Rottier, P.J. (2004). Live, attenuated coronavirus vaccines through the directed deletion of group-specific genes provide protection against feline infectious peritonitis. J. Virol. *78*, 3863–3871.

He, Y., Lu, H., Siddiqui, P., Zhou, Y., and Jiang, S. (2005). Receptor-binding domain of severe acute respiratory syndrome coronavirus spike protein contains multiple conformation-dependent epitopes that induce highly potent neutralizing antibodies. J. Immunol. *174*, 4908–4915.

He, Y., Zhou, Y., Siddiqui, P., and Jiang, S. (2004). Inactivated SARS-CoV vaccine elicits high titers of spike protein-specific antibodies that block receptor binding and virus entry. Biochem. Biophys. Res. Commun. *325*, 445–452.

Hensley, L.E., Holmes, K.V., Beauchemin, N., and Baric, R.S. (1998). Virus-receptor interactions and interspecies transfer of a mouse hepatitis virus. Adv. Exp. Med. Biol. *440*, 33–41.

Herrewegh, A.A., Smeenk, I., Horzinek, M.C., Rottier, P.J., and de Groot, R.J. (1998). Feline coronavirus type II strains 79–1683 and 79–1146 originate from a double recombination between feline coronavirus type I and canine coronavirus. J. Virol. *72*, 4508–4514.

Ho, M.S., Chen, W.J., Chen, H.Y., Lin, S.F., Wang, M.C., Di, J., Lu, Y.T., Liu, C.L., Chang, S.C., Chao, C.L., *et al.* (2005). Neutralizing antibody response and SARS severity. Emerg. Infect. Dis. *11*, 1730–1737.

Ho, T.Y., Wu, S.L., Cheng, S.E., Wei, Y.C., Huang, S.P., and Hsiang, C.Y. (2004). Antigenicity and receptor-binding ability of recombinant SARS coronavirus spike protein. Biochem. Biophys. Res. Commun. *313*, 938–947.

Ho, Y., Lin, P.H., Liu, C.Y., Lee, S.P., and Chao, Y.C. (2004). Assembly of human severe acute respiratory syndrome coronavirus-like particles. Biochem. Biophys. Res. Commun. *318*, 833–838.

Holmes, K.V. (2001). Enteric infections with coronaviruses and toroviruses. Novartis Found. Symp. *238*, 258–269; discussion 269–275.

Holmskov, U.L. (2000). Collectins and collectin receptors in innate immunity. APMIS Suppl. *100*, 1–59.

Hsiao, C.H., Wu, M.Z., Chen, C.L., Hsueh, P.R., Hsieh, S.W., Yang, P.C., and Su, I.J. (2005). Evolution of pulmonary pathology in severe acute respiratory syndrome. J. Formos. Med. Assoc. *104*, 75–81.

Hsu, H.C., Scott, D.K., and Mountz, J.D. (2005). Impaired apoptosis and immune senescence—cause or effect? Immunol. Rev. *205*, 130–146.

Hu, L.D., Zheng, G.Y., Jiang, H.S., Xia, Y., Zhang, Y., and Kong, X.Y. (2003). Mutation analysis of 20 SARS virus genome sequences: evidence for negative selection in replicase ORF1b and spike gene. Acta. Pharmacol. Sin. *24*, 741–745.

Hu, Y., Wen, J., Tang, L., Zhang, H., Zhang, X., Li, Y., Wang, J., Han, Y., Li, G., Shi, J., *et al.* (2003). The M protein of SARS-CoV: basic structural and immunological properties. Genomics Proteomics Bioinformatics *1*, 118–130.

Hui, D.S., Wong, P.C., and Wang, C. (2003). SARS: clinical features and diagnosis. Respirology *8 Suppl*, S20–24.

Ip, W.K., Chan, K.H., Law, H.K., Tso, G.H., Kong, E.K., Wong, W.H., To, Y.F., Yung, R.W., Chow, E.Y., Au, K.L., *et al.* (2005). Mannose-binding lectin in severe acute respiratory syndrome coronavirus infection. J. Infect. Dis. *191*, 1697–1704.

Ito, N., Mossel, E.C., Narayanan, K., Popov, V.L., Huang, C., Inoue, T., Peters, C.J., and Makino, S. (2005). Severe acute respiratory syndrome coronavirus 3a protein is a viral structural protein. J. Virol. *79*, 3182–3186.

Jin, H., Xiao, C., Chen, Z., Kang, Y., Ma, Y., Zhu, K., Xie, Q., Tu, Y., Yu, Y., and Wang, B. (2005). Induction of Th1 type response by DNA vaccinations with N, M, and E genes against SARS-CoV in mice. Biochem. Biophys. Res. Commun. *328*, 979–986.

Kan, B., Wang, M., Jing, H., Xu, H., Jiang, X., Yan, M., Liang, W., Zheng, H., Wan, K., Liu, Q., *et al.* (2005). Molecular evolution analysis and geographic investigation of severe acute respiratory syndrome coronavirus-like virus in palm civets at an animal market and on farms. J. Virol. *79*, 11892–11900.

Kapadia, S.U., Rose, J.K., Lamirande, E., Vogel, L., Subbarao, K., and Roberts, A. (2005). Long-term protection from SARS coronavirus infection conferred by a single immunization with an attenuated VSV-based vaccine. Virology *340*, 174–182.

Keng, C.T., Zhang, A., Shen, S., Lip, K.M., Fielding, B.C., Tan, T.H., Chou, C.F., Loh, C.B., Wang, S., Fu, J., *et al.* (2005). Amino acids 1055 to 1192 in the S2 region of severe acute respiratory syndrome coronavirus S protein induce neutralizing antibodies: implications for the development of vaccines and antiviral agents. J. Virol. *79*, 3289–3296.

Kim, T.W., Lee, J.H., Hung, C.F., Peng, S., Roden, R., Wang, M.C., Viscidi, R., Tsai, Y.C., He, L., Chen, P.J., *et al.* (2004). Generation and characterization of DNA vaccines targeting the nucleocapsid protein of severe acute respiratory syndrome coronavirus. J. Virol. *78*, 4638–4645.

Kishore, U., Bernal, A.L., Kamran, M.F., Saxena, S., Singh, M., Sarma, P.U., Madan, T., and Chakraborty, T. (2005). Surfactant proteins SP-A and SP-D in human health and disease. Arch. Immunol. Ther. Exp. (Warsz) *53*, 399–417.

Kuo, L., and Masters, P.S. (2003). The small envelope protein E is not essential for murine coronavirus replication. J. Virol. *77*, 4597–4608.

Lai, S.C., Chong, P.C., Yeh, C.T., Liu, L.S., Jan, J.T., Chi, H.Y., Liu, H.W., Chen, A., and Wang, Y.C. (2005). Characterization of neutralizing monoclonal antibodies recognizing a 15-residues epitope on the spike protein HR2 region of severe acute respiratory syndrome coronavirus (SARS-CoV). J. Biomed. Sci. *12*, 711–727.

Larson, H.E., Reed, S.E., and Tyrrell, D.A. (1980). Isolation of rhinoviruses and coronaviruses from 38 colds in adults. J. Med. Virol. *5*, 221–229.

Lassnig, C., Sanchez, C.M., Egerbacher, M., Walter, I., Majer, S., Kolbe, T., Pallares, P., Enjuanes, L., and Muller, M. (2005). Development of a transgenic mouse model susceptible to human coronavirus 229E. Proc. Natl. Acad. Sci. USA *102*, 8275–8280.

Lau, S.K., Woo, P.C., Li, K.S., Huang, Y., Tsoi, H.W., Wong, B.H., Wong, S.S., Leung, S.Y., Chan, K.H., and Yuen, K.Y. (2005). Severe acute respiratory syndrome coronavirus-like virus in Chinese horseshoe bats. Proc. Natl. Acad. Sci. USA *102*, 14040–14045.

Law, H.K., Cheung, C.Y., Ng, H.Y., Sia, S.F., Chan, Y.O., Luk, W., Nicholls, J.M., Peiris, J.S., and Lau, Y.L. (2005). Chemokine up-regulation in SARS-coronavirus-infected, monocyte-derived human dendritic cells. Blood *106*, 2366–2374.

Law, P.T., Wong, C.H., Au, T.C., Chuck, C.P., Kong, S.K., Chan, P.K., To, K.F., Lo, A.W., Chan, J.Y., Suen, Y.K., *et al.* (2005). The 3a protein of severe acute respiratory syndrome-associated coronavirus induces apoptosis in Vero E6 cells. J. Gen. Virol. *86*, 1921–1930.

Li, F., Li, W., Farzan, M., and Harrison, S.C. (2005). Structure of SARS coronavirus spike receptor-binding domain complexed with receptor. Science *309*, 1864–1868.

Li, R.W., Leung, K.W., Sun, F.C., and Samaranayake, L.P. (2004). Severe Acute Respiratory Syndrome (SARS) and the GDP. Part I: Epidemiology, virology, pathology and general health issues. Br. Dent. J. *197*, 77–80.

Li, W., Moore, M.J., Vasilieva, N., Sui, J., Wong, S.K., Berne, M.A., Somasundaran, M., Sullivan, J.L., Luzuriaga, K., Greenough, T.C., *et al.* (2003). Angiotensin-converting enzyme 2 is a functional receptor for the SARS coronavirus. Nature *426*, 450–454.

Li, W., Shi, Z., Yu, M., Ren, W., Smith, C., Epstein, J.H., Wang, H., Crameri, G., Hu, Z., Zhang, H., *et al.* (2005). Bats are natural reservoirs of SARS-like coronaviruses. Science *310*, 676–679.

Li, W., Zhang, C., Sui, J., Kuhn, J.H., Moore, M.J., Luo, S., Wong, S.K., Huang, I.C., Xu, K., Vasilieva, N., *et al.* (2005). Receptor and viral determinants of SARS-coronavirus adaptation to human ACE2. EMBO J. *24*, 1634–1643.

Lin, M.T., Hinton, D.R., Marten, N.W., Bergmann, C.C., and Stohlman, S.A. (1999). Antibody prevents virus reactivation within the central nervous system. J. Immunol. *162*, 7358–7368.

Lip, K.M., Shen, S., Yang, X., Keng, C.T., Zhang, A., Oh, H.L., Li, Z.H., Hwang, L.A., Chou, C.F., Fielding, B.C., *et al.* (2006). Monoclonal antibodies targeting the HR2 domain and the region immediately upstream of the HR2 of the S protein neutralize in vitro infection of severe acute respiratory syndrome coronavirus. J. Virol. *80*, 941–950.

Liu, S.J., Leng, C.H., Lien, S.P., Chi, H.Y., Huang, C.Y., Lin, C.L., Lian, W.C., Chen, C.J., Hsieh, S.L., and Chong, P. (2006). Immunological characterizations of the nucleocapsid protein based SARS vaccine candidates. Vaccine *24*, 3100–3108.

Liu, W., Fontanet, A., Zhang, P.H., Zhan, L., Xin, Z.T., Baril, L., Tang, F., Lv, H., and Cao, W.C. (2006). Two-year prospective study of the humoral immune response of patients with severe acute respiratory syndrome. J. Infect. Dis. *193*, 792–795.

Marshall, E., and Enserink, M. (2004). Medicine. Caution urged on SARS vaccines. Science *303*, 944–946.

Martina, B.E., Haagmans, B.L., Kuiken, T., Fouchier, R.A., Rimmelzwaan, G.F., Van Amerongen, G., Peiris, J.S., Lim, W., and Osterhaus, A.D. (2003). Virology: SARS virus infection of cats and ferrets. Nature *425*, 915.

Masters, P.S., and Rottier, P.J. (2005). Coronavirus reverse genetics by targeted RNA recombination. Curr. Top. Microbiol. Immunol. *287*, 133–159.

Mizgerd, J.P. (2006). Lung infection--a public health priority. PLoS Med. *3*, e76.

Mo, H., Zeng, G., Ren, X., Li, H., Ke, C., Tan, Y., Cai, C., Lai, K., Chen, R., Chan-Yeung, M., and Zhong, N. (2006). Longitudinal profile of antibodies against SARS-coronavirus in SARS patients and their clinical significance. Respirology *11*, 49–53.

Moss, W.J., and Polack, F.P. (2001). Immune responses to measles and measles vaccine: challenges for measles control. Viral Immunol. *14*, 297–309.

Murasko, D.M., Bernstein, E.D., Gardner, E.M., Gross, P., Munk, G., Dran, S., and Abrutyn, E. (2002). Role of humoral and cell-mediated immunity in protection from influenza disease after immunization of healthy elderly. Exp. Gerontol. *37*, 427–439.

Nelson, C.A., Pekosz, A., Lee, C.A., Diamond, M.S., and Fremont, D.H. (2005). Structure and intracellular targeting of the SARS-coronavirus Orf7a accessory protein. Structure (Camb) *13*, 75–85.

Nicholls, J.M., Butany, J., Poon, L.L., Chan, K.H., Beh, S.L., Poutanen, S., Peiris, J.S., and Wong, M. (2006). Time course and cellular localization of SARS-CoV nucleoprotein and RNA in lungs from fatal cases of SARS. PLoS Med. *3*, e27.

Nicholls, J.M., Poon, L.L., Lee, K.C., Ng, W.F., Lai, S.T., Leung, C.Y., Chu, C.M., Hui, P.K., Mak, K.L., Lim, W., *et al.* (2003). Lung pathology of fatal severe acute respiratory syndrome. Lancet *361*, 1773–1778.

Norrby, E., Enders-Ruckle, G., and Meulen, V. (1975). Differences in the appearance of antibodies to structural components of measles virus after immunization with inactivated and live virus. J. Infect. Dis. *132*, 262–269.

Okabayashi, T., Kariwa, H., Yokota, S.I., Iki, S., Indoh, T., Yokosawa, N., Takashima, I., Tsutsumi, H., and Fujii, N. (2006). Cytokine regulation in SARS coronavirus infection compared to other respiratory virus infections. J. Med. Virol. *78*, 417–424.

Okada, M., Takemoto, Y., Okuno, Y., Hashimoto, S., Yoshida, S., Fukunaga, Y., Tanaka, T., Kita, Y., Kuwayama, S., Muraki, Y., *et al.* (2005). The development of vaccines against SARS corona virus in mice and SCID-PBL/hu mice. Vaccine *23*, 2269–2272.

Olsen, C.W. (1993). A review of feline infectious peritonitis virus: molecular biology, immunopathogenesis, clinical aspects, and vaccination. Vet. Microbiol. *36*, 1–37.

Olsen, C.W., Corapi, W.V., Ngichabe, C.K., Baines, J.D., and Scott, F.W. (1992). Monoclonal antibodies to the spike protein of feline infectious peritonitis virus mediate antibody-dependent enhancement of infection of feline macrophages. J. Virol. *66*, 956–965.

Ortego, J., Escors, D., Laude, H., and Enjuanes, L. (2002). Generation of a replication-competent, propagation-deficient virus vector based on the transmissible gastroenteritis coronavirus genome. J. Virol. *76*, 11518–11529.

Pasternak, A.O., van den Born, E., Spaan, W.J., and Snijder, E.J. (2001). Sequence requirements for RNA strand transfer during nidovirus discontinuous subgenomic RNA synthesis. EMBO J. *20*, 7220–7228.

Pavlovic-Lazetic, G.M., Mitic, N.S., Tomovic, A.M., Pavlovic, M.D., and Beljanski, M.V. (2005). SARS-CoV genome polymorphism: a bioinformatics study. Genomics Proteomics Bioinformatics *3*, 18–35.

Pewe, L., Zhou, H., Netland, J., Tangudu, C., Olivares, H., Shi, L., Look, D., Gallagher, T., and Perlman, S. (2005). A severe acute respiratory syndrome-associated coronavirus-specific protein enhances virulence of an attenuated murine coronavirus. J. Virol. *79*, 11335–11342.

Poon, L.L., Chu, D.K., Chan, K.H., Wong, O.K., Ellis, T.M., Leung, Y.H., Lau, S.K., Woo, P.C., Suen, K.Y., Yuen, K.Y., *et al.* (2005). Identification of a novel coronavirus in bats. J. Virol. *79*, 2001–2009.

Qin, E., Shi, H., Tang, L., Wang, C., Chang, G., Ding, Z., Zhao, K., Wang, J., Chen, Z., Yu, M., *et al.* (2006). Immunogenicity and protective efficacy in monkeys of purified inactivated Vero-cell SARS vaccine. Vaccine *24*, 1028–1034.

Rasschaert, D., Duarte, M., and Laude, H. (1990). Porcine respiratory coronavirus differs from transmissible gastroenteritis virus by a few genomic deletions. J. Gen. Virol. *71*, 2599–2607.

Roberts, A., Paddock, C., Vogel, L., Butler, E., Zaki, S., and Subbarao, K. (2005). Aged BALB/c mice as a model for increased severity of severe acute respiratory syndrome in elderly humans. J. Virol. *79*, 5833–5838.

Roberts, A., Thomas, W.D., Guarner, J., Lamirande, E.W., Babcock, G.J., Greenough, T.C., Vogel, L., Hayes, N., Sullivan, J.L., Zaki, S., *et al.* (2006). Therapy with a severe acute respiratory syndrome-associated coronavirus-neutralizing human monoclonal antibody reduces disease severity and viral burden in golden Syrian hamsters. J. Infect. Dis. *193*, 685–692.

Roberts, A., Vogel, L., Guarner, J., Hayes, N., Murphy, B., Zaki, S., and Subbarao, K. (2005). Severe acute respiratory syndrome coronavirus infection of golden Syrian hamsters. J. Virol. *79*, 503–511.

Rota, P.A., Oberste, M.S., Monroe, S.S., Nix, W.A., Campagnoli, R., Icenogle, J.P., Penaranda, S., Bankamp, B., Maher, K., Chen, M.H., *et al.* (2003). Characterization of a novel coronavirus associated with severe acute respiratory syndrome. Science *300*, 1394–1399.

Rottier, P.J. (1999). The molecular dynamics of feline coronaviruses. Vet. Microbiol. *69*, 117–125.

Rottier, P.J., Nakamura, K., Schellen, P., Volders, H., and Haijema, B.J. (2005). Acquisition of macrophage tropism during the pathogenesis of feline infectious peritonitis is determined by mutations in the feline coronavirus spike protein. J. Virol. *79*, 14122–14130.

Schaad, M.C., and Baric, R.S. (1994). Genetics of mouse hepatitis virus transcription: evidence that subgenomic negative strands are functional templates. J. Virol. *68*, 8169–8179.

Schickli, J.H., Zelus, B.D., Wentworth, D.E., Sawicki, S.G., and Holmes, K.V. (1997). The murine coronavirus mouse hepatitis virus strain A59 from persistently infected murine cells exhibits an extended host range. J. Virol. *71*, 9499–9507.

See, R.H., Zakhartchouk, A.N., Petric, M., Lawrence, D.J., Mok, C.P., Hogan, R.J., Rowe, T., Zitzow, L.A., Karunakaran, K.P., Hitt, M.M., *et al.* (2006). Comparative evaluation of two severe acute respiratory syndrome (SARS) vaccine candidates in mice challenged with SARS coronavirus. J. Gen. Virol. *87*, 641–650.

Sestak, K., Lanza, I., Park, S.K., Weilnau, P.A., and Saif, L.J. (1996). Contribution of passive immunity to porcine respiratory coronavirus to protection against transmissible gastroenteritis virus challenge exposure in suckling pigs. Am. J. Vet. Res. *57*, 664–671.

Shi, S.Q., Peng, J.P., Li, Y.C., Qin, C., Liang, G.D., Xu, L., Yang, Y., Wang, J.L., and Sun, Q.H. (2006). The expression of membrane protein augments the specific responses induced by SARS-CoV nucleocapsid DNA immunization. Mol. Immunol. *43*, 1791–1798.

Simons, R.J., and Reynolds, H.Y. (1990). Altered immune status in the elderly. Semin. Respir. Infect. *5*, 251–259.

Sims, A.C., Baric, R.S., Yount, B., Burkett, S.E., Collins, P.L., and Pickles, R.J. (2005). Severe acute respiratory syndrome coronavirus infection of human ciliated airway epithelia: role of ciliated cells in viral spread in the conducting airways of the lungs. J. Virol. *79*, 15511–15524.

Snijder, E.J., Bredenbeek, P.J., Dobbe, J.C., Thiel, V., Ziebuhr, J., Poon, L.L., Guan, Y., Rozanov, M., Spaan, W.J., and Gorbalenya, A.E. (2003). Unique and conserved features of genome and proteome of SARS-coronavirus, an early split-off from the coronavirus group 2 lineage. J. Mol. Biol. *331*, 991–1004.

Song, H.D., Tu, C.C., Zhang, G.W., Wang, S.Y., Zheng, K., Lei, L.C., Chen, Q.X., Gao, Y.W., Zhou, H.Q., Xiang, H., *et al.* (2005). Cross-host evolution of severe acute respiratory syndrome coronavirus in palm civet and human. Proc. Natl. Acad. Sci. USA *102*, 2430–2435.

Sperry, S.M., Kazi, L., Graham, R.L., Baric, R.S., Weiss, S.R., and Denison, M.R. (2005). Single-amino-acid substitutions in open reading frame (ORF) 1b-nsp14 and ORF 2a proteins of the coronavirus mouse hepatitis virus are attenuating in mice. J. Virol. *79*, 3391–3400.

Spiegel, M., Pichlmair, A., Martinez-Sobrido, L., Cros, J., Garcia-Sastre, A., Haller, O., and Weber, F. (2005). Inhibition of Beta interferon induction by severe acute respiratory syndrome coronavirus suggests a two-step model for activation of interferon regulatory factor 3. J. Virol. *79*, 2079–2086.

Spruth, M., Kistner, O., Savidis-Dacho, H., Hitter, E., Crowe, B., Gerencer, M., Bruhl, P., Grillberger, L., Reiter, M., Tauer, C., *et al.* (2006). A double-inactivated whole virus candidate SARS coronavirus vaccine stimulates neutralising and protective antibody responses. Vaccine *24*, 652–661.

St-Jean, J.R., Desforges, M., Almazan, F., Jacomy, H., Enjuanes, L., and Talbot, P.J. (2006). Recovery of a neurovirulent human coronavirus OC43 from an infectious cDNA clone. J. Virol. *80*, 3670–3674.

Stohlman, S.A., Bergmann, C.C., Lin, M.T., Cua, D.J., and Hinton, D.R. (1998). CTL effector function within the central nervous system requires CD4+ T cells. J. Immunol. *160*, 2896–2904.

Subbarao, K., McAuliffe, J., Vogel, L., Fahle, G., Fischer, S., Tatti, K., Packard, M., Shieh, W.J., Zaki, S., and Murphy, B. (2004). Prior infection and passive transfer of neutralizing antibody prevent replication of severe acute respiratory syndrome coronavirus in the respiratory tract of mice. J. Virol. *78*, 3572–3577.

Sui, J., Li, W., Roberts, A., Matthews, L.J., Murakami, A., Vogel, L., Wong, S.K., Subbarao, K., Farzan, M., and Marasco, W.A. (2005). Evaluation of human monoclonal antibody 80R for immunoprophylaxis of severe acute respiratory syndrome by an animal study, epitope mapping, and analysis of spike variants. J. Virol. *79*, 5900–5906.

Sussman, M.A., Shubin, R.A., Kyuwa, S., and Stohlman, S.A. (1989). T-cell-mediated clearance of mouse hepatitis virus strain JHM from the central nervous system. J. Virol. *63*, 3051–3056.

Tan, Y.J., Fielding, B.C., Goh, P.Y., Shen, S., Tan, T.H., Lim, S.G., and Hong, W. (2004). Overexpression of 7a, a protein specifically encoded by the severe acute respiratory syndrome coronavirus, induces apoptosis via a caspase-dependent pathway. J. Virol. *78*, 14043–14047.

Tan, Y.J., Teng, E., Shen, S., Tan, T.H., Goh, P.Y., Fielding, B.C., Ooi, E.E., Tan, H.C., Lim, S.G., and Hong, W. (2004). A novel severe acute respiratory syndrome coronavirus protein, U274, is transported to the cell surface and undergoes endocytosis. J. Virol. *78*, 6723–6734.

Tang, L., Zhu, Q., Qin, E., Yu, M., Ding, Z., Shi, H., Cheng, X., Wang, C., Chang, G., Zhu, Q., *et al.* (2004). Inactivated SARS-CoV vaccine prepared from whole virus induces a high level of neutralizing antibodies in BALB/c mice. DNA Cell Biol. *23*, 391–394.

Temperton, N.J., Chan, P.K., Simmons, G., Zambon, M.C., Tedder, R.S., Takeuchi, Y., and Weiss, R.A. (2005). Longitudinally profiling neutralizing antibody response to SARS coronavirus with pseudotypes. Emerg. Infect. Dis. *11*, 411–416.

Thiel, V., Herold, J., Schelle, B., and Siddell, S.G. (2001). Infectious RNA transcribed in vitro from a cDNA copy of the human coronavirus genome cloned in vaccinia virus. J. Gen. Virol. *82*, 1273–1281.

Thiel, V., Ivanov, K.A., Putics, A., Hertzig, T., Schelle, B., Bayer, S., Weissbrich, B., Snijder, E.J., Rabenau, H., Doerr, H.W., *et al.* (2003). Mechanisms and enzymes involved in SARS coronavirus genome expression. J. Gen. Virol. *84*, 2305–2315.

Tirado, S.M., and Yoon, K.J. (2003). Antibody-dependent enhancement of virus infection and disease. Viral Immunol. *16*, 69–86.

Tong, S., Lingappa, J.R., Chen, Q., Shu, B., LaMonte, A.C., Cook, B.T., Birge, C., Chern, S.W., Liu, X., Galloway, R., *et al.* (2004). Direct sequencing of SARS-coronavirus S and N genes from clinical specimens shows limited variation. J. Infect. Dis. *190*, 1127–1131.

Traggiai, E., Becker, S., Subbarao, K., Kolesnikova, L., Uematsu, Y., Gismondo, M. R., Murphy, B.R., Rappuoli, R., and Lanzavecchia, A. (2004). An efficient method to make human monoclonal antibodies from memory B cells: potent neutralization of SARS coronavirus. Nat. Med. *10*, 871–875.

Tseng, C.T., Perrone, L.A., Zhu, H., Makino, S., and Peters, C.J. (2005). Severe acute respiratory syndrome and the innate immune responses: modulation of effector cell function without productive infection. J. Immunol. *174*, 7977–7985.

van der Hoek, L., Pyrc, K., Jebbink, M.F., Vermeulen-Oost, W., Berkhout, R.J., Wolthers, K.C., Wertheim-van Dillen, P.M., Kaandorp, J., Spaargaren, J., and Berkhout, B. (2004). Identification of a new human coronavirus. Nat. Med. *10*, 368–373.

van der Hoek, L., Sure, K., Ihorst, G., Stang, A., Pyrc, K., Jebbink, M.F., Petersen, G., Forster, J., Berkhout, B., and Uberla, K. (2005). Croup is associated with the novel coronavirus NL63. PLoS Med. *2*, e240.

VanCott, J.L., Brim, T.A., Lunney, J.K., and Saif, L.J. (1994). Contribution of antibody-secreting cells induced in mucosal lymphoid tissues of pigs inoculated with respiratory or enteric strains of coronavirus to immunity against enteric coronavirus challenge. J. Immunol. *152*, 3980–3990.

Wang, B., Chen, H., Jiang, X., Zhang, M., Wan, T., Li, N., Zhou, X., Wu, Y., Yang, F., Yu, Y., *et al.* (2004). Identification of an HLA-A*0201-restricted CD8+ T-cell epitope SSp-1 of SARS-CoV spike protein. Blood *104*, 200–206.

Wang, C.H., Liu, C.Y., Wan, Y.L., Chou, C.L., Huang, K.H., Lin, H.C., Lin, S.M., Lin, T.Y., Chung, K.F., and Kuo, H.P. (2005). Persistence of lung inflammation and lung cytokines with high-resolution CT abnormalities during recovery from SARS. Respir. Res. *6*, 42.

Wang, Z., Yuan, Z., Matsumoto, M., Hengge, U.R., and Chang, Y.F. (2005). Immune responses with DNA vaccines encoded different gene fragments of severe acute respiratory syndrome coronavirus in BALB/c mice. Biochem. Biophys. Res. Commun. *327*, 130–135.

Wang, Z.G., Zheng, Z.H., Shang, L., Li, L.J., Cong, L.M., Feng, M.G., Luo, Y., Cheng, S.Y., Zhang, Y.J., Ru, M.G., *et al.* (2005). Molecular evolution and multilocus sequence typing of 145 strains of SARS-CoV. FEBS Lett. *579*, 4928–4936.

Ware, L.B., and Matthay, M.A. (2000). The acute respiratory distress syndrome. N. Engl. J. Med. *342*, 1334–1349.

Weingartl, H., Czub, M., Czub, S., Neufeld, J., Marszal, P., Gren, J., Smith, G., Jones, S., Proulx, R., Deschambault, Y., *et al.* (2004). Immunization with modified vaccinia virus Ankara-based recombinant vaccine against severe acute respiratory syndrome is associated with enhanced hepatitis in ferrets. J. Virol. *78*, 12672–12676.

Woo, P.C., Lau, S.K., Tsoi, H.W., Chen, Z.W., Wong, B.H., Zhang, L., Chan, J.K., Wong, L.P., He, W., Ma, C., *et al.* (2005). SARS coronavirus spike polypeptide DNA vaccine priming with recombinant spike polypeptide from Escherichia coli as booster induces high titer of neutralizing antibody against SARS coronavirus. Vaccine *23*, 4959–4968.

Woo, P.C., Lau, S.K., Tsoi, H.W., Huang, Y., Poon, R.W., Chu, C.M., Lee, R.A., Luk, W.K., Wong, G.K., Wong, B.H., *et al.* (2005). Clinical and molecular epidemiological features of coronavirus HKU1-associated community-acquired pneumonia. J. Infect. Dis. *192*, 1898–1907.

Woo, P.C., Lau, S.K., Wong, B.H., Chan, K.H., Chu, C.M., Tsoi, H.W., Huang, Y., Peiris, J.S., and Yuen, K.Y. (2004). Longitudinal profile of immunoglobulin G (IgG), IgM, and IgA antibodies against the severe acute respiratory syndrome (SARS) coronavirus nucleocapsid protein in patients with pneumonia due to the SARS coronavirus. Clin. Diagn. Lab. Immunol. *11*, 665–668.

Wu, D., Tu, C., Xin, C., Xuan, H., Meng, Q., Liu, Y., Yu, Y., Guan, Y., Jiang, Y., Yin, X., *et al.* (2005). Civets are equally susceptible to experimental infection by two different severe acute respiratory syndrome coronavirus isolates. J. Virol. *79*, 2620–2625.

Xu, X., and Gao, X. (2004). Immunological responses against SARS-coronavirus infection in humans. Cell. Mol. Immunol. *1*, 119–122.

Yan, H., Xiao, G., Zhang, J., Hu, Y., Yuan, F., Cole, D.K., Zheng, C., and Gao, G.F. (2004). SARS coronavirus induces apoptosis in Vero E6 cells. J. Med. Virol. *73*, 323–331.

Yang, Z.Y., Huang, Y., Ganesh, L., Leung, K., Kong, W.P., Schwartz, O., Subbarao, K., and Nabel, G.J. (2004). pH-dependent entry of severe acute respiratory syndrome coronavirus is mediated by the spike glycoprotein and enhanced by dendritic cell transfer through DC-SIGN. J. Virol. *78*, 5642–5650.

Yang, Z.Y., Kong, W.P., Huang, Y., Roberts, A., Murphy, B.R., Subbarao, K., and Nabel, G.J. (2004). A DNA vaccine induces SARS coronavirus neutralization and protective immunity in mice. Nature *428*, 561–564.

Yang, Z.Y., Werner, H.C., Kong, W.P., Leung, K., Traggiai, E., Lanzavecchia, A., and Nabel, G.J. (2005). Evasion of antibody neutralization in emerging severe acute respiratory syndrome coronaviruses. Proc. Natl. Acad. Sci. USA *102*, 797–801.

Yount, B., Curtis, K.M., Fritz, E.A., Hensley, L.E., Jahrling, P.B., Prentice, E., Denison, M.R., Geisbert, T.W., and Baric, R.S. (2003). Reverse genetics with a full-length infectious cDNA of severe acute respiratory syndrome coronavirus. Proc. Natl. Acad. Sci. USA *100*, 12995–13000.

Yount, B., Denison, M.R., Weiss, S.R., and Baric, R.S. (2002). Systematic assembly of a full-length infectious cDNA of mouse hepatitis virus strain A59. J. Virol. *76*, 11065–11078.

Yount, B., Roberts, R.S., Sims, A.C., Deming, D., Frieman, M.B., Sparks, J., Denison, M.R., Davis, N., and Baric, R.S. (2005). Severe acute respiratory syndrome coronavirus group-specific open reading frames encode nonessential functions for replication in cell cultures and mice. J. Virol. *79*, 14909–14922.

Yu, D., MD, Li, H., Xu, R., MPH, He, J., Lin, J., Li, L., Li, W., Xu, H., Huang, S., and Huang, J. (2003) Prevalence of IgG antibody to SARS-associated coronavirus in animal traders--Guangdong Province, China, 2003. MMWR Morb. Mortal. Wkly. Rep. *52*, 986–987.

Yuan, X., Shan, Y., Zhao, Z., Chen, J., and Cong, Y. (2005). G0/G1 arrest and apoptosis induced by SARS-CoV 3b protein in transfected cells. Virol. J. *2*, 66.

Yuan, X., Wu, J., Shan, Y., Yao, Z., Dong, B., Chen, B., Zhao, Z., Wang, S., Chen, J., and Cong, Y. (2006). SARS coronavirus 7a protein blocks cell cycle progression at G0/G1 phase via the cyclin D3/pRb pathway. Virology *346*, 74–85.

Zakhartchouk, A.N., Liu, Q., Petric, M., and Babiuk, L.A. (2005). Augmentation of immune responses to SARS coronavirus by a combination of DNA and whole killed virus vaccines. Vaccine *23*, 4385–4391.

Zakhartchouk, A.N., Viswanathan, S., Mahony, J.B., Gauldie, J., and Babiuk, L.A. (2005). Severe acute respiratory syndrome coronavirus nucleocapsid protein expressed by an adenovirus vector is phosphorylated and immunogenic in mice. J. Gen. Virol. *86*, 211–215.

Zeng, F., Chow, K.Y., Hon, C.C., Law, K.M., Yip, C.W., Chan, K.H., Peiris, J.S., and Leung, F.C. (2004). Characterization of humoral responses in mice immunized with plasmid DNAs encoding SARS-CoV spike gene fragments. Biochem. Biophys. Res. Commun. *315*, 1134–1139.

Zeng, R., Yang, R.F., Shi, M.D., Jiang, M.R., Xie, Y.H., Ruan, H.Q., Jiang, X.S., Shi, L., Zhou, H., Zhang, L., *et al.* (2004). Characterization of the 3a protein of SARS-associated coronavirus in infected vero E6 cells and SARS patients. J. Mol. Biol. *341*, 271–279.

Zhang, C.H., Guo, Z.M., Zheng, H.Y., Lu, J.H., Wang, Y.F., Yan, X.G., Zhao, Y., Du, X.W., Zhang, X., Fang, L., *et al.* (2004). Humoral immune responses in rabbits induced by an experimental inactivated severe acute respiratory syndrome coronavirus vaccine prepared from F69 strain. Chin. Med. J. (Engl) *117*, 1625–1629.

Zhi, Y., Kobinger, G.P., Jordan, H., Suchma, K., Weiss, S.R., Shen, H., Schumer, G., Gao, G., Boyer, J.L., Crystal, R.G., and Wilson, J.M. (2005). Identification of murine CD8 T cell epitopes in codon-optimized SARS-associated coronavirus spike protein. Virology *335*, 34–45.

Zhou, J., Wang, W., Zhong, Q., Hou, W., Yang, Z., Xiao, S.Y., Zhu, R., Tang, Z., Wang, Y., Xian, Q., *et al.* (2005). Immunogenicity, safety, and protective efficacy of an inactivated SARS-associated coronavirus vaccine in rhesus monkeys. Vaccine *23*, 3202–3209.

Zhou, Z., Post, P., Chubet, R., Holtz, K., McPherson, C., Petric, M., and Cox, M. (2006). A recombinant baculovirus-expressed S glycoprotein vaccine elicits high titers of SARS-associated coronavirus (SARS-CoV) neutralizing antibodies in mice. Vaccine *24*, 3624–3631.

Zhu, M. (2004). SARS Immunity and Vaccination. Cell. Mol. Immunol. *1*, 193–198.

Zhu, M.S., Pan, Y., Chen, H.Q., Shen, Y., Wang, X.C., Sun, Y.J., and Tao, K.H. (2004). Induction of SARS-nucleoprotein-specific immune response by use of DNA vaccine. Immunol. Lett. *92*, 237–243.

Ziegler, T., Matikainen, S., Ronkko, E., Osterlund, P., Sillanpaa, M., Siren, J., Fagerlund, R., Immonen, M., Melen, K., and Julkunen, I. (2005). Severe acute respiratory syndrome coronavirus fails to activate cytokine-mediated innate immune responses in cultured human monocyte-derived dendritic cells. J. Virol. *79*, 13800–13805.

SARS and Other "New" Coronaviruses

14

Leo L.M. Poon

Abstract

SARS coronavirus is a novel human pathogen identified in 2003. Its emergence caused severe impacts on human health. This chapter will summarize the commonly used diagnostic methods for SARS. The origin of this novel virus will be discussed. Besides, other novel coronaviruses found in the post-SARS era will also be reviewed.

SARS coronavirus

Severe acute respiratory syndrome (SARS) is a new infectious respiratory disease in this century. This disease first started in Guangdong Province, China, in November 2002 and was officially recognized by World Health Organization (WHO) in February 2003. After its introduction to Hong Kong in mid February 2003, the virus spread across Vietnam, Singapore, Canada, and elsewhere within weeks. In the early stage of the SARS pandemic, other pathogens such as chlamydia and paramyxoviruses were claimed to be the causative agent. The identification of H5N1 in early 2003 also made one to hypothesize that the outbreak was caused by a pandemic avian influenza virus. However, subsequent investigations soon discarded the above hypotheses. On 17 March 2003, an international collaborative research network was set up by the WHO to identify the etiology of SARS. By the end of March 2003, researchers from Hong Kong, Germany, Canada and the USA identified a novel coronavirus as the agent responsible for the disease (Drosten *et al.*, 2003; Ksiazek *et al.*, 2003; Peiris *et al.*, 2003a). Since then, enormous efforts were taken to contain and diagnose the disease. Unfortunately, the pandemic did not cease until July 2003. In this outbreak, over 8000 SARS patients were reported and about 10% of the patients died from the disease (Poon *et al.*, 2004).

All previously known human coronaviruses are mainly associated with common cold. Thus, the identification of a coronavirus causing a severe viral pneumonia was out of the expectations from most clinical virologists. But, with the tremendous efforts made by colleagues from the WHO collaborative network, this initial query was soon fallen into ruin. First, viral RNA or seroconversion against this virus was found in the majority of the patients (Peiris *et al.*, 2003b). By contrast, no evidence of previous exposure of this pathogen was found from healthy individuals. In addition, cynomolgus macaques (*Macaca fascicularis*) infected with this pathogen shown clinical presentations similar to SARS-like illness (Fouchier *et al.*, 2003). Thus, this novel virus fulfills Koch's postulates and confirmed to be the primary etiological agent of SARS in mid April 2003.

The genome of coronavirus is a single, positive-stranded RNA. All coronaviruses have 5 major open reading frames (ORFs) which encode the replicase, spike (S), envelope (E), membrane (M) and nucleocapsid (N) proteins. Based on antigenical and genetical studies, these viruses can be subdivided into 3 groups (groups 1–3). All previously known human coronaviruses are classified into group 1 and group 2. Interestingly, antigenical and genetical studies of SARS coronavirus yielded inconsistent results. This virus was initially shown to cross-react with group 1 coronaviruses (Ksiazek *et al.* 2003). However, partial viral sequences deduced from initial studies indicated that this novel virus is genetically distinct from all previously known coronaviruses (Drosten *et al.*, 2003; Ksiazek *et al.*, 2003; Peiris *et al.*, 2003a). Further characterizations of the full genome of SARS coronavirus also suggested that this virus is distantly related to coronavirus groups 2 and 3 (Rota *et al.* 2003; Marra *et al.* 2003; Zeng *et al.*, 2003). Subsequent studies proposed that this virus is an early split off of the group 2 coronaviruses (Snijder *et al.*, 2003). Currently, SARS coronavirus is classified as a group 2b virus in the family. However, unlike other group 2 members, its viral genome does not contain a hemagglutinin-esterase gene. In addition, this virus also has some features which are similar to group 3 coronaviruses (Stavrinides and Guttman, 2004). These all point to the fact that SARS coronavirus is very distinct from other group 2 members.

Clinical diagnosis of SARS

At the early stage of the global SARS outbreak, three independent groups isolated the virus from SARS patients (Drosten *et al.*, 2003; Ksiazek *et al.*, 2003; Peiris *et al.*, 2003a). The identification of this virus and its sequence led to the development of molecular tests for clinical diagnosis of SARS (Poon *et al.*, 2004). In these investigations, two of the groups used a random RT-PCR approach to obtain SARS coronavirus sequences (Drosten *et al.*, 2003; Peiris *et al.*, 2003a). The other group used degenerated primers which are specific for group 2 coronaviruses in their studies (Ksiazek *et al.*, 2003). Interestingly, viral sequences revealed in these studies were all derived from the highly conserved ORF1b region of the replicase gene. Therefore, all the first generation molecular diagnostic tests for SARS coronavirus were almost focused on detecting the ORF1b sequences of the viral genome.

As the majority of SARS patients were seroconverted in the second week after disease onset, serological tests might not be a practical approach for early SARS diagnosis (Peiris *et al.*, 2003b). PCR-based tests seemed to be a more promising approach for identifying SARS patients early after onset of the disease. Several RT-PCR assays were reported to be useful for the detection of SARS coronavirus since the SARS outbreak. However, the sensitivity of these tests varies, and standardization of the tests has not been unified. It is impossible to have a fair and direct comparison of these tests. This problem is further complicated by the fact that details of clinical specimens in some of these studies were not reported. For the purpose of the review, only studies with detailed information of tested specimens will be discussed.

Results from four large-scaled investigations were highly similar (Chan *et al.* 2004; Tsang *et al.*, 2003; Tang *et al.* 2004; Wu *et al.*, 2004). Using conventional RT-PCR methods, the positivity rate of SARS coronavirus in respiratory samples collected from the first two days of the disease were low (about 35–65%). This might be due to the fact that the majority of these clinical samples was collected from the upper respiratory tract. The detection rates of SARS coronavirus in these studied cohorts increased as the disease progressed and peaked at around day 9–11 of the illness (range from 75 to 100%) (Chan *et al.*, 2004). In general, the detection rates of SARS coronavirus in lower respiratory samples were much higher than those determined from upper respiratory tract (Chan *et al.*, 2004). The positivity rate in respiratory specimens gradually decrease as the disease progressed, but viral RNA was still detectable in some respiratory samples collected after day 30 of the disease.

Viral RNA was also detected in non-respiratory samples such as stool, blood, cerebrospinal fluid and urine (Hung *et al.*, 2003; Lau *et al.*, 2004; Li *et al.*, 2003; Yam *et al.*, 2003). The detection of viral RNA in these specimens indicated that this disease did not only affect the respiratory system. In the SARS cohorts in Hong Kong, a significant number of SARS patients developed severe diarrhea. Of these non-respiratory samples, stool specimens had the highest proportion of positive RT-PCR results (Chan *et al.*, 2004). This observation agreed with other studies that stool samples might be alternative samples for SARS diagnosis. However, one should notice that viral RNA is not detectable in stool samples in the early phase of the disease, indicating that stool samples might not be ideal for early diagnosis of SARS.

The low sensitivity of conventional RT-PCR assays towards samples collected early after disease onset prompted many groups to develop better molecular detection assays. Amongst these assays, quantitative RT-PCR was the most commonly used approach to achieve a better sensitivity. With quantitative RT-PCR technology, the detection rate of SARS was greatly enhanced and the amount of viral RNA in clinical samples could be determined. Of various clinical specimens, samples from low respiratory tract were shown to contain the highest amount of viral RNA. For nasopharyngeal aspirate (NPA) samples, viral loads were shown to be low in the first few days of disease and only hundreds copies/ml of viral RNA were found within the first 3 days after disease onset (Poon *et al.*, 2003). The viral load of SARS coronavirus in NPA was found to peak at day 10 after disease onset and declined afterward (Poon *et al.*, 2004). The decline of vial load correlates with the synthesis of anti-SARS coronavirus antibodies (Peiris *et al.*, 2003b). Whether the decrease of viral load is related to the production of anti-SARS coronavirus antibodies remains to be elucidated.

Coronaviruses are known to generate subgenomic mRNAs for the synthesis of structural viral proteins. Due to the genomic organization, all subgenomic mRNA molecules contain the N gene of SARS coronavirus. Thus, by targeting the N gene sequence, one might expect we could have a more sensitive test for SARS diagnosis. Though this hypothesis was supported by work from animal models and *in vitro* viral cultures (Kuiken *et al.*, 2003; Drosten *et al.*, 2004), it could not be confirmed in several independent studies on SARS patients (Poon *et al.*, 2004). Further studies also demonstrated that the majority of viral RNA in clinical specimens is of genomic origin, suggesting that there is no advantage of targeting the N gene for clinical diagnosis. The reason for the low abundance of subgenomic mRNA is not clear. It is possible that these subgenomic mRNA molecules are rapidly degraded. Nonetheless, it is useful for a diagnostic laboratory to have multiple molecular assays that target to different regions of the viral genome for confirmatory purposes.

Apart from targeting subgenomic mRNA molecules, there are other ways that could enhance the detection rate of SARS coronavirus early after disease onset. We reasoned that the limited success of early diagnosis of SARS coronavirus was mainly due to the apparent low amount of SARS coronavirus in clinical samples. By extracting RNA from large volumes of nasopharyngeal aspirates and applying quantitative real-time RT-PCR technologies, we were able to detected SARS coronavirus in more than 80% of samples collected from day 1–3 after disease onset with 100% specificity (Poon *et al.*, 2003).

Other strategies were also adopted to enhance the detection rate of SARS coronavirus. Firstly, the detection rate could be enhanced by testing multiple serial clinical samples collected from the same patients (Yam *et al.*, 2003). Secondly, nested RT-PCR was also developed for better detection of the disease (Jiang *et al.*, 2004). However, one should be aware that this approach might increase the risk of having cross contamination.

In initial studies, blood samples were thought to be not useful for SARS diagnosis. However, subsequence studies on plasma and serum samples from SARS patients suggested that this might not be the case. Ng *et al.* (2003) demonstrated that SARS coronavirus was present in more than 50% of plasma and 78% of serum samples from SARS patients during the first week of illness. By using multiple serial plasma samples, 79% of SARS patients could be identified by quantitative RT-PCR methods. Further studies demonstrated that SARS coronavirus could replicate in mononuclear cells of peripheral blood from SARS patient early after disease onset (Li *et al.*, 2003). Whether the viral RNA originated from infected mononuclear cells or the respiratory system should be investigated. These encouraging results suggested that blood might be a good alternative for early diagnosis. However, one should be aware that the duration of viremia was mainly restricted to the first week of the illness. In addition, in a non-SARS epidemic period when most of respiratory infections are caused by other pathogens, respiratory samples are still the most practical clinical samples for routine clinical diagnosis.

Tsang *et al.* (2003) demonstrated that patients with positive RT-PCR results on admission had adverse outcomes in terms of survival, intensive care unit (ICU) admission, and requirement for assisted ventilation. Other data further indicated that viral load could be an independent predictor of mortality (Cheng *et al.*, 2004). For example, at day 10 after disease onset, the viral load from patients who died from the disease was about 4.5 orders of magnitude higher than the one from patients who eventually recovered. Besides, viral load was also associated with the requirement of intensive care and the occurrence of diarrhea (Ng *et al.*, 2003; Cheng *et al.*, 2004). Thus, quantitative PCR assay might be an important prognostic indicator for clinical treatments. In particular, if increased viral loads correlate with the immunopathological events of SARS, it is important to control the viral loads in the early phase of the illness.

Serological diagnosis using immunofluoresence or virus neutralization remains the gold standard for the confirmation of SARS. However, seroconversion (including IgM or IgA) occurs during the second week of illness; therefore, serology can only provide a retrospective confirmation of diagnosis. To increase the throughput of this indirect immunofluorescence method, several groups had converted this approach into ELISA plate format. In these ELISA assays, infected cells or recombinant viral proteins were used to detect the seroconversion of SARS patients (Ksiazek *et al.*, 2003; Woo *et al.*, 2004). An initial study using SARS coronavirus cell culture extract-based ELISA showed no cross-reaction with human coronavirus (HCoV) OC43 and HCoV 229E positive serum samples (Ksiazek *et al.*, 2003). However, antigenic cross-reactivity between SARS coronavirus and HCoV 229E and HCoV OC43 was subsequently demonstrated in immunized animals and human serum (Che *et al.*, 2005). Despite ELISA assays based on recombinant proteins have some cross-reactions, they should be considered for routine screening. The ELISA method for diagnosis has several advantages. It is easy to standardize, adaptable to large-scale testing, providing objective endpoints, and technically less demanding.

Viral antigen detection methods (particularly directed against the N protein) had been developed and evaluated. Initial studies suggested that this approach is much less sensitive than molecular detection methods. However, recent studies demonstrated an antigen capture ELISA that could achieve a high N protein detection rate in acute-phase sera of patients with SARS (Di *et al.*, 2005), suggesting that viral antigen detection methods might be useful for early diagnosis of SARS.

Apart from using indirect immunofluorescence approaches, neutralization assay was also used for the detection of protective antibody from SARS patients. In our initial study, sera of post-SARS patients (> 180 days after onset of clinical syndrome) were tested for protective antibody by neutralization test. More than 75% patients have sustainable good antibody titers (≥ 1:160). Only less than 4 percent of SARS patients were found not to possess detectable levels of protective antibody (K.H. Chan, The University of Hong Kong, unpublished data). In our recent follow up studies on those serologically positive SARS patients, 7 out of 11 serum samples collected at 7 months post infection were still found to contain IgM against SARS coronavirus (Chan *et al.*, 2005). Besides, neutralizing antibody and IgG antibody titers were also found to remain stable over this period.

SARS coronavirus in animals

During the SARS epidemic, a number of speculations were raised to explain the origin of this novel coronavirus. This virus was proposed to be of human, animal or even extraterrestrial origin. It was also suggested that this virus was a natural recombinant of a group 2 and a group 3 coronavirus. Others also proposed that it was a genetically modified biological weapon. But the zoonotic theory was soon supported by the isolation of SARS-like coronaviruses in Himalayan palm civets (*Paguma larvata*) in Guangdong province (Guan *et al.*, 2003). These animal viruses were also detected in other mammalian species in live animal markets. Serological studies revealed that a significant portion of traders in wild live animal markets in Guangdong Province had antibodies against SARS coronavirus (Guan *et al.*, 2003). Interestingly, none of these animal workers had developed SARS during the Guangdong Province outbreak, suggesting that most animal SARS-like coronaviruses could only cause asymptomatic infection. The reasons accounting for differences between SARS coronavirus and animal SARS-like coronavirus infections in humans are unknown. This might be due to subtle genetic differences between these viruses. Phylogenetic analysis indicated that these animal viruses were highly similar to human SARS coronavirus (> 99% sequence homology). Sequence analysis revealed that all the animal virus isolates retained an "additional" 29-nucleotide sequence that was not found in most human virus isolates. Studies on recombinant SARS coronaviruses indicated that such deletion on human isolates had little impacts on *in vitro* growth or RNA synthesis (Yount *et al.*, 2005). However, it is not known whether this 29-nucleotide deletion would modulate the pathogenesis or human-to-human transmission efficiency.

The initial discovery of SARS coronavirus in *Paguma larvata* and other animals in live animal markets suggested there were massive spillovers of this virus in live animal markets. However, several lines of evidence suggested that these animals might not be the natural reservoirs of the virus. First, the relative prevalence of the animal SARS-like coronaviruses in animal farms was much lower than in animal markets (Tu *et al.*, 2004). Besides, sequence analyses revealed the mutation rate of the SARS-like coronavirus in the palm civets to be relatively high (Kan *et al.*, 2005; Song *et al.*, 2005). These analyses indicated that the virus did not (yet) fully adapt to civets. Furthermore, results from experimental infections of civets also suggested that these animals might not be the natural host of the virus (Wu *et al.*, 2005). Thus, the above findings suggested that civets in the animal markets might have acquired the virus from another animal species. Nonetheless, it is clear that the civets amplify and perpetuate the circulation of the virus within the live animal market settings.

The above observations prompted several extensive investigations to identify the natural reservoir of this virus. Early human cases occurred in the Guangdong province, which is geographically adjacent to Hong Kong. We reasoned that the natural host of SARS coronavirus might also reside in Hong Kong. A surveillance study of coronaviruses in wild animals in Hong Kong was therefore launched after the discovery of SARS coronavirus in civets (Poon *et al.*, 2005). In this particular study, several wild civets were sampled and tested for coronaviruses. None of these wild civets was positive for the virus by both serological and molecular tests, supporting the hypothesis that civets might not be a natural carrier of SARS coronavirus.

Interestingly, a novel group 2b coronavirus, closely related to SARS coronavirus, was recently identified in bats (*Rhinolophus spp*) by two independent groups (Lau *et al.*, 2005; Li *et al.*, 2005). Similar to the virus isolated from civets, this bat virus was not able to growth in cell culture systems. Serological tests indicated that this virus might be endemic in these bat species. In one of these studies, more than half of *Rhinolophus sinicus* were found to have been previously exposed to this virus. In both of the above studies, only fecal, but not respiratory samples were positive for this novel virus, suggesting this virus might have an enteric tropism. However, no obvious signs of disease were reported. Interestingly, 42% of serum samples collected from these bats were able to neutralize human SARS coronavirus in cell cultures. This suggested the human and bat SARS coronaviruses are antigentically similar. These SARS-like coronaviruses had only about 90% nucleotide sequence identity to SARS coronavirus, suggesting that the viruses found in humans and civets in 2003 were not directly transmitted from bat species. In particular, the S1 domain of the S protein of the SARS-like coronavirus was only about 64% identical to the corresponding region in SARS coronavirus. As the S1 domain was found to be responsible for receptor binding, it is unlikely that this bat SARS-like coronavirus could use human ACE2 as receptor for human infection. This might further suggest that this virus is not readily transmitted to humans. Nonetheless, both serological and phylogenetical data all suggested that human and bat SARS coronaviruses are closely related viruses. It is possible that the bat SARS-like coronavirus represents a distant ancestor of the human SARS coronavirus. Alternatively, the human and bat SARS coronaviruses were derived from the same precursor. Further characterization of these novel bat SARS-like coronaviruses might help to understand the ecology of SARS coronavirus and animal SARS-like coronaviruses.

Novel coronaviruses in bats

Before the SARS outbreak in 2003, most coronaviruses were isolated from humans or domestic animals. This bias is presumably because viral investigations are often driven by disease outbreaks in the above populations. By contrast, relatively little is known about the prevalence of coronaviruses in wild animal species. More importantly, the SARS outbreak also highlighted the potential threats posed by coronaviruses endemic in wild animals. This prompted us to launch a survey of the prevalence of coronaviruses in wild animals in Hong Kong (Poon *et al.*, 2005). In this surveillance study, wild mammalian, avian and reptile species caught from natural reservoirs were sampled and about three hundred samples were examined by PCR assays specific for known coronaviruses. Of these samples, a novel group 1 coronavirus was identified in three bent-winged bat species (*Miniopterus magnater, Miniopterus pusillus* and *Miniopterus schreibersii*). This virus is phylogenetically close to porcine epidemic diarrhea virus. Both fecal and respiratory samples were positive for the virus. In particular, more than half of fecal samples from *Miniopterus pusillus* were positive for the virus. These observations implied that this virus might predominantly have an enteric tropism. In our on-going surveillance study, we managed to detect this virus in different *Miniopterus* populations from four consecutive years. Interestingly, further genetic analysis suggested that there are indeed several different novel species of coronaviruses in *Miniopterus* bats (Chu *et al.*, 2006). This study suggests that bent-winged bats appear to be the natural reservoir for at least two of these novel group 1 coronaviruses.

Our results implied that there is a diversity of hitherto unrecognized coronaviruses that circulate in bats (Poon *et al.*, 2005, Chu *et al.*, 2006). Recent studies on other bat species have confirmed this hypothesis (Woo *et al.*, 2006; Tang *et al.*, 2006). In these investigations, several group 1 and group 2 coronaviruses were identified in different bat species. The genomes of 4 of these viruses were completed sequenced (Tang *et al.*, 2006). Phylogenetic studies revealed that these viruses are genetically diverse. Further investigation is required to elucidate the persistence of the infection, the tissue tropism and possible pathogenicity of these viruses. Besides, further surveillance of bat coronaviruses in different geographical regions may reveal a more comprehensive picture of how bat coronaviruses, or even coronaviruses in general, have evolved.

Novel group 3 coronaviruses in birds

Recently, a Norwegian surveillance study reported three novel coronaviruses in wild graylag geese (*Anser anser*), feral pigeons (*Columbia livia*) and mallards (*Anas platyrhynchos*) (Jonassen *et al.*, 2005). Interestingly, using degenerated primers that could cross react with coronaviruses from three different groups, only group 3 viruses were identified. These results suggested that group 3 coronaviruses were fundamental different from the viruses of the other two groups. These novel avian viruses were found in cloacal samples of these animals. About 25% of fecal samples from goose were positive for the geese coronavirus, suggesting that the virus might spread by the fecal-oral transmission route. However, as cloacal swabs were the only studied samples in the investigation, it is possible that some of these viruses might be transmitted by respiratory or other routes.

A novel group 2 virus in dogs

In 2003, a novel group 2 coronavirus was identified by Erles *et al.* (2003). This virus is genetically closely related to bovine coronavirus and HCoV OC43. Unlike the group 1 canine enteric coronavirus, this new canine virus was found to be associated with respiratory diseases in dogs. In this study, 30% of the studied dogs were found to be seropositive to this virus and more than half of the dogs with mild cough were positive for this virus in PCR assays. These findings indicated that this novel virus is a very common respiratory pathogen in dog populations.

Novel group 1 coronavirus in humans: HCoV NL63

Soon after the discovery of SARS, a novel human coronavirus, HCoV NL63, was discovered in infants with respiratory illness (van der Hoek *et al.*, 2004; Fouchier *et al.*, 2004). Details descriptions of this virus will be covered in the chapter by van der Hoek in this book.

A novel group 2 coronavirus in humans: HCoV HKU1

The SARS outbreak prompted us to launch a retrospective screening study of coronaviruses in human respiratory samples. Using conserved coronavirus-specific primers, another novel human coronavirus virus (HCoV-HKU1) was identified (Woo *et al.*, 2005a). The virus was first detected from a nasopharyngeal aspirate samples from a 71-year-old man with fever and productive cough. This novel virus is genetically distinct from HCoV OC43, the other known human group 2 coronavirus. Phylogenetic studies suggested that this virus might be a recombinant of other group 2 coronaviruses (Woo *et al.*, 2005b). Subsequent studies demonstrated that this virus could be found also on other continents (Sloots *et al.*, 2005; Vabret *et al.*, 2006), suggesting that this virus is a common human pathogen around the globe. Epidemiological studies suggested that this virus was associated with respiratory illnesses and that about 2.4% of patients with community-acquired pneumonia were found to be positive for this virus (Woo *et al.*, 2005c). Recently, this virus was detected stools samples from patients with acute enteric disease, suggesting HCoV-HKU1 might also cause enteric illness (Vabret *et al.*, 2006).

Further perspectives

SARS was the first pandemic in this millennium. The disease had severe impacts on the health care, the economy and the tourist industry in many countries. With the experiences that we learned from many studies, we are in a better position in identifying SARS or other "novel" coronaviruses in the future. Further work on developing assays with better sensitivities for diagnostic purpose is a must. Standardization on RT-PCR assays and clinical samples for diagnosis might also allow us to develop a unified protocol for SARS diagnosis. In a non-pandemic situation, suspected SARS samples should be tested with caution. Ideally, serial multiple samples should be collected from suspected patients. Positive samples should be confirmed by an independent molecular test. In addition, to avoid having false negative results, assays with internal positive control should be encouraged to be used for clinical applications.

The outbreak also led to the search for novel coronaviruses in humans and animals. Within the last four years, at least two new human coronaviruses and several new animal coronaviruses were identified. These discoveries have opened up a new horizon in this field. In particular, the discoveries of animal coronaviruses also reflect the lack of understanding of coronaviruses in wildlife. These findings also suggested that wildlife is a huge reservoir for other coronaviruses that are yet to be identified. It is not know whether these "novel" viruses might have zoonotic potential. However, given the catastrophic consequences of the SARS epidemic, further investigations on viruses in wildlife and the ecology of their hosts should be encouraged.

Acknowledgments

The work is supported by funds from the Research Grant Council of Hong Kong (HKU 7356/03 M), the VOC SARS Research funds from the University of Hong Kong, and European Research Project SARS-DTV (contract no: SP22-CT-2004).

References

Chan, K.H., Poon, L.L., Cheng, V.C., Guan, Y., Hung, I.F., Kong, J., Yam, L.Y., Seto, W.H., Yuen, K.Y., and Peiris, J.S. (2004). Detection of SARS coronavirus in patients with suspected SARS. Emerg. Infect. Dis. *10*, 294–299.

Chan, K.H., Cheng, V.C., Woo, P.C., Lau, S.K., Poon, L.L., Guan, Y., Seto, W.H., Yuen, K.Y., and Peiris, J.S. (2005). Serological responses in patients with severe acute respiratory syndrome coronavirus infection and cross-reactivity with human coronaviruses 229E, OC43, and NL63. Clin. Diagn. Lab. Immunol. *12*, 1317–1321.

Che, X.Y., Qiu, L.W., Liao, Z.Y., Wang, Y.D., Wen, K., Pan, Y.X., Hao, W., Mei, Y.B., Cheng, V.C., and Yuen, K.Y. (2005). Antigenic cross-reactivity between severe acute respiratory syndrome-associated coronavirus and human coronaviruses 229E and OC43. J. Infect. Dis. *191*, 2033–2037.

Cheng, V.C., Hung, I.F., Tang, B.S., Chu, C.M., Wong, M.M., Chan, K.H., Wu, A.K., Tse, D.M., Chan, K.S., Zheng, B.J., *et al.* (2004). Viral replication in the nasopharynx is associated with diarrhea in patients with severe acute respiratory syndrome. Clin. Infect. Dis. *38*, 467–475.

Chu, D.K., Poon L.L., Chan, K.H., Chen, H., Guan, Y., Yuen, K.Y., Peiris, J.S. (2006). Coronaviruses in bent-winged bats (Miniopterus spp.). J. Gen. Virol. *87*, 2461–2466.

Di, B., Hao, W., Gao, Y., Wang, M., Wang, Y.D., Qiu, L.W., Wen, K., Zhou, D.H., Wu, X.W., Lu, E.J., *et al.* (2005). Monoclonal antibody-based antigen capture enzyme-linked immunosorbent assay reveals high sensitivity of the nucleocapsid protein in acute-phase sera of severe acute respiratory syndrome patients. Clin. Diagn. Lab. Immunol. *12*, 135–140.

Drosten, C., Gunther, S., Preiser, W., van der Werf, S., Brodt, H.R., Becker, S., Rabenau, H., Panning, M., Kolesnikova, L., Fouchier, R.A., *et al.* (2003). Identification of a novel coronavirus in patients with severe acute respiratory syndrome. N. Engl. J. Med. *348*, 1967–1976.

Drosten, C., Chiu, L.L., Panning, M., Leong, H.N., Preiser, W., Tam, J.S., Gunther, S., Kramme, S., Emmerich, P., Ng, W.L., *et al.* (2004). Evaluation of advanced reverse transcription-PCR assays and an alternative PCR target region for detection of severe acute respiratory syndrome-associated coronavirus. J. Clin. Microbiol. *42*, 2043–2047.

Erles, K., Toomey, C., Brooks, H.W., and Brownlie, J. (2003). Detection of a group 2 coronavirus in dogs with canine infectious respiratory disease. Virology *310*, 216–223.

Fouchier, R.A., Kuiken, T., Schutten, M., van Amerongen, G., van Doornum, G.J., van den Hoogen, B.G., Peiris, M., Lim, W., Stohr, K., and Osterhaus, A.D. (2003). Aetiology: Koch's postulates fulfilled for SARS virus. Nature *423*, 240.

Fouchier, R.A., Hartwig, N.G., Bestebroer, T.M., Niemeyer, B., de Jong, J.C., Simon, J.H., and Osterhaus, A.D. (2004). A previously undescribed coronavirus associated with respiratory disease in humans. Proc. Natl. Acad. Sci. USA *101*, 6212–6216.

Guan, Y., Zheng, B.J., He, Y.Q., Liu, X.L., Zhuang, Z.X., Cheung, C.L., Luo, S.W., Li, P.H., Zhang, L.J., Guan, Y.J., *et al.* (2003). Isolation and characterization of viruses related to the SARS coronavirus from animals in southern China. Science *302*, 276–278.

Hung, E.C., Chim, S.S., Chan, P.K., Tong, Y.K., Ng, E.K., Chiu, R.W., Leung, C.B., Sung, J.J., Tam, J.S., and Lo, Y.M. (2003). Detection of SARS coronavirus RNA in the cerebrospinal fluid of a patient with severe acute respiratory syndrome. Clin. Chem. *49*, 2108–2109.

Jiang, S.S., Chen, T.C., Yang, J.Y., Hsiung, C.A., Su, I.J., Liu, Y.L., Chen, P.C., and Juang, J.L. (2004). Sensitive and quantitative detection of severe acute respiratory syndrome coronavirus infection by real-time nested polymerase chain reaction. Clin. Infect. Dis. *38*, 293–296.

Jonassen, C.M., Kofstad, T., Larsen, I.L., Lovland, A., Handeland, K., Follestad, A., and Lillehaug, A. (2005). Molecular identification and characterization of novel coronaviruses infecting graylag geese (Anser anser), feral pigeons (Columbia livia) and mallards (Anas platyrhynchos). J. Gen. Virol. *86*, 1597–1607.

Kan, B., Wang, M., Jing, H., Xu, H., Jiang, X., Yan, M., Liang, W., Zheng, H., Wan, K., Liu, Q., *et al.* (2005). Molecular evolution analysis and geographic investigation of severe acute respiratory syndrome coronavirus-like virus in palm civets at an animal market and on farms. J.Virol. *79*, 11892–11900.

Ksiazek, T.G., Erdman, D., Goldsmith, C.S., Zaki, S.R., Peret, T., Emery, S., Tong, S., Urbani, C., Comer, J.A., Lim, W., *et al.* (2003). A novel coronavirus associated with severe acute respiratory syndrome. N. Engl. J. Med. *348*, 1953–1966.

Kuiken, T., Fouchier, R.A., Schutten, M., Rimmelzwaan, G.F., van Amerongen, G., van Riel, D., Laman, J.D., de Jong, T., van Doornum, G., Lim, W., *et al.* (2003). Newly discovered coronavirus as the primary cause of severe acute respiratory syndrome. Lancet *362*, 263–270.

Lau, K.K., Yu, W.C., Chu, C.M., Lau, S.T., Sheng, B., and Yuen, K.Y. (2004). Possible central nervous system infection by SARS coronavirus. Emerg. Infect. Dis. *10*, 342–344.

Lau, S.K., Woo, P.C., Li, K.S., Huang, Y., Tsoi, H.W., Wong, B.H., Wong, S.S., Leung, S.Y., Chan, K.H., and Yuen, K.Y. (2005). Severe acute respiratory syndrome coronavirus-like virus in Chinese horseshoe bats. Proc. Natl. Acad. Sci. USA *102*, 14040–14045.

Li, L., Wo, J., Shao, J., Zhu, H., Wu, N., Li, M., Yao, H., Hu, M., and Dennin, R.H. (2003). SARS-coronavirus replicates in mononuclear cells of peripheral blood (PBMCs) from SARS patients. J. Clin. Virol. *28*, 239–244.

Li, W., Shi, Z., Yu, M., Ren, W., Smith, C., Epstein, J.H., Wang, H., Crameri, G., Hu, Z., Zhang, H., *et al.* (2005). Bats are natural reservoirs of SARS-like coronaviruses. Science *310*, 676–679.

Marra, M.A., Jones, S.J., Astell, C.R., Holt, R.A., Brooks-Wilson, A., Butterfield, Y.S., Khattra, J., Asano, J.K., Barber, S.A., Chan, S.Y., *et al.* (2003). The Genome sequence of the SARS-associated coronavirus. Science *300*, 1399–1404.

Ng, E.K., Hui, D.S., Chan, K.C., Hung, E.C., Chiu, R.W., Lee, N., Wu, A., Chim, S.S., Tong, Y.K., Sung, J.J., *et al.* (2003). Quantitative analysis and prognostic implication of SARS coronavirus RNA in the plasma and serum of patients with severe acute respiratory syndrome. Clin. Chem. *49*, 1976–1980.

Peiris, J.S., Lai, S.T., Poon, L.L., Guan, Y., Yam, L.Y., Lim, W., Nicholls, J., Yee, W.K., Yan, W.W., Cheung, M.T., *et al.* (2003a). Coronavirus as a possible cause of severe acute respiratory syndrome. Lancet *361*, 1319–1325.

Peiris, J.S., Chu, C.M., Cheng, V.C., Chan, K.S., Hung, I.F., Poon, L.L., Law, K.I., Tang, B.S., Hon, T.Y., Chan, C.S., *et al.* (2003b). Clinical progression and viral load in a community outbreak of coronavirus-associated SARS pneumonia: a prospective study. Lancet *361*, 1767–1772.

Poon, L.L., Chan, K.H., Wong, O.K., Yam, W.C., Yuen, K.Y., Guan, Y., Lo, Y.M., and Peiris, J.S. (2003). Early diagnosis of SARS coronavirus infection by real time RT-PCR. J. Clin. Virol. *28*, 233–238.

Poon, L.L., Guan, Y., Nicholls, J.M., Yuen, K.Y., and Peiris, J.S. (2004). The aetiology, origins, and diagnosis of severe acute respiratory syndrome. Lancet Infect. Dis. *4*, 663–671.

Poon, L.L., Chu, D.K., Chan, K.H., Wong, O.K., Ellis, T.M., Leung, Y.H., Lau, S.K., Woo, P.C., Suen, K.Y., Yuen, K.Y., *et al.* (2005). Identification of a novel coronavirus in bats. J. Virol. *79*, 2001–2009.

Rota, P.A., Oberste, M.S., Monroe, S.S., Nix, W.A., Campagnoli, R., Icenogle, J.P., Penaranda, S., Bankamp, B., Maher, K., Chen, M.H., *et al.* (2003). Characterization of a novel coronavirus associated with severe acute respiratory syndrome. Science *300*, 1394–1399.

Sloots, T.P., McErlean, P., Speicher, D.J., Arden, K.E., Nissen, M.D., and Mackay, I.M. (2006). Evidence of human coronavirus HKU1 and human bocavirus in Australian children. J. Clin. Virol. *35*, 99–102.

Snijder, E.J., Bredenbeek, P.J., Dobbe, J.C., Thiel, V., Ziebuhr, J., Poon, L.L., Guan, Y., Rozanov, M., Spaan, W.J., and Gorbalenya, A.E. (2003). Unique and conserved features of genome and proteome of SARS-coronavirus, an early split-off from the coronavirus group 2 lineage. J. Mol. Biol. *331*, 991–1004.

Song, H.D., Tu, C.C., Zhang, G.W., Wang, S.Y., Zheng, K., Lei, L.C., Chen, Q.X., Gao, Y.W., Zhou, H.Q., Xiang, H., *et al.* (2005). Cross-host evolution of severe acute respiratory syndrome coronavirus in palm civet and human. Proc. Natl. Acad. Sci. USA *102*, 2430–2435.

Stavrinides, J., and Guttman, D.S. (2004). Mosaic evolution of the severe acute respiratory syndrome coronavirus. J. Virol. *78*, 76–82.

Tang, P., Louie, M., Richardson, S.E., Smieja, M., Simor, A.E., Jamieson, F., Fearon, M., Poutanen, S.M., Mazzulli, T., Tellier, R., *et al.* (2004). Interpretation of diagnostic laboratory tests for severe acute respiratory syndrome: the Toronto experience. CMAJ *170*, 47–54.

Tang, X.C., Zhang, J.X., Zhang, S.Y., Wang, P., Fan, X.H., Li, L.F., Li, G., Dong, B.Q., Liu, W., Cheung, C.L., *et al.* (2006). Prevalence and genetic diversity of coronaviruses in bats from china. J. Virol. *80*, 7481–7490.

Tsang, O.T., Chau, T.N., Choi, K.W., Tso, E.Y., Lim, W., Chiu, M.C., Tong, W.L., Lee, P.O., Lam, B.H., Ng, T.K., *et al.* (2003). Coronavirus-positive nasopharyngeal aspirate as predictor for severe acute respiratory syndrome mortality. Emerg. Infect. Dis. *9*, 1381–1387.

Tu, C., Crameri, G., Kong, X., Chen, J., Sun, Y., Yu, M., Xiang, H., Xia, X., Liu, S., Ren, T., *et al.* (2004). Antibodies to SARS coronavirus in civets. Emerg. Infect. Dis. *10*, 2244–2248.

Vabret, A., Dina, J., Gouarin, S., Petitjean, J., Corbet, S., and Freymuth, F. (2006). Detection of the new human coronavirus HKU1: a report of 6 cases. Clin. Infect. Dis. *42*, 634–639.

van der Hoek, L., Pyrc, K., Jebbink, M.F., Vermeulen-Oost, W., Berkhout, R.J., Wolthers, K.C., Wertheim-van Dillen, P.M., Kaandorp, J., Spaargaren, J., and Berkhout, B. (2004). Identification of a new human coronavirus. Nat. Med. *10*, 368–373.

Woo, P.C., Lau, S.K., Wong, B.H., Tsoi, H.W., Fung, A.M., Chan, K.H., Tam, V.K., Peiris, J.S., and Yuen, K.Y. (2004). Detection of specific antibodies to severe acute respiratory syndrome (SARS) coronavirus nucleocapsid protein for serodiagnosis of SARS coronavirus pneumonia. J. Clin. Microbiol. *42*, 2306–2309.

Woo, P.C., Lau, S.K., Chu, C.M., Chan, K.H., Tsoi, H.W., Huang, Y., Wong, B.H., Poon, R.W., Cai, J.J., Luk, W.K., *et al.* (2005a). Characterization and complete genome sequence of a novel coronavirus, coronavirus HKU1, from patients with pneumonia. J. Virol. *79*, 884–895.

Woo, P.C., Lau, S.K., Huang, Y., Tsoi, H.W., Chan, K.H., and Yuen, K.Y. (2005b). Phylogenetic and recombination analysis of coronavirus HKU1, a novel coronavirus from patients with pneumonia. Arch. Virol. *150*, 2299–2311.

Woo, P.C., Lau, S.K., Tsoi, H.W., Huang, Y., Poon, R.W., Chu, C.M., Lee, R.A., Luk, W.K., Wong, G.K., Wong, B.H., *et al.* (2005c). Clinical and molecular epidemiological features of coronavirus HKU1-associated community-acquired pneumonia. J. Infect. Dis. *192*, 1898–1907.

Woo, P.C., Lau, S.K., Li, K.S., Poon, R.W., Wong, B.H., Tsoi, H.W., Yip, B.C., Huang, Y., Chan, K.H., Yuen, K.Y. (2006). Molecular diversity of coronaviruses in bats. Virology *351*, 180–187.

Wu, D., Tu, C., Xin, C., Xuan, H., Meng, Q., Liu, Y., Yu, Y., Guan, Y., Jiang, Y., Yin, X., *et al.* (2005). Civets are equally susceptible to experimental infection by two different severe acute respiratory syndrome coronavirus isolates. J. Virol. *79*, 2620–2625.

Wu, H.S., Chiu, S.C., Tseng, T.C., Lin, S.F., Lin, J.H., Hsu, Y.H., Wang, M.C., Lin, T.L., Yang, W.Z., Ferng, T.L., *et al.* (2004). Serologic and molecular biologic methods for SARS-associated coronavirus infection, Taiwan. Emerg. Infect. Dis. *10*, 304–310.

Yam, W.C., Chan, K.H., Poon, L.L., Guan, Y., Yuen, K.Y., Seto, W.H., and Peiris, J.S. (2003). Evaluation of reverse transcription-PCR assays for rapid diagnosis of severe acute respiratory syndrome associated with a novel coronavirus. J. Clin. Microbiol. *41*, 4521–4524.

Yount, B., Roberts, R.S., Sims, A.C., Deming, D., Frieman, M.B., Sparks, J., Denison, M.R., Davis, N., and Baric, R.S. (2005). Severe acute respiratory syndrome coronavirus group-specific open reading frames encode nonessential functions for replication in cell cultures and mice. J. Virol. *79*, 14909–14922.

Zeng, F.Y., Chan, C.W., Chan, M.N., Chen, J.D., Chow, K.Y., Hon, C.C., Hui, K.H., Li, J., Li, V.Y., Wang, C.Y., *et al.* (2003). The complete genome sequence of severe acute respiratory syndrome coronavirus strain HKU-39849 (HK-39). Exp. Biol. Med. (Maywood) *228*, 866–873.

Human Coronavirus NL63, a Long Lost Brother

15

Krzysztof Pyrc and Lia van der Hoek

Abstract

Over the last decade a number of novel human respiratory viruses have emerged or have been recognized. A novel group I coronavirus, human coronavirus NL63 (HCoV-NL63) was identified in 2004 in the Netherlands in children and adults with acute respiratory illness. Worldwide analysis of clinical specimen indicated that this virus has widely spread within the human population. It is observed mostly in young children, elderly people and immunocompromised patients with upper and lower acute respiratory tract disease (1–10% of all respiratory infections). This chapter will discuss current knowledge on HCoV-NL63 replication, pathogenesis, disease burden and advances in treatment strategies.

Identification of HCoV-NL63

In January 2003, a 7-month child was admitted to a hospital in Amsterdam with coryza, conjunctivitis and fever (van der Hoek *et al.*, 2004). Chest radiography revealed a typical view of bronchilitis. A nasopharyngeal aspirate sample was collected five days after onset of disease and subsequently tested for known respiratory pathogens. Diagnostic tests for respiratory syncytial virus, adenovirus, influenza viruses A and B, rhinovirus, enterovirus and known coronaviruses were negative. The clinical sample was inoculated onto cell culture (human fetal lung fibroblasts, tertiary monkey kidney cells (tMK; Cynomolgus monkey) and HeLa cells). A cytophatic effect (CPE) was detected on tMK cells at day eight after inoculation. The observed CPE was diffuse, with a refractive appearance followed by cell detachment, and this CPE was more pronounced when the virus was passaged onto a monkey kidney cell line (LLC-MK2). Acid lability and chloroform sensitivity tests indicated that the virus was most likely enveloped, and did not belong to the picornavirus group. The sample was analyzed with a novel virus discovery method based on amplification fragment length polymorphism: VIDISCA (Virus Discovery cDNA-AFLP), which led to the identification of the novel human pathogen HCoV-NL63. The virus was completely sequenced and appeared to be distinct from any known member of the coronavirus family. The highest nucleotide identity was noticed with the human coronavirus (HCoV) 229E. The similarity between these two viruses was about 65% on the nucleotide level. Soon after the first publication, another group from the Netherlands reported discovery of the same human coronavirus (Fouchier *et al.*, 2004). This report describes identification of the virus in a culture, which originated from a nose swab collected in 1988 from an 8-month-old boy suffering from pneumonia. The patient sample was primarily inoculated on tMK cells and resulted in CPE onset on day 7 with ~50% of cell death on day 13. After initial isolation on tMK cells, the virus replicated in Vero cells with CPE appearance from day 4 post-inoculation onward. The virus was first investigated by electron microscopy that revealed the presence of coronavirus-like particles with an average size of 140 nm (range, 95–275 nm) and average envelope projections of 20 nm (range, 12–24 nm). Amplification by RAP-PCR yielded coronavirus PCR fragments. Interestingly, one year after the publications of both Dutch groups the team of J. Kahn described the identification of essentially the same virus and presented a novel name: New Haven (NH) coronavirus (Esper *et al.*, 2005b). However, the partial sequences deposited in GenBank show 94%—100% similarity between the prototype HCoV-NL63 strain and isolates from New Haven. In 2005 HCoV-NL63 was advised as the proper species name by the coronavirus study group of the International Committee on Taxonomy of Viruses (ICTV) executive committee.

Analysis of the early coronavirus reports suggests that the virus might have been observed previously. The research on human coronaviruses started in 1965 when B814 coronavirus was isolated from an adult with respiratory illness (Tyrrell and Bynoe, 1965). In subsequent reports from the 1960s several additional coronaviruses isolates were obtained that could be propagated in human embryonic tracheal organ culture. Of these, only HCoV-229E and HCoV-OC43 were studied during the last decades, whereas the other isolates were lost for follow-up studies (B814, HCoV-OC16, HCoV-OC37, and HCoV-OC48 (Tyrrell and Bynoe, 1965; McIntosh *et al.*, 1967)). Some of them might have been HCoV-NL63 isolates. Furthermore, in the early 1980s various isolates of human coronaviruses were described that were obtained by cell culture or organ culture (HCoV strains AD, PA, LP, KI, PR, TO, RO, GI, and HO) (Larson *et al.*, 1980; MacNaughton *et al.*, 1981; Reed, 1984). Some of these isolates were termed group II isolates because they were serologically related to HCoV-OC43 (strains RO, GI and HO), whereas others were assigned group I isolates (strains AD, PA, LP, KI, PR, and TO). Of these group I viruses, LP, KI, PR and TO could be cultured in MRC5 continuous cells, but strains AD and PA could only be cultured in human fetal tracheal and nasal organ cultures. This difference in culture characteristics resembles the difference in cell tropism also noticed for HCoV-229E and HCoV-NL63. This raises the possibility that some of these group I strains were in fact HCoV-NL63 isolates. Unfortunately, isolates AD and PA were also lost for further study (Phil Minor, personal communications). Therefore, the earliest HCoV-NL63 isolate remains the NL-strain that was obtained from a patient in 1988 (Fouchier *et al.*, 2004).

Molecular biology of HCoV-NL63

The HCoV-NL63 genome

The genome encompasses 27535 to 27553 nucleotides (depending on the virus strain). It is predicted to contain seven functional open reading frames and untranslated regions at its 5′ and 3′ ends. The two large 5′ terminal ORFs 1a and 1b encode non-structural proteins that are required for viral replication. The remaining ORFs in the 3′ part of the genome encode the four structural proteins spike (S), envelope (E), membrane (M) and nucleocapsid (N) and one accessory ORF3 protein (Table 15.1) (Pyrc *et al.*, 2004).

Table 15.1 HCoV-NL63 genomic organization

ORF	Gene	Protein expressed	Position in genome (nt)	Length (nt)	Length (aa)	Frame
1	1a		287–12439	12153	4050	+2
1,2	1a/1b		287–20475	20190	6729	+2/+1
		nsp1	287–616	330	110	+2
		nsp2	617–2980	2364	788	+2
		nsp3; PL^{pro}	2981–7672	4692	1564	+2
		nsp4	7673–9103	1431	477	+2
		nsp5; M^{pro}	9104–10012	909	303	+2
		nsp6	10013–10849	837	279	+2
		nsp7	10850–11098	249	83	+2
		nsp8	11099–11683	585	195	+2
		nsp9	11684–12010	327	109	+2
		nsp10	12011–12415	405	135	+2
		nsp12; RdRp	12416–15195	2781	927	+2/+1
		nsp13; MBD, NTPase, Helicase	15196–16986	1791	597	+1
		nsp14; ExoN	16987–18540	1554	518	+1
		nsp15; NendoU	18541–19572	1032	344	+1
		nsp16; 2′-O-MT	19573–20472	900	300	+1
3	S	Spike (S)	20472–24542	4071	1356	+3
4	ORF3	ORF3	24542–25219	678	225	+2
5	E	Envelope (E)	25200–25433	234	77	+3
6	M	Membrane (M)	25442–26122	681	226	+2
7	N	Nucleocapsid (N)	26133–27266	1134	377	+3

Phylogenetic analysis

HCoV-NL63 is closely related to group I coronaviruses according to phylogenetic analyses (Figure 15.1) (van der Hoek *et al.*, 2004). Similarity is the highest with HCoV-229E and porcine epidemic diarrhea virus (PEDV), 65% and 61%, respectively. These three viruses cluster with a recently described bat coronavirus (Poon *et al.*, 2005) in a subgroup within group I (Ib), excluding transmissible gastroenteritis virus (TGEV), Canine enteric coronavirus (CCoV) and Feline infectious peritonitis virus (FIPV) (Table 15.2). But also within the HCoV-NL63 species different phylogenetic subgroups can be identified. Most phylogenetic analyses have focused on a small region in the ORF 1a. Several groups have sequenced this region of field isolates and found 2 genetic clusters (Moes *et al.*, 2005; van der Hoek *et al.*, 2004; Arden *et al.*, 2005; Bastien *et al.*, 2005a). This clustering does

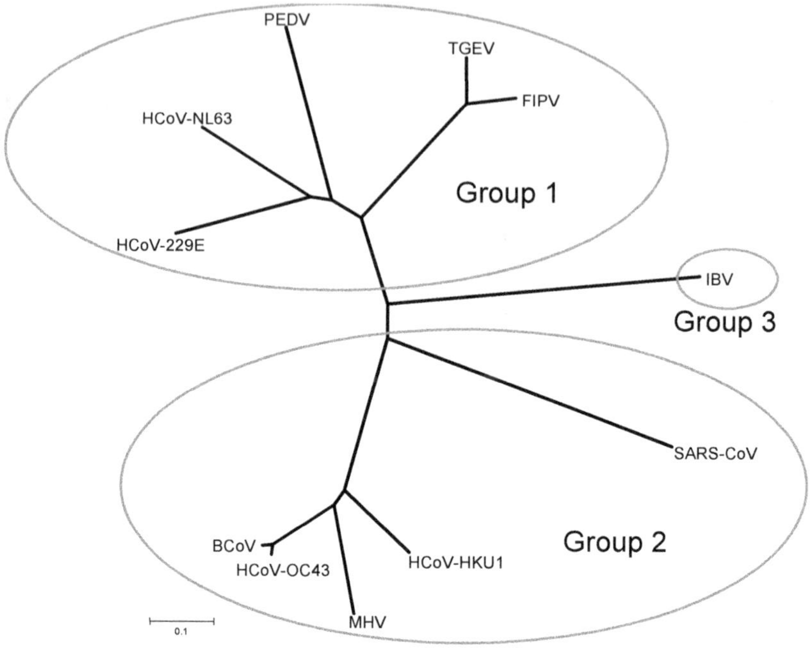

Figure 15.1 Phylogenetic analysis of the *Coronaviridae* family. In the analysis full genome sequences were employed and analysis was done with the Neighbor-joining method and Kimura 2-parameter model.

not correlate with the geographic location of the patients (Figure 15.2). Also in the spike gene two genetic clusters are observed. For those patients for which both regions have been sequenced a discordant clustering was found (Moes *et al.*, 2005) indicating that recombination between HCoV-NL63 isolates occurs frequently (Figure 15.2; indicated

Table 15.2 Nucleotide and amino acid similarity between HCoV-NL63 and other coronaviruses

Group	Species	1a	1b	S	E	M	N
Ia	TGEV	53/38[a]	69/71	53/42	54/30	49/44	48/35
	FIPV	53/38	69/70	53/40	51/27	45/39	47/35
Ib	HCoV-229E	63/57	76/81	56/55	59/47	64/61	52/43
	PEDV	58/50	71/77	55/44	54/42	65/65	47/38
IIa	MHV	46/23	59/53	46/21	50/15	48/33	37/20
	HCoV-HKU1	49/23	62/53	48/22	45/18	51/33	38/21
	HCoV-OC43	48/23	59/52	47/22	41/18	49/33	38/21
IIb	SARS-CoV	47/23	59/55	46/21	46/18	45/29	43/24
III	IBV	46/22	60/52	43/23	34/10	49/24	41/20

a Percentage of nucleotide/amino acid identity.

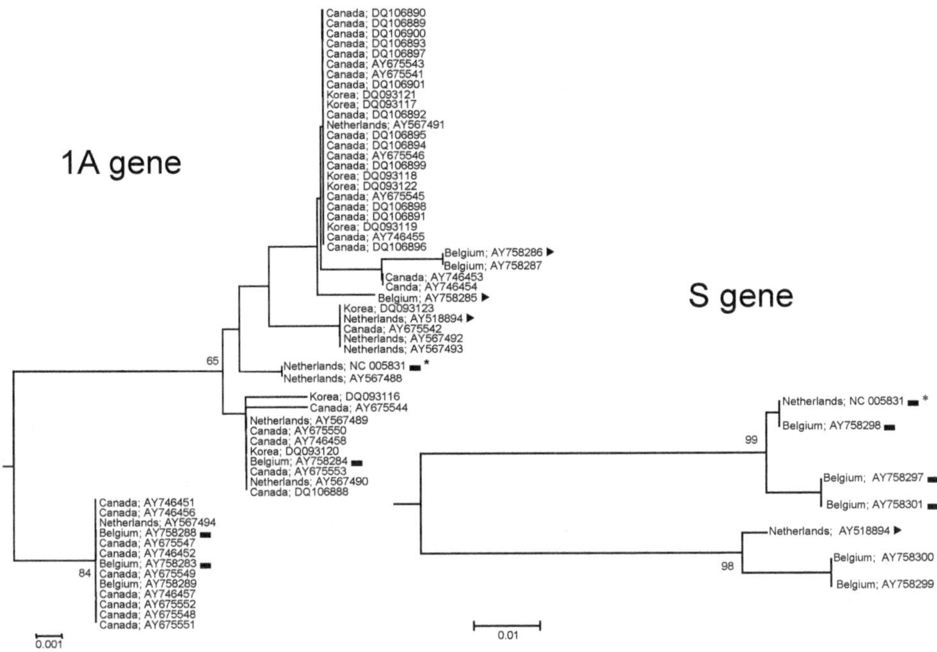

Figure 15.2 Phylogenetic analysis of HCoV-NL63 isolates. Analysis includes the partial sequence of the PL^{pro} gene (1a) and S gene. Name of each taxon consists of its GenBank accession number and country of origin. The index strain that was discovered first is marked with an asterisk. The symbols ▶ and ▬ were added to illustrate the discordance in phylogenetic clustering at the 1a gene and the S gene. Analysis was done with the Neighbor-joining method and Kimura 2-parameter model.

with ▶ and ▬). At present four full genome sequences of HCoV-NL63 are available (Amsterdam 1, NL, A57 and A496). They share ~99% similarity on the nucleotide level, but the 5′ region of the 1a gene and the 5′ part of the S gene are the most variable among the different isolates. Analyses of the full-length sequences further supports the theory that recombination between HCoV-NL63 isolates occurs (Pyrc *et al.*, 2006b). Another region with a unique feature is the M gene. Analysis of the sequence suggests that HCoV-NL63 M is a hybrid protein, which is partially related to 229E M (N-terminus) and partially to PEDV M (C-terminus). This suggest that even interspecies recombination might have taken place (Pyrc *et al.*, 2005).

The HCoV-NL63 genome contains a unique N-terminal region of the S protein (aa position 17–196). This region shows no significant similarity with any other coronavirus or cellular sequence (van der Hoek *et al.*, 2004). The origin of the insert is not known, and one might speculate that HCoV-229E lost the 5′ region of its S protein, similar to what has been described for porcine respiratory coronavirus and TGEV (Laude *et al.*, 1995). This region is also most diverse between HCoV-NL63 isolates on both nucleotide and amino acid level. Furthermore, it contains many glycosylation sites and is immunodominant (Pyrc and van der Hoek, unpublished data). Its exact function in virus replication is however unknown.

Transcription

Coronavirus replication is a complex, not yet fully understood mechanism (Sawicki and Sawicki, 1990; Sawicki and Sawicki, 1998). The 5′ end of the HCoV-NL63 genomic RNA contains the untranslated leader sequence of 72 nucleotides that ends with the leader Transcription Regulation Sequence (TRS) element. This leader TRS is very similar to sequences that can be found upstream of each open reading frame (body TRSs). The RNA-dependent RNA polymerase (RdRp) has been proposed to pause after a body TRS of a particular gene during minus strand synthesis, subsequently switching to the leader TRS and thus adding a common leader sequence to each minus strand subgenomic (sg) mRNA. This discontinuous transcription mechanism is based on base-pairing of the nascent minus strand copy RNA with the plus strand leader TRS. Inspection of sg mRNA junctions of HCoV-NL63 shows that they are composed of the part of the HCoV-NL63 genome that is directly downstream of a particular body TRS, and a 5′ end derived from the leader sequence (Pyrc *et al.*, 2004). The most conserved TRS region of HCoV-NL63 was defined by multiple sequence alignment as AACUAAA. This core sequence is conserved in almost all sg mRNA. The only exception is the E gene sg mRNA that contains the sub-optimal TRS core AACUAUA. A 13-nucleotide sequence upstream of the E gene core sequence harbors perfect homology to the leader sequence and it was suggested that the upstream sequence compensates for the absence of an optimal TRS core during discontinuous (–) strand synthesis (Pyrc *et al.*, 2004). HCoV-NL63 generates five distinct sg mRNAs during HCoV-NL63 replication, encoding S, ORF3 protein, E, M and N (Figure 15.3). The 1a/ab polyprotein is supposed to be translated from the genomic RNA (Pyrc *et al.*, 2004). Northern blot analysis revealed that the transcription levels of sg mRNAs of HCoV-NL63 are inversely related to the distance between the body TRS and the 3′ end of the genome (Pyrc *et al.*, 2004).

Translation

The sg mRNAs of S, E, M, N and ORF3 protein are monocistronic, so they all encode for a single protein. A completely different situation is observed for the large replicase gene. This gene encodes the polyproteins 1a and 1ab. Expression of polyprotein 1ab requires ribosomal frameshifting. Ribosomal frameshifting in coronaviruses was the first described non-retroviral example of translational frameshifting in higher eukaryotes (Brierley *et al.*, 1989). It is a mechanism of translational regulation in which a directed change of translational reading frame allows the synthesis of a single protein from two overlapping ORFs. In coronaviruses, frameshifting is thought to regulate the level of replicative proteins (Brierley *et al.*, 1989; Brierley *et al.*, 1987). Coronaviral ribosomal frameshifting is induced by a complex RNA structure called a pseudoknot, consisting of a hairpin of which the loop interacts with a sequence further downstream, and a slippery site upstream of the hairpin (Brierley *et al.*, 1989). The pseudoknot structure in HCoV-NL63 was found to possess an extended structure and was therefore called an "elaborated" pseudoknot, similarly to the structure found in HCoV-229E, PEDV and TGEV (Baranov *et al.*, 2005; Mathews *et al.*, 1999; Herold and Siddell, 1993; Pyrc *et al.*, 2005). In the HCoV-NL63 genome, the ribosomal frameshift site is located inside the RNA-dependent RNA-polymerase gene. The predicted "elaborated" pseudoknot consists of a hairpin with a highly conserved 11 basepair stem, and a loop of 8 nucleotides that interacts with a sequence 167 nucleotides

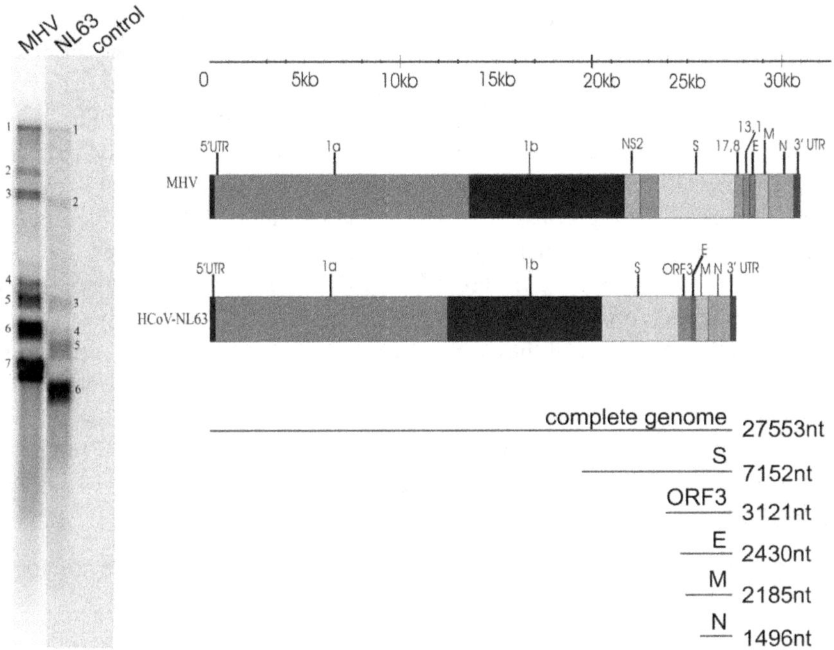

Figure 15.3 Transcripts of HCoV-NL63. The left panel shows the Northern blot analysis of HCoV-NL63 RNA in infected LLC-MK2 cells. RNA of HCoV-NL63 (NL63 lane) was compared with RNA of MHV strain A59 (MHV lane). Non-infected LLC-MK2 cells are included as a negative control (control lane). The right panel shows the MHV and HCoV-NL63 genome organization and the HCoV-NL63 sg-mRNAs. Reprinted from Pyrc *et al.* (2004).

downstream (with 5 nucleotides basepairing) that is presented on a hairpin structure. The heptanucleotide slippery UUUAAAC sequence is present just upstream of the first hairpin at position nt 12433 (Pyrc *et al.*, 2005).

Post-translational modification

Similar to all other coronaviruses, HCoV-NL63 is supposed to employ post-translational proteolytic processing as a key regulatory mechanism in the expression of its replicative proteins (Denison and Perlman, 1986) (see Table 15.1 for an overview). In the genome of HCoV-NL63 three proteinases can be predicted in the 5′ region of the 1a/1ab polyproteins. First, two papain-like proteinases (PL^{pro}s) are encoded in non-structural protein (nsp) 3. The HCoV-NL63 PL^{pro} domains, $PL1^{pro}$ and $PL2^{pro}$, are both expected to have catalytic activity (Pyrc *et al.*, 2005). Cleavage of polyproteins 1a/1ab is predicted between nsp1|nsp2, nsp2|nsp3 and nsp3|nsp4, releasing the functional papain-like proteinase protein (nsp3) molecule by auto-cleavage (Gao *et al.*, 2003). Analysis of the HCoV-NL63 cleavage sites indicates that HCoV-NL63-PL^{pro}s cut between two small amino acids with short uncharged side chains, similar to homologous enzymes from other species from the *Coronaviridae* family (Lim and Liu, 1998).

The second proteinase of HCoV-NL63 resides in nsp5. It has a predicted serine-like proteinase activity and is designated chymotrypsin-like proteinase or "the main proteinase"

(M^{pro}) to stress its dominant function. On the level of primary and secondary structure, HCoV-NL63 M^{pro} is closely related to M^{pro} of HCoV-229E. The majority of non-structural proteins in the polyproteins 1a/1ab are released by M^{pro}. The order of cleavage sites is conserved among coronaviruses and appears to depend on the accessibility of specific sites in the context of the polyprotein (Pinon *et al.*, 1999). Analysis of the cleavage sites of HCoV-NL63 M^{pro} reveals an interesting cleavage site change between nsp13 and nsp14. In contrast to most coronaviruses, this cleavage site of HCoV-NL63 contains a histidine residue at position P1 instead of glutamine (Ziebuhr and Siddell, 1999; Hegyi and Ziebuhr, 2002; Fan *et al.*, 2004; Pyrc *et al.*, 2005). This was observed in four different HCoV-NL63 isolates (Amsterdam 1, A57, A496 and NL), and might suggest altered substrate specificity or the absence of the cleavage site. For instance, alteration of the consensus M^{pro} cleavage sequence LQ|(A,S,G) for M^{pro} derived from other coronaviruses resulted in significantly reduced cleavage efficiencies in most cases (Ziebuhr and Siddell, 1999; Fan *et al.*, 2004; Hegyi *et al.*, 2002). Surprisingly, a recent report about HCoV-HKU1 suggest that the HCoV-HKU1 genome harbors the same alteration in nsp13|nsp14 cleavage site (Woo *et al.*, 2005).

Entry to the target cell

Cell tropism and receptor usage

The S protein of a coronavirus is a type I membrane glycoprotein that is incorporated into the virion as a surface protein. It mediates attachment and entry into host cells (Cavanagh, 1983; Hofmann *et al.*, 2005) and is synthesized as a 180- to 200-kDa protein. The protein can be cleaved by host-derived proteases (e.g. furin) (Huang *et al.*, 2006; Simmons *et al.*, 2005; de Haan *et al.*, 2004) into two, non-covalently associated subunits S1 and S2. The HCoV-NL63 S protein resembles other group I viral fusion proteins, with its short C-terminal tail inside the virion, single transmembrane domain, rod-like S2 domain and located on top of the S2 domain a globular head consisting of the S1 region (Pyrc *et al.*, 2005). Receptor binding is mediated by the S1 subunit, while the membrane-anchored S2 portion is required for fusion of viral and cellular membranes.

The receptor specificity among coronaviruses varies between species. This variability is especially visible among group II viruses: MHV uses CAECAM-1 (Williams *et al.*, 1991), HCoV-OC43 and BCoV use O-acetylated sialic acid (Vlasak *et al.*, 1988) and SARS-CoV uses angiotensin converting enzyme 2 (ACE2) (Li *et al.*, 2003). Before the discovery of HCoV-NL63 it was generally thought that all group I coronaviruses use CD13 (also known as aminopeptidase N) as receptor, because the representatives of that group (HCoV-229E, porcine, feline and canine coronaviruses) engage CD13 for cell entry (Oh *et al.*, 2003; Oh *et al.*, 2004; Delmas *et al.*, 1992; Yeager *et al.*, 1992; Tresnan *et al.*, 1996; Benbacer *et al.*, 1997). Studies on HCoV-NL63 cell tropism reveals that the virus, unlike its closest relative HCoV-229E, can be propagated on LLC-MK2, Vero-E6 and Huh-7 cells (van der Hoek *et al.*, 2004; Fouchier *et al.*, 2004; Hofmann *et al.*, 2005). These cells all express ACE2, suggesting that HCoV-NL63 and SARS-CoV might use the same receptor. Further analysis engaging the pseudotyping retroviral system confirmed that HCoV-NL63 does not use CD13, but indeed uses ACE2 as the receptor for cell entry (Hofmann *et al.*, 2005).

The ACE2 molecule was first identified as a homolog of angiotensin-converting enzyme with zinc metalloproteinase activity (Tipnis *et al.*, 2000). The activity of ACE2 differs from ACE (Danilczyk *et al.*, 2003), but both enzymes are key players of the renin–angiotensin system, thereby regulating cardiac and vascular functions (Crackower *et al.*, 2002). There is now an increasing body of evidence supporting the role for the renin–angiotensin system in acute lung injury. It is thought that ACE influences the vascular permeability, air-vessel interface and pneumocyte viability and indeed, increased levels of ACE in the respiratory tract were found in patients with an acute respiratory distress syndrome (ARDS) (Idell *et al.*, 1987). ACE2 acts in the opposite way and protects against lung injury. It was suggested that binding of SARS-CoV to ACE2 downregulates ACE2 levels, thus aggravating the lung injury and produce lung edema during infection (Kuba *et al.*, 2005). HCoV-NL63 also binds to ACE2, but unlike SARS-CoV, an HCoV-NL63 infection is generally not associated with ARDS. Therefore the use of ACE2 as a viral receptor by itself does not invariably result in aggravating acute lung injury.

Cell entry

Enveloped viruses employ two strategies to enter their target cells. One involves glycoprotein mediated engagement of receptors on the cell surface, which triggers fusion of the viral and cellular membranes (Doms, 2000). Alternatively, receptor binding can trigger endocytosis followed by transport into an endocytic compartment, where the low pH triggers the fusion (Sieczkarski and Whittaker, 2002). Inhibitors of vacuolar acidification are well-established tools to assess if the fusion of the membranes is pH dependent (Drose and Altendorf, 1997; Drose *et al.*, 1993; Perez and Carrasco, 1994). Treatment with ammonium chloride or bafilomycin, an inhibitor of vacuolar proton pumps (Bowman *et al.*, 1988), demonstrated that HCoV-NL63 S-mediated infection of hepatocytes (Huh-7 cells) depends on low pH (50% inhibition), but to lower extent than SARS-CoV (97% inhibition) (Hofmann *et al.*, 2006). Also results of I-Chueh Huang *et al.* shows that treatment with NH_4Cl poorly inhibits HCoV-NL63, compared to a strong inhibition of SARS-CoV (Huang *et al.*, 2006), suggesting that the entry into an acidified environment of the endosome is not essential to HCoV-NL63. These results suggest that HCoV-NL63 may use the same route of entry as MHV, via direct fusion of the membranes rather than the one used by HCoV-229E, via alveolar mediated endocytosis. Additionally, I-Chueh Huang *et al.* showed that, unlike SARS-CoV, HCoV-NL63 does not require cathepsin L activity for infection of LLC-MK2 cells (Huang *et al.*, 2006). The question of the entry pathway and S protein intracellular processing of HCoV-NL63 needs further study.

The spike proteins of coronaviruses exhibit a characteristic membrane fusion mechanism that is driven by conformational changes in the S protein, with rearrangements and interactions of heptad repeat regions (HR1 and HR2) located in the S2 domain. Association of the HR1 and HR2 domains into a hetero-hexamer brings the fusion peptide that is located near the N-terminus of HR1 in close proximity to the transmembrane domain, thereby facilitating membrane fusion. HCoV-NL63 S also contains the HR1 and HR2 regions with typical seven-residue periodicity. Characteristic for group I viruses are the two additional repeats in both HR regions and also HCoV-NL63 contains these additional repeats. HR2 of HCoV-NL63 is located adjacent to the transmembrane domain, and HR1 is about 170 amino acids towards the N-terminus. These regions associate tightly into the hetero-hexamer and are essential for HCoV-NL63 entry (Pyrc *et al.*, 2006a).

Infection

Prevalence and disease association

The first publication on HCoV-NL63 described several infected patients thus illustrating that infection by the virus is not a rare event (van der Hoek *et al.*, 2004). Among 614 respiratory specimens collected between December 2002 and April 2003, 7 individuals were positive for HCoV-NL63 with an incidence in the winter months up to 7%. All individuals that were HCoV-NL63 positive had upper or lower respiratory tract disease (URTI and LRTI) or both (van der Hoek *et al.*, 2004). The report from Fouchier describes HCoV-NL63 in a clinical sample derived from 1988 and four additional samples out of 139 respiratory specimens (2.9%) collected from November 2000 to January 2002 (Fouchier *et al.*, 2004). Since then, several groups reported HCoV-NL63 in patients suffering from respiratory tract disease (Figure 15.4). In Australia, HCoV-NL63 could be detected in 16 of 766 patients with acute respiratory disease (2%) (Arden *et al.*, 2005). A Japanese study on 118 children under the age of 2 that suffered from bronchiolitis and were admitted to the hospital revealed 3 positive patients (2.5%), and a second Japanese study presented 5 HCoV-NL63 infections among 419 specimens collected from children with respiratory infections (1.2%) (Suzuki *et al.*, 2005). Amongst 279 hospitalized children (with URTI and LRTI) in Belgium, 7 children were identified with HCoV-NL63 (2.3%) (Moes *et al.*, 2005). In France, 28 of 300 patients with URTI and LRTI under the age of 20 contained HCoV-NL63 RNA (9.3%) (Vabret *et al.*, 2005) and in a Swiss cohort

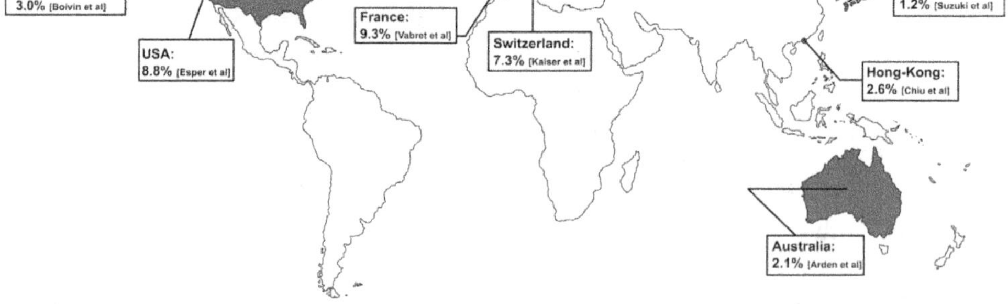

Figure 15.4 The worldwide distribution of HCoV-NL63. Countries with documented HCoV-NL63 infections are indicated in grey. Blank regions represent countries that did not search for HCoV-NL63 infection. There are no reports on the absence of HCoV-NL63 in certain countries. In the boxes the frequency of HCoV-NL63 infection is indicated.

of neonates, HCoV-NL63 was identified in 6 of 82 cases (7%) during their first episode of LRTI (Kaiser *et al.*, 2005). Three studies describe HCoV-NL63 infection in Canada (Bastien *et al.*, 2005a; Bastien *et al.*, 2005b; Boivin *et al.*, 2005). Bastien *et al.* reported 45 HCoV-NL63-positive patients of a total of 1765 specimens (2.5%). Seven of these Canadian patients were involved in an outbreak of acute respiratory-tract infection in a personal care home for elderly (Bastien *et al.*, 2005b). The third Canadian survey reported a frequency of 3% (12 of 396 tested specimens). In this study HCoV-NL63 infection was also investigated in healthy control subjects (Boivin *et al.*, 2005). The virus was present in 1.7% (3 of 177 tested specimens) in the healthy control group illustrating that HCoV-NL63 infection might be also asymptomatic or, alternatively, that some control individuals are shedding the virus after a symptomatic phase of HCoV-NL63 infection (see below). HCoV-NL63 was also detected in Hong Kong in 15 of 587 children (2.6%) (Chiu *et al.*, 2005). These Hong Kong children participated in a prospective clinical and virological study on children under the age of 18 with acute respiratory tract infection. The study group represented the population, such that it could be estimated that HCoV-NL63 infection causes more than 200 hospital admissions each year per 100000 children under the age of 6 in Hong Kong. A similar study was performed on the PRI.DE cohort, which is a population-based retrospective German multicenter study on pediatric respiratory infections (van der Hoek *et al.*, 2005). The study included young children from pediatric practices and hospitals with clinical signs of laryngotracheitis (croup), bronchitis, bronchiolitis, pneumonia, or apnea. Of the 949 PRI.DE samples, 49 (5.2%) were positive for HCoV-NL63. Various clinical diagnoses of lower respiratory tract disease were observed for the HCoV-NL63-positive patients, including croup, bronchitis, bronchiolitis, and pneumonia. A high frequency of co-infections in HCoV-NL63-positive samples (59%) was observed. However, in children with no other pathogen incidence of croup was high, suggesting that HCoV-NL63 is one of the pathogens causing croup in early childhood. Croup is an inflammation of the trachea and is characterized by a loud barking couch that usually worsens at night.

Respiratory coronavirus infections occur more often in the winter and spring than in the summer and fall. Already in the first report about HCoV-NL63 the winter preference of HCoV-NL63 was presented. Additional studies from several countries with moderate climate (Belgium, Australia, Japan, Canada, Germany and France) confirmed that observation (Figure 15.5). Only in Hong Kong, HCoV-NL63 showed a spring-summer peak of activity, whereas in the same study HCoV-OC43 was mostly found in the winter season. This indicates that the seasonality of HCoV-NL63 in tropical and subtropical regions is not restricted to the winter season.

HCoV-NL63 versus other coronaviruses, frequency of infection

Until recently only human coronaviruses 229E and OC43 were thoroughly studied. The third human coronavirus, SARS-CoV, was identified during the 2002–2003 outbreak of severe acute respiratory syndrome (SARS). Although SARS-CoV is the most pathogenic human coronavirus and responsible for ~800 fatal cases (10% mortality), it was eradicated within one season due to highly effective global public health response. To compare the burden of disease caused by the non-SARS-CoV coronaviruses, the frequency of HCoV-NL63 infection was compared with the frequency of HCoV-229E and HCoV-OC43

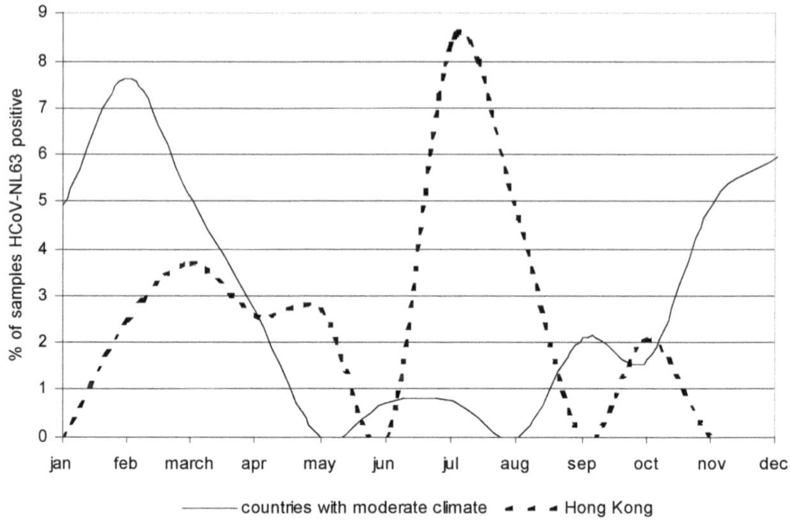

Figure 15.5 Seasonality of HCoV-NL63 infection. Studies in moderate climate (Germany, Canada, Belgium, France) and sub-tropical climate (Hong Kong) were included. The values on the Y axis represent the percentage of HCoV-NL63 positive samples.

infection. The screening of a Belgian sample collection for HCoV-OC43, HCoV-229E and HCoV-NL63 led to identification of 7 patients with HCoV-NL63 infection (2.3%), 7 patients with HCoV-OC43 infection (2.3%) and one patient with HCoV-229E infection (0.3%). Similar results were obtained in our laboratory for patients with respiratory disease that were diagnosed in the Public Health Laboratory at the Municipal Health Service of Amsterdam. Of the 63 respiratory samples that were tested for the 3 human coronaviruses in the winter of 2002–2003, 3 samples were positive for HCoV-NL63 (4.7%), 3 for HCoV-OC43 (4.7%) and no HCoV-229E was detected (M. van Zon, S. Bruisten, J. Spaargaren and L. van der Hoek unpublished results).

The population based study of S. Chiu *et al.* identified 26 children with coronaviral infection (4.4% of 587 patients). Of them 15 were infected with HCoV-NL63 (2.6%), 9 by HCoV-OC43 (1.5%) and HCoV-229E was found only in 2 patients (0.3%) (Chiu *et al.*, 2005).

In a French study HCoV-NL63 was detected in 9.3% of 300 patients, while HCoV-OC43 was detected in 1.2% of samples and no HCoV-229E positive sample was identified (Vabret *et al.*, 2005). A Swiss study on coronavirus infection detected 13 positive patients (16%) (Kaiser *et al.*, 2005). Among these, 6 samples were positive for HCoV-NL63 (7.3%), 5 samples were HCoV-OC43 positive (6%) and HCoV-229E was found in 3 cases (3.7%).

Combining these results suggests that the frequency of infection with different coronaviruses follows the following rule: HCoV-NL63 ≥ HCoVOC43 > HCoV-229E. In all reports HCoV-NL63 and HCoV-OC43 were found most frequently.

Table 15.3 Overview of clinical symptoms of HCoV-NL63 infected patients

Country of origin	No. of samples	Clinical symptoms: Fever	Cough	Running nose	URTI	LRTI	Co-infection
The Netherlands (Fouchier *et al.*, 2004)	4	100%	100%	100%	N.D.	N.D.	N.A.*
The Netherlands (van der Hoek *et al.*, 2004)	8	N.D.	N.D.	N.D.	38%	38%	25%
Belgium (Moes *et al.*, 2005)	7	86%	57%	29%	43%	57%	N.A.
Switzerland (Kaiser *et al.*, 2005)	6	83%	100%	100%	67%	100%	33%
Germany (van der Hoek *et al.*, 2005)	49	43%	N.D.	N.D.	0%	100%	59%
France (Vabret *et al.*, 2005)	28	61%	N.D.	39%	22%	39%	N.A.
USA (Esper *et al.*, 2005b)	79	48%	64%	61%	N.D.	N.D.	12%
Canada (Bastien *et al.*, 2005a)	19	53%	47%	11%	26%	21%	N.A.
Canada (Bastien *et al.*, 2005b)	26	31%	54%	31%	42%	58%	4%
Canada (Boivin *et al.*, 2005)	12	58%	100%	100%	17%	83%	75%
Japan (Ebihara *et al.*, 2005a)	3	100%	100%	100%	33%	100%	N.A.
Japan (Suzuki *et al.*, 2005)	5	100%	80%	80%	80%	20%	N.A.
Hong Kong (Chiu *et al.*, 2005)	15	20%	67%	60%	27%	20%	33%
Australia (Arden *et al.*, 2005)	16	69%	81%	56%	19%	81%	38%

* Not applicable because all or part of the clinical samples was selected for absence of other respiratory virus.

Co-infection with a second respiratory virus

Several studies showed that HCoV-NL63 infection is often combined with a second respiratory virus. The frequency of co-infection might reach even 75% (Table 15.3). In the PRI.DE cohort, double infections of HCoV-NL63 and a second respiratory virus was apparent in 29 of the 49 HCoV-NL63 positive samples (59%) with the highest co-infection frequency with RSV-A (41%), followed by PIV3 (10%) and RSV-B (8%). On the contrary, the other population based study from Hong Kong reported 15 HCoV-NL63 positive patients but double infection was rather rare (20%) and the most common second respiratory virus was influenza A. These data suggest that there is no preference for the second pathogen, but that the frequency of a co-infection with a specific agent correlates with its co-seasonality in the population. Several other reports also show that HCoV-NL63 is frequently present with a second virus, although the numbers are not representative because often a selection for samples that are negative for other viral pathogens was performed. The direct impact of co-infection with other respiratory viruses on the disease severity is not known, but doubly infected patients are more often hospitalized then single HCoV-NL63 infected patients (van der Hoek *et al.*, 2005). During a co-infection the viral load of HCoV-NL63 is significantly lowered, which may illustrate the influence of innate immune responses triggered by one virus on the replication of HCoV-NL63 (Chiu *et al.*, 2005; van der Hoek *et al.*, 2005).

Virus shedding

The study of S. Chiu shows that the HCoV-NL63 load in respiratory specimens correlates inversely with the time after onset of symptoms (Chiu *et al.*, 2005). While the mean viral load on day zero of infection was estimated to be $\sim 2 \times 10^8$ copies/ml, it decreased linearly to the level of 3×10^4 copies/ml on day 3. Although it is known that a coronavirus infection is generally cleared within a week it has been reported that coronaviruses can also be shed for a long period (Herrewegh *et al.*, 1997; Liu *et al.*, 2004). For HCoV-NL63 the situation seems not much different. Three weeks after an initial HCoV-NL63 infection, the virus was still detectable in respiratory specimens of 50% of the previously infected children. At this time all children were free of symptoms (Kaiser *et al.*, 2005). This prolonged shedding may explain HCoV-NL63 detection in symptom-free control persons (Boivin *et al.*, 2005).

Sero-prevalence

Coronaviruses are responsible for 10 to 30% of common cold cases each winter season. This means that during a lifetime virtually everybody will experience an infection with these viruses, and everybody will consequently carry antibodies against these viruses. This is true for HCoV-NL63. Adults carry neutralizing antibodies that can inhibit HCoV-NL63 S mediated infection (Hofmann *et al.*, 2005). In contrast, strong neutralization of HCoV-229E S-harboring pseudotypes was observed with only a minority of adult sera. These results were confirmed in our laboratory. We showed that sera from healthy adult donors all neutralize HCoV-NL63 (Pyrc *et al.*, 2006a). Further evidence for the high frequency of specific HCoV-NL63 antibodies in adult sera was delivered by Western blot analysis. The unique N-terminal part of the HCoV-NL63 S protein was expressed and tested for recognition by human sera from adults. All tested adult sera (n=20) recognized the HCoV-NL63 S protein (Pyrc and van der Hoek, unpublished data). Sera from infants poorly neutralized infection driven by both HCoV-NL63 S and HCoV-229E S using the HCoV-NL63 pseudotyping system (Hofmann *et al.*, 2005). Screening of a larger number of samples with an HCoV-NL63-specific-ELISA will allow a more precise determination of the average age at which HCoV-NL63 antibodies are acquired.

HCoV-NL63 and Kawasaki disease

In 2005 an association between HCoV-NL63 infection and Kawasaki disease was reported (Esper *et al.*, 2005a). Kawasaki disease is a systemic vasculitis in childhood that may result in aneurysms of the coronary arteries. In the developed world, Kawasaki disease is the most common cause of acquired heart disease in children. The evidence that Kawasaki disease is linked with an infectious agent includes temporal clustering and marked seasonality, geographic clustering (Chang *et al.*, 2004; Burns *et al.*, 2005) and age distribution (Chang *et al.*, 2004). Previously, various infectious agents have been suggested, including bacteria, allergens and some viruses (Chou *et al.*, 2005; Takahashi *et al.*, 2004; Nigro *et al.*, 1994). Coronary complications can be reduced significantly by the use of intravenous immunoglobulin (IVIG) therapy combined with oral aspirin (Kato *et al.*, 1979; Leung, 1989). The study on coronavirus infection and Kawasaki disease included 11 children with Kawasaki disease and 22 control children sampled in the same week as the Kawasaki patients. Eight of the children with Kawasaki disease were infected by HCoV-NL63 (73%), while in the control group of 22 children only one single HCoV-NL63 positive patient was identified (4.5%). Soon thereafter several groups reported lack of association between HCoV-NL63 infection and Kawasaki disease (Chang *et al.*, 2006; Shimizu *et al.*, 2005; Ebihara *et al.*, 2005b). Differences in geographic region, definition of cases, selection bias and PCR methods might explain to some extent the differences between J. Kahn's group and others. To solve this matter, a sensitive immunological assay is needed to investigate whether seroconversion to HCoV-NL63 occurs during Kawasaki disease.

Treatment strategies

All published data point to HCoV-NL63 as being one of the most prevalent pathogenic human coronavirus, associated with acute respiratory disease and croup in children. The development of therapeutic agents that are able to efficiently inhibit viral replication, and may supply a useful therapy for acute respiratory illness of children and immunocompromised patients is thus indispensable. Combined with fast diagnostic tools to identify HCoV-NL63 infection, the viral inhibitors would provide an alternative to the regular treatment. Several inhibitors are known to reduce replication of at least some coronaviruses including HCoV-NL63 (Yang *et al.*, 2005; Pyrc *et al.*, 2006a). These inhibitors act at various steps of the coronavirus replication cycle, e.g. receptor binding, membrane fusion, transcription and post-translational processing. Unfortunately, up to date there is no animal model for HCoV-NL63 available that would allow validation of the findings *in vivo*.

As mentioned above, virtually all sera from adults with RTI or from healthy adults carry neutralizing antibodies. Thus, it is not surprising that intravenous immunoglobulins (IVIG), consisting for 95% of pooled human IgGs isolated from serum of healthy donors, inhibit HCoV-NL63 replication. IVIG is approved as an intravenously delivered drug by the Food and Drug Administration and successfully used to treat several diseases, mostly primary immune deficiencies and autoimmune neuromuscular disorders, but also respiratory diseases (e.g. RSV; Hemming *et al.*, 1987) and Kawasaki disease (Gruber *et al.*, 1987; Leung, 1989). The mode of action of this agent in HCoV-NL63 infection is most probably based on its potency to neutralize the virus. The anti-HCoV-NL63 potential, low cytotoxicity and availability of the drug makes it a perfect candidate for human trials. A second target for inhibition of HCoV-NL63 replication concerns membrane fusion. As mentioned in above, the S proteins of coronaviruses are class I fusion proteins and therefore membrane fusion follows a characteristic path. The peptide corresponding to the HCoV-NL63 HR2 domain associates tightly into a complex with HCoV-NL63 HR1 (Pyrc *et al.*, 2006a). Addition of HR2 peptide during HCoV-NL63 infection of LLC-MK2 cells efficiently blocks the replication of the virus (Pyrc *et al.*, 2006a), illustrating that the peptide interferes with the natural HR1-HR2 interaction during membrane fusion and that HCoV-NL63 HR2 peptide is a suitable candidate to treat HCoV-NL63 infection. The third target for virus inhibition is the viral genome itself. HCoV-NL63 is a single stranded RNA virus, and thus vulnerable to inhibition by siRNA. We explored the effectivity of RNAi mechanism with two synthetic siRNAs targeting the S gene. Transfection of these oligoribonucleotides at the nanomolar concentration resulted in profound decrease in virus yield. The effectiveness of siRNA against respiratory track illnesses in a therapeutic setting was demonstrated recently in the mouse and monkey model with intranasal administration of naked siRNA or with transfection reagents targeting RSV, PIV and SARS-CoV (Li *et al.*, 2005; Bitko *et al.*, 2005; Zhang *et al.*, 2005). Inhaled siRNA in low doses might offer a fast, potent and easily administrable antiviral tool against HCoV-NL63 infection in humans. Another very interesting antiviral approach was suggested by Yang *et al.* with a broad inhibitor affecting the function of the coronavirus M^{pro}. Different host receptors for cellular entry, poorly conserved structural proteins, and the high mutation and recombination rates of coronaviruses pose a significant problem in the development of wide-spectrum anti-CoV drugs and vaccines. M^{pro} is a key enzyme during post-translational processing and all coronaviral M^{pro}s share a highly conserved substrate-recognition pocket. Designed on this conserved structural region mechanism-based irreversible inhibitors were tested in an *in vitro* assay and appeared to selectively inhibit several coronaviral M^{pro}s including HCoV-NL63 M^{pro} (Yang *et al.*, 2005).

Yet another approach in inhibition of HCoV-NL63, engages ribonucleoside analogues interfering with virus replication (Pyrc *et al.*, 2006a). Two pyrimidine nucleoside analogues (β-D-N^4-hydroxycitidine and 6-azauridine) showed a high HCoV-NL63-inhibitory potential. The antiviral mechanism of these analogues is still unclear, but likely these nucleoside analogues are incorporated in the new nascent strand during transcription and result in chain termination. Alternatively, several pyrimidine ribonucleoside analogs, including 6-azauridine, act as anti-metabolites, exerting the pharmacological effects in their monophosphate forms by inhibiting uridine 5′-monophosphate synthase and thereby interfering with UTP metabolism. Additionally, incorporation of the nucleoside analogues may change the processivity and fidelity of viral transcription. This results in an increased mutagenicity rate forcing the virus into "error catastrophe" as described previously for ribavirin. Nucleoside analogues are known for inhibition of several viruses, including HIV, pestivirus, hepacivirus, flavivirus, Hepatitis C virus, West Nile virus, FIPV and SARS-CoV infections.

Concluding remarks

Until the autumn of 2002, coronaviruses were generally considered as common cold viruses with low pathogenicity. The SARS-CoV outbreak of 2002–2003 attracted the attention to this family of viruses and subsequent identification of additional members of the *Coronaviridae* family should therefore not come as a surprise. HCoV-NL63 with its high incidence and worldwide spread seems to be one of the clinically most relevant coronaviruses. Studies on the incidence of HCoV-NL63 infection report a frequency of 1–10% of all respiratory infections in humans. The virus was diagnosed most frequently in children, elderly people and immunocompromised patients. Typical for an HCoV-NL63 infection is the frequent occurrence of croup, but also lower respiratory tract infections that require hospitalization and relatively mild symptoms like fever, cough, sore throat and rhinitis are observed.

An analysis of receptor usage and cell tropism of HCoV-NL63 revealed that the virus binds to and uses angiotensin-converting enzyme (ACE2), the receptor that is also used by SARS-CoV. In contrast to SARS-CoV, HCoV-NL63 is generally not associated with acute respiratory distress syndrome. Therefore, the use of ACE2 as a viral receptor is by itself not sufficient to cause aggravating acute lung injury. ACE2 is highly expressed in organs that are affected during Kawasaki disease. Furthermore, IVIG, the reagent that is used to treat Kawasaki disease, is also an effective inhibitor of HCoV-NL63 replication. Combined with the high frequency of HCoV-NL63 infection in Kawasaki disease, at least in one study, these findings suggest a link between Kawasaki disease and HCoV-NL63 infection. However, this tempting hypothesis could not be confirmed by other research groups.

The high prevalence of HCoV-NL63 infection, together with very high co-infection frequencies with other respiratory viruses, increases the risk of viral recombination. Additionally, the fact that HCoV-NL63 shares the receptor with SARS-CoV provides the opportunity for co-infection of the same cell, which is a prerequisite for recombination. The emergence of highly pathogenic recombinant forms of HCoV-NL63 is thus not impossible. Fortunately there are several antivirals that have the capacity to inhibit HCoV-NL63 infection.

During a short period a large body of knowledge on HCoV-NL63 has been gained. However, the lack of a cDNA molecular clone and the lack of an appropriate animal model system hamper future research on this virus. These topics are among the main goals for future HCoV-NL63 research.

References

Arden, K.E., Nissen, M.D., Sloots, T.P., and Mackay, I.M. (2005). New human coronavirus, HCoV-NL63, associated with severe lower respiratory tract disease in Australia. J. Med. Virol. *75*, 455–462.

Baranov, P.V., Henderson, C.M., Anderson, C.B., Gesteland, R.F., Atkins, J.F., and Howard, M.T. (2005). Programmed ribosomal frameshifting in decoding the SARS-CoV genome. Virology *332*, 498–510.

Bastien, N., Anderson, K., Hart, L., Van Caeseele, P., Brandt, K., Milley, D., Hatchette, T., Weiss, E.C., and Li, Y. (2005a). Human coronavirus NL63 infection in Canada. J. Infect. Dis. *191*, 503–506.

Bastien, N., Robinson, J.L., Tse, A., Lee, B.E., Hart, L., and Li, Y. (2005b). Human coronavirus NL-63 infections in children: a 1-year study. J. Clin. Microbiol. *43*, 4567–4573.

Benbacer, L., Kut, E., Besnardeau, L., Laude, H., and Delmas, B. (1997). Interspecies aminopeptidase-N chimeras reveal species-specific receptor recognition by canine coronavirus, feline infectious peritonitis virus, and transmissible gastroenteritis virus. J. Virol. *71*, 734–737.

Bitko, V., Musiyenko, A., Shulyayeva, O., and Barik, S. (2005). Inhibition of respiratory viruses by nasally administered siRNA. Nat. Med. *11*, 50–55.

Boivin, G., Baz, M., Cote, S., Gilca, R., Deffrasnes, C., Leblanc, E., Bergeron, M.G., Dery, P., and De Serres, G. (2005). Infections by human coronavirus-NL in hospitalized children. Pediatr. Infect. Dis. J *24*, 1045–1048.

Bowman, E.J., Siebers, A., and Altendorf, K. (1988). Bafilomycins: a class of inhibitors of membrane ATPases from microorganisms, animal cells, and plant cells. Proc. Natl. Acad. Sci. USA *85*, 7972–7976.

Brierley, I., Boursnell, M.E., Binns, M.M., Bilimoria, B., Blok, V.C., Brown, T.D., and Inglis, S.C. (1987). An efficient ribosomal frame-shifting signal in the polymerase-encoding region of the coronavirus IBV. EMBO J. *6*, 3779–3785.

Brierley, I., Digard, P., and Inglis, S.C. (1989). Characterization of an efficient coronavirus ribosomal frameshifting signal: requirement for an RNA pseudoknot. Cell *57*, 537–547.

Burns, J.C., Cayan, D.R., Tong, G., Bainto, E.V., Turner, C.L., Shike, H., Kawasaki, T., Nakamura, Y., Yashiro, M., and Yanagawa, H. (2005). Seasonality and temporal clustering of Kawasaki syndrome. Epidemiology *16*, 220–225.

Cavanagh, D. (1983). Coronavirus IBV: structural characterization of the spike protein. J. Gen. Virol. *64*, 2577–2583.

Chang, L.Y., Chang, I.S., Lu, C.Y., Chiang, B.L., Lee, C.Y., Chen, P.J., Wang, J.T., Ho, H.N., Chen, D.S., and Huang, L.M. (2004). Epidemiologic features of Kawasaki disease in Taiwan, 1996–2002. Pediatrics *114*, e678-e682.

Chang, L.Y., Chiang, B.L., Kao, C.L., Wu, M.H., Chen, P.J., Berkhout, B., Yang, H.C., and Huang, L.M. (2006). Lack of association between infection with a novel human coronavirus (HCoV), HCoV-NH, and Kawasaki Disease in Taiwan. J. Infect. Dis. *193*, 283–286.

Chiu, S.S., Chan, K.H., Chu, K.W., Kwan, S.W., Guan, Y., Poon, L.L., and Peiris, J.S. (2005). Human coronavirus NL63 infection and other coronavirus infections in children hospitalized with acute respiratory disease in Hong Kong, China. Clin. Infect. Dis. *40*, 1721–1729.

Chou, C.T., Chang, J.S., Ooi, S.E., Huo, A.P., Chang, S.J., Chang, H.N., and Tsai, C.Y. (2005). Serum anti-Yersinia antibody in Chinese patients with Kawasaki disease. Arch. Med. Res. *36*, 14–18.

Crackower, M.A., Sarao, R., Oudit, G.Y., Yagil, C., Kozieradzki, I., Scanga, S.E., Oliveira-dos-Santos, A.J., da Costa, J., Zhang, L., Pei, Y., Scholey, J., Ferrario, C.M., Manoukian, A.S., Chappell, M.C., Backx, P.H., Yagil, Y., and Penninger, J.M. (2002). Angiotensin-converting enzyme 2 is an essential regulator of heart function. Nature *417*, 822–828.

Danilczyk, U., Eriksson, U., Crackower, M.A., and Penninger, J.M. (2003). A story of two ACEs. J. Mol. Med. *81*, 227–234.

de Haan, C.A., Stadler, K., Godeke, G.J., Bosch, B.J., and Rottier, P.J. (2004). Cleavage inhibition of the murine coronavirus spike protein by a furin-like enzyme affects cell–cell but not virus-cell fusion. J. Virol. *78*, 6048–6054.

Delmas, B., Gelfi, J., L'Haridon, R., Vogel, L.K., Sjostrom, H., Noren, O., and Laude, H. (1992). Aminopeptidase N is a major receptor for the entero-pathogenic coronavirus TGEV. Nature *357*, 417–420.

Denison, M.R., and Perlman, S. (1986). Translation and processing of mouse hepatitis virus virion RNA in a cell-free system. J. Virol. *60*, 12–18.

Doms, R.W. (2000). Beyond receptor expression: the influence of receptor conformation, density, and affinity in HIV-1 infection. Virology *276*, 229–237.

Drose, S., and Altendorf, K. (1997). Bafilomycins and concanamycins as inhibitors of V-ATPases and P-ATPases. J. Exp. Biol. *200*, 1–8.

Drose, S., Bindseil, K.U., Bowman, E.J., Siebers, A., Zeeck, A., and Altendorf, K. (1993). Inhibitory effect of modified bafilomycins and concanamycins on P- and V-type adenosinetriphosphatases. Biochemistry *32*, 3902–3906.

Ebihara, T., Endo, R., Ma, X., Ishiguro, N., and Kikuta, H. (2005a). Detection of human coronavirus NL63 in young children with bronchiolitis. J. Med. Virol. *75*, 463–465.

Ebihara, T., Endo, R., Ma, X., Ishiguro, N., and Kikuta, H. (2005b). Lack of association between New Haven coronavirus and Kawasaki disease. J. Infect. Dis. *192*, 351–352.

Esper, F., Shapiro, E.D., Weibel, C., Ferguson, D., Landry, M.L., and Kahn, J.S. (2005a). Association between a novel human coronavirus and Kawasaki disease. J. Infect. Dis. *191*, 499–502.

Esper, F., Weibel, C., Ferguson, D., Landry, M.L., and Kahn, J.S. (2005b). Evidence of a novel human coronavirus that is associated with respiratory tract disease in infants and young children. J. Infect. Dis. *191*, 492–498.

Fan, K., Wei, P., Feng, Q., Chen, S., Huang, C., Ma, L., Lai, B., Pei, J., Liu, Y., Chen, J., and Lai, L. (2004). Biosynthesis, purification, and substrate specificity of severe acute respiratory syndrome coronavirus 3C-like proteinase. J. Biol. Chem. *279*, 1637–1642.

Fouchier, R.A., Hartwig, N.G., Bestebroer, T.M., Niemeyer, B., de Jong, J.C., Simon, J.H., and Osterhaus, A.D. (2004). A previously undescribed coronavirus associated with respiratory disease in humans. Proc. Natl. Acad. Sci. USA *101*, 6212–6216.

Gao, F., Ou, H.Y., Chen, L.L., Zheng, W.X., and Zhang, C.T. (2003). Prediction of proteinase cleavage sites in polyproteins of coronaviruses and its applications in analyzing SARS-CoV genomes. FEBS Lett. *553*, 451–456.

Gruber, W.C., Wilson, S.Z., Throop, B.J., and Wyde, P.R. (1987). Immunoglobulin administration and ribavirin therapy: efficacy in respiratory syncytial virus infection of the cotton rat. Pediatr. Res. *21*, 270–274.

Hegyi, A., Friebe, A., Gorbalenya, A.E., and Ziebuhr, J. (2002). Mutational analysis of the active centre of coronavirus 3C-like proteases. J. Gen. Virol. *83*, 581–593.

Hegyi, A., and Ziebuhr, J. (2002). Conservation of substrate specificities among coronavirus main proteases. J. Gen. Virol. *83*, 595–599.

Hemming, V.G., Rodriguez, W., Kim, H.W., Brandt, C.D., Parrott, R.H., Burch, B., Prince, G.A., Baron, P.A., Fink, R.J., and Reaman, G. (1987). Intravenous immunoglobulin treatment of respiratory syncytial virus infections in infants and young children. Antimicrob. Agents Chemother. *31*, 1882–1886.

Herold, J., and Siddell, S.G. (1993). An "elaborated" pseudoknot is required for high frequency frameshifting during translation of HCV 229E polymerase mRNA. Nucleic Acids Res. *21*, 5838–5842.

Herrewegh, A.A., Mahler, M., Hedrich, H.J., Haagmans, B.L., Egberink, H.F., Horzinek, M.C., Rottier, P.J., and de Groot, R.J. (1997). Persistence and evolution of feline coronavirus in a closed cat-breeding colony. Virology *234*, 349–363.

Hofmann, H., Pyrc, K., van der Hoek, L., Geier, M., Berkhout, B., and Pohlmann, S. (2005). Human coronavirus NL63 employs the severe acute respiratory syndrome coronavirus receptor for cellular entry. Proc. Natl. Acad. Sci. USA *102*, 7988–7993.

Hofmann, H., Simmons, G., Rennekamp, A.J., Chaipan, C., Gramberg, T., Heck, E., Geier, M., Wegele, A., Marzi, A., Bates, P., Pohlmann, S. (2006). Highly conserved regions within the spike proteins of human coronaviruses 229E and NL63 determine recognition of their respective cellular receptors. J. Virol. *80*, 8639–8652.

Huang, I.C., Bosch, B.J., Li, F., Li, W., Lee, K.H., Ghiran, S., Vasilieva, N., Dermody, T.S., Harrison, S.C., Dormitzer, P.R., Farzan, M., Rottier, P.J., and Choe, H. (2006). SARS coronavirus, but not human coronavirus NL63, utilizes cathepsin L to infect ACE2-expressing cells. J. Biol. Chem. *281*, 3198–3203.

Idell, S., Kueppers, F., Lippmann, M., Rosen, H., Niederman, M., and Fein, A. (1987). Angiotensin converting enzyme in bronchoalveolar lavage in ARDS. Chest *91*, 52–56.

Kaiser, L., Regamey, N., Roiha, H., Deffernez, C., and Frey, U. (2005). Human coronavirus NL63 associated with lower respiratory tract symptoms in early life. Pediatr. Infect. Dis. J. *24*, 1015–1017.

Kato, H., Koike, S., and Yokoyama, T. (1979). Kawasaki disease: effect of treatment on coronary artery involvement. Pediatrics *63*, 175–179.

Kuba, K., Imai, Y., Rao, S., Gao, H., Guo, F., Guan, B., Huan, Y., Yang, P., Zhang, Y., Deng, W., Bao, L., Zhang, B., Liu, G., Wang, Z., Chappell, M., Liu, Y., Zheng, D., Leibbrandt, A., Wada, T., Slutsky, A.S., Liu, D., Qin, C., Jiang, C., and Penninger, J.M. (2005). A crucial role of angiotensin converting enzyme 2 (ACE2) in SARS coronavirus-induced lung injury. Nat. Med. *11*, 875–879.

Larson, D.J., Solorzano, R.F., Morehouse, L.G., and Olson, L.D. (1980). Mild transmissible gastroenteritis in pigs suckling vaccinated sows. J. Am. Vet. Med. Assoc. *176*, 539–542.

Laude, H., Godet, M., Bernard, S., Gelfi, J., Duarte, M., and Delmas, B. (1995). Functional domains in the spike protein of transmissible gastroenteritis virus. Adv. Exp. Med. Biol. *380*, 299–304.

Leung, D.Y. (1989). The immunologic effects of IVIG in Kawasaki disease. Int. Rev. Immunol. *5*, 197–202.

Li, B.J., Tang, Q., Cheng, D., Qin, C., Xie, F.Y., Wei, Q., Xu, J., Liu, Y., Zheng, B.J., Woodle, M.C., Zhong, N., and Lu, P.Y. (2005). Using siRNA in prophylactic and therapeutic regimens against SARS coronavirus in Rhesus macaque. Nat. Med. *11*, 944–951

Li, W., Moore, M.J., Vasilieva, N., Sui, J., Wong, S.K., Berne, M.A., Somasundaran, M., Sullivan, J.L., Luzuriaga, K., Greenough, T.C., Choe, H., and Farzan, M. (2003). Angiotensin-converting enzyme 2 is a functional receptor for the SARS coronavirus. Nature *426*, 450–454.

Lim, K.P., and Liu, D.X. (1998). Characterization of the two overlapping papain-like proteinase domains encoded in gene 1 of the coronavirus infectious bronchitis virus and determination of the C-terminal cleavage site of an 87-kDa protein. Virology *245*, 303–312.

Liu, W., Tang, F., Fontanet, A., Zhan, L., Zhao, Q.M., Zhang, P.H., Wu, X.M., Zuo, S.Q., Baril, L., Vabret, A., Xin, Z.T., Shao, Y.M., Yang, H., and Cao, W.C. (2004). Long-term SARS coronavirus excretion from patient cohort, China. Emerg. Infect. Dis. *10*, 1841–1843.

MacNaughton, M.R., Madge, M.H., and Reed, S.E. (1981). Two antigenic groups of human coronaviruses detected by using enzyme-linked immunosorbent assay. Infect. Immun. *33*, 734–737.

Mathews, D.H., Sabina, J., Zuker, M., and Turner, D.H. (1999). Expanded sequence dependence of thermodynamic parameters improves prediction of RNA secondary structure. J. Mol. Biol. *288*, 911–940.

McIntosh, K., Becker, W.B., and Chanock, R.M. (1967). Growth in suckling-mouse brain of "IBV-like" viruses from patients with upper respiratory tract disease. Proc. Natl. Acad. Sci. USA *58*, 2268–2273.

Moes, E., Vijgen, L., Keyaerts, E., Zlateva, K., Li, S., Maes, P., Pyrc, K., Berkhout, B., van der Hoek, L., and Van Ranst, M. (2005). A novel pancoronavirus RT-PCR assay: frequent detection of human coronavirus NL63 in children hospitalized with respiratory tract infections in Belgium. BMC Infect. Dis. *5*, 6.

Nigro, G., Zerbini, M., Krzysztofiak, A., Gentilomi, G., Porcaro, M.A., Mango, T., and Musiani, M. (1994). Active or recent parvovirus B19 infection in children with Kawasaki disease. Lancet *343*, 1260–1261.

Oh, J.S., Song, D.S., and Park, B.K. (2003). Identification of a putative cellular receptor 150 kDa polypeptide for porcine epidemic diarrhea virus in porcine enterocytes. J. Vet. Sci. *4*, 269–275.

Oh, J.S., Song, D.S., Yang, J.S., Song, J.Y., Yoo, H.S., Jang, Y.S., and Park, B.K. (2004). Effect of soluble porcine aminopeptidase N on antibody production against porcine epidemic diarrhea virus. J. Vet. Sci. *5*, 353–357.

Perez, L., and Carrasco, L. (1994). Involvement of the vacuolar H(+)-ATPase in animal virus entry. J. Gen. Virol. *75*, 2595–2606.

Pinon, J.D., Teng, H., and Weiss, S.R. (1999). Further requirements for cleavage by the murine coronavirus 3C-like proteinase: identification of a cleavage site within ORF1b. Virology *263*, 471–484.

Poon, L.L., Chu, D.K., Chan, K.H., Wong, O.K., Ellis, T.M., Leung, Y.H., Lau, S.K., Woo, P.C., Suen, K.Y., Yuen, K.Y., Guan, Y., and Peiris, J.S. (2005). Identification of a novel coronavirus in bats. J. Virol. *79*, 2001–2009.

Pyrc, K., Berkhout, B., and van der Hoek, L. (2005). Molecular characterization of human coronavirus NL63. In Recent Research Developments in Infection & Immunity Vol. 3, S.G.Pandalai, ed. Transworld Research Network), pp. 25–48.

Pyrc, K., Bosch, B.J., Berkhout, B., Jebbink, M.F., Dijkman, R., Rottier, P., and van der Hoek, L. (2006a). Inhibition of HCoV-NL63 infection at early stages of the replication cycle. Antimicrob. Agents Chemother. *50*, 2000–2008.

Pyrc, K., Dijkman, R., Deng, L.C., Jebbink, M.F., Berkhout, B., and van der Hoek, L. (2006b). Mosaic structure of human coronavirus NL63, one thousand years of evolution. J. Mol. Biol. *364*, 964–973.

Pyrc, K., Jebbink, M.F., Berkhout, B., and van der Hoek, L. (2004). Genome structure and transcriptional regulation of human coronavirus NL63. Virol. J. *1*, 7.

Reed, S.E. (1984). The behaviour of recent isolates of human respiratory coronavirus in vitro and in volunteers: evidence of heterogeneity among 229E-related strains. J. Med. Virol. *13*, 179–192.

Sawicki, S.G., and Sawicki, D.L. (1990). Coronavirus transcription: subgenomic mouse hepatitis virus replicative intermediates function in RNA synthesis. J. Virol. *64*, 1050–1056.

Sawicki, S.G., and Sawicki, D.L. (1998). A new model for coronavirus transcription. Adv. Exp. Med. Biol. *440*, 215–219.

Shimizu, C., Shike, H., Baker, S.C., Garcia, F., Hoek, L., Kuijpers, T.W., Reed, S.L., Rowley, A.H., Shulman, S.T., Talbot, H.K., Williams, J.V., and Burns, J.C. (2005). Human Coronavirus NL63 Is Not Detected in the Respiratory Tracts of Children with Acute Kawasaki Disease. J. Infect. Dis. *192*, 1767–1771.

Sieczkarski, S.B., and Whittaker, G.R. (2002). Dissecting virus entry via endocytosis. J. Gen. Virol. *83*, 1535–1545.

Simmons, G., Gosalia, D.N., Rennekamp, A.J., Reeves, J.D., Diamond, S.L., and Bates, P. (2005). Inhibitors of cathepsin L prevent severe acute respiratory syndrome coronavirus entry. Proc. Natl. Acad. Sci. USA *102*, 11876–11881.

Suzuki, A., Okamoto, M., Ohmi, A., Watanabe, O., Miyabayashi, S., and Nishimura, H. (2005). Detection of human coronavirus-NL63 in children in Japan. Pediatr. Infect. Dis. J. *24*, 645–646.

Takahashi, K., Oharaseki, T., Wakayama, M., Yokouchi, Y., Naoe, S., and Murata, H. (2004). Histopathological features of murine systemic vasculitis caused by Candida albicans extract--an animal model of Kawasaki disease. Inflamm. Res. *53*, 72–77.

Tipnis, S.R., Hooper, N.M., Hyde, R., Karran, E., Christie, G., and Turner, A.J. (2000). A human homolog of angiotensin-converting enzyme. Cloning and functional expression as a captopril-insensitive carboxypeptidase. J. Biol. Chem. *275*, 33238–33243.

Tresnan, D.B., Levis, R., and Holmes, K.V. (1996). Feline aminopeptidase N serves as a receptor for feline, canine, porcine, and human coronaviruses in serogroup I. J. Virol. *70*, 8669–8674.

Tyrrell, D.A., and Bynoe, M.L. (1965). Cultivation of a novel type of common-cold virus in organ cultures. Br. Med. J. *5448*, 1467–1470.

Vabret, A., Mourez, T., Dina, J., van der Hoek, L., Gouarin, S., Petitjean, J., Brouard, J., and Freymuth, F. (2005). Human coronavirus NL63, France. Emerg. Infect. Dis. *11*, 1225–1229.

van der Hoek, L., Pyrc, K., Jebbink, M.F., Vermeulen-Oost, W., Berkhout, R.J., Wolthers, K.C., Wertheim-van Dillen, P.M., Kaandorp, J., Spaargaren, J., and Berkhout, B. (2004). Identification of a new human coronavirus. Nat. Med. *10*, 368–373.

van der Hoek, L., Sure, K., Ihorst, G., Stang, A., Pyrc, K., Jebbink, M.F., Petersen, G., Forster, J., Berkhout, B., and Uberla, K. (2005). Croup is associated with the novel coronavirus NL63. PLoS. Med. 2, e240.

Vlasak, R., Luytjes, W., Spaan, W., and Palese, P. (1988). Human and bovine coronaviruses recognize sialic acid-containing receptors similar to those of influenza C viruses. Proc. Natl. Acad. Sci. USA *85*, 4526–4529.

Williams, R.K., Jiang, G.S., and Holmes, K.V. (1991). Receptor for mouse hepatitis virus is a member of the carcinoembryonic antigen family of glycoproteins. Proc. Natl. Acad. Sci. USA *88*, 5533–5536.

Woo, P.C., Huang, Y., Lau, S.K., Tsoi, H.W., and Yuen, K.Y. (2005). In silico analysis of ORF1ab in coronavirus HKU1 genome reveals a unique putative cleavage site of coronavirus HKU1 3C-like protease. Microbiol. Immunol. *49*, 899–908.

Yang, H., Xie, W., Xue, X., Yang, K., Ma, J., Liang, W., Zhao, Q., Zhou, Z., Pei, D., Ziebuhr, J., Hilgenfeld, R., Yuen, K.Y., Wong, L., Gao, G., Chen, S., Chen, Z., Ma, D., Bartlam, M., and Rao, Z. (2005). Design of Wide-Spectrum Inhibitors Targeting Coronavirus Main Proteases. PLoS. Biol. 3, e324.

Yeager, C.L., Ashmun, R.A., Williams, R.K., Cardellichio, C.B., Shapiro, L.H., Look, A.T., and Holmes, K.V. (1992). Human aminopeptidase N is a receptor for human coronavirus 229E. Nature 357, 420–422.

Zhang, W., Yang, H., Kong, X., Mohapatra, S., Juan-Vergara, H., Hellermann, G., Behera, S., Singam, R., Lockey, R.F., and Mohapatra, S.S. (2005). Inhibition of respiratory syncytial virus infection with intranasal siRNA nanoparticles targeting the viral NS1 gene. Nat. Med. *11*, 56–62.

Ziebuhr, J., and Siddell, S.G. (1999). Processing of the human coronavirus 229E replicase polyproteins by the virus-encoded 3C-like proteinase: identification of proteolytic products and cleavage sites common to pp1a and pp1ab. J. Virol. *73*, 177–185.

16 Current Status of Antiviral Severe Acute Respiratory Syndrome Coronavirus Research

Els Keyaerts, Leen Vijgen, and Marc Van Ranst

Abstract

Before the emergence of the severe acute respiratory syndrome associated coronavirus (SARS-CoV), no efforts were put into the search for antivirals against coronaviruses. The rapid transmission and high mortality rate made SARS a global threat for which no efficacious therapy was available and empirical strategies had to be used to treat the patients. New insights into the field of the SARS-CoV genome structure and pathogenesis revealed novel potential anti-coronavirus targets. Several proteins encoded by the SARS-CoV could be considered as targets for therapeutic intervention: the spike protein, the main protease, the NTPase/helicase, the RNA dependent RNA polymerase and different other viral protein-mediated processes. Different animal models have now been established to enable the evaluation of potential anti-SARS-CoV drugs *in vivo*. The development of effective drugs against SARS-CoV may also provide new strategies for the prevention or treatment of other coronavirus diseases in animals or humans. In this chapter, we review and discuss experimental and clinical data about possible coronavirus antivirals.

Introduction

Before the emergence of severe acute respiratory syndrome coronavirus (SARS-CoV), most of the contributions to coronavirus research came from the veterinary field, where coronaviruses are known to cause devastating epizootics of respiratory or enteric diseases in livestock and poultry. In human medicine however, coronaviruses had somewhat been neglected since they were considered as almost unimportant harmless common cold viruses. As a result of the recent appearance of SARS-CoV, coronaviruses have attained a more prominent position in studies of antiviral strategies. The search for inhibitors of viral replication depends on the understanding of the events taking place at the molecular level during viral infection. All the essential steps during the viral life cycle are potential targets for antiviral drugs (Figure 16.1). This chapter includes a summary of the therapies given during the SARS-epidemic, followed by an overview of possible antiviral drugs which show *in vitro* inhibitory effects against SARS-CoV. A summary of compounds reported for their antiviral activity against SARS-CoV is given in Table 16.1.

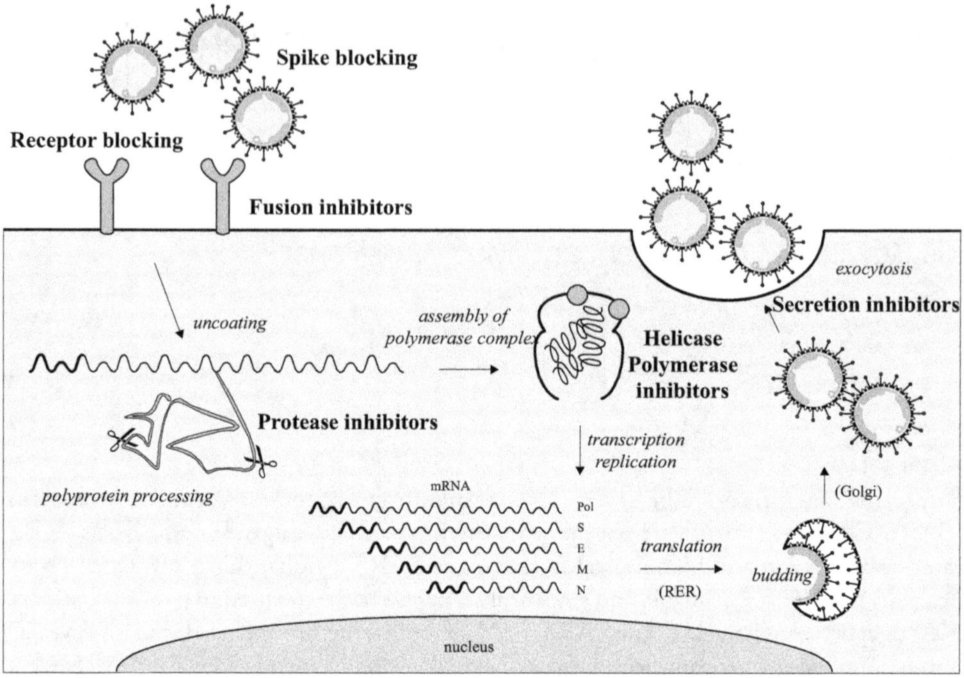

Figure 16.1 Possible antiviral intervention of the coronavirus life cycle. A coronavirus infection is initiated by binding of the viral spike protein to a specific receptor on the cell membrane. Subsequently, conformational changes are induced in the spike protein, that initiate fusion of the viral envelope with the host cell membrane. Molecules that block binding to the receptor or inhibit the receptor-induced conformational changes are possible antiviral drugs. Protease inhibitors can prevent the cleavage of the SARS-CoV polymerase protein and can in this way inhibit succeeding viral RNA synthesis. Another option is to target the coronavirus RNA synthesis, which occurs in a virus-specific cytoplasmic compartment bordered by a double membrane. Small interfering RNA or other inhibitors can block the discontinuous RNA transcription. Alternatively, nucleoside inhibitors can be designed to block specifically the SARS-CoV replication without damaging the cell. Coronaviruses are released by exocytosis, so inhibitors of secretion can be tested for antiviral activity

Therapy during the SARS epidemic

During the epidemic in 2003, treatment of SARS was empirical due to the limited understanding of this new disease, and included broad spectrum antibiotics, antiviral agents, interferons, corticosteroids, and human immunoglobulins.

Ribavirin

Ribavirin (1-β-D-ribofuranosyl-1*H*-1,2,4-triazole-3-carboxamide) is a purine nucleoside analogue with *in vitro* activity against various RNA viruses, including orthomyxoviruses (influenza A and B), paramyxoviruses (measles, respiratory syncytial virus), arenaviruses (lassa, junin, etc.) and coronaviruses (De Clercq, 2004; Sidwell *et al.*, 1987; Sidwell *et al.*, 1977; Weiss and Oostrom-Ram, 1989). The principal target for ribavirin, in its 5′-

Table 16.1 Overview of the effectiveness of different anti-SARS-CoV drugs with there 50% inhibitory concentration (IC_{50}) or 50% effective concentration (EC_{50}), 50% cytotoxic concentration (CC_{50}) and selectivity index

Compound	EC_{50} or IC_{50}	CC_{50}	SI	Assay	Target	Reference
Ribavirin	20–80 μg/ml	> 200 μg/ml	2.5–10	a	Nucleoside analogue	Saijo *et al.*, 2005
	2.2–9.4μg/ml	nd	nc	a	Nucleoside analogue	Morgenstern *et al.*, 2005
	50–200 μg/ml	> 1000 μg/ml	5–80	a	Nucleoside analogue	Chen *et al.*, 2004a
	> 1000 μg/ml	> 1000 μg/ml	nc	a	Nucleoside analogue	Cinatl *et al.*, 2003a
	10000 μg/ml*	nd	nc	a	Nucleoside analogue	Tan *et al.*, 2004
Lopinavir	4–8 μg/ml	32 μg/ml	4–8	a	Nucleoside analogue	Chen *et al.*, 2004a
Nelfinavir	48 nm	14.5 μM	302	a	Post entry	Yamamoto *et al.*, 2004
IFN-α	5000 IU	> 10000 IU	> 2	a	Unknown	Chen *et al.*, 2004a
IFN-α	25 U/ml	nd	nc	a	Unknown	Zheng *et al.*, 2004a
IFN-α	4950 IU/ml	> 10000 IU/ml	> 2	a	Unknown	Cinatl *et al.*, 2003b
Alferon	0.8 IU/ml	nd	nc	a	Unknown	Tan *et al.*, 2004
IFN-α-2b	> 1000 IU/ml	nd	nc	a	Unknown	Stroher *et al.*, 2004
IFN alfacon 1	0.001 μg/ml	> 1μg/ml	1000	a	Unknown	Paragas *et al.*, 2005
IFN β 1a	10000 IU	> 10000 IU	nc	a	Unknown	Chen *et al.*, 2004a
IFN β 1a	14 IU/ml	nd	nc	a	Unknown	Zheng *et al.*, 2004a
IFN-β 1a pretreatment	50 IU/ml	nd	nc	a	Unknown	Hensley *et al.*, 2004
IFN-β 1a posttreatment	500 IU/ml	nd	nc	a	Unknown	Hensley *et al.*, 2004
IFN-β	95 IU/ml	> 10000 IU/ml	> 105	a	Unknown	Cinatl *et al.*, 2003b
Betaferon	0.2 IU/ml	nd	nc	a	Unknown	Tan *et al.*, 2004

Table 16.1 continued

Compound	EC_{50} or IC_{50}	CC_{50}	SI	Assay	Target	Reference
IFN-γ	2500 IU/ml	> 10000 IU/ml	> 4	a	Unknown	Cinatl *et al.*, 2003b
Multiferon	2 IU/ml	nd	nc	a	Unknown	Tan *et al.*, 2004
NAAE	57 μM	nd	nc	b	ACE2	Huentelman *et al.*, 2004
193 amino acid fragment S protein	< 10 nm	nd	nc	a	S protein	Wong *et al.*, 2004
TGG	4.5 μM	1080 μM	240	a	S protein	Yi *et al.*, 2004
Luteolin	10.6 μM	1550 μM	14.6	a	S protein	Yi *et al.*, 2004
HR121	4.13 μM	nd	nc	c	S protein-HR	Ni *et al.*, 2005
HR212	0.95 μM	nd	nc	c	S protein-HR	Ni *et al.*, 2005
HR1-1	0.14 μM	nd	nc	c	S protein-HR	Yuan *et al.*, 2004
HR2-18	1.19 μM	nd	nc	c	S protein-HR	Yuan *et al.*, 2004
CP-1	19 μM	nd	nc	a	S protein-HR	Liu *et al.*, 2004
Synthetic HR2	0.5–5 nm	nd	nc	a	S protein-HR	Zhu *et al.*, 2004
HR2-38	66.2 nm	nd	nc	a	S protein-HR	Zhu *et al.*, 2004
HR2-44	500 nm	nd	nc	a	S protein-HR	Zhu *et al.*, 2004
sHR2-1	43 μM	nd	nc	a	S protein-HR	Bosch *et al.*, 2004
sHR2-2	24 μM	nd	nc	a	S protein-HR	Bosch *et al.*, 2004
sHR2-8	17 μM	nd	nc	a	S protein-HR	Bosch *et al.*, 2004
sHR2-9	34 μM	nd	nc	a	S protein-HR	Bosch *et al.*, 2004
SP-4	4.30 nmol	nd	nc	c	S protein	Ho *et al.*, 2006

SP-8	6.99 nmol	nd	nc	c	S protein	Ho *et al.*, 2006
SP-10	1.88 nmol	nd	nc	c	S protein	Ho *et al.*, 2006
FL-078	4.5 μM**	nd	nc	d	3CLpro	Bacha *et al.*, 2004
FL-101	16 μM**	nd	nc	d	3CLpro	Bacha *et al.*, 2004
FL-106	6 μM**	nd	nc	d	3CLpro	Bacha *et al.*, 2004
FL-136	6 μM**	nd	nc	d	3CLpro	Bacha *et al.*, 2004
FL-166	40 nm**	nd	nc	d	3CLpro	Bacha *et al.*, 2004
MAC-5576	0.5 μM**	nd	nc	e	3CLpro	Blanchard *et al.*, 2004
MAC-8120	4.3 μM**	nd	nc	e	3CLpro	Blanchard *et al.*, 2004
MAC-13985	7 μM**	nd	nc	e	3CLpro	Blanchard *et al.*, 2004
MAC-22272	2.6 μM**	nd	nc	e	3CLpro	Blanchard *et al.*, 2004
MAC-30731	7 μM**	nd	nc	e	3CLpro	Blanchard *et al.*, 2004
CS11	23 μM	nd	nc	a	3CLpro	Dooley *et al.*, 2006
Sinigrin	217 μM	> 10000 μM	> 46	a	3CLpro	Lin *et al.*, 2005
Indigo	752 μM	7375 μM	9.8	a	3CLpro	Lin *et al.*, 2005
Aloe emodin	366 μM	11592 μM	31.7	a	3CLpro	Lin *et al.*, 2005
Hesperetin	8.3 μM	2718 μM	327.1	a	3CLpro	Lin *et al.*, 2005
MP576	7 μM	> 50 μM	> 7	a	3CLpro	Kao *et al.*, 2004a
MP521	14 μM	> 50 μM	> 4	a	3CLpro	Kao *et al.*, 2004a
Compound	EC_{50} or IC_{50}	CC_{50}	SI	Assay	Target	Reference
Dipeptidomimetic α,β-unsat. ester	0.18 μM	> 200 μM	> 1000	a	3CLpro	Shie *et al.*, 2005b

Table 16.1 continued

Compound	EC_{50} or IC_{50}	CC_{50}	SI	Assay	Target	Reference
Hexachlorophene	4 μM**	nd	nc	e	3CLpro	Liu *et al.*, 2005a
Calmidazolium	61 μM**	nd	nc	d	3CLpro	Liu *et al.*, 2005b
C9521	178 μM**	nd	nc	d	3CLpro	Liu *et al.*, 2005b
195623	84 μM**	nd	nc	d	3CLpro	Liu *et al.*, 2005b
Isatin derivates	0.95–17.50 μM	nd	nc	e	3CLpro	Chen *et al.*, 2005b
Cinanserin	25 μM*	nd	nc	f	Unknown	Chen *et al.*, 2005a
Mizoribine	13.5–45.7 μM	> 771 μM	17–57	a	Nucleoside analogue	Saijo *et al.*, 2005
Aurintricarboxylic acid	0.2 mg/ml	37.5 mg/ml	187	a	Nucleoside analogue	He *et al.*, 2004
Bananin	< 10 μM	> 300 μM	> 30	a	Helicase	Tanner *et al.*, 2005
HE602	6 μM	nd	nd	a	Helicase	Kao *et al.*, 2004b
SNAP	222 μM	587.7 μM	3	a	Unknown	Keyaerts *et al.*, 2004a
Chloroquine	8.8 μM	266 μM	30	a	Unknown	Keyaerts *et al.*, 2004b
	4.4 μM	nd	nc	a	Unknown	Vincent *et al.*, 2005
Glycyrrhizin	> 400 μg/ml	> 400 μg/ml	nc	a	Unknown	Chen *et al.*, 2004a
	300 μg/ml	> 20000 μg/ml	> 67	a	Unknown	Cinatl *et al.*, 2003a
Pyridine N-oxide derivatives (157)	4.2 μg/ml	> 100 μg/ml	> 23.3	a	Unknown	Balzarini *et al.*, 2006b

Plant lectin UDA	1.3 μg/ml	> 100 μg/ml	> 77	a	Unknown	Keyaerts *et al.*, 2007
6-azauridine	16.8 μg/ml	104 μg/ml	6	a	Unknown	Cinatl *et al.*, 2003a
Pyrazofurin	4.2 μg/ml	52 μg/ml	12	a	Unknown	Cinatl *et al.*, 2003a
Niclosamide	1–3 μM	250 μM	83	a	Unknown	Wu *et al.*, 2004a
Baicalin	25–50 μg/ml	> 100 μg/ml	2–4	a	Unknown	Chen *et al.*, 2004a
Valinomycin	0.85 μM	68 μM	80	a	Unknown	Wu *et al.*, 2004b
Aescin	6.0 μM	15 μM	2.5	a	Unknown	Wu *et al.*, 2004b
Reserpine	3.4 μM	25 μM	7.3	a	Unknown	Wu *et al.*, 2004b
Blancoxanthone	3 μg/ml	nd	nc	a	Unknown	Shen *et al.*, 2005
Pyranojacareubin	15 μg/ml	nd	nc	a	Unknown	Shen *et al.*, 2005
Iminocyclitol derivates	5–10 μM	> 10 μM	nc	a	α-Glucosidase	Liang *et al.*, 2006
Rimantadine	32 μg/ml	64 μg/ml	2	a	Unknown	Chen *et al.*, 2004a
1,2,4-triazole analogue (17a)	21 μM	nd	nc	a	Unknown	Cho *et al.*, 2006
1,2,3-triazole analogue (17c)	47 μM	nd	nc	a	Unknown	Cho *et al.*, 2006
Geldanamycin	0.91 μM	350 μM	385	a	Unknown	Li *et al.*, 2004

Note: nd: not done, nc: not calculatable, a: cell-based assay; b: ACE2 enzyme activity assay; c: HIV-luc/SARS pseudotyped virus entry inhibition assay; d: enzymatic activity assay; e: 3CLpro activity measurement using FRET; f: HCoV-229E replicon assay; * complete inhibition of CPE with 1000 plaque forming units per well; ** K_i or dissociation constant for inhibitor binding. 3CLpro: main protease, HR heptat repeat, S-protein: spike protein.

monophosphate form, is inosine monophosphate dehydrogenase, which is required for de novo synthesis of guanosine triphosphate. In its 5′-triphosphate form, ribavirin can also interfere with the viral RNA polymerase and the 5′-capped oligonucleotide primer formation required for transcription of the viral RNA genome (De Clercq, 2004). Ribavirin can be given orally at doses of 800–1200 mg per day for treatment of hepatitis C virus infections, and intravenously or as a small size droplet aerosol (solution of 20 mg/ml) for the treatment of RSV infections.

Due to its broad antiviral spectrum, ribavirin was chosen as an empirical treatment during the SARS-CoV epidemic in 2003. Subsequent clinical experience showed that ribavirin did not appear to have a significant effect. A randomized clinical study of 190 patients from the initial outbreak in the Guangdong province findings indicates that ribavirin provided no benefit in the resolution of symptoms or survival (Zhao *et al.*, 2003).

Initially, *in vitro* studies of ribavirin demonstrated no anti-SARS-CoV effect (Cinatl *et al.*, 2003a; Stroher *et al.*, 2004; Tan *et al.*, 2004). It was suggested that this lack of activity is due to an insufficient phosphorylation of ribavirin to its active triphosphorylated form in Vero cells (Morgenstern *et al.*, 2005). More recent experiments demonstrated anti-SARS-CoV activity of ribavirin in fetal rhesus kidney 4 cells in concentrations of 50 to 100 μg/ml (Chen *et al.*, 2004a). These concentrations are above the mean plasma levels which are about 24 μg/ml after intravenous administration of 1000 mg ribavirin (Koren *et al.*, 2003).

HIV-1 protease inhibitors

Lopinavir is an HIV protease inhibitor and is co-formulated (Kaletra®) with ritonavir, an agent that effectively inhibits the cytochrome CYP3A metabolic pathway to boost plasma lopinavir levels (Cooper *et al.*, 2003). The findings of recent studies (Chan *et al.*, 2003; Chu *et al.*, 2004; Chen *et al.*, 2004a) suggest that Kaletra®, when combined with ribavirin may be effective as an antiviral agent against SARS-CoV. The early use of kaletra may decrease the initial viral load during the viral replication phase and hence improve the subsequent immunopathological lung damage.

Nelfinavir, an HIV protease inhibitor, has also been tested for its *in vitro* activity against the SARS-CoV and has yielded contradictory observations. One study found strong inhibitory activity at clinical attainable, non-toxic concentrations, but interestingly, this study could not observe an anti-SARS-CoV activity of lopinavir *in vitro* (Yamamoto *et al.*, 2004). Another *in vitro* study did not observe any significant SARS-CoV inhibition of nelfinavir (Tan *et al.*, 2004).

Interferons

Interferons (IFNs) play a key role in the first line defense against virus infections by activation of the innate immune system. In addition to these direct antiviral effects, IFN-α and -β also enhance the expression of class I and class II MHC molecules on the surface of infected cells, which enhances the presentation of viral antigens on specific immune cells. Interferons are classified according to their cellular origin and the type of receptors they bind to. IFN-α and -β are classified as type I since they bind to IFN cell surface receptors type 1. Although they have different binding affinities, their biological effects are similar.

IFN-γ is a type II IFN, since it binds only to type 2 receptors. Leukocytes produce IFN-α, fibroblasts produce IFN-β, and activated T-cells and natural killer cells produce IFN-γ.

The effectiveness of IFNs against coronavirus-infection was demonstrated *in vitro* and *in vivo* for different coronaviruses, including human coronaviruses, murine hepatitis virus (MHV), avian infectious bronchitis virus (IBV) and transmissible gastroenteritis virus (TGEV) (Sperber and Hayden, 1989; Fuchizaki *et al.*, 2003; Pei *et al.*, 2001; Jordan and Derbyshire, 1995). Different studies describe the protective effect of interferon given intranasally before and after virus challenge of volunteers with HCoV-229E (Turner *et al.*, 1986; Higgins *et al.*, 1983; Tyrrell, 1986).

The *in vitro* antiviral activity of IFNs on SARS-CoV replication has been reported by several studies (Chen *et al.*, 2004a; Cinatl *et al.*, 2003b; Sainz *et al.*, 2004; Stroher *et al.*, 2004; Tan *et al.*, 2004; Zheng *et al.*, 2004a; Spiegel *et al.*, 2005; Hensley *et al.*, 2004). The majority of studies describe the antiviral activity of type I IFNs, with varying activity depending on the cell type that was used for the *in vitro* infection (Cinatl *et al.*, 2003b; Hensley *et al.*, 2004; Spiegel *et al.*, 2005; Stroher *et al.*, 2004; Tan *et al.*, 2004). Although IFN-γ seemed to exhibit only a limited *in vitro* anti-SARS-CoV effect, two reports suggested a synergistic effect of IFN-β and IFN-γ (Scagnolari *et al.*, 2004; Sainz *et al.*, 2004).

Documentation about IFN treatment of SARS patients is limited. One study in a cohort of 190 SARS-patients in Guangdong reported that early administration of high-dose steroids and non-invasive ventilation appeared to offer the best supportive treatment, but the use of INF-α did not result in an obvious beneficial effect (Zhao *et al.*, 2003). A preliminary, uncontrolled study in Toronto reported that treatment of SARS-patients with IFN alfacon1, a second-generation cytokine that was engineered to contain the most frequently occurring amino acids among the non-allelic IFN-α subtypes, was well tolerated and of therapeutic benefit (Paragas *et al.*, 2005; Loutfy *et al.*, 2003). A cell-based study confirms the anti-SARS-CoV activity of alfacon1 with a 50% effective concentration EC_{50} of 0.001 μg/ml (Paragas *et al.*, 2005). *In vitro* studies suggest a synergistically increased SARS-CoV replication inhibition by a combination of IFN-β and ribavirin (Morgenstern *et al.*, 2005; Chen *et al.*, 2004a). However, this synergistic effect between type I IFNs and ribavirin was not observed in a third study (Tan *et al.*, 2004).

Immunomodulatory therapy: corticosteroids and immunoglobulins

During the immunopathological phase of the disease an immunomodulatory agent was usually given to control the over-exuberant host response, and corticosteroids have been most commonly used for this purpose in the 2003 outbreak (Jones *et al.*, 2004). Before the SARS-CoV was identified as the causative agent of SARS a standard treatment protocol was developed by So and colleagues (So *et al.*, 2003). The protocol allowed commencement of high, but not pulsed, dose methyl-prednisolone therapy only after worsening of clinical and radiographical parameters after a trial of antibiotics, with subsequent tapering over the next three weeks should there be satisfactory response. Patients with a poor response to corticosteroids were treated with human gamma immunoglobulins. However, randomized controlled trials are however required to confirm the efficacy of immunoglobulins. Several centers in Hong Kong used plasma from convalescent patients to treat the more severe cases apparently not responding to other treatments (Wong *et al.*, 2003).

Entry and fusion inhibitors

Entry inhibitors form a novel class of anti-HIV drugs that act by blocking the viral ability to infect the cell. Enfuvirtide (T-20), an HIV fusion inhibitor, is the first entry inhibitor approved by the food and drug administration (FDA). Most entry inhibitors under clinical development are directed against enveloped viruses. Viral entry of enveloped viruses, including coronaviruses involves two major steps (see Figure 16.1). First, the virus attaches to its specific receptor on the cell surface, and subsequently, the receptor binding induces the viral-envelope protein to undergo conformational changes that mediate fusion between the viral and cellular membranes (Doms, 2004). The coronavirus spike glycoprotein, responsible for these two steps in the coronavirus entry, is translated as a large polypeptide and is in most coronaviruses subsequently cleaved into two functional subunits S1 and S2. The S1 domain contains the binding domain for the host cell receptor and S2 is the transmembrane subunit that mediates fusion between the viral and cellular membranes. Since each step of the viral-entry pathway is a potential target for antiviral agents, entry inhibitors fall into several categories, depending on which step they target. Recently, Simmons and colleagues demonstrated that SARS-CoV infection results from a unique, three-step process: receptor binding and induced conformational changes in the spike glycoprotein followed by cathepsin L proteolysis within endosomes. The requirement for cathepsin L proteolysis identifies a previously unrecognized class of inhibitors for SARS-CoV infection (Simmons *et al.*, 2005). Also a role of calpains in SARS-CoV infection is suggested (Cinatl *et al.*, 2005; Barnard *et al.*, 2004).

Receptor blocking

The first category of entry inhibitors includes compounds that bind to viral receptors. The identification of angiotensin-converting enzyme-2 (ACE2) as a functional receptor for the SARS-CoV has led to crystallographic studies of the native and inhibitor-bound forms of ACE2. Structural based screening of approximately 140000 small molecules by *in silico* molecular docking led to the identification of N-(2-aminoethyl)-1-aziridine-ethanamine (NAAE), an ACE2 inhibitor that also effectively blocks the SARS-CoV spike protein-mediated cell fusion (Huentelman *et al.*, 2004). Another small-molecule inhibitor MLN-4760 was reported to induce large conformational changes upon binding to the ACE2 catalytic site, but it remains to be elucidated if these conformational changes are able to prevent virus binding (Towler *et al.*, 2004). Earlier, mutations of ACE2 zinc-coordinating residues were already reported to have no effect on syncytium formation (Li *et al.*, 2003; Xiao *et al.*, 2003). A major drawback of using receptor inhibitors as antiviral drugs is the ability of some viruses to use alternative cellular receptors. Jeffers and colleagues identified CD209L (L-SIGN), a human cellular glycoprotein, as an alternative receptor for SARS-CoV (Jeffers *et al.*, 2004). Secondly, unlike viral enzyme targets where a limited copy number of catalytic molecules must be inhibited, virus receptors probably exist in large numbers on the cell. Thus receptor-targeted therapy may require large amounts of drugs to saturate the available targets.

Spike blocking

The second category of entry inhibitors includes compounds that bind to the virus and prevent it from interacting with its receptors. A 193 amino acid fragment (amino acid residues 318 to 510) on the SARS-CoV spike protein has been identified as the ACE2-binding domain. The 193-residue fragment showed antiviral activity through blocking S-protein-mediated infection (Wong *et al.*, 2004). Two neutralizing MAbs against S1 were reported, one MAb with an epitope within the receptor binding domain and the other MAb binds externally to the receptor-binding domain. When the MAbs were applied prophylactically, they were able to protect mice against SARS-CoV challenge. Clinical trials are planned with a human MAb directed against an epitope within the receptor-binding domain (Greenough *et al.*, 2005). Prophylactic use of the human MAb 80R provided protection in a SARS-CoV mouse model (Sui *et al.*, 2004; Sui *et al.*, 2005). Furthermore, three human neutralizing MAb against the spike glycoprotein were selected from semi-synthetic antibody phage display libraries (van den Brink *et al.*, 2005). The MAb with the highest potency (CR3014) was further tested in a ferret SARS-CoV model. Prophylactic treatment with the MAb at 10 mg/kg reduced SARS-CoV replication in the lungs of infected ferrets by 3.3 logs, completely prevented the development of SARS-CoV induced macroscopic lung pathology and abolished shedding of virus in pharyngeal secretions (Ter Meulen *et al.*, 2004).

Taken together, human neutralizing MAbs are effective prophylactic SARS-CoV entry inhibitors and when they are given shortly after infection, MAbs can reduce the severity of clinical symptoms. However, prophylaxis strategy based on neutralizing MAbs will only be effective together with genotyping of the infectious strain, as amino acid substitutions in the earlier discussed 80R epitope are already described for SARS-like CoVs from civet cats (Sui *et al.*, 2005). Furthermore, molecular characterization of the SARS-CoV spike glycoprotein has revealed genetic diversity among isolates. Functional effects of spike mutations have been determined by analyzing their affinity for the ACE2 receptor, and their sensitivity to MAb neutralization with viral pseudotypes. Although minor differences among strains transmitted during human outbreaks in early 2003 were found, substantial functional changes were detected in the spike protein derived from a case in late 2003 and from two palm civets. These functional changes led to a minor dependence on the ACE-2 receptor and a marked resistance to Ab inhibition. Moreover, Abs that neutralized most human spike glycoproteins enhanced entry mediated by the civet virus spike glycoproteins (Yang *et al.*, 2005b).

Besides these MAb, small-molecule inhibitors that accomplish the same feat have been reported. Two small molecules, tetra-O-galloyl-β-d-glucose (TGG) and luteolin interfere with the entry of SARS-CoV into its host cells by binding avidly with the surface spike protein. TGG exhibits prominent anti-SARS-CoV activity with a 50% effective concentration of 4.5 μM and a selectivity index of 240 (Yi *et al.*, 2004). Synthetic peptides targeting the spike gene were also reported. SARS-CoV infectivity was reduced over 10000-fold through pre-incubation with two of these peptides, while it was completely inhibited in the presence of three peptides. Molecular modeling of the SARS-CoV peplomer suggests that three of these antiviral peptides map to the interfaces between the three monomers (Zheng *et al.*, 2005). Recently, three small peptides derived from the spike protein were reported to inhibit the binding of spike protein to ACE2 (Ho *et al.*, 2006).

Fusion inhibitors

The final category of entry inhibitors prevents the conformational changes needed for membrane fusion. Such a compound, enfuvirtide is licensed by the FDA as an inhibitor of HIV membrane fusion. After binding of the HIV gp120 to CD4 and a co-receptor, gp41 changes its conformation via heptad repeat region 1 (HR1) and HR2 interaction to form a six-helix bundle structure, which brings the viral and cellular membranes into close proximity, leading to membrane fusion. At the intermediate state, soluble synthesized HR1, HR2 or homologues can bind the exposed HR1 or HR2 region and block the formation of the six-helix bundle, thus inhibiting virus fusion with the target cell (Jiang *et al.*, 2002). The S2 domain of the SARS-CoV spike glycoprotein is a typical class I viral fusion protein (Bosch *et al.*, 2003). Similar to HIV-1 gp41, the SARS-CoV S2 domain contains HR1 and HR2 sequences, which tend to form a coiled-coil structure (Liu *et al.*, 2004).

Different synthetic and recombinant peptides corresponding to sequences of SARS-CoV spike protein HR1 and HR2 regions have been reported to interfere with conformational changes, resulting in inhibition of SARS-CoV fusion with the target cell (Ni *et al.*, 2005; Liu *et al.*, 2004; Yuan *et al.*, 2004; Bosch *et al.*, 2004; Zhu *et al.*, 2004). It is a matter of discussion whether the HIV-1 gp41 HR2-derived peptide T20 (enfuvirtide) could be a possible SARS-CoV inhibitor given the similarities between the two viruses. Veiga and colleagues tested the possibility of interaction between both T20 and the related peptide T-1249 and the SARS-CoV spike protein HR1- and HR2-derived peptides. Their biophysical data show a significant interaction between a SARS-CoV HR1-derived peptide and T20. However, the interaction is too moderate to be able to provide therapeutic efficiency (Veiga *et al.*, 2006). Recently, MAbs were raised against a SARS-CoV spike protein fragment, encompassing the HR2-domain. Binding sites of these MAbs were mapped to four linear epitopes. Two of them were located within the HR2 region and two immediately upstream of the HR2 domain. MAbs targeting these epitopes showed *in vitro* neutralizing activities and were able to inhibit cell–cell membrane fusion (Lip *et al.*, 2006).

SARS-CoV enzymes as targets for antiviral agents

Virus-encoded proteins that are essential for the viral life cycle and exhibit low similarity to host proteins are attractive drug targets. Three viral enzymes were suggested as targets for drug discovery: the proteases, the RNA-dependent RNA polymerase and the helicase (Snijder *et al.*, 2003).

Main protease inhibitors

Due to its key function in the viral life cycle, the coronavirus protease is an excellent target for the development of antivirals. The coronavirus main protease is a cysteine protease, which is frequently called 3C-like protease (3CLpro) because its substrate specificity is similar to that observed for picornavirus 3C protease (Gorbalenya *et al.*, 1989). Several crystal structures of coronavirus 3CLpro reported from TGEV, HCoV 229E and SARS-CoV revealed a common feature in 3CLpro: two chymotrypsin-like β-domains, which are responsible for the catalytic reaction, and a third α-helical domain with unclear biological function (Anand *et al.*, 2002; Anand *et al.*, 2003; Yang *et al.*, 2003). The active site of SARS-CoV contains a catalytic dyad defined by His41 and Cys145, which is similar to the arrangement found in other coronavirus proteases. The major quaternary structure of

SARS-CoV 3CLpro at neutral pH is a dimer and the dimeric interface was suggested as a target for structural-based drug design (Chou *et al.*, 2004; Fan *et al.*, 2004; Shi *et al.*, 2004). Different techniques are used to study possible 3CLpro inhibitors.

A theoretical homology model of the SARS-CoV 3CL protease based on sequence homology with TGEV as well as the solved crystal structure have both been used to dock substrate mimics and for virtual screening of collections of synthetic and natural compounds, as well as approved antiviral drugs to evaluate their ability to inhibit the 3CL protease (Toney *et al.*, 2004; Xiong *et al.*, 2003; Jenwitheesuk and Samudrala, 2003; Sirois *et al.*, 2004; Rajnarayanan *et al.*, 2004; Liu and Zhou, 2005; Zhang *et al.*, 2005; Du *et al.*, 2005; Kaeppler *et al.*, 2005). It has been proposed that the rhinovirus 3C protease inhibitor AG7088 may be modified so that it would be useful for SARS therapy. Docking studies of KZ7008, a derivative of AG7088 to the SARS-CoV main protease have been reported (Chou *et al.*, 2003; Anand *et al.*, 2003). Zhang and colleagues studied possible other inhibitors in anti-SARS drug development and suggest a link of SARS-CoV to other viruses, including human rhinovirus, poliovirus, arteritis virus, hepatitis A virus, hepatitis C virus, and dengue virus (Zhang and Yap, 2004a). Other compounds identified as potential inhibitors of the 3CL protease from these studies include the HIV-1 protease inhibitor L-700417 (Jenwitheesuk and Samudrala, 2003) and the natural compound sabadinine (Toney *et al.*, 2004), however the *in vitro* or *in vivo* inhibitory potency of these compounds has not yet been verified.

Several *in vitro* screenings have tested commercially available compound libraries for their ability to inhibit replication of SARS-CoV (Bacha *et al.*, 2004; Blanchard *et al.*, 2004; Hsu *et al.*, 2004; Kao *et al.*, 2004a; Wu *et al.*, 2004b; Lin *et al.*, 2005; Wu *et al.*, 2004a; Liu *et al.*, 2005b; Chen *et al.*, 2005a; Yamamoto *et al.*, 2004; Tan *et al.*, 2004; Zhang and Yap, 2004b). The HIV-1 protease inhibitors indinavir, saquinavir, ritonavir, lopinavir, TYA5, TYB5, and KNI-272 were reported to be unable to inhibit SARS-CoV replication *in vitro* (Tan *et al.*, 2004; Yamamoto *et al.*, 2004). Another HIV-1 protease inhibitor, nelfinavir, was shown to inhibit SARS-CoV replication with an EC_{50} of 48 nm (Yamamoto *et al.*, 2004). However lack of complete inhibition of viral replication with compound concentrations up to 10 μM was observed in another study (Tan *et al.*, 2004). Cinanserin, a serotonin antagonist was reported to inhibit the catalytic SARS-CoV 3CLpro activity with IC_{50} values of 5 μM (Chen *et al.*, 2005a). Calmidazolium, an calmodulin-antagonist was reported to be effective (Liu *et al.*, 2005b). Bifunctional boronic acid inhibitors were designed to target the serine cluster near the catalytic residues (Bacha *et al.*, 2004).

Peptide analogues and organic compounds derived from known rhinovirus 3 protease inhibitors were also synthesized and some of them were found to be active towards the SARS-CoV 3CL protease (Kao *et al.*, 2004b; Wu *et al.*, 2004b; Jain *et al.*, 2004; Chen *et al.*, 2005b; Lee *et al.*, 2005; Du *et al.*, 2005; Martina *et al.*, 2005; Dooley *et al.*, 2006; Ghosh *et al.*, 2005; Yang *et al.*, 2005a; Shie *et al.*, 2005b; Shie *et al.*, 2005a).

To be cleavable by the SARS-CoV proteinase the octapeptide must have a good fit for binding to the active site. However, such a peptide can be modified in its scissile bond with some simple routine procedure so that it still binds to the active site but will completely lose its cleavability. Such a molecule can be compared to a key that can be inserted into a lock but can neither open the lock nor be pulled out from it, spontaneously becoming an ideal competitive inhibitor against the SARS-CoV proteinase (Chou *et al.*, 2003). Several of these octapeptide inhibitors have been reported (Gan *et al.*, 2005; Du *et al.*, 2004; Du *et al.*, 2005; Fan *et al.*, 2005; Lee *et al.*, 2005). Recently, Wei and colleagues reported that the N-terminal octapeptide (N-finger) acts as a dimerization inhibitor of SARS-CoV 3CL protease (Wei *et al.*, 2006).

Papain-like protease inhibitors

Coronaviruses encode one or two accessory cysteine proteases, called coronavirus papain-like proteases (PLpro) that recognize and process one to three sites in the amino-terminal half of the replicase polyprotein during assembly of the viral replication complex. Most coronaviruses harbor two such papain-like protease domains, PL1pro and PL2pro, whereas SARS-CoV and IBV utilize only one, which is equivalent to PL2pro (Snijder *et al.*, 2003). By using a fluorescence based inhibitor-screening platform, it was observed that the zinc ion and its conjugates inhibit the enzymatic activity of the SARS-CoV PL2pro (Han *et al.*, 2005).

Polymerase inhibitors

Molecular modeling revealed that the SARS-CoV RNA dependent RNA polymerase (RdRp) does not contain a hydrophilic pocket in which non-nucleoside inhibitors can bind (Xu *et al.*, 2003). Many nucleoside analogues were screened, and the most selective nucleoside analogue inhibitor for SARS-CoV is Beta-D-N4-hydroxycytidine, with an EC_{90} of 6 μM by virus yield reduction assay (Barnard *et al.*, 2004). Mizoribine has an IC_{50} between 3.5 and 16 μg/ml (13.5- 45.7 μM), depending on the SARS-CoV strain that was used (Saijo *et al.*, 2005). Aurintricarboxylic acid (ATA), a general inhibitor of nucleases was also suggested to inhibit the SARS-CoV RdRp (He *et al.*, 2004; Yap *et al.*, 2005). Ribavirin, a broad-spectrum nucleoside analogue effective in the treatment of several viral infections, was discussed earlier in this chapter.

Helicase inhibitors

RNA helicases are a diverse class of enzymes that use the energy of NTP hydrolysis to unwind duplex RNA. Because of the vital role of its enzymatic activities for viral replication, the helicase may represent a suitable target for antiviral therapy of coronavirus infections (Ivanov *et al.*, 2004; Davidson and Siddell, 2003). However, given the huge number of cellular helicases, the development of selective coronavirus helicase inhibitors can be expected to be a major challenge. Two studies have reported on helicase inhibitors. First, bananins are reported to be effective inhibitors of the ATPase activity of the SARS-CoV helicase with IC_{50} values in the range of 0.5 to 3 μM. In a cell culture system of SARS-CoV, bananin exhibited an EC_{50} of less than 10 μM and a CC_{50} of over 300 μM (Tanner *et al.*, 2005). Second, high throughput screening of a chemical library identified a small-molecule inhibitor (HE602) that targets the helicase with an EC_{50} below 10 μM (Kao *et al.*, 2004b).

Nitric oxide

Nitric oxide (NO) has complex and diverse functions in physiological and pathophysiological phenomena. It is a key molecule in the pathogenesis of infectious diseases. NO biosynthesis, particularly through the expression of an inducible NO synthetase (iNOS) occurs in a variety of microbial infections. NO has been reported to exert antiviral effects against several DNA viruses such as murine poxvirus, herpes simplex virus, and Epstein-Barr virus, and some RNA viruses, such as coxsackievirus, poliovirus, Japanese encephalitis virus, vesicular stomatitis virus and murine coronavirus (Lane *et al.*, 1997; Torre *et al.*, 2002; Bi and Reiss, 1995; Lin *et al.*, 1997).

We tested two NO donors, S-nitroso-N-acetylpenicillamine (SNAP) and sodium nitroprusside (SNP) *in vitro* for their effect on the SARS-CoV replication. The survival rate of SARS-CoV infected cells was greatly increased by the treatment with SNAP, and the concentration of this compound needed to inhibit the viral cytopathic effect to 50% was 222 μM, with a selectivity index of 3 (Keyaerts *et al.*, 2004a). No anti-SARS-CoV effect could be detected for SNP. Later on, our results were confirmed by a Swedish group, which also reported on the anti-SARS-CoV capacity of SNAP (Akerstrom *et al.*, 2005). Hitherto, the antiviral mechanism by which NO inhibits SARS-CoV replication is not elucidated. For coxsackievirus, it is demonstrated that NO is capable of nitrosylating the cysteine residue in the active site of viral protease 3C, inhibiting protease activity and interrupting the viral life cycle (Saura *et al.*, 1999). The SARS-CoV main protease also has a cysteine residue in its catalytic dyad (as discussed earlier). Akerstrom and colleagues reported inhibition by NO of both viral proteins and RNA synthesis. NO inactivation of the 3CL protease could explain this reduction in both viral RNA and viral protein levels. A rescue trial in Beijing suggested that inhalation of NO gas may be of benefit to SARS patients (Chen *et al.*, 2004b). In a small group of SARS patients, inhalation of NO improved the arterial oxygenation and clinical signs in patients with severe SARS. Furthermore they observed an improvement of chest radiograph findings. As a result of the improved arterial oxygenation, the non-invasive ventilatory pressure support could be ended. After termination of the therapy, the increased oxygen saturation remained, thus not only an immediate improvement was seen, but also a lasting effect on the disease itself. These results encourage the use of inhalation of NO in a possible next outbreak.

Chloroquine

The SARS-CoV spike glycoprotein mediates viral entry through pH-dependent endocytosis, and infection is sensitive to lysosomotropic agents that change endosomal pH (Simmons *et al.*, 2004). Savarino and colleagues hypothesized that chloroquine might be of some use for the clinical management of SARS (Savarino *et al.*, 2003). Chloroquine is a clinically approved drug effective against malaria, and is known to elicit antiviral effects against several viruses, including HIV type 1 (HIV-1) (Tsai *et al.*, 1990; Pardridge *et al.*, 1998; Savarino *et al.*, 2001), hepatitis B virus (Kouroumalis and Koskinas, 1986), herpes simplex virus type 1 (Singh *et al.*, 1996) and HCoV-229E (Blau and Holmes, 2001). The antiviral effects of chloroquine against HIV-1 replication are currently being tested in clinical trials (Savarino *et al.*, 2003). Besides a direct antiviral effect, chloroquine also has immunomodulatory activity, suppressing the production and release of tumor necrosis factor and interleukin 6, which mediate the inflammatory complications of several viral diseases (Savarino *et al.*, 2003). Our group reported that chloroquine exerted anti-SARS-CoV activity with an EC_{50} of 8.8 μM (Keyaerts *et al.*, 2004b). Interestingly, time-of addition experiments pointed out that chloroquine was equally active when added during virus adsorption or at 1 hr after infection. At later time points, a gradual loss of the antiviral activity of chloroquine was seen. This indicates that the compound does likely not only interfere with the early steps of viral replication, i.e. attachment or penetration. In 2005, Vincent and colleagues reported that chloroquine is effective in preventing SARS-CoV infection in cell culture when the drug is added 24h prior to the infection (Vincent *et al.*, 2005). In addition, they found that effective anti-SARS concentrations of chloroquine and NH_4Cl impaired the terminal glycosylation of ACE2, which may result in reduced binding affinities between SARS-CoV and its receptor. The mechanism of antiviral action of chloroquine is not yet elucidated but is possibly a multitarget mechanism, depending of the time-point at which the drug is added. When added during and shortly after infection, chloroquine probably affects the endosome-mediated fusion, and when the drug is given after this first target, it can still act on later stages of the viral life cycle, as reported for other viruses (Savarino *et al.*, 2003).

RNA interference

RNA interference is a sequence-specific RNA degradation process in the cytoplasm of eukaryotic cells that is induced by double-stranded RNA. RNA interference induced by small interfering RNA (siRNA) can inhibit the expression of viral antigens, and thereby provides a new approach to the therapy of pathogenic viruses and has been reported for numerous viruses, including several important human pathogens such as human immunodeficiency virus type 1, hepatitis C virus, hepatitis B virus, dengue virus, poliovirus and influenza virus A (Haasnoot *et al.*, 2003; Wu and Chan, 2006).

Screening of 48 chemically synthesized siRNA candidates targeting elements throughout the entire SARS-CoV genome revealed four effective siRNA duplexes (Zheng *et al.*, 2004b). Two of these promising siRNAs were further evaluated *in vivo* (Li *et al.*, 2005). The siRNAs provided relief from SARS-CoV infection-induced fever, diminished SARS-CoV viral levels and reduced acute diffuse alveoli damage. The accumulated siRNA dosages of 10 to 40 mg/kg did not show any sign of siRNA-induced toxicity. Tao and colleagues demonstrated that siRNA, which targets the SARS-CoV N gene sequence, efficiently reduced the level of transcription of the corresponding viral RNA and ultimately resulted in the reduction of the N protein levels (Tao *et al.*, 2005). Other siRNA duplexes were reported to reduce target gene expression over 80% at a concentration of 60 nm *in vitro* in a sequence specific manner (Shi *et al.*, 2005). RNA interference targeting the coronavirus RNA dependent RNA polymerase using short hairpin RNA expression plasmids can specifically inhibit expression of extraneous coronavirus RdRp *in vitro* (Lu *et al.*, 2004).

Recently, peptide-conjugated antisense morpholino oligomers, designed to bind to specific sequences of the SARS-CoV genome, proved to be a selective tool to manipulate SARS-CoV gene expression (Neuman *et al.*, 2005).

Other antiviral SARS-CoV compounds

Different groups set up an antiviral assay to screen their libraries for drugs with antiviral activity against the SARS-CoV (Hertzig *et al.*, 2004; Ivens *et al.*, 2005; Keyaerts *et al.*, 2004c). A variety of products were found to exert anti-coronavirus activity *in vitro* with a good selectivity index.

Glycyrrhizin

Glycyrrhizin was one of the first compounds reported for its anti-SARS-CoV activity with an EC_{50} of 365μM in cell culture (Cinatl *et al.*, 2003a). To improve the EC_{50}, glycyrrhizin derivatives were tested and revealed that the introduction of an *N*-acetylglycosamine into the glycoside chain resulted in a 10-fold increased anti-SARS-CoV activity compared to glycyrrhizin. The sugar moiety seemed to be essential for antiviral properties as the glycyrrhizin aglycone did not exert anti-SARS-CoV effects (Hoever *et al.*, 2005; Cinatl *et al.*, 2005). The mechanism of antiviral action of glycyrrhizin is still unclear. Glycyrrhizin may influence the viral replication by up-regulation of expression of iNOS and production of NO (Yi *et al.*, 1996). Preliminary results indeed showed that glycyrrhizin induces NO synthase in Vero cells and that virus replication is inhibited when a nitrous oxide donor was added to the culture medium (Cinatl *et al.*, 2003a).

Pyridine N-oxide derivatives

Our group reported the antiviral effect of a wide variety of pyridine N-oxide derivatives against feline infectious peritonitis virus (FIPV) and SARS-CoV in cell culture. The pyridine-oxide part was found indispensable for anti-coronavirus activity. One of the most active and selective compounds was shown to inhibit FIPV replication at the transcriptional level (Balzarini *et al.*, 2006b).

Plant lectins

Furthermore, we observed that plant lectins show antiviral activity against FIPV and SARS-CoV (Keyaerts *et al.*, 2007). Plant lectins have been reported as HIV fusion inhibitors. Besides inhibiting HIV infection, they also prevent transmission by blocking cell-to-cell contact. Mannose- and N-acetylglucosamine-specific agglutinins are reported to exert an inhibitory effect on HIV, cytomegalovirus infection, respiratory syncytial virus infection and influenza A virus infection *in vitro*. Since the spike glycoprotein of coronaviruses is heavily glycosylated, one may expect coronavirus infectivity and virus-cell fusion to be inhibited by those lectins that are specific for the glycans present in the spike glycoprotein.

Niclosamide

Niclosamide, an existing antihelminthic drug was reported to abolish the viral antigen synthesis at a concentration of 1.56 μM (Wu *et al.*, 2004a). The mechanism of action is currently not known, but the authors would exclude that viral attachment, entry, or the main protease was targeted.

Chinese traditional medicine

Numerous small molecules of herbal origin were shown to have antiviral activity against SARS-CoV. Baicalin, a flavonoid derived from Scutellaria baicalensis, a mint family member, inhibited SARS-CoV at *in vivo* achievable concentrations (Chen *et al.*, 2004a). Aescin (the major active compound from the horse chestnut tree), reserpine (a naturally occurring alkaloid produced by several members of the genus Rauwolfia), and ginsenoside-Rb1 (one of the pharmacologically active components of the traditional Chinese herb Panax ginseng) were reported to posses inhibitory effects towards SARS-CoV replication at non-toxic concentrations (Wu *et al.*, 2004b). Additionally, blancoxanthone, isolated from the roots of the Calophyllum blancoi tree exhibited significant anti-coronavirus activity (Shen *et al.*, 2005).

Aglycon derivates of glycopeptide antibiotics

Several glycopeptide derivatives modified with hydrophobic substituents showed selective antiviral activity against FIPV and SARS-CoV in cell culture at non-toxic concentrations (Balzarini *et al.*, 2006a).

Other compounds

Geldanamycin, an antibiotic targeting the ADP/ATP binding site of heat shock protein 90 (Hsp90) inhibited SARS-CoV replication with an EC50 of 0.91 μM (Li *et al.*, 2004).

A five-membered iminocyclitol derivative was found to be a potent and selective inhibitor of the α-glucosidase glycoprotein-processing and was further derivatized to antiviral agents against Japanese encephalitis virus, dengue virus and SARS-CoV (Liang *et al.*, 2006).

Cho and colleagues reported on two newly synthesized cyclopentenyl carbocyclic nucleosides (1,2,4-triazole analogue (17a) and 1,2,3-triazole analogue (17c)) exhibiting antiviral activity against SARS-CoV with EC_{50} of 47 μM and 21 μM respectively (Cho *et al.*, 2006).

Conclusions

At this moment no consensus about the best treatment strategy for SARS can be provided. Numerous products are reported to have antiviral activity *in vitro* against SARS-CoV. It will be important to test promising antiviral compounds in animal models. Although the initial severe acute respiratory syndrome epidemic has been controlled by conventional measures, the SARS-CoV or another yet unknown animal coronavirus can be (re)-introduced in the human population. The development of effective drugs against SARS-CoV may also be useful in the treatment of other human coronavirus-induced diseases and economically important coronavirus diseases of poultry, livestock and companion animals.

Acknowledgments

We would like to thank all the colleagues of the laboratory of Clinical & Epidemiological Virology, Department of Microbiology and Immunology, Rega Institute for Medical Research, University of Leuven, Belgium, for helpful comments and discussion. We would like to thank Piet Maes for the graphical assistance. This work is part of the activities of the Euro-Asian SARS-DTV Network (SP22-CT-2004–511064) supported by the European Commission specific research and technological development program "Integrating and Strengthening the European Research Area." This work was supported by a postdoctoral fellowship of the Research Fund K.U.Leuven to Leen Vijgen.

References

Akerstrom, S., Mousavi-Jazi, M., Klingstrom, J., Leijon, M., Lundkvist, A., and Mirazimi, A. (2005). Nitric oxide inhibits the replication cycle of severe acute respiratory syndrome coronavirus. J. Virol. *79*, 1966–1969.

Anand, K., Palm, G.J., Mesters, J.R., Siddell, S.G., Ziebuhr, J., and Hilgenfeld, R. (2002). Structure of coronavirus main proteinase reveals combination of a chymotrypsin fold with an extra alpha-helical domain. EMBO J. *21*, 3213–3224.

Anand, K., Ziebuhr, J., Wadhwani, P., Mesters, J.R., and Hilgenfeld, R. (2003). Coronavirus main proteinase (3CLpro) structure: basis for design of anti-SARS drugs. Science *300*, 1763–1767.

Bacha, U., Barrila, J., Velazquez-Campoy, A., Leavitt, S.A., and Freire, E. (2004). Identification of novel inhibitors of the SARS coronavirus main protease 3CLpro. Biochemistry *43*, 4906–4912.

Balzarini, J., Keyaerts, E., Vijgen, L., De Clercq, E., Printsevskaya, S., Preobrazhenskaya, M., and Van Ranst, M. (2006a). Inhibitory activity of vancomycin, eremomycin and teicoplanin aglycon derivatives against feline and human (i.e. SARS) coronaviruses. Antiviral Res. *72*, 20–33.

Balzarini, J., Keyaerts, E., Vijgen, L., Vandermeer, F., Stevens, M., De Clercq, E., Egberink, H., and Van Ranst, M. (2006b). Pyridine N-oxide derivatives are inhibitory to the human SARS and feline infectious peritonitis coronavirus in cell culture. J. Antimicrob. Chemother. *57*, 472–81.

Barnard, D.L., Hubbard, V.D., Burton, J., Smee, D.F., Morrey, J.D., Otto, M.J., and Sidwell, R.W. (2004). Inhibition of severe acute respiratory syndrome-associated coronavirus (SARSCoV) by calpain inhibitors and beta-D-N4-hydroxycytidine. Antivir. Chem. Chemother. *15*, 15–22.

Bi, Z., and Reiss, C.S. (1995). Inhibition of vesicular stomatitis virus infection by nitric oxide. J. Virol. *69*, 2208–2213.

Blanchard, J.E., Elowe, N.H., Huitema, C., Fortin, P.D., Cechetto, J.D., Eltis, L.D., and Brown, E.D. (2004). High-throughput screening identifies inhibitors of the SARS coronavirus main proteinase. Chem. Biol. *11*, 1445–1453.

Blau,D., and Holmes,K.V. (2001). Human Coronavirus HCoV-229E enters susceptible cells via the endocytic pathway. In *The Nidoviruses, Coronaviruses and Arteriviruses*, E Lavi, ed. (New York: Kluwer), pp. 193–197.

Bosch, B.J., Martina, B.E., van der, Z.R., Lepault, J., Haijema, B.J., Versluis, C., Heck, A.J., De, G.R., Osterhaus, A.D., and Rottier, P.J. (2004). Severe acute respiratory syndrome coronavirus (SARS-CoV) infection inhibition using spike protein heptad repeat-derived peptides. Proc. Natl. Acad. Sci. USA *101*, 8455–8460.

Bosch, B.J., van der, Z.R., de Haan, C.A., and Rottier, P.J. (2003). The coronavirus spike protein is a class I virus fusion protein: structural and functional characterization of the fusion core complex. J. Virol. *77*, 8801–8811.

Chan, K.S., Lai, S.T., Chu, C.M., Tsui, E., Tam, C.Y., Wong, M.M., Tse, M.W., Que, T.L., Peiris, J.S., Sung, J., Wong, V.C., and Yuen, K.Y. (2003). Treatment of severe acute respiratory syndrome with lopinavir/ritonavir: a multicentre retrospective matched cohort study. Hong Kong Med. J. *9*, 399–406.

Chen, F., Chan, K.H., Jiang, Y., Kao, R.Y., Lu, H.T., Fan, K.W., Cheng, V.C., Tsui, W.H., Hung, I.F., Lee, T.S., Guan, Y., Peiris, J.S., and Yuen, K.Y. (2004a). *In vitro* susceptibility of 10 clinical isolates of SARS coronavirus to selected antiviral compounds. J. Clin. Virol. *31*, 69–75.

Chen, L., Gui, C., Luo, X., Yang, Q., Gunther, S., Scandella, E., Drosten, C., Bai, D., He, X., Ludewig, B., Chen, J., Luo, H., Yang, Y., Yang, Y., Zou, J., Thiel, V., Chen, K., Shen, J., Shen, X., and Jiang, H. (2005a). Cinanserin is an inhibitor of the 3C-like proteinase of severe acute respiratory syndrome coronavirus and strongly reduces virus replication *in vitro*. J. Virol. *79*, 7095–7103.

Chen, L., Liu, P., Gao, H., Sun, B., Chao, D., Wang, F., Zhu, Y., Hedenstierna, G., and Wang, C.G. (2004b). Inhalation of nitric oxide in the treatment of severe acute respiratory syndrome: a rescue trial in Beijing. Clin. Infect. Dis. *39*, 1531–1535.

Chen, L.R., Wang, Y.C., Lin, Y.W., Chou, S.Y., Chen, S.F., Liu, L.T., Wu, Y.T., Kuo, C.J., Chen, T.S., and Juang, S.H. (2005b). Synthesis and evaluation of isatin derivatives as effective SARS coronavirus 3CL protease inhibitors. Bioorg. Med. Chem. Lett. *15*, 3058–3062.

Cho, J.H., Bernard, D.L., Sidwell, R.W., Kern, E.R., and Chu, C.K. (2006). Synthesis of Cyclopentenyl Carbocyclic Nucleosides as Potential Antiviral Agents Against Orthopoxviruses and SARS. J. Med. Chem. *49*, 1140–1148.

Chou, C.Y., Chang, H.C., Hsu, W.C., Lin, T.Z., Lin, C.H., and Chang, G.G. (2004). Quaternary Structure of the Severe Acute Respiratory Syndrome (SARS) Coronavirus Main Protease. Biochemistry *43*, 14958–14970.

Chou, K.C., Wei, D.Q., and Zhong, W.Z. (2003). Binding mechanism of coronavirus main proteinase with ligands and its implication to drug design against SARS. Biochem. Biophys. Res. Commun. *308*, 148–151.

Chu, C.M., Cheng, V.C., Hung, I.F., Wong, M.M., Chan, K.H., Chan, K.S., Kao, R.Y., Poon, L.L., Wong, C.L., Guan, Y., Peiris, J.S., and Yuen, K.Y. (2004). Role of lopinavir/ritonavir in the treatment of SARS: initial virological and clinical findings. Thorax *59*, 252–256.

Cinatl, J., Michaelis, M., Hoever, G., Preiser, W., and Doerr, H.W. (2005). Development of antiviral therapy for severe acute respiratory syndrome. Antiviral Res. *66*, 81–97.

Cinatl, J., Morgenstern, B., Bauer, G., Chandra, P., Rabenau, H., and Doerr, H.W. (2003a). Glycyrrhizin, an active component of liquorice roots, and replication of SARS-associated coronavirus. Lancet *361*, 2045–2046.

Cinatl, J., Morgenstern, B., Bauer, G., Chandra, P., Rabenau, H., and Doerr, H.W. (2003b). Treatment of SARS with human interferons. Lancet *362*, 293–294.

Cooper, C.L., van Heeswijk, R.P., Gallicano, K., and Cameron, D.W. (2003). A review of low-dose ritonavir in protease inhibitor combination therapy. Clin. Infect. Dis. *36*, 1585–1592.

Davidson, A., and Siddell, S. (2003). Potential for antiviral treatment of severe acute respiratory syndrome. Curr. Opin. Infect. Dis. *16*, 565–571.

De Clercq, E. (2004). Antiviral drugs in current clinical use. J. Clin. Virol. *30*, 115–133.

Doms, R.W. (2004). Viral entry denied. N. Engl. J. Med. *351*, 743–744.

Dooley, A.J., Shindo, N., Taggart, B., Park, J.G., and Pang, Y.P. (2006). From genome to drug lead: Identification of a small-molecule inhibitor of the SARS virus. Bioorg. Med. Chem. Lett. *16*, 830–833.

Du, Q., Wang, S., Wei, D., Sirois, S., and Chou, K.C. (2005). Molecular modeling and chemical modification for finding peptide inhibitor against severe acute respiratory syndrome coronavirus main proteinase. Anal. Biochem. *337*, 262–270.

Du, Q.S., Wang, S.Q., Zhu, Y., Wei, D.Q., Guo, H., Sirois, S., and Chou, K.C. (2004). Polyprotein cleavage mechanism of SARS CoV Mpro and chemical modification of the octapeptide. Peptides *25*, 1857–1864.

Fan, K., Ma, L., Han, X., Liang, H., Wei, P., Liu, Y., and Lai, L. (2005). The substrate specificity of SARS coronavirus 3C-like proteinase. Biochem. Biophys. Res. Commun. *329*, 934–940.

Fan, K., Wei, P., Feng, Q., Chen, S., Huang, C., Ma, L., Lai, B., Pei, J., Liu, Y., Chen, J., and Lai, L. (2004). Biosynthesis, purification, and substrate specificity of severe acute respiratory syndrome coronavirus 3C-like proteinase. J. Biol. Chem. *279*, 1637–1642.

Fuchizaki, U., Kaneko, S., Nakamoto, Y., Sugiyama, Y., Imagawa, K., Kikuchi, M., and Kobayashi, K. (2003). Synergistic antiviral effect of a combination of mouse interferon-alpha and interferon-gamma on mouse hepatitis virus. J. Med. Virol. *69*, 188–194.

Gan, Y.R., Huang, H., Huang, Y.D., Rao, C.M., Zhao, Y., Liu, J.S., Wu, L., and Wei, D.Q. (2005). Synthesis and activity of an octapeptide inhibitor designed for SARS coronavirus main proteinase. Peptides.

Ghosh, A.K., Xi, K., Ratia, K., Santarsiero, B.D., Fu, W., Harcourt, B.H., Rota, P.A., Baker, S.C., Johnson, M.E., and Mesecar, A.D. (2005). Design and synthesis of peptidomimetic severe acute respiratory syndrome chymotrypsin-like protease inhibitors. J. Med. Chem. *48*, 6767–6771.

Gorbalenya, A.E., Donchenko, A.P., Blinov, V.M., and Koonin, E.V. (1989). Cysteine proteases of positive strand RNA viruses and chymotrypsin-like serine proteases. A distinct protein superfamily with a common structural fold. FEBS Lett. *243*, 103–114.

Greenough, T.C., Babcock, G.J., Roberts, A., Hernandez, H.J., Thomas, W.D., Jr., Coccia, J.A., Graziano, R.F., Srinivasan, M., Lowy, I., Finberg, R.W., Subbarao, K., Vogel, L., Somasundaran, M., Luzuriaga, K., Sullivan, J.L., and Ambrosino, D.M. (2005). Development and characterization of a severe acute respiratory syndrome-associated coronavirus-neutralizing human monoclonal antibody that provides effective immunoprophylaxis in mice. J. Infect. Dis. *191*, 507–514.

Haasnoot, P.C., Cupac, D., and Berkhout, B. (2003). Inhibition of virus replication by RNA interference. J. Biomed. Sci. *10*, 607–616.

Han, Y.S., Chang, G.G., Juo, C.G., Lee, H.J., Yeh, S.H., Hsu, J.T., and Chen, X. (2005). Papain-like protease 2 (PLP2) from severe acute respiratory syndrome coronavirus (SARS-CoV): expression, purification, characterization, and inhibition. Biochemistry *44*, 10349–10359.

He, R., Adonov, A., Traykova-Adonova, M., Cao, J., Cutts, T., Grudesky, E., Deschambaul, Y., Berry, J., Drebot, M., and Li, X. (2004). Potent and selective inhibition of SARS coronavirus replication by aurintricarboxylic acid. Biochem. Biophys. Res. Commun. *320*, 1199–1203.

Hensley, L.E., Fritz, L.E., Jahrling, P.B., Karp, C.L., Huggins, J.W., and Geisbert, T.W. (2004). Interferon-beta 1a and SARS coronavirus replication. Emerg. Infect. Dis. *10*, 317–319.

Hertzig, T., Scandella, E., Schelle, B., Ziebuhr, J., Siddell, S.G., Ludewig, B., and Thiel, V. (2004). Rapid identification of coronavirus replicase inhibitors using a selectable replicon RNA. J. Gen. Virol. *85*, 1717–1725.

Higgins, P.G., Phillpotts, R.J., Scott, G.M., Wallace, J., Bernhardt, L.L., and Tyrrell, D.A. (1983). Intranasal interferon as protection against experimental respiratory coronavirus infection in volunteers. Antimicrob. Agents Chemother. *24*, 713–715.

Ho, T.Y., Wu, S.L., Chen, J.C., Wei, Y.C., Cheng, S.E., Chang, Y.H., Liu, H.J., and Hsiang, C.Y. (2006). Design and biological activities of novel inhibitory peptides for SARS-CoV spike protein and angiotensin-converting enzyme 2 interaction. Antiviral Res. *69*, 70–76.

Hoever, G., Baltina, L., Michaelis, M., Kondratenko, R., Baltina, L., Tolstikov, G.A., Doerr, H.W., and Cinatl, J., Jr. (2005). Antiviral activity of glycyrrhizic acid derivatives against SARS-coronavirus. J. Med. Chem. *48*, 1256–1259.

Hsu, J.T., Kuo, C.J., Hsieh, H.P., Wang, Y.C., Huang, K.K., Lin, C.P., Huang, P.F., Chen, X., and Liang, P.H. (2004). Evaluation of metal-conjugated compounds as inhibitors of 3CL protease of SARS-CoV. FEBS Lett. *574*, 116–120.

Huentelman, M.J., Zubcevic, J., Hernandez Prada, J.A., Xiao, X., Dimitrov, D.S., Raizada, M.K., and Ostrov, D.A. (2004). Structure-Based Discovery of a Novel Angiotensin-Converting Enzyme 2 Inhibitor. Hypertension *44*:903–906.

Ivanov, K.A., Thiel, V., Dobbe, J.C., van der, M.Y., Snijder, E.J., and Ziebuhr, J. (2004). Multiple enzymatic activities associated with severe acute respiratory syndrome coronavirus helicase. J. Virol. *78*, 5619–5632.

Ivens, T., Van den, E.C., Van, A.K., Nijs, E., Dams, G., Bettens, E., Ohagen, A., Pauwels, R., and Hertogs, K. (2005). Development of a homogeneous screening assay for automated detection of antiviral agents active against severe acute respiratory syndrome-associated coronavirus. J. Virol. Methods *129*, 56–63.

Jain, R.P., Pettersson, H.I., Zhang, J., Aull, K.D., Fortin, P.D., Huitema, C., Eltis, L.D., Parrish, J.C., James, M.N., Wishart, D.S., and Vederas, J.C. (2004). Synthesis and evaluation of keto-glutamine analogues as potent inhibitors of severe acute respiratory syndrome 3CLpro. J. Med. Chem. *47*, 6113–6116.

Jeffers, S.A., Tusell, S.M., Gillim-Ross, L., Hemmila, E.M., Achenbach, J.E., Babcock, G.J., Thomas, W.D., Jr., Thackray, L.B., Young, M.D., Mason, R.J., Ambrosino, D.M., Wentworth, D.E., Demartini, J.C., and Holmes, K.V. (2004). CD209L (L-SIGN) is a receptor for severe acute respiratory syndrome coronavirus. Proc. Natl. Acad. Sci. USA *101*, 15748–15753.

Jenwitheesuk, E., and Samudrala, R. (2003). Identifying inhibitors of the SARS coronavirus proteinase. Bioorg. Med. Chem. Lett. *13*, 3989–3992.

Jiang, S., Zhao, Q., and Debnath, A.K. (2002). Peptide and non-peptide HIV fusion inhibitors. Curr. Pharm. Des *8*, 563–580.

Jones, B.M., Ma, E.S., Peiris, J.S., Wong, P.C., Ho, J.C., Lam, B., Lai, K.N., and Tsang, K.W. (2004). Prolonged disturbances of *in vitro* cytokine production in patients with severe acute respiratory syndrome (SARS) treated with ribavirin and steroids. Clin. Exp. Immunol. *135*, 467–473.

Jordan, L.T., and Derbyshire, J.B. (1995). Antiviral action of interferon-alpha against porcine transmissible gastroenteritis virus. Vet. Microbiol. *45*, 59–70.

Kaeppler, U., Stiefl, N., Schiller, M., Vicik, R., Breuning, A., Schmitz, W., Rupprecht, D., Schmuck, C., Baumann, K., Ziebuhr, J., and Schirmeister, T. (2005). A new lead for nonpeptidic active-site-directed inhibitors of the severe acute respiratory syndrome coronavirus main protease discovered by a combination of screening and docking methods. J. Med. Chem. *48*, 6832–6842.

Kao, R.Y., To, A.P., Ng, L.W., Tsui, W.H., Lee, T.S., Tsoi, H.W., and Yuen, K.Y. (2004a). Characterization of SARS-CoV main protease and identification of biologically active small molecule inhibitors using a continuous fluorescence-based assay. FEBS Lett. *576*, 325–330.

Kao, R.Y., Tsui, W.H., Lee, T.S., Tanner, J.A., Watt, R.M., Huang, J.D., Hu, L., Chen, G., Chen, Z., Zhang, L., He, T., Chan, K.H., Tse, H., To, A.P., Ng, L.W., Wong, B.C., Tsoi, H.W., Yang, D., Ho, D.D., and Yuen, K.Y. (2004b). Identification of novel small-molecule inhibitors of severe acute respiratory syndrome-associated coronavirus by chemical genetics. Chem. Biol. *11*, 1293–1299.

Keyaerts, E., Vijgen, L., Chen, L., Maes, P., Hedenstierna, G., and Van Ranst, M. (2004a). Inhibition of SARS-coronavirus infection *in vitro* by S-nitroso-N-acetylpenicillamine, a nitric oxide donor compound. Int. J. Infect. Dis. *8*, 223–226.

Keyaerts, E., Vijgen, L., Maes, P., Neyts, J., and Van Ranst, M. (2004b). *In vitro* inhibition of severe acute respiratory syndrome coronavirus by chloroquine. Biochem. Biophys. Res. Commun. *323*, 264–268.

Keyaerts, E., Vijgen, L., Neyts, J., De Clercq, E., Balzarini, J., and Van Ranst, M. (2004c). A colorimetric cell culture assay for the identification of SARS coronavirus inhibitors. Antiviral Res. *62*, A76.

Keyaerts, E.,- Vijgen, L., Pannecouque, C., Van Damme, E., Peumans, W., Egberink, H., Balzarini, J., and Van Ranst, M. (2007). Plant lectins are potent inhibitors of coronaviruses by interfering with two targets in the viral replication cycle. Antivir. Res. doi:10.1016/j.antiviral.2007.03.003

Koren, G., King, S., Knowles, S., and Phillips, E. (2003). Ribavirin in the treatment of SARS: A new trick for an old drug? CMAJ. *168*, 1289–1292.

Kouroumalis, E.A., and Koskinas, J. (1986). Treatment of chronic active hepatitis B (CAH B) with chloroquine: a preliminary report. Ann. Acad. Med. Singapore *15*, 149–152.

Lane, T.E., Paoletti, A.D., and Buchmeier, M.J. (1997). Disassociation between the *in vitro* and *in vivo* effects of nitric oxide on a neurotropic murine coronavirus. J. Virol. *71*, 2202–2210.

Lee, T.W., Cherney, M.M., Huitema, C., Liu, J., James, K.E., Powers, J.C., Eltis, L.D., and James, M.N. (2005). Crystal structures of the main peptidase from the SARS coronavirus inhibited by a substrate-like aza-peptide epoxide. J. Mol. Biol. *353*, 1137–1151.

Li, B.J., Tang, Q., Cheng, D., Qin, C., Xie, F.Y., Wei, Q., Xu, J., Liu, Y., Zheng, B.J., Woodle, M.C., Zhong, N., and Lu, P.Y. (2005). Using siRNA in prophylactic and therapeutic regimens against SARS coronavirus in Rhesus macaque. Nat. Med. *11*, 944–951.

Li, W., Moore, M.J., Vasilieva, N., Sui, J., Wong, S.K., Berne, M.A., Somasundaran, M., Sullivan, J.L., Luzuriaga, K., Greenough, T.C., Choe, H., and Farzan, M. (2003). Angiotensin-converting enzyme 2 is a functional receptor for the SARS coronavirus. Nature *426*, 450–454.

Li, Y.H., Tao, P.Z., Liu, Y.Z., and Jiang, J.D. (2004). Geldanamycin, a ligand of heat shock protein 90, inhibits the replication of herpes simplex virus type 1 *in vitro*. Antimicrob. Agents Chemother. *48*, 867–872.

Liang, P.H., Cheng, W.C., Lee, Y.L., Yu, H.P., Wu, Y.T., Lin, Y.L., and Wong, C.H. (2006). Novel five-membered iminocyclitol derivatives as selective and potent glycosidase inhibitors: new structures for antivirals and osteoarthritis. Chembiochemistry *7*, 165–173.

Lin, C.W., Tsai, F.J., Tsai, C.H., Lai, C.C., Wan, L., Ho, T.Y., Hsieh, C.C., and Chao, P.D. (2005). Anti-SARS coronavirus 3C-like protease effects of Isatis indigotica root and plant-derived phenolic compounds. Antiviral Res. *68*, 36–42.

Lin, Y.L., Huang, Y.L., Ma, S.H., Yeh, C.T., Chiou, S.Y., Chen, L.K., and Liao, C.L. (1997). Inhibition of Japanese encephalitis virus infection by nitric oxide: antiviral effect of nitric oxide on RNA virus replication. J. Virol. *71*, 5227–5235.

Lip, K.M., Shen, S., Yang, X., Keng, C.T., Zhang, A., Oh, H.L., Li, Z.H., Hwang, L.A., Chou, C.F., Fielding, B.C., Tan, T.H., Mayrhofer, J., Falkner, F.G., Fu, J., Lim, S.G., Hong, W., and Tan, Y.J. (2006). Monoclonal antibodies targeting the HR2 domain and the region immediately upstream of the HR2 of the S protein neutralize *in vitro* infection of severe acute respiratory syndrome coronavirus. J. Virol. *80*, 941–950.

Liu, B., and Zhou, J. (2005). SARS-CoV protease inhibitors design using virtual screening method from natural products libraries. J. Comput. Chem. *26*, 484–490.

Liu, S., Xiao, G., Chen, Y., He, Y., Niu, J., Escalante, C.R., Xiong, H., Farmar, J., Debnath, A.K., Tien, P., and Jiang, S. (2004). Interaction between heptad repeat 1 and 2 regions in spike protein of SARS-associated coronavirus: implications for virus fusogenic mechanism and identification of fusion inhibitors. Lancet *363*, 938–947.

Liu, Y.C., Huang, V., Chao, T.C., Hsiao, C.D., Lin, A., Chang, M.F., and Chow, L.P. (2005a). Screening of drugs by FRET analysis identifies inhibitors of SARS-CoV 3CL protease. Biochem. Biophys. Res. Commun. *333*, 194–199.

Liu, Z., Huang, C., Fan, K., Wei, P., Chen, H., Liu, S., Pei, J., Shi, L., Li, B., Yang, K., Liu, Y., and Lai, L. (2005b). Virtual screening of novel noncovalent inhibitors for SARS-CoV 3C-like proteinase. J. Chem. Inf. Model. *45*, 10–17.

Loutfy, M.R., Blatt, L.M., Siminovitch, K.A., Ward, S., Wolff, B., Lho, H., Pham, D.H., Deif, H., LaMere, E.A., Chang, M., Kain, K.C., Farcas, G.A., Ferguson, P., Latchford, M., Levy, G., Dennis, J.W., Lai, E.K., and Fish, E.N. (2003). Interferon alfacon-1 plus corticosteroids in severe acute respiratory syndrome: a preliminary study. JAMA *290*, 3222–3228.

Lu, A., Zhang, H., Zhang, X., Wang, H., Hu, Q., Shen, L., Schaffhausen, B.S., Hou, W., and Li, L. (2004). Attenuation of SARS coronavirus by a short hairpin RNA expression plasmid targeting RNA-dependent RNA polymerase. Virology *324*, 84–89.

Martina, E., Stiefl, N., Degel, B., Schulz, F., Breuning, A., Schiller, M., Vicik, R., Baumann, K., Ziebuhr, J., and Schirmeister, T. (2005). Screening of electrophilic compounds yields an aziridinyl peptide as new active-site directed SARS-CoV main protease inhibitor. Bioorg. Med. Chem. Lett. *15*, 5365–5369.

Morgenstern, B., Michaelis, M., Baer, P.C., Doerr, H.W., and Cinatl, J., Jr. (2005). Ribavirin and interferon-beta synergistically inhibit SARS-associated coronavirus replication in animal and human cell lines. Biochem. Biophys. Res. Commun. *326*, 905–908.

Neuman, B.W., Stein, D.A., Kroeker, A.D., Churchill, M.J., Kim, A.M., Kuhn, P., Dawson, P., Moulton, H.M., Bestwick, R.K., Iversen, P.L., and Buchmeier, M.J. (2005). Inhibition, escape, and attenuated growth of severe acute respiratory syndrome coronavirus treated with antisense morpholino oligomers. J. Virol. *79*, 9665–9676.

Ni, L., Zhu, J., Zhang, J., Yan, M., Gao, G.F., and Tien, P. (2005). Design of recombinant protein-based SARS-CoV entry inhibitors targeting the heptad-repeat regions of the spike protein S2 domain. Biochem. Biophys. Res. Commun. *330*, 39–45.

Paragas, J., Blatt, L.M., Hartmann, C., Huggins, J.W., and Endy, T.P. (2005). Interferon alfacon1 is an inhibitor of SARS-corona virus in cell-based models. Antiviral Res. *66*, 99–102.

Pardridge, W.M., Yang, J., and Diagne, A. (1998). Chloroquine inhibits HIV-1 replication in human peripheral blood lymphocytes. Immunol. Lett. *64*, 45–47.

Pei, J., Sekellick, M.J., Marcus, P.I., Choi, I.S., and Collisson, E.W. (2001). Chicken interferon type I inhibits infectious bronchitis virus replication and associated respiratory illness. J. Interferon Cytokine Res. *21*, 1071–1077.

Rajnarayanan, R.V., Dakshanamurthy, S., and Pattabiraman, N. (2004). "Teaching old drugs to kill new bugs": structure-based discovery of anti-SARS drugs. Biochem. Biophys. Res. Commun. *321*, 370–378.

Saijo, M., Morikawa, S., Fukushi, S., Mizutani, T., Hasegawa, H., Nagata, N., Iwata, N., and Kurane, I. (2005). Inhibitory effect of mizoribine and ribavirin on the replication of severe acute respiratory syndrome (SARS)-associated coronavirus. Antiviral Res. *66*, 159–163.

Sainz, B., Mossel, E.C., Peters, C.J., and Garry, R.F. (2004). Interferon-beta and interferon-gamma synergistically inhibit the replication of severe acute respiratory syndrome-associated coronavirus (SARS-CoV). Virology *329*, 11–17.

Saura, M., Zaragoza, C., McMillan, A., Quick, R.A., Hohenadl, C., Lowenstein, J.M., and Lowenstein, C.J. (1999). An antiviral mechanism of nitric oxide: inhibition of a viral protease. Immunity. *10*, 21–28.

Savarino, A., Boelaert, J.R., Cassone, A., Majori, G., and Cauda, R. (2003). Effects of chloroquine on viral infections: an old drug against today's diseases? Lancet Infect. Dis. *3*, 722–727.

Savarino, A., Gennero, L., Sperber, K., and Boelaert, J.R. (2001). The anti-HIV-1 activity of chloroquine. J. Clin. Virol. *20*, 131–135.

Scagnolari, C., Vicenzi, E., Bellomi, F., Stillitano, M.G., Pinna, D., Poli, G., Clementi, M., Dianzani, F., and Antonelli, G. (2004). Increased sensitivity of SARS-coronavirus to a combination of human type I and type II interferons. Antivir. Ther. *9*, 1003–1011.

Shen, Y.C., Wang, L.T., Khalil, A.T., Chiang, L.C., and Cheng, P.W. (2005). Bioactive pyranoxanthones from the roots of Calophyllum blancoi. Chem. Pharm. Bull. (Tokyo) *53*, 244–247.

Shi, J., Wei, Z., and Song, J. (2004). Dissection study on the severe acute respiratory syndrome 3C-like protease reveals the critical role of the extra domain in dimerization of the enzyme: defining the extra domain as a new target for design of highly specific protease inhibitors. J. Biol. Chem. *279*, 24765–24773.

Shi, Y., Yang, d.H., Xiong, J., Jia, J., Huang, B., and Jin, Y.X. (2005). Inhibition of genes expression of SARS coronavirus by synthetic small interfering RNAs. Cell Res. *15*, 193–200.

Shie, J.J., Fang, J.M., Kuo, C.J., Kuo, T.H., Liang, P.H., Huang, H.J., Yang, W.B., Lin, C.H., Chen, J.L., Wu, Y.T., and Wong, C.H. (2005a). Discovery of potent anilide inhibitors against the severe acute respiratory syndrome 3CL protease. J. Med. Chem. *48*, 4469–4473.

Shie, J.J., Fang, J.M., Kuo, T.H., Kuo, C.J., Liang, P.H., Huang, H.J., Wu, Y.T., Jan, J.T., Cheng, Y.S., and Wong, C.H. (2005b). Inhibition of the severe acute respiratory syndrome 3CL protease by peptidomimetic alpha,beta-unsaturated esters. Bioorg. Med. Chem. *13*, 5240–5252.

Sidwell, R.W., Huffman, J.H., Call, E.W., Warren, R.P., Radov, L.A., and Murray, R.J. (1987). Inhibition of murine hepatitis virus infections by the immunomodulator 2,3,5,6,7,8-hexahydro-2-phenyl-8,8-dimethoxy-imidazo[1,2a)pyridine (PR-879–317A). Antimicrob. Agents Chemother. *31*, 1130–1134.

Sidwell, R.W., Huffman, J.H., Campbell, N., and Allen, L.B. (1977). Effect of ribavirin on viral hepatitis in laboratory animals. Ann. N. Y. Acad. Sci. *284*, 239–246.

Simmons, G., Gosalia, D.N., Rennekamp, A.J., Reeves, J.D., Diamond, S.L., and Bates, P. (2005). Inhibitors of cathepsin L prevent severe acute respiratory syndrome coronavirus entry. Proc. Natl. Acad. Sci. USA *102*, 11876–11881.

Simmons, G., Reeves, J.D., Rennekamp, A.J., Amberg, S.M., Piefer, A.J., and Bates, P. (2004). Characterization of severe acute respiratory syndrome-associated coronavirus (SARS-CoV) spike glycoprotein-mediated viral entry. Proc. Natl. Acad. Sci. USA *101*, 4240–4245.

Singh, A.K., Sidhu, G.S., Friedman, R.M., and Maheshwari, R.K. (1996). Mechanism of enhancement of the antiviral action of interferon against herpes simplex virus-1 by chloroquine. J. Interferon Cytokine Res. *16*, 725–731.

Sirois, S., Wei, D.Q., Du, Q., and Chou, K.C. (2004). Virtual screening for SARS-CoV protease based on KZ7088 pharmacophore points. J. Chem. Inf. Comput. Sci. *44*, 1111–1122.

Snijder, E.J., Bredenbeek, P.J., Dobbe, J.C., Thiel, V., Ziebuhr, J., Poon, L.L., Guan, Y., Rozanov, M., Spaan, W.J., and Gorbalenya, A.E. (2003). Unique and conserved features of genome and proteome of SARS-coronavirus, an early split-off from the coronavirus group 2 lineage. J. Mol. Biol. *331*, 991–1004.

So, L.K., Lau, A.C., Yam, L.Y., Cheung, T.M., Poon, E., Yung, R.W., and Yuen, K.Y. (2003). Development of a standard treatment protocol for severe acute respiratory syndrome. Lancet *361*, 1615–1617.

Sperber, S.J., and Hayden, F.G. (1989). Comparative susceptibility of respiratory viruses to recombinant interferons-alpha 2b and -beta. J. Interferon Res. *9*, 285–293.

Spiegel, M., Pichlmair, A., Martinez-Sobrido, L., Cros, J., Garcia-Sastre, A., Haller, O., and Weber, F. (2005). Inhibition of Beta interferon induction by severe acute respiratory syndrome coronavirus suggests a two-step model for activation of interferon regulatory factor 3. J. Virol. *79*, 2079–2086.

Stroher, U., DiCaro, A., Li, Y., Strong, J.E., Aoki, F., Plummer, F., Jones, S.M., and Feldmann, H. (2004). Severe acute respiratory syndrome-related coronavirus is inhibited by interferon-alpha. J. Infect. Dis. *189*, 1164–1167.

Sui, J., Li, W., Murakami, A., Tamin, A., Matthews, L.J., Wong, S.K., Moore, M.J., Tallarico, A.S., Olurinde, M., Choe, H., Anderson, L.J., Bellini, W.J., Farzan, M., and Marasco, W.A. (2004). Potent neutralization of severe acute respiratory syndrome (SARS) coronavirus by a human mAb to S1 protein that blocks receptor association. Proc. Natl. Acad. Sci. USA *101*, 2536–2541.

Sui, J., Li, W., Roberts, A., Matthews, L.J., Murakami, A., Vogel, L., Wong, S.K., Subbarao, K., Farzan, M., and Marasco, W.A. (2005). Evaluation of human monoclonal antibody 80R for immunoprophylaxis of severe acute respiratory syndrome by an animal study, epitope mapping, and analysis of spike variants. J. Virol. *79*, 5900–5906.

Tan, E.L., Ooi, E.E., Lin, C.Y., Tan, H.C., Ling, A.E., Lim, B., and Stanton, L.W. (2004). Inhibition of SARS coronavirus infection *in vitro* with clinically approved antiviral drugs. Emerg. Infect. Dis. *10*, 581–586.

Tanner, J.A., Zheng, B.J., Zhou, J., Watt, R.M., Jiang, J.Q., Wong, K.L., Lin, Y.P., Lu, L.Y., He, M.L., Kung, H.F., Kesel, A.J., and Huang, J.D. (2005). The adamantane-derived bananins are potent inhibitors of the helicase activities and replication of SARS coronavirus. Chem. Biol. *12*, 303–311.

Tao, P., Zhang, J., Tang, N., Zhang, B.Q., He, T.C., and Huang, A.L. (2005). Potent and specific inhibition of SARS-CoV antigen expression by RNA interference. Chin Med. J. (Engl.) *118*, 714–719.

Ter Meulen, J., Bakker, A.B., van den Brink, E.N., Weverling, G.J., Martina, B.E., Haagmans, B.L., Kuiken, T., de, K.J., Preiser, W., Spaan, W., Gelderblom, H.R., Goudsmit, J., and Osterhaus, A.D. (2004). Human monoclonal antibody as prophylaxis for SARS coronavirus infection in ferrets. Lancet *363*, 2139–2141.

Toney, J.H., Navas-Martin, S., Weiss, S.R., and Koeller, A. (2004). Sabadinine: a potential non-peptide anti-severe acute-respiratory-syndrome agent identified using structure-aided design. J. Med. Chem. *47*, 1079–1080.

Torre, D., Pugliese, A., and Speranza, F. (2002). Role of nitric oxide in HIV-1 infection: friend or foe? Lancet Infect. Dis. *2*, 273–280.

Towler, P., Staker, B., Prasad, S.G., Menon, S., Tang, J., Parsons, T., Ryan, D., Fisher, M., Williams, D., Dales, N.A., Patane, M.A., and Pantoliano, M.W. (2004). ACE2 X-ray structures reveal a large hinge-bending motion important for inhibitor binding and catalysis. J. Biol. Chem. *279*, 17996–18007.

Tsai, W.P., Nara, P.L., Kung, H.F., and Oroszlan, S. (1990). Inhibition of human immunodeficiency virus infectivity by chloroquine. AIDS Res. Hum. Retroviruses *6*, 481–489.

Turner, R.B., Felton, A., Kosak, K., Kelsey, D.K., and Meschievitz, C.K. (1986). Prevention of experimental coronavirus colds with intranasal alpha-2b interferon. J. Infect. Dis. *154*, 443–447.

Tyrrell, D.A. (1986). The efficacy and tolerance of intranasal interferons: studies at the Common Cold Unit. J. Antimicrob. Chemother. *18 Suppl B*, 153–156.

van den Brink, E.N., Ter, M.J., Cox, F., Jongeneelen, M.A., Thijsse, A., Throsby, M., Marissen, W.E., Rood, P.M., Bakker, A.B., Gelderblom, H.R., Martina, B.E., Osterhaus, A.D., Preiser, W., Doerr, H.W., de, K.J., and Goudsmit, J. (2005). Molecular and biological characterization of human monoclonal antibodies binding to the spike and nucleocapsid proteins of severe acute respiratory syndrome coronavirus. J. Virol. *79*, 1635–1644.

Veiga, S., Yuan, Y., Li, X., Santos, N.C., Liu, G., and Castanho, M.A. (2006). Why are HIV-1 fusion inhibitors not effective against SARS-CoV? Biophysical evaluation of molecular interactions. Biochim. Biophys. Acta *1760*, 55–61.

Vincent, M.J., Bergeron, E., Benjannet, S., Erickson, B.R., Rollin, P.E., Ksiazek, T.G., Seidah, N.G., and Nichol, S.T. (2005). Chloroquine is a potent inhibitor of SARS coronavirus infection and spread. Virol. J. *2*, 69.

Wei, P., Fan, K., Chen, H., Ma, L., Huang, C., Tan, L., Xi, D., Li, C., Liu, Y., Cao, A., and Lai, L. (2006). The N-terminal octapeptide acts as a dimerization inhibitor of SARS coronavirus 3C-like proteinase. Biochem. Biophys. Res. Commun. *339*, 865–872.

Weiss, R.C., and Oostrom-Ram, T. (1989). Inhibitory effects of ribavirin alone or combined with human alpha interferon on feline infectious peritonitis virus replication *in vitro*. Vet. Microbiol. *20*, 255–265.

Wong, S.K., Li, W., Moore, M.J., Choe, H., and Farzan, M. (2004). A 193-amino acid fragment of the SARS coronavirus S protein efficiently binds angiotensin-converting enzyme 2. J. Biol. Chem. *279*, 3197–3201.

Wong, V.W., Dai, D., Wu, A.K., and Sung, J.J. (2003). Treatment of severe acute respiratory syndrome with convalescent plasma. Hong. Kong. Med. J. *9*, 199–201.

Wu, C.J., and Chan, Y.L. (2006). Antiviral applications of RNAi for coronavirus. Expert. Opin. Investig. Drugs *15*, 89–97.

Wu, C.J., Jan, J.T., Chen, C.M., Hsieh, H.P., Hwang, D.R., Liu, H.W., Liu, C.Y., Huang, H.W., Chen, S.C., Hong, C.F., Lin, R.K., Chao, Y.S., and Hsu, J.T. (2004a). Inhibition of severe acute respiratory syndrome coronavirus replication by niclosamide. Antimicrob. Agents Chemother. *48*, 2693–2696.

Wu, C.Y., Jan, J.T., Ma, S.H., Kuo, C.J., Juan, H.F., Cheng, Y.S., Hsu, H.H., Huang, H.C., Wu, D., Brik, A., Liang, F.S., Liu, R.S., Fang, J.M., Chen, S.T., Liang, P.H., and Wong, C.H. (2004b). Small molecules targeting severe acute respiratory syndrome human coronavirus. Proc. Natl. Acad. Sci. USA *101*, 10012–10017.

Xiao, X., Chakraborti, S., Dimitrov, A.S., Gramatikoff, K., and Dimitrov, D.S. (2003). The SARS-CoV S glycoprotein: expression and functional characterization. Biochem. Biophys. Res. Commun. *312*, 1159–1164.

Xiong, B., Gui, C.S., Xu, X.Y., Luo, C., Chen, J., Luo, H.B., Chen, L.L., Li, G.W., Sun, T., Yu, C.Y., Yue, L.D., Duan, W.H., Shen, J.K., Qin, L., Shi, T.L., Li, Y.X., Chen, K.X., Luo, X.M., Shen, X., Shen, J.H., and Jiang, H.L. (2003). A 3D model of SARS_CoV 3CL proteinase and its inhibitors design by virtual screening. Acta Pharmacol. Sin. *24*, 497–504.

Xu, X., Liu, Y., Weiss, S., Arnold, E., Sarafianos, S.G., and Ding, J. (2003). Molecular model of SARS coronavirus polymerase: implications for biochemical functions and drug design. Nucleic Acids Res. *31*, 7117–7130.

Yamamoto, N., Yang, R., Yoshinaka, Y., Amari, S., Nakano, T., Cinatl, J., Rabenau, H., Doerr, H.W., Hunsmann, G., Otaka, A., Tamamura, H., Fujii, N., and Yamamoto, N. (2004). HIV protease inhibitor nelfinavir inhibits replication of SARS-associated coronavirus. Biochem. Biophys. Res. Commun. *318*, 719–725.

Yang, H., Xie, W., Xue, X., Yang, K., Ma, J., Liang, W., Zhao, Q., Zhou, Z., Pei, D., Ziebuhr, J., Hilgenfeld, R., Yuen, K.Y., Wong, L., Gao, G., Chen, S., Chen, Z., Ma, D., Bartlam, M., and Rao, Z. (2005a). Design of wide-spectrum inhibitors targeting coronavirus main proteases. PLoS. Biol. *3*, e324.

Yang, H., Yang, M., Ding, Y., Liu, Y., Lou, Z., Zhou, Z., Sun, L., Mo, L., Ye, S., Pang, H., Gao, G.F., Anand, K., Bartlam, M., Hilgenfeld, R., and Rao, Z. (2003). The crystal structures of severe acute respiratory syndrome virus main protease and its complex with an inhibitor. Proc. Natl. Acad. Sci. USA *100*, 13190–13195.

Yang, Z.Y., Werner, H.C., Kong, W.P., Leung, K., Traggiai, E., Lanzavecchia, A., and Nabel, G.J. (2005b). Evasion of antibody neutralization in emerging severe acute respiratory syndrome coronaviruses. Proc. Natl. Acad. Sci. USA *102*, 797–801.

Yap, Y., Zhang, X., Andonov, A., and He, R. (2005). Structural analysis of inhibition mechanisms of aurintricarboxylic acid on SARS-CoV polymerase and other proteins. Comput. Biol. Chem. *29*, 212–219.

Yi, H., Nakashima, I., and Isobe, K. (1996). Enhancement of nitric oxide production from activated macrophages by glycyrrhizin. Am. J. Chin Med. *24*, 271–278.

Yi, L., Li, Z., Yuan, K., Qu, X., Chen, J., Wang, G., Zhang, H., Luo, H., Zhu, L., Jiang, P., Chen, L., Shen, Y., Luo, M., Zuo, G., Hu, J., Duan, D., Nie, Y., Shi, X., Wang, W., Han, Y., Li, T., Liu, Y., Ding, M., Deng, H., and Xu, X. (2004). Small molecules blocking the entry of severe acute respiratory syndrome coronavirus into host cells. J. Virol. *78*, 11334–11339.

Yuan, K., Yi, L., Chen, J., Qu, X., Qing, T., Rao, X., Jiang, P., Hu, J., Xiong, Z., Nie, Y., Shi, X., Wang, W., Ling, C., Yin, X., Fan, K., Lai, L., Ding, M., and Deng, H. (2004). Suppression of SARS-CoV entry by peptides corresponding to heptad regions on spike glycoprotein. Biochem. Biophys. Res. Commun. *319*, 746–752.

Zhang, X.W., and Yap, Y.L. (2004a). Exploring the binding mechanism of the main proteinase in SARS-associated coronavirus and its implication to anti-SARS drug design. Bioorg. Med. Chem. *12*, 2219–2223.

Zhang, X.W., and Yap, Y.L. (2004b). Old drugs as lead compounds for a new disease? Binding analysis of SARS coronavirus main proteinase with HIV, psychotic and parasite drugs. Bioorg. Med. Chem. *12*, 2517–2521.

Zhang, X.W., Yap, Y.L., and Altmeyer, R.M. (2005). Generation of predictive pharmacophore model for SARS-coronavirus main proteinase. Eur. J. Med. Chem. *40*, 57–62.

Zhao, Z., Zhang, F., Xu, M., Huang, K., Zhong, W., Cai, W., Yin, Z., Huang, S., Deng, Z., Wei, M., Xiong, J., and Hawkey, P.M. (2003). Description and clinical treatment of an early outbreak of severe acute respiratory syndrome (SARS) in Guangzhou, PR China. J. Med. Microbiol. *52*, 715–720.

Zheng, B., He, M.L., Wong, K.L., Lum, C.T., Poon, L.L., Peng, Y., Guan, Y., Lin, M.C., and Kung, H.F. (2004a). Potent inhibition of SARS-associated coronavirus (SCOV) infection and replication by type I interferons (IFN-alpha/beta) but not by type II interferon (IFN-gamma). J. Interferon Cytokine Res. *24*, 388–390.

Zheng, B.J., Guan, Y., Hez, M.L., Sun, H., Du, L., Zheng, Y., Wong, K.L., Chen, H., Chen, Y., Lu, L., Tanner, J.A., Watt, R.M., Niccolai, N., Bernini, A., Spiga, O., Woo, P.C., Kung, H.F., Yuen, K.Y., and Huang, J.D. (2005). Synthetic peptides outside the spike protein heptad repeat regions as potent inhibitors of SARS-associated coronavirus. Antivir. Ther. *10*, 393–403.

Zheng, B.J., Guan, Y., Tang, Q., Du, C., Xie, F.Y., He, M.L., Chan, K.W., Wong, K.L., Lader, E., Woodle, M.C., Lu, P.Y., Li, B., and Zhong, N. (2004b). Prophylactic and therapeutic effects of small interfering RNA targeting SARS-coronavirus. Antivir. Ther. *9*, 365–374.

Zhu, J., Xiao, G., Xu, Y., Yuan, F., Zheng, C., Liu, Y., Yan, H., Cole, D.K., Bell, J.I., Rao, Z., Tien, P., and Gao, G.F. (2004). Following the rule: formation of the 6-helix bundle of the fusion core from severe acute respiratory syndrome coronavirus spike protein and identification of potent peptide inhibitors. Biochem. Biophys. Res. Commun. *319*, 283–288.

Index

R

S

T

U

V

X

Z

www.ingramcontent.com/pod-product-compliance
Lightning Source LLC
LaVergne TN
LVHW011909200926
848484LV00002B/574

9781904455165